The Schrödinger-Virasoro Algebra

Theoretical and Mathematical Physics

The series founded in 1975 and formerly (until 2005) entitled *Texts and Monographs in Physics* (TMP) publishes high-level monographs in theoretical and mathematical physics. The change of title to *Theoretical and Mathematical Physics* (TMP) signals that the series is a suitable publication platform for both the mathematical and the theoretical physicist. The wider scope of the series is reflected by the composition of the editorial board, comprising both physicists and mathematicians.

The books, written in a didactic style and containing a certain amount of elementary background material, bridge the gap between advanced textbooks and research monographs. They can thus serve as basis for advanced studies, not only for lectures and seminars at graduate level, but also for scientists entering a field of research.

For further volumes:
http://www.springer.com/series/720

Jérémie Unterberger
Claude Roger

The Schrödinger-Virasoro Algebra

Mathematical structure and dynamical Schrödinger symmetries

With a Foreword by Malte Henkel

Springer

Jérémie Unterberger
Université Henri Poincaré
Institut Elie Cartan
Vandoeuvre-les-Nancy
France
unterberger@iecn.u-nany.fr

Claude Roger
Université Lyon I
Faculté des Sciences et Technologies
Département de Mathématiques
Boulevard du 11 novembre 1918, 43
69622 Villeurbanne Cedex
France
roger@math.univ-lyon1.fr

ISSN 1864-5879 e-ISSN 1864-5887
ISBN 978-3-642-26959-2 ISBN 978-3-642-22717-2 (eBook)
DOI 10.1007/978-3-642-22717-2
Springer Heidelberg Dordrecht London New York

Softcover reprint of the hardcover 1st edition 2012

Printed on acid-free paper

Springer is part of Springer Science+Business Media (www.springer.com)

Foreword

"A physicist needs that his equations should be mathematically sound, ... he should not neglect quantities unless they are small."

P.A.M. Dirac, Europhys. News **8**, 1-4 (1977)

"Mathematics, if used thoughtfully, is almost always useful – and occasionally essential – to progress in theoretical physics."

R. Geroch, *Mathematical physics* (1985)

"L.D. Landau n'exigait pas qu'un théoricien sachc 'beaucoup' de mathématiques, mais il fallait être capable, en prenant une quelconque intégrale indéfinie de résoudre n'importe quelle équation différentielle ..."

E.M. Lifchitz, *Biographie de L.D. Landau* (1969)

"Was aber dieser Formalismus eigentlich bedeutete, war keineswegs klar. Die Mathematik war, wie es öfters vorkommt, klüger als das sinngebende Denkcn."

M. Born et H. Born *Der Luxus des Gewissens* (1969)

"Symmetry ... is one idea by which man through the ages has tried to comprehend and create order, beauty and perfection. ... Whenever you have to do with a structure-endowed entity, try to determine its group of automorphisms. You can expect to gain a deep insight."

H. Weyl, *Symmetry* (1952)

Symmetries, using a word derived from the Greek $\sigma\upsilon\mu\mu\eta\tau\rho\iota\tilde{\alpha}$ – "with equal measure", have since its invention by the ancients played an ever increasing rôle. Here, I shall comment, from a physicist's point of view, on applications of a particular class of symmetries in physics, namely scale-symmetries and their extensions. Especially, in what follows we shall be interested in *continuous* symmetries, formally described by *Lie groups* (discovered some five quarters of a century ago). More specifically, this book, which was grown from a long article by its authors on the geometry and representation theory of the *Schrödinger Lie group* and its Lie algebra,[1] considers *space-time symmetries*. Arguably, the most famous example of such a symmetry principle in physics is Einstein's celebrated theory of relativity.[2] In essence, the (special) *principle of relativity* asserts that under a change of time and space coordinates given by an element of the Poincaré Lie group, the laws which describe physical systems should remain formally invariant or, as physicists say, should transform *co-variantly*. This space-time symmetry requirement is enough to fix the structure of (special) relativity and provides a general framework for all physical theories. Contact with the material world as studied through experiments is achieved by identifying the invariant velocity, whose existence is predicted by Poincaré-covariance alone, with the measured speed of light.[3]

From an æsthetic point of view, it is then natural to ask whether co-variance under a larger set of dynamical symmetry transformations is a sensible *démarche* in physics. Indeed, soon after the formulation of the special theory of relativity, it was shown by Bateman and by Cunningham that Maxwell's equations for a free electromagnetic field in the vacuum admit the *conformal group* as a dynamical symmetry.[4] The conformal group includes the Poincaré group as a subgroup. Conformal transformations in a d-dimensional Euclidean space are local scale-transformations, which modify locally length scales $\boldsymbol{r} \mapsto \lambda(\boldsymbol{r})\boldsymbol{r}$ with a space-dependent local rescaling factor $\lambda(\boldsymbol{r})$ but such that angles are kept unchanged. Of particular significance is the case of two spatial dimensions, since in that case the Lie algebra of conformal transformations is infinite-dimensional, i.e. it is isomorphic to the direct sum of analytic and anti-analytic transformations, respectively, of a single complex variable z or $\bar{z}$. A well-known basis of infinitesimal generators is the set $\{\ell_n, \bar{\ell}_n\}_{n\in\mathbb{Z}}$ with $\ell_n := -z^{n+1}\partial_z$ and $\bar{\ell}_n := -\bar{z}^{n+1}\partial_{\bar{z}}$. They generate a commuting pair of algebras[5] $\mathrm{Vect}(S^1) \oplus \mathrm{Vect}(S^1)$ of vector fields on the circle

$$[\ell_n, \ell_{n'}] = (n - n')\ell_{n+n'} \;\;,\;\; [\ell_n, \bar{\ell}_{n'}] = 0 \;\;,\;\; [\bar{\ell}_n, \bar{\ell}_{n'}] = (n - n')\bar{\ell}_{n+n'} \qquad (1)$$

with $n, n' \in \mathbb{Z}$. For a long time, conformal invariance, although a well-known property of free field-theories (associated e.g. to the Klein-Gordon or to the free Dirac equation), did not play a very important role in physics. This perception only began to change when in 1970 Polyakov pointed out that the long-established scale-invariance of many-body systems at the so-called *critical point* of a second-order

[1] C. Roger and J. Unterberger, Ann. Henri Poincaré **7**, 1477 (2006).

[2] A. Einstein, Ann. der Physik **17**, 891 (1905).

[3] Since the beginning of the 1990s, most articles in theoretical physics are freely available electronically, see http://fr.arxiv.org/ and then the categories `hep-th`, `math-ph` or `cond-mat`. The references quoted here try to give an entry to the literature, are usually the earliest reference(s) known to me and are not meant to be complete in any sense.

[4] E. Cunningham, Proc. London Math. Soc. **8**, 77 (1909); H. Bateman, Proc. London Math. Soc. **8**, 223 (1910).

[5] First defined by E. Cartan in 1909 and later studied by others over fields of characteristic $p > 0$, in which case it is spanned by the ℓ_n with $-1 \leq n \leq p - 2$, and known as *Witt algebra*.

phase transition naturally extends to conformal invariance.[6] Indeed, in physical systems, the cooperative properties are encoded by the *energy-momentum* (or stress-energy) *tensor*. For large classes of sufficiently local interactions, it may be verified that combined with the habitually assumed properties of translation- and rotation-invariance, the tracelessness of the energy-momentum tensor which follows from scale-invariance, is sufficient for conformal invariance of the model.[7] The form of the two- and three-point correlation functions C_{12} and C_{123} is fixed through their covariance under projective transformations, i.e. $\mathscr{L}^{(2)}C_{12} = 0$ and $\mathscr{L}^{(3)}C_{123} = 0$, where $\mathscr{L}^{(2,3)}$ is the two- or three-body operator of the (projective) conformal *Ward identities*, derived from the generators in $\langle \ell_{\pm1,0}, \bar{\ell}_{\pm1,0} \rangle$.[6]

The enormous impact of conformal invariance for the understanding of *two-dimensional* phase transitions was first realized in the ground-breaking work by Belavin, Polyakov and Zamolodchikov,[8] who considered the mode expansion of the two non-vanishing components of the complex energy-momentum tensor, $T(z) = \sum_{n\in\mathbb{Z}} L_n z^{-n-2}$ and $\bar{T}(\bar{z}) = \sum_{n\in\mathbb{Z}} \bar{L}_n \bar{z}^{-n-2}$. They showed that the effects of interactions and fluctuations can be condensed into a single real constant c, the *central charge* (or *conformal anomaly number*), since the components $L_n, \bar{L}_n$ generate a pair of commuting *Virasoro algebras*

$$
\begin{aligned}
[L_n, L_{n'}] &= (n-n')L_{n+n'} + \frac{c}{12}(n^3-n)\delta_{n+n',0} \\
[L_n, \bar{L}_{n'}] &= 0 \\
[\bar{L}_n, \bar{L}_{n'}] &= (n-n')\bar{L}_{n+n'} + \frac{c}{12}(n^3-n)\delta_{n+n',0}
\end{aligned}
\tag{2}
$$

with $n, n' \in \mathbb{Z}$. This is a *central extension* of the Lie algebra Vect(S^1) (see (1)), parametrized by the constant c. (The Virasoro algebra actually originates in attempts to quantify the motion of a bosonic string, consistently with the principle of relativity). The most simple objects in a conformal field-theory are the *primary operator* $\phi(z)$, which are defined as follows. Consider the mode expansion of a field operator $\phi(z) = \sum_{m\in\mathbb{Z}} \phi_m z^{-m-\mu}$, where the (real) constant $\mu = \mu_\phi$ is called the *conformal weight* of the field operator $\phi = \phi_\mu$. By definition, the modes $\phi_{\mu,m}$ of a primary operator ϕ_μ satisfy

$$
[L_n, \phi_{\mu,m}] = (n(\mu-1)-m)\,\phi_{\mu,n+m}. \tag{3}
$$

[6] A.M. Polyakov, Sov. Phys. JETP Lett. **12**, 381 (1970).

[7] C.G. Callan, S. Coleman and R. Jackiw, Ann. of Phys. **59**, 42 (1970). However, since then systems with local interactions which are scale-invariant, but not conformally invariant have been found. See J. Polchinsky, Nucl. Phys. **B303**, 226 (1988) and V. Riva and J.L. Cardy, Phys. Lett. **B622**, 339 (2005) for specific examples.

[8] A.A. Belavin, A.M. Polyakov and A.B. Zamolodchikov, Nucl. Phys. **B241**, 333 (1984).

Field operators depending on both z and $\bar{z}$ are built from tensor products $\phi_{\mu,\bar{\mu}}(z,\bar{z}) = \phi_\mu(z)\otimes\bar{\phi}_{\bar{\mu}}(\bar{z})$. A *conformal physical model* is constituted from a set of primary operators ϕ, together with all so-called *secondary operator* $L_{-n_1}L_{-n_2}\cdots\bar{L}_{-\bar{n}_2}\bar{L}_{-\bar{n}_1}\phi$, where $n_i, \bar{n}_i \in \mathbb{N}$.

In certain cases, it is possible to classify the admissible primary operators of a conformal physical model. In order to see how this comes about, one considers a quantum mechanical formulation of the theory characterized by a ground state $|0\rangle$, which by assumption is unique. Working in the context of so-called *radial quantization* (where the ordering of operators $\phi_1(z_1), \phi_2(z_2)$ in products of operators is canonically $\phi_1(z_1)\phi_2(z_2)$ if $|z_1| > |z_2|$), one associates to each primary operator $\phi = \phi_\mu(z)$ a state $|\mu\rangle := \phi_\mu(0)|0\rangle$. These states arise as highest-weight states in the representation theory of the Virasoro algebra (indeed, they are characterized algebraically by $L_0|\mu\rangle = \mu|\mu\rangle$ and $L_n|\mu\rangle = 0\ \forall n > 0$) and may be classified with the help of the celebrated *Kac formula*.[9] Unitary representations belong to one of the two following series:[10]

$$\begin{aligned} &\text{case 1:} \qquad c \geq 1 \qquad \mu \geq 0 \\ &\text{case 2:} \qquad c = c_m \qquad \mu = \mu_{r,s} \end{aligned} \tag{4}$$

$$c_m = 1 - \tfrac{6}{m(m+1)}\ ,\ \mu_{r,s} = \frac{[r(m+1) - sm]^2 - 1}{4m(m+1)}$$

where $m = 2, 3, 4\ldots$, $1 \leq r \leq m-1$ and $1 \leq s \leq m$. The admissible central charges $c_m, m \geq 2$ correspond to simple models of statistical physics at their critical point. For example, $c = \frac{1}{2}$ $(m = 3)$ corresponds to the *Ising model* which is one of the paradigmatic models in physics for the description of the phase transition between ferromagnetic and paramagnetic phases, $c = \frac{7}{10}$ $(m = 4)$ corresponds to the *tri-critical point* of the *diluted Ising model*, $c = \frac{4}{5}$ $(m = 5)$ to the three-states *Potts model* used e.g. for the description of adsorption processes and so on. The central charge of a given model may be calculated explicitly, for example from the *Casimir effect*. Systems belonging to the discrete series (case 2 above) are called *unitary minimal models* and have the special property that for each admissible value of c, there exists only a *finite* number of conformal weights $\mu_{r,s}$, which in turn correspond to physical observables.[11] For example, in the Ising model, there are three primary

---

[9] B.L. Feigin and D.B. Fuks, Funct. Anal. Appl. **16**, 114 (1982); **17**, 241 (1983); V.G. Kac and A.K. Raina, *Bombay lectures on highest weight representations of infinite-dimensional Lie algebras*, World Scientific (Singapore 1987).

[10] P. Goddard, A. Kent and D. Olive, Comm. Math. Phys. **103**, 105 (1986); D. Friedan, Z. Qiu and S. Shenker, Comm. Math. Phys. **107**, 535 (1986).

[11] More generally, (non-unitary) *minimal models* or *rational conformal field-theories* are those conformal physical models which possess a finite number of primary operators. Their conformal weights are given by a generalised Kac formula and are rational numbers. Physical applications include two-dimensional turbulence, percolation and polymers.

operators with conformal weights $\mu_{1,1} = \mu_{2,3} = 0$, $\mu_{1,2} = \mu_{2,2} = \frac{1}{16}$ and $\mu_{1,3} = \mu_{2,1} = \frac{1}{2}$. Physically, they describe the identity operator $\mathbf{1}$, the magnetisation density σ and the energy density ϵ, respectively. The primary operators of a minimal model form an associative and commutative *operator product algebra*, related to the short-distance expansion of the operator product $\phi_1(z_1)\phi_2(z_2)$ as $|z_1 - z_2| \to 0$. For example, the operator product algebra for the Ising model is $\sigma \cdot \sigma = \mathbf{1} + \epsilon$, $\epsilon \cdot \epsilon = \mathbf{1}$, $\sigma \cdot \epsilon = \sigma$ and $\mathbf{1} \cdot \chi = \chi$, where $\chi = \mathbf{1}, \sigma, \epsilon$. These operators are often realized as *vertex operators*.

In minimal models, the known conformal weights allow to calculate easily any of the critical exponents which were introduced for a phenomenological description of the system's behaviour close to a critical point. Besides, any n-point correlation function between primary operators satisfies a system of linear partial differential equations from which these correlators may be explicitly found. By the additional requirement of *modular invariance*,[12] a classification of the modular invariant partition functions can be derived,[13] which goes a long way towards a classification of the universality classes described by minimal models. The physics literature abounds with applications of these results.[14]

* * *

Given the enormous practical utility of conformal invariance for the description of phase transitions *at* equilibrium, what about dynamical symmetries in time-dependent phenomena? Is it possible to use ideas and methods borrowed from conformal invariance? From a physical point of view, one may consider the time-dependent behaviour of *equilibrium* systems at a critical point, leading to a study of *equilibrium critical dynamics*. More generally, *non-equilibrium critical dynamics* looks at the time-dependent scaling behaviour of a physical system at a critical point, far from thermal equilibrium. Finally, it is also conceivable that the full space-time-dependence of a physical system may be scale-invariant even if the stationary, time-independent behaviour does not have a scale-invariance.

In many cases, one considers the *algebraic* rescaling of time and space $t \mapsto \lambda^z t$ and $\boldsymbol{r} \mapsto \lambda \boldsymbol{r}$. Here, z is called the *dynamical exponent*.

One of the most simple examples of this kind is given by the *diffusion equation* $(2\mathcal{M}\partial_t - \Delta_r)\psi(t, \boldsymbol{r}) = 0$, where Δ_r denotes the spatial Laplacian and $D = (2\mathcal{M})^{-1}$ is called the diffusion constant. Alternatively, if one writes $\mathcal{M} = -im$, where the real constant m is the mass of the particle, one obtains the free *Schrödinger equation*. As first realized by Lie at the end of the nineteenth century [15]

---

[12] J.L. Cardy, Nucl. Phys. **B270**, 186 (1986); **B275**, 200 (1986).

[13] A. Cappelli, C. Itzykson and J.-B. Zuber, Nucl. Phys. **B280**, 445 (1987); Comm. Math. Phys. **113**, 1 (1987).

[14] For introductions, see e.g. M. Schottenloher, *A mathematical introduction to conformal field-theory*, 2nd ed., Springer (Heidelberg 2008); P. di Francesco, P. Mathieu and D. Sénéchal, *Conformal field-theory*, Springer (Heidelberg 1997); M. Henkel, *Conformal invariance and critical phenomena*, Springer (Heidelberg 1999).

[15] In his Königsberg *Vorlesungen über Dynamik* in 1842-43, Jacobi noted *en passant* that the classical equations of motion of the n-body problem (with an inverse-square potential) are invariant

and rediscovered by physicists several times some 40 years ago, the set of space-time transformations

$$t \mapsto t' = \frac{\alpha t + \beta}{\gamma t + \delta}\ ,\ \boldsymbol{r} \mapsto \boldsymbol{r}' = \frac{\mathscr{R}\boldsymbol{r} + \boldsymbol{v}t + \boldsymbol{a}}{\gamma t + \delta}\ ;\ \alpha\delta - \beta\gamma = 1 \tag{5}$$

generates a new solution of the diffusion/Schrödinger equation from a known one. The above set of continuous transformations, parametrized by $\alpha, \beta, \gamma, \delta, \boldsymbol{a}, \boldsymbol{v}$ and the rotation matrix $\mathscr{R} \in SO(d)$, forms a Lie group which is now usually called the *Schrödinger group* $Sch(d)$.[16] In distinction to the conformal group, the Schrödinger group is the semi-direct product of $SL(2,\mathbb{R})$ by the Galilei group and hence is not semi-simple; it admits a non-trivial central extension. It follows that the solutions $\psi = \psi(t, \boldsymbol{r})$ of the diffusion/Schrödinger equation transform under *projective* representations of the Schrödinger group

$$\psi(t,\boldsymbol{r}) \mapsto (T_g\psi)(t,\boldsymbol{r}) = f_g\left(g^{-1}(t,\boldsymbol{r})\right)\psi\left(g^{-1}(t,\boldsymbol{r})\right) \tag{6}$$

where $g \in Sch(d)$ and the companion function f_g reads explicitly[17]

$$f_g(t,\boldsymbol{r}) = (\gamma t+\delta)^{-1/2}\exp\left[-\frac{\mathscr{M}}{2}\frac{\gamma \boldsymbol{r}^2+2\mathscr{R}\boldsymbol{r}\cdot(\gamma\boldsymbol{a}-\delta\boldsymbol{v})+\gamma\boldsymbol{a}^2-\delta t\boldsymbol{v}^2+2\gamma\boldsymbol{a}\cdot\boldsymbol{v}}{\gamma t+\delta}\right] \tag{7}$$

The complexity of this expression should be a more than sufficient motivation instead to concentrate on the *Schrödinger* Lie *algebra* $\mathfrak{sch}(d) := \mathrm{Lie}(Sch(d))$, which may be obtained from the generators

$$\begin{aligned}
L_n &:= -t^{n+1}\partial_t - \frac{n+1}{2}t^n\boldsymbol{r}\cdot\nabla_{\boldsymbol{r}} - \frac{\mathscr{M}}{4}n(n+1)\boldsymbol{r}^2 - \frac{x}{2}(n+1)t^n \\
Y_m^{(j)} &:= -t^{m+1/2}\nabla_{r_j} - \left(m+\frac{1}{2}\right)\mathscr{M}t^{m-1/2}r_j \\
M_p &:= -t^p\mathscr{M} \\
R_p^{(i,j)} &:= t^p\left(r_i\nabla_{r_j} - r_j\nabla_{r_i}\right) = -R_p^{(j,i)}
\end{aligned} \tag{8}$$

---

under the elements of the Schrödinger group. In A. Clebsch and E. Lottner, editors, *Gesammelte Werke von C.G.J. Jacobi*, 18–31, Berlin, 1866/1884, Akademie der Wissenschaften.

[16] High-energy physicists often refer to it as *non-relativistic conformal group* (although this name is inappropriate, since $Sch(d)$ cannot be obtained as a non-relativistic limit of a conformal group, see p. xv below) and certain mathematical schools call it a *generalized Galilei group*. It describes space-time scaling with a dynamical exponent $z = 2$.

[17] The mathematical literature often distinguishes carefully between the cases $\mathscr{M} \neq 0$ and $\mathscr{M} = 0$ and considers the former as a central extension of the latter. We shall always assume $\mathscr{M} \neq 0$, unless explicitly stated otherwise, since this is the case relevant for most physical applications.

with $n = \pm 1, 0$, $m = \pm\frac{1}{2}$, $p = 0$ and $i, j = 1, \ldots, d$. Here, x denotes the scaling dimension (it is the analogue of the conformal weights $\mu, \bar{\mu}$) of the wave function on which these generators act. The non-vanishing commutators read explicitly

$$\begin{aligned}
[L_n, L_{n'}] &= (n - n')L_{n+n'} \\
\left[L_n, Y_m^{(j)}\right] &= \left(\frac{n}{2} - m\right) Y_{n+m}^{(j)} \\
[L_n, M_{n'}] &= -n' M_{n+n'}, \quad \left[L_n, R_{n'}^{(i,j)}\right] = -n' R_{n+n'}^{(i,j)} \\
\left[Y_m^{(i)}, Y_{m'}^{(j)}\right] &= \delta_{i,j}(m - m')M_{m+m'}, \quad \left[Y_m^{(k)}, R_n^{(i,j)}\right] = \delta_{i,k} Y_{n+m}^{j} - \delta_{j,k} Y_{n+m}^{(i)} \\
\left[R_n^{(i,j)}, R_{n'}^{(k,\ell)}\right] &= \delta_{j,k} R_{n+n'}^{(i,\ell)} + \delta_{i,\ell} R_{n+n'}^{(j,k)} - \delta_{j,\ell} R_{n+n'}^{(i,k)} - \delta_{i,k} R_{n+n'}^{(j,\ell)}
\end{aligned} \tag{9}$$

Note that the Galilei algebra generated by the $Y_{\frac{1}{2}}^{(j)}$ has been centrally extended. The value of the extra so-called *mass generator* M_0, namely, the *mass* parameter $\mathscr{M}$, plays the same rôle as the central charge c in the study of two-dimensional conformal symmetry.

Comparison with the conformal commutation relations (1) suggested[18] to extend the generators (8) to four *infinite* families by extending indices $n \in \mathbb{Z}$ and $m \in \mathbb{Z} + \frac{1}{2}$. The commutators (9) remain valid with $n, n' \in \mathbb{Z}$ and $m, m' \in \mathbb{Z} + \frac{1}{2}$. In this way, we have obtained an infinite-dimensional Lie algebra (which contains the algebra $\mathrm{Vect}(S^1)$ as a sub-algebra), the mathematical theory of which is the object of study of this book. This Lie algebra is called herein *Schrödinger–Virasoro algebra* and is denoted by $\mathfrak{sv}(d)$, while formulas (9) for $n \in \mathbb{Z}$ and $p \in \mathbb{Z} + \frac{1}{2}$ generate a representation of $\mathfrak{sv}(d)$ called *vector-field representation* and denoted by $d\pi_{x/2}$. By analogy with (3), we may construct 'conformal primary operators' $L(t)$, $Y^{(j)}(t)$, $M(t)$ and $R^{(i,j)}(t)$, whose 'conformal weights', respectively, have the values 2, $\frac{3}{2}$, 1 and 1.

A comparison between (2) and (9) suggests looking for central extensions of $\mathfrak{sv}(d)$. The mathematical theory presented in this book allows this to be done systematically; see Chap. 7. Physical applications of this still remain to be constructed.

We shall briefly comment on Schrödinger Ward identities below, in the context of the non-relativistic AdS/CFT correspondence and of physical ageing phenomena.

Schrödinger symmetry has been found in a large variety of physical situations. We now briefly list some of them.

---

[18] M. Henkel, J. Stat. Phys. **75**, 1023 (1994) and Nucl. Phys. **B641**, 405 (2002) for the cases $d = 1$ and $d = 2$. The construction of the generators L_n was motivated by the similarity of their time-dependent part $-t^{n+1}\partial_t$ with the generators of the algebra $\mathrm{Vect}(S^1)$. Given $L_{-1,0}$ and $Y_{-1/2}$ and requiring that $[L_n, L_{n'}] = (n - n')L_{n+n'}$, the form of the other generators is already fixed.

1. *Free diffusion and free Schrödinger equation*. The Schrödinger group in one spatial dimension is one of the earliest examples of a physically relevant Lie group.[19] Since the free diffusion/Schrödinger equation may be seen as the Euler-Lagrange equation of motion of free non-relativistic free-field theory, the Schrödinger-invariance of that field-theory has been one of the physicists' methods for rediscovering the Schrödinger group.[20] In that context, the emphasis shifts from the consideration of the properties of solutions of the diffusion/Schrödinger equation to the properties of the energy-momentum tensor. For example, it is possible, for large classes of sufficiently local theories, to show that Galilei-invariance together with dynamical scale-invariance (with $z = 2$) are enough to imply full Schrödinger invariance,[21] analogously to what has long been known for many conformally invariant systems.

   This may be naturally generalized from the so far considered scalar case to the *Lévy-Leblond field* with spin $\frac{1}{2}$ (see Chap. 8) and further to higher spin fields.

   At first sight, there seems to be an important difference between the Schrödinger and diffusion equations, in that the solutions of the first are complex-valued (necessary for consistency with Galilei-invariance), while those of the second are real-valued. We note that the Galilei-invariance of the diffusion equation can be understood as follows: the diffusion equation is to be considered just one of a pair of equations. The second equation involves, in the context of the physical response-field formalism (both at and far from equilibrium, see below on p. xix), the so-called "response field" $\widetilde{\psi} = \widetilde{\psi}(t, \boldsymbol{r})$. Now the doublet $\begin{pmatrix} \psi \\ \widetilde{\psi} \end{pmatrix}$ plays the same rôle under Galilei-transformation as the pair $\begin{pmatrix} \psi \\ \psi^* \end{pmatrix}$ in the usual Schrödinger equation. In this way, one may have a real-valued solution $\psi = \psi(t, \boldsymbol{r})$ of the diffusion equation such that Galilei- and Schrödinger-invariance are indeed satisfied.
2. *Schrödinger equation with a potential*. The dynamical symmetries of the equation $\left(-2im\partial_t - \Delta_r - V(t, \boldsymbol{r})\right)\psi = 0$ were analyzed in detail by U. Niederer. For example, for a freely falling particle or a harmonic oscillator, the dynamical symmetry group of the associated Schrödinger equation is isomorphic to $Sch(d)$, but the representations to be used are different from the one of the free Schrödinger equation. In fact, the potentials $V = V(t, \boldsymbol{r})$ admitting a non-trivial

---

[19] S. Lie, Arch. Math. Nat. Vid. (Kristiania) **6**, 328 (1882); S. Lie, *Vorlesungen über Differentialgleichungen mit bekannten infinitesimalen Transformationen*, Teubner (Leipzig 1891).

[20] H.A. Kastrup, Nucl. Phys. **B7**, 545 (1968); U. Niederer, Helv. Phys. Acta **45**, 802 (1972); C.R. Hagen, Phys. Rev. **D5**, 377 (1972); G. Burdet and M. Perrin, Lett. Nuovo Cim. **4**, 651 (1972); R. Jackiw, Phys. Today, **25**, 23 (1972).

[21] M. Henkel and J. Unterberger, Nucl. Phys. **B660**, 407 (2003).

dynamical symmetry group, which can be as large as *Sch*(d), are characterized as solutions of a certain differential equation.[22]

Many Schrödinger equations with a potential admit non-trivial dynamical symmetries which are subgroups of *Sch*(d).

The study of various representations of the Schrödinger algebra and of the Schrödinger–Virasoro algebra is one of the major themes in this book, see Chaps. 3–5 for the general algebraic theory, Chap. 6 for vertex algebra representations and Chaps. 8–10 for a detailed study of the action of the Schrödinger–Virasoro algebra on Schrödinger operators.

3. *Noiseless Burgers equation.* This equation, in one spatial dimension, reads $\partial_t u + \frac{1}{2}\partial_x u^2 - \nu\partial_x^2 u = f$ and is used for simplified descriptions of turbulence or shocks of a fluid flow $u = u(t,x)$ dependent on an external force $f = f(t,x)$. If one has $f = 0$, the unforced Burgers equation is an example of a non-linear Schrödinger-invariant equation. However, the solutions $u = u(t,x)$ do *not* transform according to (6), (7) with $\mathcal{M} = (2\nu)^{-1}$, as one might have conjectured by comparing with the linear Schrödinger equation. Rather, one needs different representations of $\mathfrak{sch}(1)$, with $\mathcal{M} = 0$, and which also include additive terms (and not merely multiplicative ones as in (6)) in the transformation of u.[23] Furthermore, these representations may be extended to representations of $\mathfrak{sv}(1)$ which relate Burgers' equations with different external forces.[24]

   On the other hand, Schrödinger-invariance is broken down to Galilei-invariance, with a Lie algebra spanned by $\langle L_{-1}, Y_{\pm 1/2}\rangle$, if f is taken to be a *random* force with a Gaussian distribution.[24]
4. *Euler's equation of a perfect fluid.* Euler's hydrodynamical equations for a perfect, viscous fluid are given by $\partial_t \rho + \nabla\cdot(\rho \boldsymbol{v}) = 0$ and $\rho\big(\partial_t + \boldsymbol{v}\cdot\nabla\big)\boldsymbol{v} = -\nabla p + \rho\nu\Delta\boldsymbol{v}$, where $\rho = \rho(t,\boldsymbol{r})$ is the fluid density whose flow is described by the velocity field $\boldsymbol{v} = \boldsymbol{v}(t,\boldsymbol{r})$ and the pressure $p = p(t,\boldsymbol{r})$. If one further has a so-called *polytropic equation of state* $p \sim \rho^{\gamma_p}$ with *polytropic exponent* $\gamma_p = 1+2/(d+1)$, then Euler's equations are Schrödinger-invariant.[25] Here, the so-called 'special' transformation (parametrized in (5) by γ) relates collapsing and exploding solutions. This has been applied to relate supernova *ex*plosions to laboratory experiments on plasma *im*plosions.[26]

---

[22] U. Niederer, Helv. Phys. Acta **47**, 167 (1974).

[23] U. Niederer, Helv. Phys. Acta **51**, 220 (1978). Herein, the Schrödinger-invariance for several other generalised non-linear diffusion equations in one spatial dimension is also proven.

[24] E. Ivashkevich, J. Phys. **A30**, L525 (1997).

[25] This result has been known to Russian mathematicians since the 1960s; see L.V. Ovsiannikov, *Group analysis of differential equations*, Academic Press (London 1982) and was independently rediscovered by physicists, see M. Hassaïne and P. Horváthy, Ann. of Phys. **282**, 218 (2000) and Phys. Lett. **A279**, 215 (2001); L. O'Raifeartaigh and V.V. Sreedhar, Ann. of Phys. **293**, 215 (2001).

[26] L. O'C Drury and J.T. Mendonça, Phys. Plasmas **7**, 5148 (2000).

5. *Self-interacting theories and non-linear Schrödinger equations.* An equilibrium field-theoretical description of a system at a multi-critical point of order p may be given in terms of a Landau-Ginzburg functional $\mathscr{H}[\phi] = \int \mathrm{d}\boldsymbol{r}\, \left(\frac{1}{2}(\nabla\phi)^2 + g\phi^{2+2p}\right)$, where the field ϕ describes the space-time-dependent order parameter. Here $p = 1$ corresponds to the usual critical point, $p = 2$ to a tri-critical point and so on. The equilibrium critical dynamics of such a system is often captured by a *time-dependent Ginzburg-Landau equation* of motion of the form $\partial_t\phi = -\frac{\delta\mathscr{H}}{\delta\phi} = \Delta\phi + 2(p+1)g\phi^{1+2p}$.

   By definition, the system is at its *characteristic dimension* $d = d^{(c)}$, where the coupling constant g is dimensionless. A dimensional analysis of the Janssen-de Dominicis functional $\mathscr{J}[\widetilde{\phi}, \phi] := \int \mathrm{d}t\mathrm{d}\boldsymbol{r}\, \left[\widetilde{\phi}\partial_t\phi + \widetilde{\phi}\frac{\delta\mathscr{H}}{\delta\phi}\right]$ shows that this is achieved here for $p = 2/d$. It is mathematical textbook knowledge that the time-dependent Ginzburg-Landau equation is Schrödinger-invariant if and only if $d = d^{(c)}$.[27]

   Generalising this argument to the *Gross-Pitaevskii equation* (with a complex-valued order-parameter ϕ), applications of Schrödinger-invariance to the Bose-Einstein condensation have been discussed.[28]

   From a physical point of view, it looks unnatural to have Schrödinger-invariance only for $d = d^{(c)}$. Alternatively, one might consider that since in general g is dimensionful, it should transform under scale and special Schrödinger transformations. Generalized generators for the Schrödinger algebra taking this into account may readily be constructed and in this generalised sense, the time-dependent Ginzburg-Landau (or Gross-Pitaevskii) equations are Schrödinger-invariant for any d.[29]

6. *Matter fields interacting with a Chern-Simons gauge field.* Consider the gauged non-linear Schrödinger equation in $d = 2$ space dimensions

$$\mathrm{i}\partial_t\psi = \left(-\frac{1}{2m}(\nabla - \mathrm{i}e\boldsymbol{A})^2 + eA^0 - \Lambda\psi^*\psi\right)\psi \tag{10}$$

   where $(A^0, \boldsymbol{A})$ is an electromagnetic vector potential in $(1+2)$ dimensions and Λ is a constant. The electromagnetic density $\rho = \psi^*\psi$ and current $\boldsymbol{J} = \frac{1}{2\mathrm{i}m}\left[\psi^*(\nabla - \mathrm{i}e\boldsymbol{A})\psi - \psi\left((\nabla - \mathrm{i}e\boldsymbol{A})\psi\right)^*\right]$ are defined as usual. However, the electric and magnetic fields do not satisfy Maxwell's equations but rather obey the field-current relations

$$B = \nabla \wedge \boldsymbol{A} = -\frac{e}{\kappa}\rho \;,\; \boldsymbol{E} = -\nabla A^0 - \partial_t\boldsymbol{A} = \frac{e}{\kappa}\, {}^*\boldsymbol{J} \tag{11}$$

---

[27] See e.g. C.D. Boyer, R.T. Sharp and P. Winternitz, J. Math. Phys. **17**, 1439 (1976); W.I. Fushchich, W.M. Shtelen and N.I. Serov, *Symmetry analysis and exact solutions of equations of nonlinear mathematical physics*, Kluwer (Dordrecht 1993).

[28] P. Ghosh, Phys. Lett. **A308**, 411 (2003).

[29] S. Stoimenov and M. Henkel, Nucl. Phys. **B723**, 205 (2005).

with the abbreviation ${}^*J^i = \epsilon^{ij} J_j$ where ϵ^{ij} is the totally antisymmetric tensor and $i, j = 1, 2$. For a given value of Λ, the static solutions can be interpreted as Chern-Simons vortices.[30]

7. *Conditional symmetries of differential equations.* To find symmetries of certain differential equations, one may relax the requirements and look instead for so-called *non-classical* or *conditional symmetries* which are valid *modulo* an auxiliary condition.[31] For example,[32] one may show in this way that the Schrödinger–Virasoro algebra $\mathfrak{sv}(2)$ with $\mathcal{M} = 0$ is a conditional dynamical symmetry of the non-linear equation

$$\det \begin{pmatrix} u_t & u_{x_1} & u_{x_2} \\ u_{tx_1} & u_{x_1x_1} & u_{x_1x_2} \\ u_{tx_2} & u_{x_2x_1} & u_{x_2x_2} \end{pmatrix} = \frac{\partial}{\partial x_1}\left(\frac{u_{x_1}}{u^2}\right) + \frac{\partial}{\partial x_2}\left(\frac{u_{x_2}}{u^2}\right), \tag{12}$$

with the abbreviations $u_t = \partial_t u$, $u_{x_j} = \partial u/\partial x_j$ etc. Here, the auxiliary condition is the *Monge-Ampère equation* $u_{x_1x_1}u_{x_2x_2} - u_{x_1x_2}^2 = 0$. Generalisations exist to any spatial dimension $d \geq 2$ and to arbitrary dynamical exponents z.

8. *Relationship with conformal algebras.* In d spatial dimensions, the Schrödinger algebra $\mathfrak{sch}(d)$ may be considered as a sub-algebra of the complexified conformal algebra $\mathfrak{conf}(d+2)$ in $d+2$ dimensions.[33] A constructive way to see this is to consider the mass parameter $\mathcal{M}$ as an additional coordinate. By way of Fourier-transform, the projective factor in the transformation law (6) becomes a transformation law in the 'dual' coordinate ζ and the projective representation becomes a true representation.[34] Simply stated, the Schrödinger operator $2\mathcal{M}\partial_t - \Delta_r$ becomes through the formal substitution $\mathcal{M} \rightsquigarrow \partial/\partial_\zeta$ the differential operator $2\partial_\zeta\partial_t - \Delta_r$, which by a further rotation in the coordinates (t, ζ) becomes a Klein-Gordon operator. Considering all possible transformations of ζ compatible with $\mathfrak{sch}(d)$, we arrive at the conformal algebra $\mathfrak{conf}(d+2)$, as explained in more details in Chap. 2. [35]

   We mention two physical applications of this construction:[35] (i) the physically required causality of the response functions of systems undergoing ageing far from equilibrium (see 12. below) can be derived from the condition that the response functions transform co-variantly under the conformal algebra $\mathfrak{conf}(d+2)$; (ii) re-introducing the *speed of light c* as a dimensionful constant

---

[30] R. Jackiw and S.-Y. Pi, Phys. Rev. **D42**, 3500 (1990); C. Duval, P. Horváthy and L. Palla, Phys. Lett. **B325**, 39 (1994); M. Hassaïne and P. Horváthy, Phys. Lett. **A279**, 215 (2001).

[31] G.W. Bluman and J.D. Cole, J. Math. Mech. **18**, 1025 (1969); W.I. Fushchych, W.M. Shtelen and M.I. Serov, Dopov. Akad. Nauk Ukr. Ser. **A9**, 17 (1988); D. Levi and P. Winternitz, J. Phys. **A22**, 2915 (1989).

[32] R. Cherniha and M. Henkel, J. Math. Anal. Appl. **298**, 487 (2004).

[33] G. Burdet, M. Perrin and P. Sorba, Comm. Math. Phys. **34**, 85 (1973).

[34] This observation was made, for the Galilei group, by D. Giulini, Ann. of Phys. **249**, 222 (1996).

[35] M. Henkel and J. Unterberger, Nucl. Phys. **B660**, 407 (2003).

relating 'space' and 'time' variables allows to take the 'non-relativistic limit' $c \to \infty$ in a controlled way. This limit turns out *not* to lead to a group contraction and, surprisingly, the projected algebra $\mathfrak{conf}(d+2) \stackrel{c\to\infty}{\longrightarrow} \mathrm{CGA}(d) \not\cong \mathfrak{sch}(d)$ is not isomorphic to the Schrödinger algebra.

Furthermore, carrying out the 'non-relativistic limit' on a pair of commuting algebras $\mathfrak{vect}(S^1) \oplus \mathfrak{vect}(S^1)$ (see (1)) via a group contraction, one does *not* obtain the Schrödinger–Virasoro algebra $\mathfrak{sv}(1)$ either.[36] Rather, we have for the generators, after the contraction

$$\begin{aligned} L_n &= -t^{n+1}\partial_t - (n+1)t^n r\partial_r - (n+1)xt^n - n(n+1)\gamma t^{n-1} r \\ Y_n &= -t^{n+1}\partial_r - (n+1)\gamma t^n \end{aligned} \tag{13}$$

where x is the scaling dimension and γ a free parameter. The non-vanishing commutators are $[L_n, L_{n'}] = (n-n')L_{n+n'}$ and $[L_n, Y_{n'}] = (n-n')Y_{n+n'}$ with $n, n' \in \mathbb{Z}$. This is an infinite-dimensional extension of an algebra called either *non-relativistic* or *Conformal Galilei algebra*[37] $\mathrm{CGA}(1) \cong \langle L_{\pm1,0}, Y_{\pm1,0}\rangle \not\cong \mathfrak{sch}(1)$. From the representation (13), we read off the dynamical exponent $z = 1$ which is distinct from the Schrödinger algebra.[38]

The relationship with conformal algebras becomes important for the formulation of non-relativistic analogues of the AdS/CFT correspondence, studied intensively in string theory, see 11. below.

9. *Bargmann structures*. The extension from the Schrödinger algebra into a conformal algebra has a geometric meaning which we now briefly discuss[39] (see also Chap. 1). To illustrate, consider a free non-relativistic particle of mass m. Its Lagrange function $L = \frac{m}{2}\left(\frac{d\boldsymbol{r}}{dt}\right)^2$ is *not* invariant under a Galilei-transformation $t \mapsto t$, $\boldsymbol{r} \mapsto \boldsymbol{r} + \boldsymbol{v}t$. However, by introducing an additional 'dual' coordinate ζ, an invariant Lagrangian

$$\mathscr{L} := L + m\frac{d\zeta}{dt} = \frac{m}{2}\left(\left(\frac{d\boldsymbol{r}}{dt}\right)^2 + 2\frac{d\zeta}{dt}\right) = \frac{m}{2}g_{ab}\frac{dr^a}{dt}\frac{dr^b}{dt} \tag{14}$$

---

[36] M. Henkel, Nucl. Phys. **B641**, 405 (2002); M. Henkel, R. Schott, S. Stoimenov and J. Unterberger, math-ph/0601028. We point out that there also exist representations of $\mathrm{CGA}(1)$ with $z = 2$.

[37] P. Havas and J. Plebanski, J. Math. Phys. **19**, 482 (1978) and rediscovered independently by M. Henkel, Phys. Rev. Lett. **78**, 1940 (1997) and J. Negro, M.A. del Olmo and A. Rodríguez-Marco, J. Math. Phys. **38**, 3786 and 3810 (1997).

[38] In the context of physical ageing, the algebra $\mathrm{CGA}(d)$ is sometimes called the "altern algebra" $\mathfrak{alt}(d)$. In contrast to the Schrödinger algebra, $\mathrm{CGA}(d)$ does not arise as dynamical symmetry in hydrodynamic equations, P.-M. Zhang and P. Horváthy, Eur. Phys. J. **C65**, 607 (2010), R. Cherniha and M. Henkel, J. Math. Anal. Appl. **369**, 120 (2010). There exists a so-called 'exotic' central extension of $\mathrm{CGA}(2)$, J. Lukierski, P.C. Stichel and W.J. Zakrzewski, Phys. Lett. **A357**, 1 (2006); **B650**, 203 (2007) which does arise as conditional symmetry in certain non-linear hydrodynamical equations, R. Cherniha and M. Henkel, *ibid.*

[39] C. Duval, G. Burdet, H.P. Künzle and M. Perrin, Phys. Rev. **D31**, 1841 (1985).

may be constructed, where $a, b = -1, 0, 1, \ldots, d$ (we use the notation $r^0 = t$ and $r^{-1} = \zeta$). Of course, the invariance of $\mathscr{L}$ depends on the correct choice of the transformation law for ζ. The invariance of $\mathscr{L}$ under $Sch(d)$ corresponds to *geodesic motion* for the metric $g_{ab}\mathrm{d}r^a\mathrm{d}r^b = 2\mathrm{d}\zeta\mathrm{d}t + \mathrm{d}\boldsymbol{r}^2$. The $(d+2)$-dimensional space under consideration admits a covariantly constant Killing vector ∂_ζ and is called a *Bargmann manifold*.

For example, one may use this formalism to prove the Schrödinger-invariance of the non-relativistic n-body problem with gravitational/Coulombian interactions.[40]

The Lie algebras of conformal vector fields with a non-relativistic Newton-Cartan space-time can be classified. For flat space, the Schrödinger algebra is recovered for time-like geodesics, while for light-like geodesics, the conformal Galilei algebra is obtained.[41]

10. *Non-relativistic supersymmetries*. Often, it is of interest to extend a Lie algebra to a Lie super-algebra by introducing a graduation. In this way, besides the usual (even) generators which satisfy between them commutation relations, one may introduce additional (odd) generators which may close when so-called *anti*-commutators are formed. Structures of this kind arise naturally in many instances. As a simple example, one may observe that the Schrödinger-invariant scalar one-dimensional diffusion equation, taken together with the spin-$\frac{1}{2}$ Lévy-Leblond equation, which is also Schrödinger-invariant, is part of a supermultiplet of the so-called *super-Schrödinger algebra* with two supercharges, $\tilde{\mathfrak{s}}^{(2)} \cong \mathfrak{osp}(2|2) \ltimes \mathfrak{sh}(2|2)$, which is the semi-direct product of the orthosymplectic algebra $\mathfrak{osp}(2|2)$ and the supersymmetric extension $\mathfrak{sh}(2|2)$ of the Heisenberg algebra. If, analogously to the extension to a conformal algebra discussed above for the Schrödinger algebra, one considers the masses as further variables, then the super-Schrödinger symmetry may be extended to $\mathfrak{osp}(2|4)$.[44] As an application, we mention here super-symmetric quantum mechanics.[42] Supersymmetric extensions of $\mathfrak{sch}(d)$ with N supercharges have been systematically constructed.[43] Finally, infinite families of supersymmetric extensions of $\mathfrak{sv}(1)$ have been found through the use of extended Poisson algebras.[44] This will be described in Chap. 11.

From a less formal point of view, a motivation for looking for super-symmetric extensions of Schrödinger symmetry may come from the fact that the physically well-motivated *Fokker-Planck* or *Kramers equations*, whose

[40] C. Duval, G. Gibbons and P. Horváthy, Phys. Rev. **D43**, 3907 (1991).

[41] C. Duval and P. Horváthy, J. Phys. **A42**, 465206 (2009).

[42] J. Beckers and V. Hussin, Phys. Lett. **A118**, 319 (1986); J. Beckers, N. Debergh and A.G. Nikitin, J. Math. Phys. **33**, 152 (1992); P.K. Ghosh, Nucl. Phys. **B681**, 359 (2004).

[43] C. Duval and P. Horváthy, J. Math. Phys. **35**, 2516 (1994).

[44] M. Henkel and J. Unterberger, Nucl. Phys. **B746**, 155 (2006).

solutions give the probability distributions for the statistical description of many-body systems, are naturally supersymmetric.[45]

11. *Non-relativistic AdS/CFT correspondence*. Recently, it has been asked if there might exist physical applications of the at present intensively studied correspondence between anti-de Sitter space and conformal field-theory (*AdS/CFT correspondence*), originally discovered in a string-theory context,[46] to *non*-relativistic condensed matter problems. Using the embedding of the Schrödinger algebra into the conformal algebra described above on p. xv, a non-relativistic version of the AdS/CFT correspondence involving the Schrödinger algebra can be formulated.[47] It is impossible here to indicate all lines in this very active field and we shall only quote one illustrative example. Consider a scalar field in a background AdS metric and the solution of the associated classical equation of motion. Through the AdS/CFT dictionary,[46] one can map this onto the two-point Green function in a Schrödinger-invariant theory. The result turns out to be identical to the one obtained previously from the Schrödinger Ward identities (see also below).[48] This is not just some abtract correspondence, as shown by a recent application to the physics of cold atoms.[49]

    This appears to be a very promising field for future applications of the Schrödinger–Virasoro algebra.
12. *Ageing phenomena*. When a many-body system is brought rapidly, via a fast change ('*quench*') of its thermodynamic parameters, into the coexistence region of its phase diagram, the competition between the several equivalent thermodynamic states may lead to an extremely slow relaxation behaviour towards these stationary states, so that the relaxation time scale becomes formally infinite. Furthermore, the state of the system depends on the time since the quench, and hence time-translation-invariance no longer holds. If in addition the time-dependent observables display dynamical scaling, we say that the system undergoes *physical ageing*. Although physical ageing is habitually associated with glasses, it can also occur for non-glassy systems – a simple example being the so-called *phase-ordering kinetics* of a simple magnet (e.g. described by an Ising model) quenched from an initial high-temperature state to a very low temperature. It is well-accepted that in this situation the conditions for physical ageing are all satisfied. Furthermore, the value of the dynamical exponent z is 2 for a purely relaxational dynamics without any conservation

---

[45]G. Parisi and N. Sourlas, Nucl. Phys. **B206**, 321 (1982); J. Tailleur, S. Tanase-Nicola and J. Kurchan, J. Stat. Phys. **122**, 557 (2006).

[46]See e.g. O. Aharony, S.S. Gubser, J. Maldacena, H. Ooguri and Y. Oz, Phys. Rep. **323**, 183 (2000).

[47]E.g. C. Leiva and M.S. Plyushchay, Ann. of Phys. **307**, 372 (2003); K. Balasubramanian and J. McGreevy, Phys. Rev. Lett. **101**, 061601 (2008); D.T. Son, Phys. Rev. **D78**, 046003 (2008). See C. Duval, M. Hassaïne and P. Horváthy, Ann. of Phys. **324**, 1158 (2009) for the relationship of AdS/CFT to the geometry of Bargmann structures.

[48]D. Minic and M. Pleimling, Phys. Rev. **E78**, 061108 (2008).

[49]C.A. Fuertes and S. Moroz, Phys. Rev. **D79**, 106004 (2009).

laws.[50] This is a case for non-equilibrium space-time dynamical scaling, although the stationary equilibrium state is a trivial ordered state without any static scaling behaviour.

For a description of such phenomena from the perspective of dynamical symmetry, we first observe that only sub-algebras of the Schrödinger algebras without the time-translations generated by L_{-1} can be used. Secondly, the equation of motion of the order-parameter is a *stochastic Langevin equation* (in a physicist's notation)

$$2\mathscr{M}\partial_t\phi = \Delta_r\phi + \frac{\delta\mathscr{V}[\phi]}{\delta\phi} + \eta \tag{15}$$

where $\mathscr{V}[\phi]$ is a Ginzburg-Landau potential, and η is a 'noise' term, usually assumed to be Gaussian, centered and with variance $\langle\eta(t,\boldsymbol{r})\eta(t',\boldsymbol{r}')\rangle = 2T\delta(t-t')\delta(\boldsymbol{r}-\boldsymbol{r}')$. Here T is the temperature of the heat bath to which the system is coupled. Often, the initial state is also taken from some random ensemble and has to be averaged over. Ageing behaviour is conveniently studied physically through the two-time correlation and response functions (averaged over thermal noise and/or the initial conditions – for simplicity the spatial coordinates are suppressed)

$$C(t,s) := \langle\phi(t)\phi(s)\rangle,\quad R(t,s) := \left.\frac{\delta\langle\phi(t)\rangle}{\delta h(s)}\right|_{h=0} = \langle\phi(t)\widetilde{\phi}(s)\rangle \tag{16}$$

where, again, $\widetilde{\phi}$ is the response field associated to the order-parameter ϕ and the last relation follows from the response-field formalism.[51]

As we have already mentioned above for the Burgers equation, the noise term destroys any non-trivial symmetry. This does not mean that symmetry principles cannot be used in this case, however. Using the *response-field formalism*,[51] the effective action can often be decomposed, viz. $\mathscr{J}[\widetilde{\phi},\phi] = \mathscr{J}_0[\widetilde{\phi},\phi] + \mathscr{J}_b[\widetilde{\phi}]$, into a 'deterministic term' $\mathscr{J}_0[\widetilde{\phi},\phi] = \int \mathrm{d}t\mathrm{d}\boldsymbol{r}\,\left[\widetilde{\phi}(2\mathscr{M}\partial_t - \Delta_r)\phi + \widetilde{\phi}\frac{\delta\mathscr{V}}{\delta\phi}\right]$ which can be Schrödinger-invariant and a 'noise term', which for a purely thermal Gaussian noise takes the form $\mathscr{J}_b[\widetilde{\phi}] = T\int \mathrm{d}t\mathrm{d}\boldsymbol{r}\,\widetilde{\phi}^2$. The Galilei-invariance of the 'deterministic term' – or more precisely, the invariance under the central mass generator M_0 – leads to the celebrated *Bargmann superselection rules*,[52] which in our case state that the correlators

[50] A.J. Bray, Adv. Phys. **43**, 357 (1994).

[51] P.C. Martin, E.D. Siggia and H.A. Rose, Phys. Rev. **A8**, 423 (1973); H.-K. Janssen, Z. Phys. **B23**, 377 (1976); C.J. De Dominicis, J. Physique (Colloque) **37**, 247 (1976).

[52] V. Bargman, Ann. of Math. **56**, 1 (1954). This result depends on the validity of the transformation law (6).

$\left\langle \underbrace{\phi\phi\cdots\phi\phi}_{n \text{ times}} \underbrace{\widetilde{\phi\phi}\cdots\widetilde{\phi\phi}}_{m \text{ times}} \right\rangle_0$ vanish for $n \neq m$. Here, the index 0 indicates that the averages are calculated via functional integrals within the deterministic theory, characterized by the effective action $\mathscr{J}_0$, as follows: $\langle \mathscr{A} \rangle_0 = \int \mathscr{D}\widetilde{\phi}\mathscr{D}\phi\, \mathscr{A}[\phi] e^{-\mathscr{J}_0[\widetilde{\phi},\phi]}$. A formal perturbation series in the noise term $\mathscr{J}_b$ (which corresponds e.g. to a small-T expansion) should truncate because of the Bargmann superselection rule. A generic result is then that the two-time response $R(t,s) = \langle \phi(t)\widetilde{\phi}(s)\rangle_0$ should not depend explicitly on the noise at all! Finally, if $\mathscr{J}_0$ has a larger symmetry than mere Galilei-invariance, we can use this to constrain the form of the two-time response function by requiring that $R(t,s)$ should transform covariantly under these. This leads to the Schrödinger Ward identities, which give a system of linear differential equations for $R(t,s)$, viz. $\mathscr{X} R(t,s) = 0$ where the two-particle operators $\mathscr{X}$ are derived from the *ageing subalgebra* $\mathfrak{age}(d) \subset \mathfrak{sch}(d)$ in which the generator of time-translations, L_{-1}, has been taken out. From these, we derive the scaling form $R(t,s) = s^{-1-a} f_R(t/s)$, and further predict explicitly the scaling function $f_R(y) = f_0 y^{1+a'-\lambda_R/z}(y-1)^{-1-a'}$, where λ_R, a and a' are ageing exponents and f_0 is a normalisation constant.[53] In this way, the conceptual difficulty that stochastic equations cannot have non-trivial dynamical symmetries has been eliminated.[54]

The available evidence, including several exactly solved models, but which also comes from large-scale numerical simulations of systems for which an exact solution is not known, allows to test the predicted $f_R(y)$. The results strongly suggest that the ageing behaviour in phase-ordering kinetics should have indeed a dynamical special Schrödinger symmetry in addition to dynamical scaling. Correlators can be treated in an analogous manner.[55]

13. *Extensions.*

A recent development considers generalised predictions for co-variant two-point functions. In *logarithmic conformal theories*,, which one may obtain heuristically by replacing the conformal weights $\mu \mapsto \begin{pmatrix} \mu & 1 \\ 0 & \mu \end{pmatrix}$ by a Jordan matrix, logarithmic terms arise in the two-point functions.[56] Similarly, logarithmic extensions of both the Schrödinger and the conformal Galilei algebras have been constructed, with analogous results for the two-point functions.[57] Further

[53]Time-translation-invariance would give $a = a' = \lambda_R/z - 1$.

[54]A. Picone and M. Henkel, Nucl. Phys. **B688**, 217 (2004).

[55]M. Henkel, M. Pleimling, C. Godrèche and J.-M. Luck, Phys. Rev. Lett. **87**, 265701 (2001); M. Henkel, Nucl. Phys. **B641**, 405 (2002); M. Henkel and F. Baumann, J. Stat. Mech. P07015 (2007); M. Henkel and M. Pleimling, *Non-equilibrium phase transitions, vol 2: Ageing and dynamical scaling far from equilibrium*, Springer (2010).

[56]V. Gurarie, Nucl. Phys. **B410**, 535 (1993); M. Rahimi-Tabar *et al.*, Nucl. Phys. **B497**, 555 (1997).

[57]A. Husseiny and S. Rouhani, J. Math. Phys. at press (2010), arxiv:1001.1036.

generalisation to logarithmic representations of $\mathfrak{age}(d)$ hint at the possibility that the critical contact process (which is in the directed percolation universality class) might be described in terms of logarithmic ageing-invariance.[58] Such a result would at least be analogous to $2D$ critical percolation, where a relationship with logarithmic conformal invariance is well-established.[59]

It is currently actively investigated how to formulate, in general, similar principles for arbitrary values of the dynamical exponent z (this is needed for example when discussing non-equilibrium critical dynamics, realised after a quench to the critical temperature $T = T_c > 0$). We hope to be able to construct in this way a systematic theory, to be called *local scale-invariance*. The constraints coming from the mathematical consistency play an important rôle in its construction. Besides the value of z itself, the following ingredients might turn out to be of relevance: (a) the form of the locally scale-invariant deterministic part of the equations of motion, (b) since the equation of motion for the order-parameter is a stochastic Langevin equation, the Bargmann superselection rules must be generalizable to generic values of z, in order to be able to use[58] the symmetries of the 'deterministic part' in the analysis of the full stochastic equation.[60] (c) The causality of the response functions, recast into dispersion relations, might be a way to make contact with the Schrödinger–Virasoro algebra and its generalizations. This will have to be combined with (d) some kind of short-time expansion in order to explore possible operator product algebras.

This (incomplete) list of physical problems related to Schrödinger-invariance illustrates how much remains still to be understood about Schrödinger symmetry. The physicist's hope will be that the methods described in this book may contribute to the solution of these physical questions. On the other hand, it is to be expected that the presentation of the current state of the theory of Schrödinger symmetry given in this book will by itself stimulate further mathematical questions. To what extent this will eventually become reality might depend, and not in a small way, on the attitude of the reader.

Nancy *Malte Henkel*

---

[58] M. Henkel, arxiv:1009.4139.

[59] P. Mathieu and D. Ridout, Phys. Lett. **B657**, 120 (2007).

[60] F. Baumann and M. Henkel, work in progress; X. Durang and M. Henkel, J. Phys. **A42**, 395004 (2009).

Preface

The object of the present monography is to give an up-to-date, self-contained presentation of a recently discovered mathematical structure: the *Schrödinger–Virasoro algebra*. The study of the structure of this infinite-dimensional Lie algebra containing the Virasoro algebra, and the various contexts in which it appears naturally, will lead us to touch upon such different topics as mechanics, statistical physics, Poisson geometry, integrable systems, supergeometry, representation theory and cohomology of infinite-dimensional Lie algebras, spectral theory of Schrödinger operators.

The original motivation for introducing the Schrödinger–Virasoro algebra was the following (see the Preface for a more physically minded point of view).

There is, in the physical literature of the twentieth century, a deeply rooted belief that physical systems – macroscopic systems for statistical physicists, quantum particles and fields for high energy physicists – could and should be classified according to their symmetries.

Let us just point at two very well-known physical examples : *elementary particles* on the $(3+1)$-dimensional Minkowski space-time, and *two-dimensional conformal field theory*.

1. From the point of view of *covariant quantization*, introduced at the time of Wigner back in the 1930s [1], *elementary particles* of relativistic quantum mechanics (of positive mass, say) may be described as irreducible unitary representations of the Poincaré group $P_4 \simeq SO_0(3,1) \ltimes \mathbb{R}^4$, the group of affine isometries of Minkowski space-time, or in other words the semi-direct product of the Lorentz group of rotations and relativistic boosts by space-time translations; the physical states of a particle of mass $m > 0$ and spin $s \in \frac{1}{2}\mathbb{N}$ are in bijection with the states of the Hilbert space corresponding to the associated irreducible representation of P_4.

[1] see e.g. references and a history of the subject in: S. Weinberg. *The quantum theory of fields.*, Cambridge University Press (1996).

This *covariant quantization* was revisited by the school of Souriau in the 1960s and 1970s [2] as a particular case of *geometric quantization*. The latter is a piece of a wide program of geometrization of classical and quantum mechanics. It allows a geometric construction of a mapping from *classical observables* on a symplectic manifold $\mathcal{M}$ (the *phase space* of Hamiltonian mechanics) to operators (that is, *quantum observables*) on L^2-sections of a fiber bundle obtained from $\mathcal{M}$ by polarization. In most physical cases, there is a non-trivial Lie group of symmetries of the symplectic manifold, which is represented in this framework by unitary operators. As for the electromagnetic field, it appears as a perturbation of the underlying symplectic structure; this principle is described by Souriau as *symplectic materialism*. One of the main outcomes of this program is a general method for constructing *wave equations* that are *covariant under the action of a group of symmetries* preserving a certain particular *geometric structure*.

2. *Two-dimensional conformal field theory*, on the other hand, is an attempt at understanding the universal behaviour of two-dimensional statistical systems at *equilibrium* and at the *critical temperature*, where they undergo a second-order phase transition (see the Preface for details and references). Starting from the basic assumption of translational and rotational invariance, together with the fundamental hypothesis (confirmed by the observation of the fractal structure of the systems and the existence of long-range correlations, and made into a cornerstone of renormalization-group theory) that scale invariance holds at criticality, one is [3] naturally led to the idea that invariance under the whole conformal group $\mathrm{Conf}(d)$ should also hold. This group is known to be finite-dimensional as soon as the space dimension d is larger than or equal to three, so physicists became very interested in dimension $d = 2$, where *local conformal transformations* are given by *holomorphic or anti-holomorphic functions*. A systematic investigation of the theory of *representations* of the *Virasoro algebra* (considered as a *central extension* of the algebra of infinitesimal holomorphic transformations) in the 1980s led to introduce a class of physical models (called *unitary minimal models*), corresponding to *degenerate unitary highest weight representations* of the Virasoro algebra with central charge less than one. Miraculously, covariance alone is enough to allow the computation of the statistic correlations – or so-called *n-point functions* – for these highly constrained models.

Let us emphasize in particular the following well-known facts, to which we shall refer several times in the sequel. Covariance under projective transformations (or homographies) $z \mapsto \frac{az+b}{cz+d}$ fixes up to a constant two- and three-point functions. On the other hand, four-point functions $\langle \phi_1(z_1) \dots \phi_4(z_4) \rangle$ are fixed only up to a *scaling function* depending on the cross-ratio $\frac{(z_1-z_3)/(z_1-z_4)}{(z_2-z_3)/(z_2-z_4)}$.

[2] J.-M. Souriau. *Structure des systèmes dynamiques*. Maîtrises de mathématiques, Dunod, Paris (1970).

[3] For systems with sufficiently short-ranged interactions (see Preface for details and counterexamples).

* * *

The *Schrödinger–Virasoro algebra* was originally discovered by M. Henkel [4] in 1994 while he was trying to apply the concepts and methods of conformal field theory to models of statistical physics which either undergo a dynamics, whether in or out of equilibrium, or are no longer isotropic. The idea was that, contrary to the case of relativistic physics or conformal field theory, the different coordinates (called for convenience *time* and *space*) should not play the same rôle. Replacing *isotropic scale transformations* $\boldsymbol{r} \mapsto \lambda\boldsymbol{r}$ with *anisotropic dilatations* $(t,\boldsymbol{r}) \mapsto (\lambda^z t, \lambda\boldsymbol{r}), \lambda > 0$ with $z \neq 1$ changes fundamentally the geometry. Then natural questions arise, such as: Is there anything like conformal geometry for $z \neq 1$? Does there exist a notion of local scale invariance in low dimensions as in the case of conformal field theory?

It turns out that the answer to the first question is *positive* for $z = 2$. The geometric theory has been developed by C. Duval, H. Künzle, P. Horváthy and other authors [5]. Lorentzian geometry has to be replaced with the so-called *Newton-Cartan geometry* in $(1+d)$ dimensions, which is the right geometric framework for Newtonian (instead of relativistic) mechanics, defined by a one-form (locally written dt) representing the time-coordinate, by a metric structure on the fibers $t = \text{Cst}$, and by a (alas non unique) connection preserving these two. Various covariant wave equations may be obtained using the tools of geometric quantization, among which the two simplest ones: the *free Schrödinger equation* $\Delta_0\psi := (-2\mathrm{i}\mathcal{M}\partial_t - \Delta_r)\psi = 0$, and the *Dirac-Lévy Leblond equation*, which is associated to the Schrödinger equation in the same way as the Dirac equation to the Laplace equation $\Delta_r\psi = 0$ in the relativistic setting [6].

Contrary to the conformal case, there are several groups of symmetries associated to the Newton-Cartan geometry, among which (by increasing order with respect to inclusion) the *Galilei group*, the *Schrödinger group*, and also (by weakening the assumptions though) the *Schrödinger–Virasoro group*.

The *Galilei group* $\mathrm{Gal}(d)$ is the group generated by time translations and space rotations, and by motions with constant speed. It is the symmetry group of classical mechanics; the induced changes of frames leave invariant the equations of classical physics. It also leaves invariant the Euler equation for perfect, incompressible fluids without viscosity, obtained from Newton's equation of dynamics by a shift of point of view from Lagrangian to Eulerian mechanics.

The *Schrödinger group* $\mathrm{Sch}(d)$ is the group of *projective Lie symmetries* of the free Schrödinger equation $\Delta_0\psi(t,r) := (-2\mathrm{i}\mathcal{M}\partial_t - \Delta_r)\psi(t,r) = 0$, see Preface, including a subgroup of *time-homographies coupled with space-transformations*

[4] M. Henkel. *Schrödinger invariance and strongly anisotropic critical systems*, J. Stat. Phys. **75**, 1023 (1994).

[5] See references to articles by these authors in Chap. 1.

[6] J.-M. Lévy-Leblond. *Nonrelativistic particles and wave equations*, Comm. Math. Phys. **6**, 286 (1967).

that is isomorphic to the group of two-by-two matrices with determinant one, $SL(2,\mathbb{R})$. It contains the *Bargmann group*, which is a central extension of the Galilei group. In physical applications, the value of the central element $M_0 \in \mathfrak{sch}(d) = Lie(\mathrm{Sch}(d))$ is interpreted as the mass of the particle or the total mass of the system. The Schrödinger group also preserves a number of field equations coming either from non-relativistic condensed matter physics (e.g. Landau liquids or Bose-Einstein condensation, see 5. in the Preface) or from a direct application of the principles of geometric quantization (e.g. non-relativistic electromagnetism, see 6. in the Preface).

The *Schrödinger–Virasoro group* $\mathrm{SV}(d)$ is obtained by removing the requirement that symmetries should preserve the particular choice of connection. One then obtains *for any space dimensionality d* an infinite-dimensional Lie group, with Lie algebra $\mathfrak{sv}_d$ given by (8) in the Preface. Details are given in Chap. 1. We concentrate in this book on the particular case $d=1$. Then $\mathrm{SV} = \mathrm{SV}(1)$ is a *semi-direct product* of $\mathrm{Diff}(S^1)$ – the *group of diffeomorphisms of the circle* – by a rank-2 infinite-dimensional nilpotent group, $\mathrm{SV} \simeq \mathrm{Diff}(S^1) \ltimes H$, with typical element denoted by $(\phi;(\alpha,\beta))$, $\phi \in \mathrm{Diff}(S^1)$, $\alpha,\beta \in C^\infty(S^1)$ [7]. By centrally extending $\mathrm{Diff}(S^1)$ to the Virasoro group Vir, one may also consider the group $\widetilde{\mathrm{SV}} = \mathrm{Vir} \ltimes H$ and its Lie algebra $\widetilde{\mathfrak{sv}} = \mathfrak{vir} \ltimes \mathfrak{h}$, whose center is generated by the mass generator M_0 and by the central generator in $\mathfrak{vir}$.

Schrödinger transformations in $\mathrm{Sch} \subset \mathrm{SV}$ play the same rôle as projective transformations in conformal invariance. Computations show that covariance under Schrödinger transformations only fixes two-point functions, which are given up to a constant by the heat kernel. In three-point functions, contrary to the conformal case (see above), an arbitrary scaling function of some complicated expression in terms of the coordinates appears. The consequences of the postulate of covariance under the Schrödinger group or subgroups of it have been explored quite systematically by M. Henkel and his collaborators [8], and proved analytically or observed numerically on several models, in particular in cases where *physical ageing* sets in. A more detailed account is given in the Preface. While these are interesting, the theory is undeniably not as far-reaching as 2d-conformal field theory because the Schrödinger group is *finite-dimensional*. On the other hand, the *Schrödinger–Virasoro group* may seem at first sight to be the right candidate for local-scale invariance. However, these symmetries have not yet been observed on physical models. Maybe this approach is doomed to fail because it is even difficult to find wave equations invariant under the Schrödinger–Virasoro group [9].

* * *

---

[7]The elements in $SL(2,\mathbb{R})$ coming from $\mathrm{Sch}(1)$ represent the finite projective transformations in the $\mathrm{Diff}(S^1)$-factor.

[8]See M. Henkel, M. Pleimling. *Nonequilibrium phase transitions. Vol. 2, Ageing and dynamical scaling far from equilibrum*, Springer (2010).

[9]R. Cherniha, M. Henkel. *On non-linear partial differential equations with an infinite-dimensional conditional symmetry*, J. Math. Anal. Appl. **298**, 487 (2004).

While the above approach, developed in Chap. 1, has led, for the time being, neither to further developments nor to applications, other fruitful points of view on the Schrödinger–Virasoro group have gradually emerged.

1. The approach closest to the previous one is to see the Schrödinger–Virasoro group not as a symmetry group of a given wave equation, but as a *reparametrization group* for a given class of equations. This way of seeing things is well-known in the case of the Virasoro group: the *group of diffeomorphisms of the circle*, $\mathrm{Diff}(S^1)$, acts on the space of *Hill operators* $\partial_x^2 + u_0(x)$, $x \in \mathbb{R}/2\pi\mathbb{Z}$, otherwise known as periodic Sturm-Liouville operators on the line. This affine action, σ^{Hill}, is a left-and-right action,

$$\phi \mapsto$$
$$\left(\partial_x^2+u_0(x) \mapsto \sigma^{Hill}(\phi)(\partial_x^2+u_0(x)) := \pi_{3/2}^{Hill}(\phi)\circ(\partial_x^2+u_0(x)) \circ \pi_{-\frac{1}{2}}^{Hill}(\phi)^{-1}\right), \tag{1}$$

where $\pi_\lambda^{Hill}(\phi)u_0(x) := (\frac{d\phi}{dx})^\lambda (u_0 \circ \phi^{-1})(x)$ is the natural action of $\mathrm{Diff}(S^1)$ on the space of $(-\lambda)$-densities, also called (without specifying the weight) *tensor densities*. The orbits of this action have been classified independently by A.A. Kirillov [10] on the one hand, and by V.F. Lazutkin and T.F. Pankratova [11] on the other. A.A. Kirillov actually deduces a set of *normal forms* for the orbits from a classification up to conjugacy of all possible generators of the symmetry subgroups (also called *isotropy subgroups*) of Sturm-Liouville operators. The main characteristic of a given orbit is its *monodromy*, namely, the Floquet matrix in $SL(2,\mathbb{R})$ relating $\psi(x)$ to $\psi(x+2\pi)$, where $\psi = \begin{pmatrix}\psi_1\\ \psi_2\end{pmatrix}$ is the general solution of the Hill equation $(\partial_x^2 + u_0)\psi(x) = 0$. The eigenvalues of this matrix give the behaviour at infinity of the solutions. Another important characteristic is the oscillatory character of the solutions [12]. Generic orbits are of *type I* (*elliptic type* [13] with *oscillatory solutions*, or *hyperbolic type* with *non-oscillatory* solutions) or of *type II* (*hyperbolic type*, with *oscillatory solutions*), but there are also non-generic orbits of *type III*, with *unipotent monodromy*.

The above analysis carries over very nicely to the space of *generalized harmonic oscillators*, which are time-dependent Schrödinger operators in (1+1)-

[10] A.A. Kirillov. *Infinite-dimensional Lie groups: their orbits, invariants and representations. The geometry of moments*, Lecture Notes in Mathematics **970**, 101–123 (1982).

[11] V.F. Lazutkin, T.F. Pankratova. *Normal forms and versal deformations for Hill's equations*, Funct. Anal. Appl. **9** (4), 306–311 (1975).

[12] *Oscillatory solutions* have an infinite number of zeroes, *non-oscillatory solutions* have at most one zero.

[13] Two-by two matrices of determinant one with eigenvalues $e^{i\theta}$, $0 < \theta < \pi$, resp. $e^{\pm t}$, $t > 0$, resp. ± 1, are called *elliptic*, resp. *hyperbolic*, resp. *unipotent*.

dimensions of a very special type,

$$\mathscr{S}^{aff}_{\leq 2} := \left\{-2\mathrm{i}\mathscr{M}\partial_t - \partial_r^2 + V_0(t) + V_1(t)r + V_2(t)r^2\right\}, \qquad (2)$$

with time-dependent periodic quadratic potential. This space is preserved by the following affine action of the Schrödinger–Virasoro group,

$$\sigma_{1/4} : (\phi; (\alpha, \beta)) \mapsto \left(D \mapsto \pi_{5/4}(\phi; (\alpha, \beta)) \circ D \circ \pi_{1/4}(\phi; (\alpha, \beta))^{-1}\right), \qquad (3)$$

where the *vector-field representation* π_λ exponentiates the realization $d\pi_\lambda$ of SV found by extrapolation by M. Henkel, as explained in the Preface (formulas are recalled at the end of the Introduction). We summarize here the contents of Chap. 9. In the semi-classical limit, these operators give back the Hill operators, which explains why part of this action is equivalent to σ^{Hill}. The new feature here is the existence of an exact *invariant* I, sometimes called *Ermakov-Lewis invariant* [14] – a time-dependent second-order differential operator in the space coordinates, commuting with the Schrödinger operator –. When I has a discrete spectrum, the solutions of the Schrödinger operator are the eigenfunctions of I, multiplied by an explicit time-dependent *phase* yielding the *monodromy operator* – a bounded operator on $L^2(\mathbb{R})$ this time – ; contrary to the usual *adiabatic scheme* [15], these results are non-perturbative. Let us now come to our results:

- the orbit classification due to A.A. Kirillov carries over to the case of generalized harmonic oscillators for *generic* orbits. There also appear supplementary, *non-generic* orbits of type I or type III, due to a *resonance* between the *quadratic* and the *linear parts of the potential.* A choice of normal forms for these orbits, together with the associated isotropy subgroups, is given in Sect. 9.2.4. The isotropy subgroups are shown to be isomorphic to subgroups of the Schrödinger group or of some covering of it.
- the *invariant I* may be reconstructed from the *orbital data,* thus leading to a *unification of the algebraic, geometric and analytic methods.* One may show that I is conjugate to some simple *model operator* belonging to a finite family of operators [16].
- even when the spectrum of I is non-discrete (so that the standard adiabatic scheme is not valid), the monodromy operator is shown to be essentially conjugate to a *multiplication operator* $L^2(\Sigma) \to L^2(\Sigma),\ f(\lambda) \mapsto e^{\mathrm{i}\lambda T} f(\lambda)$,

---

[14] H.R. Lewis, W.B. Riesenfeld. *An exact quantum theory of the time-dependent harmonic oscillator and of a charged particle in a time-dependent electromagnetic field*, J. Math. Phys. **10** (8), 1458–1473 (1969).

[15] See e.g. A. Joye, *Geometric and mathematical aspects of the adiabatic theorem of quantum mechanics*, Ph. D. Thesis, Ecole Polytechnique Fédérale de Lausanne (1992).

[16] Namely, the *free Laplacian*, the *harmonic oscillator*, the *"harmonic repulsor"* and the *Airy operator*.

where $\Sigma \subset \mathbb{R}$, $\Sigma = \frac{1}{2} + \mathbb{N}$, $\mathbb{R}$ or $\mathbb{R}_+$ is the *spectrum* of I, and T is the *period* of the orbit.

2. The above action σ^{Hill} on Hill operators equivalent to the *coadjoint action* of the Virasoro group, and thus Hamiltonian for the *Kirillov-Kostant-Souriau Poisson structure* on the dual of the Virasoro algebra. The latter is the simplest of a family of compatible (i.e. mutually commuting) infinite-dimensional Poisson structures, a so-called *hierarchy*. Such structures usually come up with infinite families of Hamiltonian equations induced by commuting Hamiltonian operators, thus implying their complete integrability. This is the most efficient scheme to construct completely integrable systems. In the infinite-dimensional case, one gets *integrable partial differential equations* (rather than ordinary differential equations), the simplest one in our context being the *Korteweg-De Vries equation* $\frac{\partial u}{\partial t} = u\frac{\partial u}{\partial x} + \frac{\partial^3 u}{\partial x^3}$, which describes an *isospectral deformation* $\partial_x^2 + u_0(x) \rightsquigarrow \partial_x^2 + u(t,x)$ of a Hill operator [17]. The Poisson-Lie group Poisson bracket on the *Volterra group* integrating the *algebra of formal pseudo-differential symbols* of the type $u_{-1}(x)\partial_x^{-1} + u_{-2}(x)\partial_x^{-2} + \ldots$ leads by *symplectic reduction* to the same equation.

   Some but not all of these arguments may be reproduced in the case of the *linear action*

$$\tilde{\sigma}_\mu : (\phi; (\alpha,\beta)) \mapsto \left(D \mapsto \pi_{\mu+2}(\phi;(\alpha,\beta)) \circ D \circ \pi_\mu(\phi;(\alpha,\beta))^{-1}\right) \qquad (4)$$

   on the *linear space of time-dependent periodic Schrödinger operators*, $\mathcal{S}^{lin} := \{a(t)(-2\mathrm{i}\mathcal{M}\partial_t - \partial_r^2) + V(t,r)\}$; see Chap. 10. Note that, because the index μ has been shifted by two instead of one, compare with (3), the associated *affine space* $\mathcal{S}^{aff} := \{-2\mathrm{i}\mathcal{M}\partial_t - \partial_r^2 + V(t,r)\}$ is *not* preserved by this action. Contrary to the case of Hill operators, this action is unrelated to the coadjoint action of $\mathfrak{sv}$. On the other hand, it may be retrieved by symplectic reduction from the *current algebra* over a Volterra-type space of formal pseudo-differential symbols. Thus $\mathcal{S}^{lin}$ appears as a coadjoint orbit of a huge looped group integrating this current algebra; the $\tilde{\sigma}_\mu$-actions are in this sense coadjoint actions, and are Hamiltonian for the usual Kirillov-Kostant-Souriau bracket (see Theorem 10.2 in Sect. 10.6). This suggests of course to look for related integrable systems.

3. The above results point out to the importance of the point of view of *Poisson geometry* (see Chaps. 2, 10, 11). They rely on the embedding of the Schrödinger–Virasoro algebra as a *subquotient* of the *extended Poisson algebra* [18] on the two-dimensional torus, $\mathbb{C}[q,q^{-1}][p^{\frac{1}{2}},p^{-\frac{1}{2}}]$ with Poisson bracket $\{F,G\} =$

[17] For a beautiful introduction to the subject, see e.g. P.G. Drazin and R.S. Johnson, *Solitons: an introduction*, Cambridge Texts in Applied Mathematics, Cambridge University Press, Cambridge (1989), or the monography by L. Guieu and C. Roger cited in the bibliography.

[18] Unfortunately, the natural embedding of $\mathrm{Vect}(S^1) \subset \mathfrak{sv}$ into the Poisson algebra does not extend to $\mathfrak{sv}$.

$\frac{\partial F}{\partial p}\frac{\partial G}{\partial q} - \frac{\partial G}{\partial p}\frac{\partial F}{\partial q}$, or of its natural quantization, the algebra of *formal pseudo-differential symbols* on the line $\mathbb{C}[\xi,\xi^{-1}][[\partial_\xi^{\frac{1}{2}},\partial_\xi^{-\frac{1}{2}}]]$ with Lie bracket induced by the natural associative product of operators. This point of view may be generalized to a supersymmetric setting, yielding a large class of *superizations of the Schrödinger–Virasoro algebra* $\mathfrak{sns}^{(N)}$ called *Schrödinger-Neveu-Schwarz algebras*, containing algebras of *super-contact* vector fields (often called *super-conformal algebras* in the literature), namely, algebras of super-vector fields preserving the kernel of the super-contact form $dq + \sum_{i=1}^{N} \theta^i d\theta^i$. These arguments will be developed in Chap. 11 (see Definition 11.20 in Sect. 11.3). Similarly to what happened originally with the Schrödinger–Virasoro algebra, its N = 2 supersymmetric generalization, $\mathfrak{sns}^{(2)}$ extending the *Neveu-Schwarz algebra* [19], arises as a natural infinite-dimensional extrapolation of the symmetry generators of a *supersymmetric Schrödinger equation* (see Sect. 11.2.1), obtained from the *supersymmetric model* on super-space-time $\mathbb{R}^{(3|2)}$ with coordinates $(t,\zeta,r;\theta_1,\theta_2)$ by a Fourier transform with respect to ζ, formally $\partial_\zeta \rightsquigarrow \mathrm{i}\mathcal{M}$ (see 8. in the Preface for the analogous construction relating the Klein-Gordon equation in $(d+2)$-dimensions to the Schrödinger equation $(-2\mathrm{i}\mathcal{M}\partial_t - \Delta_r)\psi_{\mathcal{M}}(t,r) = 0$ in $(d+1)$-dimensions).

* * *

Leaving aside the above geometrical and physical aspects, which may be considered as a motivation or maybe as applications, one may also study the *algebraic properties* of this infinite-dimensional Lie algebra for its own sake, which is the subject of Chaps. 3–7.

The Schrödinger–Virasoro algebra $\mathfrak{sv}$ is a semi-direct product of the *centerless Virasoro algebra* $\mathrm{Vect}(S^1) \simeq \langle L_n, n \in \mathbb{Z}\rangle$ – seen as the Lie algebra of *smooth vector fields on the circle*, and naturally identified with the Lie algebra of the diffeomorphism group $\mathrm{Diff}(S^1)$ introduced above – by an infinite-dimensional, rank 2 nilpotent Lie algebra

$$\mathfrak{h} \simeq \left\langle Y_m, M_p, m \in \frac{1}{2} + \mathbb{Z}, p \in \mathbb{Z} \right\rangle \tag{5}$$

In other words, $\mathfrak{sv} \simeq \mathrm{Vect}(S^1) \ltimes \mathfrak{h}$ where $[\mathfrak{h},[\mathfrak{h},\mathfrak{h}]] = 0$. The generic element may be written in terms of three Laurent series [20],

$$\mathcal{L}_f + \mathcal{Y}_g + \mathcal{M}_h := \sum_{n\in\mathbb{Z}} f_n L_n + \sum_{m\in\frac{1}{2}+\mathbb{Z}} g_m Y_m + \sum_{p\in\mathbb{Z}} h_p M_p, \tag{6}$$

where $f(z) = \sum f_n z^{n+1}$, $g(z) = \sum g_m z^{m+\frac{1}{2}}$ and $h(z) = \sum h_p z^p$.

---

[19] A. Neveu and J.H. Schwarz. *Factorizable dual model of pions*, Nucl. Phys. **31**, 86 (1971).

[20] We then use *calligraphic letters* to avoid any confusion with the Laurent components.

Semi-direct products of the type $\mathrm{Vect}(S^1) \ltimes \mathscr{F}_\lambda$, where $\mathscr{F}_\lambda$ is a *tensor density module* of $\mathrm{Vect}(S^1)$ – the semi-classical analogue of a *primary operator* of weight $\lambda + 1$ in the language of conformal field theory –, have been studied in great details [21], in particular from the cohomological point of view. The Lie algebra $\mathfrak{sv}$, as a $\mathrm{Vect}(S^1)$-module, is isomorphic to $\mathrm{Vect}(S^1) \ltimes (\mathscr{F}_{\frac{1}{2}} \oplus \mathscr{F}_0)$, *but* $\mathfrak{h} \simeq \mathscr{F}_{\frac{1}{2}} \oplus \mathscr{F}_0$ (as a vector space) is *not* abelian. Hence $\mathfrak{sv}$ may be viewed as the next example in the order of increasing complexity. The techniques developed by D.B. Fuks [22] also apply to the case of $\mathfrak{sv}$ and allow a *detailed* and almost exhaustive *cohomological study* (mainly concerning *deformations* and *central extensions*) in Chap. 7. We show in particular (see Theorem 7.2 in Sect. 7.2) that there exist exactly *three independent families of deformations* of the bracket of $\mathfrak{sv}$, among which a family denoted by $\mathfrak{sv}_\epsilon$, which is isomorphic to $\mathrm{Vect}(S^1) \ltimes (\mathscr{F}_{\frac{1}{2}+\epsilon} \oplus \mathscr{F}_{2\epsilon})$ as a $\mathrm{Vect}(S^1)$-module.

Chapters 3–6 are concerned with the study of different classes of representations of $\mathfrak{sv}$, inspired by the well-developed Virasoro theory: *unitary highest-weight modules* (or *induced representations*, or *Verma modules*), *coinduced representations*, and more specifically the *coadjoint representation*. Let us comment on each of these.

1. *Unitary highest-weight modules* (see Chap. 4) are induced from a *character* of the commutative algebra $\langle L_0, M_0 \rangle$, and characterized by their *mass* and by the *conformal weight* of their highest-weight vector. As opposed to the Virasoro case, non-trivial unitary highest-weight modules are all *non-degenerate* (see Theorem 4.2 in Sect. 4.1), so the theory seems to be of little interest.
2. *Coinduced representations*, on the other hand, provide an interesting *class of representations generalizing* the $\mathrm{Vect}(S^1)$-*tensor density modules*; one finds the following explicit formulas in Chap. 5, see Theorem 5.3 in Sect. 5.1,

$$d\tilde{\rho}(\mathscr{L}_f) = \left(-f(t)\partial_t - \frac{1}{2}f'(t)r\partial_r - \frac{1}{4}f''(t)r^2\partial_\zeta\right) \otimes \mathrm{Id}_{\mathscr{H}_\rho} + f'(t)d\rho(L_0)$$
$$+\frac{1}{2}f''(t)r d\rho(Y_{\frac{1}{2}}) + \frac{1}{4}f'''(t)r^2 d\rho(M_1); \tag{7}$$

$$d\tilde{\rho}(\mathscr{Y}_f) = (-f(t)\partial_r - f'(t)r\partial_\zeta)\otimes\mathrm{Id}_{\mathscr{H}_\rho} + f'(t)d\rho(Y_{\frac{1}{2}}) + f''(t)r\, d\rho(M_1); \tag{8}$$

$$d\tilde{\rho}(\mathscr{M}_f) = -f(t)\partial_\zeta \otimes \mathrm{Id}_{\mathscr{H}_\rho} + f'(t)\, d\rho(M_1). \tag{9}$$

---

[21] T. Tsujishita, *On the continuous cohomology of the Lie algebra of vector fields*, Proc. Japan Acad. Ser. A Math. Sci. **53** (4), 134–138 (1977), and V. Ovsienko, C. Roger. *Generalizations of Virasoro group and Virasoro algebra through extensions by modules of tensor-densities on* S^1, Indag. Math. (N.S.) **9** (2), 277–288 (1998).

[22] D.B. Fuks. *Cohomology of infinite-dimensional Lie algebras*, Contemporary Soviet Mathematics, Consultants Bureau, New York (1986).

where $\rho : \mathfrak{g}_0 \to Hom(\mathcal{H}_\rho, \mathcal{H}_\rho)$ is any representation of the solvable Lie subalgebra $\mathfrak{g}_0 := \langle L_0, Y_{\frac{1}{2}}, M_1 \rangle$. The known realizations of $\mathfrak{sv}$ as a *reparametrization group* of *wave equations* – in particular, of *Schrödinger* or *Dirac-Lévy-Leblond operators*, see Chap. 8 – all belong to this family. Such explicit formulas are easily obtained by taking into consideration the *Cartan prolongation structure* of $\mathfrak{sv}$ (see Theorem 5.1 in Sect. 5.1), based on the graduation of $\mathfrak{sv}$ given by the polynomial degree of the vector fields in the vector-field representation.

3. *Vertex representations* are defined in Chap. 6. Recall first that *primary operators* for vertex representations of the Virasoro algebra are *quantum fields* $X = \sum_n X_n z^n$ realized on a *Fock space*, whose Laurent components X_n are *quantized tensor densities*, i.e. they satisfy the commutation relations $[L_n, X_m] = (\lambda n - m)X_{n+m}$ for some λ; physicists call $\lambda + 1$ the *conformal weight* of X. Regarding time-and-space dependent operators as Laurent series in the *time coordinate* leads to a natural transposition of these notions to $\mathfrak{sv}$-*primary fields*. The operators defined in Chap. 6 are *primary* with respect to some of the above-defined coinduced representations. The *polynomial fields* which have been successfully constructed (see Theorems 6.6 and 6.7 in Sect. 6.3.2) are unfortunately *degenerate*, in the sense that the action of M_0 is *nilpotent*. Hence M_0 may not be interpreted as a *mass*. We conjecture though, see Sect. 6.5, the existence of *massive fields* – on which M_0 acts as a non-trivial scalar – and compute by a formal analytic extension their *two-* and *three-point functions*, see Theorems 6.9 and 6.10. The latter are different from but closely connected to the *conformally covariant three-point functions* in three dimensions. As shown in Chap. 2, although $\mathfrak{sch} \subset \mathfrak{conf}(3)_{\mathbb{C}}$ on the one hand, and $\mathfrak{sch} \subset \mathfrak{sv}$ on the other, there seems to be no reasonable way to combine conformal *and* Schrödinger–Virasoro symmetries, which makes the above results puzzling.

   Though these preliminary results may probably be extended, it would be reasonable to try first to work out a physical context in which these quantum fields would appear, in order to gain some intuition and to go beyond what may appear for the moment as a somewhat formal exercise.

4. Finally, the *coadjoint representation* (contrary to the Virasoro case) does not belong to the above family of representations, and is studied separately in Chap. 3. As in the case of the coadjoint action of the Virasoro group, the study of the isotropy subalgebras lead to differential equations which are easily solved[23]. The reader acquainted with the classification (obtained independently by O. Mathieu on the one hand and by C. Martin and A. Piard on the other [24]) of the representations of the Virasoro algebra may wonder whether the coadjoint

[23] A.A. Kirillov, op. cit., and the monography by L. Guieu and C. Roger, op. cit.

[24] C. Martin, A. Piard. *Classification of the indecomposable bounded admissible modules over the Virasoro Lie algebra with weight spaces of dimension not exceeding two*, Comm. Math. Phys. **150**, 465–493 (1992); and O. Mathieu, *Classification of Harish-Chandra modules over the Virasoro Lie algebra*, Invent. Math. **107**, 225–234 (1992).

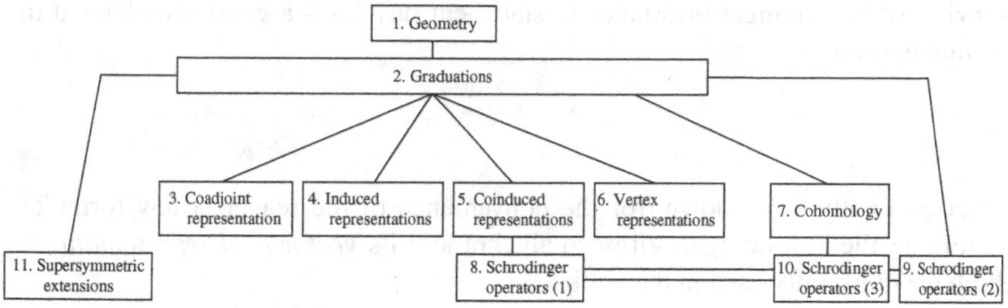

representation belongs to a new class of its own. For the moment, however, it looks isolated in the picture.

Let us give some suggestions for reading. Chapters 1 and 2 are introductive; definitions and results contained in these two chapters are frequently needed, and are quoted throughout the book. The cohomological results contained in Chap. 7 are not required for the other chapters (the deformations and central extensions used elsewhere, in particular in the chapters concerned with representation theory, are all introduced in Chap. 2), but the spectral sequence method used to compute central extensions of semi-direct products, together with Fuks' results on the Virasoro cohomology, are used once more in Chap. 10. Chapters 8–10, devoted to Schrödinger operators, have a strong thematic unity despite the differences in the methods and the language; an introductory discussion is placed at the beginning of Chap. 8, and Sect. 8.1 should be read first. Chapters 3–6 on representation theory also have a strong thematic unity, but may be read separately. Chapter 11 on supergeometry is an extension to the supersymmetric setting of Chap. 2.

The following diagram shows some possible orders of reading.

Almost all the material contained in this monograph has been published elsewhere, save for the introductory parts and Chap. 4. The first steps into Newton-Cartan geometry in Chap. 1 are a short summary of a theory developed by Duval, Horváthy, Künzle,... Chap. 9 includes a detailed account of classical results about Ermakov-Lewis quantum invariants and about Hill operators, in particular the orbit classification due to A.A. Kirillov; the classification of Hill operators by their lifted monodromy (Sect. 9.2.2), and the solution of the Hill equations and the determination of their monodromy in terms of the ξ-invariant in the isotropy subalgebra (Sect. 9.3.2) should be folklore results, but we have not found them in the literature. Apart from these, all other developments are due either to one of the authors or to both – including contributions of the author of the Preface.

The reader is assumed to have a background knowledge on the Virasoro algebra, with its applications to conformal field theory (for Chap. 6 mainly) and to integrable systems. Reading some sections of the frequently cited monograph *L'Algèbre et le Groupe de Virasoro: aspects géométriques et algébriques, généralisations* by L. Guieu and C. Roger [43] may help him find his way around. The two volumes of the (recently appeared) book *Nonequilibrium phase transitions* by M. Henkel, H. Hinrichsen, S. Lübeck and M. Pleimling [57, 58] give in particular an up-to-date overview of Schrödinger invariance in statistical physics – a good complement to this monograph.

* * *

Let us finally write down, for the convenience of the reader, a few formulas concerning the Schrödinger–Virasoro algebra and its *vector-field representations*, which are constantly used in the book.

Generators of the Schrödinger–Virasoro algebra

$$\mathfrak{sv} = \langle L_n,\, n \in \mathbb{Z}\rangle \ltimes \left\langle Y_m, M_p,\, m \in \frac{1}{2} + \mathbb{Z}, p \in \mathbb{Z}\right\rangle \simeq \mathrm{Vect}(S^1) \ltimes \mathfrak{h} \tag{10}$$

Commutation relations of the generators

$$[L_n, L_m] = (n-m)L_{n+m}; \quad [L_n, Y_m] = \left(\frac{n}{2} - m\right) Y_{n+m}, \quad [L_n, M_p] = -pM_{n+p} \tag{11}$$

and

$$[Y_{m_1}, Y_{m_2}] = (m_1 - m_2)M_{m_1+m_2},\ [Y_m, M_p] = [M_{p_1}, M_{p_2}] = 0 \tag{12}$$

with $n, p, p_1, p_2 \in \mathbb{Z}$ and $m, m_1, m_2 \in \frac{1}{2} + \mathbb{Z}$.

Vector-field representation: massive version $d\pi_\lambda$

One defines

$$d\pi_\lambda(L_n) = -t^{n+1}\partial_t - \lambda(n+1)t^n - \frac{1}{2}(n+1)t^n r\partial_r - \frac{\mathcal{M}}{4}(n+1)nt^{n-1}r^2$$

$$d\pi_\lambda(Y_m) = -t^{m+\frac{1}{2}}\partial_r - \mathcal{M}(m+\frac{1}{2})t^{m-\frac{1}{2}}r \tag{13}$$

$$d\pi_\lambda(M_n) = -\mathcal{M}t^n \tag{14}$$

where $\mathcal{M}$ is an arbitrary (non-zero) constant which plays the role of a mass parameter. These formulas agree with the original realization of $\mathfrak{sv}$ found by M. Henkel, see (8) in the Foreword, if one sets $\lambda = x/2$, where x is the *scaling dimension* of the field.

Vector-field representation: Fourier version $d\tilde{\pi}_\lambda$

$$d\tilde{\pi}_\lambda(L_n) = -t^{n+1}\partial_t - \lambda(n+1)t^n - \frac{1}{2}(n+1)t^n r\partial_r - \frac{1}{4}(n+1)nt^{n-1}r^2\partial_\zeta$$

$$d\tilde{\pi}_\lambda(Y_m) = -t^{m+\frac{1}{2}}\partial_r - \left(m+\frac{1}{2}\right)t^{m-\frac{1}{2}}r\partial_\zeta$$

$$d\tilde{\pi}_\lambda(M_n) = -t^n\partial_\zeta \tag{15}$$

Both families of representations are called *vector field representations* of $\mathfrak{sv}$. As a matter of fact, the representation $d\tilde{\pi}_\lambda$ may be deduced from $d\pi_\lambda$ by a formal Laplace transform with respect to the parameter $\mathcal{M}$.

Sometimes the scaling dimension is unimportant. In such cases we simply write sometimes $d\pi$ or $d\tilde{\pi}$.

Villeurbanne Cedex — *Claude Roger*
Vandoeuvre-les-Nancy — *Jérémie Unterberger*

Contents

Acronyms

$a\bar{b}$-theory	Boson and superboson algebra, 91
$\mathfrak{Bar}(d)$	Bargmann algebra, 4
$\mathfrak{conf}(d)$	Conformal Lie algebra in d space dimensions, Sect. 2.2
$\mathcal{D}^{aff}$	Affine space of Dirac operators, 157
D_{θ^a}	Left-invariant (super)derivatives, Sect. 11.2.1
$D\Psi D$, $D\Psi D_\xi$	algebra of extended pseudodifferential symbols, 211
$\mathcal{F}_\lambda$	Vect(S^1)-density modules, 9
$\mathfrak{fsv}$	Subalgebra of formal vector fields in $\mathfrak{sv}$, Theorem 5.1
$\mathfrak{g}$	Vect$(S^1) \ltimes \mathcal{L}_t(\widetilde{(\Psi D_r)_{\leq 1}})$, 220
G_D	Stabilizer of a Schrödinger operators D, 176
Gal(d), $\mathfrak{Gal}(d)$	Galilei group and algebra, 2
$\mathcal{L}_t(\widetilde{(\Psi D_r)_{\leq 1}})$	Looped centrally extended algebra of pseudodifferential symbols, Sect. 10.3.2
$\mathcal{M}$	Mass parameter, xxxiv
$\mathfrak{md}_\epsilon^d$	Lie algebra preserving multi-diagonal differential operators, Lemma 5.3
$\mathfrak{osp}(n\|2m)$	Orthosymplectic algebra, 254
$\mathcal{P}^{(2m\|N)}$	Super-Poisson algebra of functions on the $(2m\|N)$-supertorus, 251
$\mathcal{P}^{(2\|2)}_{\leq 2}$	Lie subalgebra of $\mathcal{P}^{(2\|2)}$ isomorphic to $\tilde{\mathfrak{s}}^{(2)}$, 252
$\mathcal{P}^{(2m\|2)}_{(2)} \subset \mathcal{P}^{(2m\|2)}$	Lie subalgebra of quadratic polynomials, isomorphic to $\mathfrak{osp}(2\|2m)$, 254
$\widetilde{\mathcal{P}}^{(2\|N)}$	Extended super-Poisson algebra, 258
$\widetilde{\mathcal{P}}^{(2\|N)}_{\leq \kappa}$	Subspace of $\widetilde{\mathcal{P}}^{(2\|N)}$, 258
S	Schwarzian derivative, Sect. 3.1
$\mathfrak{s}^{(2)}$	Algebra of Lie symmetries of the $(3\|2)$-supersymmetric model, isomorphic to $\mathfrak{osp}(2\|4)$, 250
$\tilde{\mathfrak{s}}^{(2)} \subset \mathfrak{s}^{(2)}$	$(N=2)$-super-Schrödinger algebra, 246
$\mathcal{S}^{aff}$	Affine space of Schrödinger operators, 149
Sch(d), $\mathfrak{sch}(d)$	Schrödinger Lie group and algebra in d space dimensions, 5

Sch, $\mathfrak{sch}$	Schrödinger Lie group and algebra (in one space dimension)
$\overline{\mathfrak{sch}}$	Extended Schrödinger algebra, 233
$\mathfrak{se}(3\|2)$	Super-Euclidean Lie algebra of $\mathbb{R}^{3\|2}$, 241
$\mathfrak{sgal}$	Super-Galilean Lie algebra, 244
$\mathfrak{sns}^{(N)}$	Schrödinger-Neveu-Schwarz algebra, 258
$\mathfrak{sv}$	Schrödinger–Virasoro algebra (in one space-dimension), xxxiv
$\mathfrak{sv}(0)$	Twisted Schrödinger–Virasoro algebra, 19
$\mathfrak{sv}_\epsilon$	Deformed Schrödinger–Virasoro algebra, 19
$\overline{\mathfrak{sv}}$	Extended Schrödinger–Virasoro Lie algebra, 78
$\overline{\mathfrak{sv}}_{c,\kappa,\alpha}$	Central extension of $\overline{\mathfrak{sv}}$, 79
$\mathfrak{sv}(d)$	Schrödinger–Virasoro algebra in d space-dimensions, 10
SV	Schrödinger–Virasoro group in one space-dimension, 12
$\mathscr{T}_t$	Time-shift transformation, 215
Vect(S^1)	Algebra of vector fields on the circle, 8
$\mathfrak{vir}$	Virasoro algebra, 9
$\mathscr{X}_f^{(i)}$	Generalized symmetries of free Schrödinger equation, 215
Z	Superfield, 89
δ_1, δ_2	Graduations of $\mathfrak{sv}$, Sect. 2.1
Δ_0	Free Schrödinger operator, 149
Θ	Non-local transformation, 213
$d\pi_\lambda, d\tilde{\pi}_\lambda$	Vector field representations, xxxiv
$d\pi_\lambda^\sigma$	Action of $\mathfrak{sv}$ of $\mathscr{D}^{aff}$, 158
$d\tilde{\rho}$	Coinduced representation, 68
$d\sigma_\lambda$	Action of $\mathfrak{sv}$ on $\mathscr{S}^{aff}$, 150
$\Phi(t,r,\zeta)$	$\mathfrak{sv}$- or $\overline{\mathfrak{sv}}$-primary fields, 85
$\Phi_{j,k}$	Polynomial fields, 100
${}_\alpha\Phi_{j,k}$	Generalized polynomial fields, 104
ΨD	Algebra of formal pseudodifferential symbols, 210
Ω	Operator associated to $\overline{\mathfrak{sv}}$-primary fields, 85

Chapter 1
Geometric Definitions of $\mathfrak{sv}$

The aim of this chapter is to describe a geometrical background common to physical theories that are invariant under the Schrödinger group or under various subgroups of it arising in different contexts, as explained in details in the Preface. All these are natural groups of symmetries for *Newtonian geometry*. The Schrödinger–Virasoro group subsequently is introduced as yet another group of symmetries of the same geometric structures, albeit in a weaker sense. The reader not particularly skilled in classical differential geometry may skip it without major inconvenients, since it is independent of the sequel, save for a few basic definitions and formulas concerning the Schrödinger, Virasoro and Schrödinger–Virasoro groups and algebras, in particular the definition of the *tensor-density Virasoro modules*, the abstract definition of the *Schrödinger–Virasoro algebra* in one space-dimension, together with its *vector-field representations*; but all necessary formulas have been gathered at the end of the Introduction.

1.1 From Newtonian Mechanics to the Schrödinger–Virasoro Algebra

Lorentzian geometry constitutes the geometric foundation of relativistic mechanics; its geometric formalism makes it possible to define the equations of general relativity in a coordinate-free, covariant way, in the largest possible generality.

Owing to the success of the theory of general relativity, it was natural that one should also try to geometrize Newtonian mechanics. This task was carried out within about half a century later, by several authors, including J.-M. Lévy-Leblond, C. Duval, H.P. Künzle and others (see for instance [14,22–25,76]), leading to a geometric reformulation of Newtonian mechanics on the so-called *Newton-Cartan manifolds*, and also to the discovery of new fundamental field equations for Newtonian particles.

J. Unterberger and C. Roger, *The Schrödinger-Virasoro Algebra*, Theoretical and Mathematical Physics, DOI 10.1007/978-3-642-22717-2_1,

Most of the Lie algebras and groups that will constitute our object of study in this article appear to be closely related to the Newton-Cartan geometry. That is why we chose to give a short introduction to Newton-Cartan geometry; the main objective is to lead as quickly as possible to a geometric definition of the Schrödinger group and its infinite-dimensional generalization, the *Schrödinger–Virasoro group*.

Following the usual scheme of classical differential geometry, we shall cast the subsequent definitions into the language of the theory of *G-structures* [73]. Let us recall that a *frame* on M is a collection of bases of tangent spaces at every point of M, depending smoothly on the coordinates. These frames can be stuck together to form the *frame bundle* $P(M) \stackrel{\pi}{\longrightarrow} M$, which is a principal $GL(n,\mathbb{R})$-bundle. Now, for $G \subset GL(n,\mathbb{R})$, a G-structure on M is a reduction to G of the structural group of the frame bundle. In other terms one has a principal G-bundle $P_G \stackrel{\pi}{\longrightarrow} M$ such that $P(M) = P_G \times_G GL(n,\mathbb{R})$.

Let us just mention two examples of G- structures:

- If M is orientable, then a $GL(n,\mathbb{R})^+$-structure on it is simply an orientation on M obtained by considering all direct frames with respect to the chosen orientation on M.
- In Einstein's general relativity theory, frames which are compatible with the given Lorentz metric, or gravitational field, yield a G-structure, where G is the Lorentz group.

The foremost group in Newtonian geometry is the *Galilei group*, Gal(d). It is the group of affine transformations of $(d+1)$-dimensional space-time of the following type:

$$(x,t) \longrightarrow (\mathscr{R}x + bt + c,\ t + e) \tag{1.1}$$

for $\mathscr{R} \in SO(d)$, $b \in \mathbb{R}^d$, $c \in \mathbb{R}^d$, $e \in \mathbb{R}$; it is a $\frac{(d+1)(d+2)}{2}$-dimensional group. Its linear part (obtained by setting $c = e = 0$), called *homogeneous Galilei group* and denoted by HGal(d), is a $\frac{d(d+1)}{2}$-dimensional subgroup of $GL(d+1,\mathbb{R})$.

Definition 1.1 (Galilei *G*-structure). A HGal(d)-structure on a manifold M of dimension $d+1$ is called a Galilei structure on M.

So one has a principal subbundle $P_G \subset P(M)$ of the frame bundle on M; to such a reduction is associated a pair of tensor fields (θ,γ), where $\theta \in \Omega^1(M)$ is a closed 1-form and $\gamma \in \Gamma\ (Sym(TM \otimes TM))$ a symmetric contravariant non-negative two-tensor field, whose one-dimensional kernel is generated by θ. Roughly speaking, θ measures time variations, and γ gives the Riemannian metric on the space coordinates in order to have the reduction from $GL(d)$ to $SO(d)$. So to speak, Galilei structures are to Newtonian mechanics what Lorentzian structures are to relativistic mechanics.

It is now easy to describe the local geometry of such structures: if $U \subset M$ is a sufficiently small chart one has local coordinates (t, r_i) $i = 1, ..., d$ on U such that:

$$\left[\begin{array}{l} \theta_{|U} = dt \\ \gamma_{|U} = \sum_{i,j} \gamma_{ij} dr_i \otimes dr_j \end{array}\right.$$

Of course, the global geometry can be much more complicated: one has a codimension one foliation defined by the closed one-form θ (see for example [49]) together with a smooth family of Riemannian metrics on leaves. For physical applications, one usually postulates the existence of a global time coordinate t, giving a submersion $M \xrightarrow{t} \mathbb{R}$.

Definition 1.2. A *Newton-Cartan manifold* of dimension $d+1$ is a manifold M with a Galilei structure and a torsionless connection ∇ preserving it.

In other words, the tensors (θ, γ) associated with the Galilei structure are assumed to be parallel with respect to the connection ∇, namely, $\nabla\theta = 0$ and $\nabla\gamma = 0$. Unlike the Lorentzian or Riemannian case, one has no canonical connection of Levi-Civita type. Many options are possible, so here the physics cannot be unequivocally deduced from the geometry.

Remark. The term "Newton-Cartan manifold" was used by C. Duval in [23]. The reference to Cartan reminds that this kind of geometries has been first considered by E. Cartan in the early twenties [15].

One may now consider the group of automorphisms preserving the Newton-Cartan structures, i.e. leaving the tensor fields (θ, γ) and the connection ∇ invariant; it is the natural generalization of the Poincaré group in special relativity. In the case of the standard flat Newton-Cartan manifold $\mathbb{R} \times \mathbb{R}^d$, with coordinates $(t, r_1, ..., r_d)$, $\theta = dt, \gamma = \sum_{i=1}^{d} \partial_{r_i} \otimes \partial_{r_i}$, one naturally recovers the Galilei group $\mathrm{Gal}(d)$. For later use, we shall give generators of its Lie algebra $\mathfrak{Gal}(d)$ in terms of vector fields on $\mathbb{R} \times \mathbb{R}^d$. It can be written:

$$\mathfrak{Gal}(d) = \langle L_{-1} \rangle \oplus \left\langle Y^i_{-\frac{1}{2}}, Y^i_{\frac{1}{2}} \right\rangle_{1 \le i \le d} \oplus \langle R_{ij} \rangle_{1 \le i < j \le d}, \tag{1.2}$$

including the time and space translations $L_{-1} = -\partial_t$, $Y^i_{-\frac{1}{2}} = -\partial_{r_i}$, the generators of motion with constant speed $Y^i_{\frac{1}{2}} = -t\partial_{r_i}$ and rotations $\mathcal{R}_{ij} = r_i\partial_{r_j} - r_j\partial_{r_i}$.

1.1.1 From Galilei to Schrödinger: Central Extensions and Projective Automorphisms

The use of group cohomology, and more generally homological algebra techniques, in mathematical physics, appeared in the fifties: let us mention for example the well-known work of Wigner and Inönü for deformations of Lie algebras (see for example [61] or [104]). Their approach uses group theory to describe special relativity as a deformation of non-relativistic mechanics (or, perhaps better, non-relativistic mechanics as a *contraction* of special relativity), the coefficient of deformation being $\frac{1}{c}$ where c is the speed of light. Technically, the Galilei algebra is deformed into the Poincaré algebra, which is the Lie algebra of affine isometries of Minkowski space-time. We discuss deformations in Chap. 7 below. Here we shall introduce the central extension of $\mathrm{Gal}(d)$ as considered first by Bargmann [4]; we shall give its explicit construction on the Lie algebra level. One computes easily the second cohomology space $H^2(\mathfrak{Gal}(d); \mathbb{R})$, see Chap. 7, which classifies central extensions of $\mathrm{Gal}(d)$; it is one-dimensional, generated by the following cocycle

$$C\left(Y^i_{-\frac{1}{2}}, Y^j_{\frac{1}{2}}\right) = \delta_{i,j} \quad i, j = 1...d$$

In other words, the Lie bracket of the Y-generators has been deformed into

$$\left[Y^i_{-\frac{1}{2}}, Y^j_{\frac{1}{2}}\right] := \delta_{i,j} M_0, \tag{1.3}$$

where M_0 is an extra generator in the center, i.e. commuting with all other generators. We shall denote by $\mathfrak{Bar}(d)$ the centrally extended Lie algebra, called *Bargmann algebra*. So one has a short exact sequence of Lie algebras:

$$0 \longrightarrow \mathbb{R} \longrightarrow \mathfrak{Bar}(d) \longrightarrow \mathfrak{Gal}(d) \longrightarrow 0 \tag{1.4}$$

Schur's lemma implies that for irreducible representations of $\mathfrak{Bar}(d)$, M_0 acts as multiplication by a scalar, whose value will be called the *mass* of the corresponding representation.

This terminology, together with the interpretation of the mass in terms of a two-dimensional cohomology class, has been made clear by the work of J.M. Souriau. In the symplectization of classical mechanics, a prominent rôle is played by the construction of the momentum relative to the action of the dynamical group; it is precisely the defect of equivariance of this momentum which yields a 2-cohomology class of the Lie algebra, then interpreted as the mass of the corresponding particle. [1]

We shall now enlarge the Bargmann algebra by considering projective automorphisms of Newton-Cartan structures. *Infinitesimal projective automorphisms* of the flat Newton-Cartan structure on $\mathbb{R} \times \mathbb{R}^d$ are connection-preserving vector fields X on $\mathbb{R} \times \mathbb{R}^d$ such that (denoting by $\mathscr{L}$ the Lie derivative)

$$\mathscr{L}_X dt = \lambda\, dt, \quad \mathscr{L}_X \left(\sum_{i=1}^d \partial_{r_i} \otimes \partial_{r_i}\right) = \lambda \left(\sum_{i=1}^d \partial_{r_i} \otimes \partial_{r_i}\right) \tag{1.5}$$

[1] In fact, the generator of $H^2(\mathfrak{Gal}(d); \mathbb{R})$ given above can be integrated to a group cocycle, which in turn gives the Souriau cocycle in $H^1(\mathrm{Gal}(d); \mathfrak{Gal}(d)^*)$; this cocycle is then naturally interpreted in the mechanical framework as a first integral called *momentum*, which gives precisely the total *mass* of the system. For details on the geometrical and physical arguments, see the book by Souriau [109], pp. 148-154. For generalities on the cohomological computations, see [43], Sect. 2.3.

for some function $\lambda \in C^\infty(\mathbb{R}\times\mathbb{R}^d)$. Such transformations conserve the geodesics globally up to a change of parametrization, so they form naturally a Lie subalgebra of $\mathrm{Vect}(\mathbb{R}\times\mathbb{R}^d)$, the Lie algebra of C^∞ vector fields on $\mathbb{R}\times\mathbb{R}^d$.

Proposition 1.3. *The Lie algebra of projective automorphisms of the flat Newton-Cartan structure is generated by the following vector fields.*

$$L_{-1} = -\partial_t,\ L_0 = -t\partial_t - \frac{1}{2}\sum_i r_i\partial_{r_i},\ L_1 = -t^2\partial_t - t\sum_i r_i\partial_{r_i}$$

$$Y^i_{-\frac{1}{2}} = -\partial_{r_i},\ Y^i_{\frac{1}{2}} = -t\partial_{r_i}$$

$$R^{ij} = r_i\partial_{r_j} - r_j\partial_{r_i},\quad 1\le i<j\le d \tag{1.6}$$

One is now ready to give the formal definition of the *Schrödinger Lie algebra.*

Definition 1.4. The *Schrödinger Lie algebra* in dimension d, denoted by $\mathfrak{sch}(d)$, is the Lie algebra with generators

$$L_{-1}, L_0, L_1;\quad Y^i_{-\frac{1}{2}}, Y^i_{\frac{1}{2}},\ i=1,\dots,d;\quad R^{ij},\ 1\le i<j\le d;\quad M_0.$$

The vector fields (L_{-1}, L_0, L_1) generate a copy of $\mathfrak{sl}(2,\mathbb{R})$. The brackets between $Y^i_{-\frac{1}{2}}, Y^i_{\frac{1}{2}}, R^{ij}, M_0$ are those of $\mathfrak{Bar}(d)$, and the brackets of L_{-1}, L_0, L_1 with the other generators are defined by

$$\left[L_n, Y^i_{-\frac{1}{2}}\right] = \left(\frac{n}{2}+\frac{1}{2}\right)Y^i_{n-\frac{1}{2}},\ \left[L_n, Y^i_{\frac{1}{2}}\right] = \left(\frac{n}{2}-\frac{1}{2}\right)Y^i_{n+\frac{1}{2}}\quad (n=-1,0,1); \tag{1.7}$$

$$[L_n, R^{ij}] = 0,\quad [L_n, M_0] = 0. \tag{1.8}$$

One may describe this algebra in terms of a semi-direct product. Set $\mathfrak{Bar}(d) = \mathbb{R}L_{-1}\ltimes\overline{\mathfrak{Bar}}(d)$, where $\overline{\mathfrak{Bar}}(d) = \mathrm{span}\langle Y^i_{\pm\frac{1}{2}}, R_{ij}, M_0\rangle \subset \mathfrak{Bar}(d)$ is the Lie subalgebra obtained by 'forgetting' time-translations; one then has

$$\mathfrak{sch}(d) = \mathfrak{sl}(2,\mathbb{R})\ltimes\overline{\mathfrak{Bar}}(d). \tag{1.9}$$

It may be exponentiated into a connected Lie group $\mathrm{Sch}(d) = SL(2,\mathbb{R})\ltimes\overline{Bar}(d)$, called *Schrödinger Lie group* in d space dimensions (see [98]), where $\overline{Bar}(d)$ is the Lie group naturally associated to $\overline{\mathfrak{Bar}}(d)$.

The previous proposition can then be interpreted as follows: *the Lie algebra of projective automorphisms of the flat Newton-Cartan structure defines a massless representation of the Lie algebra* $\mathfrak{sch}(d)$.

Finally, the name Schrödinger algebra is justified by the fact that $\mathfrak{sch}(d)$ acts projectively on solutions of the free Schrödinger equation $(2\mathcal{M}\partial_t - \Delta_d)\psi = 0$ where $\Delta_d := \sum_{i=1}^{d} \partial_{r_i}^2$ and $\mathcal{M} \in \mathbb{C}$ [2] is a scalar. One has:

Proposition 1.5 (Lie symmetries of the free Schrödinger equation). *The algebra of projective Lie symmetries of the free Schrödinger equation, i.e. the Lie algebra of differential operators $\mathcal{D}$ on $\mathbb{R} \times \mathbb{R}^d$ of order at most one, satisfying:*

$$(2\mathcal{M}\partial_t - \Delta_d)(\psi) = 0 \Longrightarrow (2\mathcal{M}\partial_t - \Delta_d)(\mathcal{D}\psi) = 0$$

defines a representation $d\pi^d_{d/4}$ of mass $\mathcal{M}$ of the Schrödinger algebra, with the following realization:

$$d\pi^d_{d/4}(L_{-1}) = -\partial_t,\ d\pi^d_{d/4}(L_0) = -t\partial_t - \frac{1}{2}\sum_i r_i\partial_{r_i} - \frac{d}{4},\ d\pi^d_{d/4}(L_1)$$
$$= -t^2\partial_t - t\sum_i r_i\partial_{r_i} - \frac{1}{2}\mathcal{M}r^2 - \frac{d}{2}t$$

$$d\pi^d_{d/4}\left(Y^i_{-\frac{1}{2}}\right) = -\partial_{r_i},\ d\pi^d_{d/4}\left(Y^i_{\frac{1}{2}}\right) = -t\partial_{r_i} - \mathcal{M}r_i,\ d\pi^d_{d/4}(M_0) = -\mathcal{M}$$
$$d\pi^d_{d/4}(R^{ij}) = r_i\partial_{r_j} - r_j\partial_{r_i},\quad 1 \le i < j \le d. \tag{1.10}$$

Let us mention, anticipating on Sect. 1.2, that realizing instead L_0 by the operator $-t\partial_t - \frac{1}{2}\sum_i r_i\partial_{r_i} - \lambda$ and L_1 by $-t^2\partial_t - t\sum_i r_i\partial_{r_i} - \frac{1}{2}\mathcal{M}r^2 - 2\lambda t$ ($\lambda \in \mathbb{R}$), leads to a family of representations $d\pi^d_\lambda$ of $\mathfrak{sch}(d)$. This accounts for the parameter $d/4$ in our definition of $d\pi^d_{d/4}$. The parameter 2λ may be interpreted physically as the *scaling dimension* x of the field on which $\mathfrak{sch}(d)$ acts, $\lambda = d/4$ for solutions of the free Schrödinger equation in d space dimensions (see Preface).

We shall also frequently use the realization $d\tilde{\pi}^d_\lambda$ of the Schrödinger algebra given by the Laplace transform of the above generators with respect to the mass, which is formally equivalent to replacing the parameter $\mathcal{M}$ in the above formulas by ∂_ζ. This simple transformation leads to a representation $d\tilde{\pi}^d_\lambda$ of $\mathfrak{sch}(d)$; let us consider the particular case $\lambda = 0$ for simplicity. Then $d\tilde{\pi}^d := d\tilde{\pi}^d_0$ gives a realization of $\mathfrak{sch}(d)$ by vector fields on $\mathbb{R}^{d+2}$ with coordinates t, r_i $(i = 1, \ldots, d)$ and ζ. Let us write the action of the generators for further reference (see [54]):

[2] On an *analytic* level, one should distinguish between the *heat equation* ($\mathcal{M}$ real) and the *Schrödinger equation* ($\mathcal{M}$ imaginary). On an *algebraic* level though, this distinction is irrelevant. The present convention, used everywhere *except* in Chaps. 8–10, prevents unwanted $\sqrt{-1}$-factors.

$$d\tilde{\pi}^d(L_{-1}) = -\partial_t, d\tilde{\pi}^d(L_0) = -t\partial_t - \frac{1}{2}\sum_i r_i\partial_{r_i},$$

$$d\tilde{\pi}^d(L_1) = -t^2\partial_t - t\sum_i r_i\partial_{r_i} - \frac{1}{2}r^2\partial_\zeta$$

$$d\tilde{\pi}^d\left(Y^i_{-\frac{1}{2}}\right) = -\partial_{r_i}, d\tilde{\pi}^d\left(Y^i_{\frac{1}{2}}\right) = -t\partial_{r_i} - r_i\partial_\zeta, d\tilde{\pi}^d(M_0) = -\partial_\zeta$$

$$d\tilde{\pi}^d(R^{ij}) = r_i\partial_{r_j} - r_j\partial_{r_i}, \quad 1 \le i < j \le d. \tag{1.11}$$

In other words, the action $d\tilde{\pi}^d$ is conjugate to $d\pi^d$ through the (formal) Laplace transform

$$\psi(\mathcal{M};t,r) \longrightarrow \tilde{\psi}(\zeta;t,r) = \int \psi(\mathcal{M};t,r)e^{\mathcal{M}\zeta}\, d\mathcal{M}.$$

So, according to the context, one may use either the representation by differential operators on $\mathbb{R}^{d+1}$ of order one, or the representation by vector fields on $\mathbb{R}^{d+2}$. Both points of view prove to be convenient.

1.1.2 From Schrödinger to Schrödinger–Virasoro

The last step which will lead us finally from the Schrödinger Lie algebra to the *Schrödinger–Virasoro* Lie algebra consists in a certain sense in forgetting about the Newton-Cartan geometry and requiring only the covariance of the tensors (θ, γ). Namely, one has the following in the case of the usual structure on $\mathbb{R} \times \mathbb{R}^d$:

Proposition 1.6. *The Lie algebra of vector fields X on $\mathbb{R} \times \mathbb{R}^d$ - not necessarily preserving the connection, such that (1.5) are verified, is generated by the following set of transformations:*

(i) $$\mathcal{L}_f = -f(t)\partial_t - \frac{1}{2}f'(t)\sum_{i=1}^d r_i\partial_{r_i} \quad \text{(Virasoro – like transformations)}$$

(ii) $$\mathcal{Y}^i_{g_i} = -g_i(t)\partial_{r_i} \quad \text{(time – dependent space translations)}$$

(iii) $$\mathcal{R}^{ij}_{h_{ij}} = -h_{ij}(t)(r_i\partial_{r_j} - r_j\partial_{r_i}),\ 1 \le i < j \le d$$
(time – dependent space rotations)

where f, g_i, h_{ij} are arbitrary functions of t.

Proof. Put $r = (r_1, \ldots, r_d)$. Let $X = f(t,r)\partial_t + \sum_{i=1}^d g_i(t,r)\partial_{r_i}$ verifying conditions (1.5), and let $\mathcal{L}_X = [X, .]$ be the Lie derivation with respect to the vector field X. Then $d(\mathcal{L}_X dt) = d(\lambda dt) = 0$, so $\lambda = -f'(t)$ is a function of time only. Hence

$$\mathcal{L}_X\left(\sum_{j=1}^{d} \partial_{r_j}^2\right) = -2\partial_{r_j} g_i \sum_{i,j} \partial_{r_j} \otimes \partial_{r_i}$$
$$= -2\sum_j \partial_{r_j} g_j \partial_{r_j} \otimes \partial_{r_j} - 2\sum_{i \neq j} \partial_{r_j} g_i \partial_{r_j} \otimes \partial_{r_i}$$

so $\partial_{r_j} g_i = -\partial_{r_i} g_j$ if $i \neq j$, which gives the time-dependent rotations, and $2\partial_{r_j} g_j = f'$ for $j = 1, \ldots, d$, which gives the Virasoro-like transformations and the time-dependent translations. □

Note that the Lie algebra of Proposition 1.5 corresponds to $f(t) = 1, t, t^2$; $g_i(t) = 1, t$ and $h_{ij}(t) = 1$.

One easily sees that the $\mathcal{R}^{ij}_{h_{ij}}$ generate the *current algebra* $C^\infty(\mathbb{R}, \mathfrak{so}(d))$ on $\mathfrak{so}(d)$, while the $\mathcal{R}^{ij}_{h_{ij}}$ and the $\mathcal{Y}^i_{g_i}$ generate together the current algebra on the Euclidean Lie algebra $\mathfrak{eucl}(d) = \mathfrak{so}(d) \ltimes \mathbb{R}^d$. The transformations $\mathcal{L}_f$ generate a copy of the Lie algebra of tangent vector fields on $\mathbb{R}$, denoted by Vect($\mathbb{R}$), namely,

$$[\mathcal{L}_f, \mathcal{L}_g] = \mathcal{L}_{\{f,g\}} \tag{1.12}$$

where $\{f, g\} = f'g - fg'$. So this Lie algebra can be described algebraically as a semi-direct product $\mathrm{Vect}(\mathbb{R}) \ltimes C^\infty(\mathbb{R}, \mathfrak{eucl}(d))$. In our realization, it is embedded as a subalgebra of $\mathrm{Vect}(\mathbb{R} \times \mathbb{R}^d)$.

For both topological and algebraic reasons, we shall from now on compactify the time coordinate t. So we shall work on $S^1 \times \mathbb{R}^d$, and $\mathrm{Vect}(\mathbb{R}) \ltimes C^\infty(\mathbb{R}, \mathfrak{eucl}(d))$ is replaced by $\mathrm{Vect}(S^1) \ltimes C^\infty(S^1, \mathfrak{eucl}(d))$, where $\mathrm{Vect}(S^1)$ is the *centerless Virasoro algebra*.

It may be the right place to recall some well-known facts about the Virasoro algebra, that we shall use throughout the present monography.

Definition 1.7 (algebra of vector fields on the circle). Let $\mathrm{Vect}(S^1)$ be the Lie algebra of C^∞-vector fields on the circle.

We represent an element of $\mathrm{Vect}(S^1)$ by the vector field $f(z)\partial_z$, where $f \in \mathbb{C}[z, z^{-1}]$ is a Laurent polynomial. Vector field brackets $[f(z)\partial_z, g(z)\partial_z] = (fg' - f'g)\partial_z$, may equivalently be rewritten in the basis $(\ell_n)_{n\in\mathbb{Z}}$, $\ell_n = -z^{n+1}\partial_z$ (also called Fourier components), which yields $[\ell_n, \ell_m] = (n - m)\ell_{n+m}$. Notice the unusual choice of signs, justified (among other arguments) by the anteriority of [50] on our subject.

The Lie algebra $\mathrm{Vect}(S^1)$ has only one non-trivial central extension (see [43] or [64] for instance, and also the introduction to Chap. 7 for some explanations if needed), given by the so-called *Virasoro cocycle c* defined by

$$c(f\partial_z, g\partial_z) = \int_{S^1} f'''(z)g(z)\, dz, \tag{1.13}$$

or, in Fourier components,

$$c(\ell_n, \ell_m) = \delta_{n+m,0}(n+1)n(n-1). \tag{1.14}$$

The resulting centrally extended Lie algebra, called *Virasoro algebra*, will be denoted by $\mathfrak{vir}$. Explicitly, the new bracket reads

$$[\ell_n, \ell_m]_{\mathfrak{vir}} = (n-m)\ell_{n+m} + \delta_{n+m,0}(n+1)n(n-1)K, \tag{1.15}$$

where K is an extra generator in the center (as for the above central extension of the Galilei algebra).

The Lie algebra $\mathrm{Vect}(S^1)$ has a one-parameter family of representations $\mathcal{F}_\lambda$, $\lambda \in \mathbb{R}$, called *tensor-density modules*.

Definition 1.8 (tensor-density modules). We denote by $\mathcal{F}_\lambda$ the representation of $\mathrm{Vect}(S^1)$ on $\mathbb{C}[z, z^{-1}]$ given by[3]

$$\ell_n.z^m = (\lambda n - m)z^{n+m}, \quad n, m \in \mathbb{Z}. \tag{1.16}$$

An element of $\mathcal{F}_\lambda$ is naturally understood as a $(-\lambda)$-density $\phi(z)dz^{-\lambda}$, acted by $\mathrm{Vect}(S^1)$ as

$$f(z)\partial_z.\phi(z)dz^{-\lambda} = (f\phi' - \lambda f'\phi)(z)dz^{-\lambda}. \tag{1.17}$$

In the bases $\ell_n = -z^n\partial_z$ and $a_m = z^m dz^{-\lambda}$, one gets $\ell_n.a_m = (\lambda n - m)a_{n+m}$.

Replacing formally t by the compactified variable z in the formulas of Proposition 1.6, and putting $f(z) = -z^{n+1}$, $g_i(z) = -z^{n+\frac{1}{2}}$, $h_{ij}(z) = -z^n$, one gets a realization of $\mathrm{Vect}(S^1) \ltimes C^\infty(S^1, \mathfrak{eucl}(d))$ as a Lie subalgebra of $\mathrm{Vect}(S^1 \times \mathbb{R}^d)$ generated by $L_n, Y^i_{m_i}, R^{ij}_{p_{ij}}$ (with integer indices n and p_{ij} and half-integer indices m_i), with the following set of relations:

$$\begin{gathered}
[L_n, L_p] = (n-p)L_{n+p} \\
\left[L_n, Y^i_m\right] = \left(\frac{n}{2} - m\right)Y^i_{n+m}, \quad \left[L_n, R^{ij}_p\right] = -pR^{ij}_{n+p} \\
\left[Y^i_m, Y^j_{m'}\right] = 0, \left[R^{ij}_p, Y^k_m\right] = \delta_{j,k}Y^i_{m+p} - \delta_{i,k}Y^j_{m+p} \\
\left[R^{ij}_n, R^{kl}_p\right] = \delta_{j,k}R^{il}_{n+p} + \delta_{i,l}R^{jk}_{n+p} - \delta_{j,l}R^{ik}_{n+p} - \delta_{i,k}R^{jl}_{n+p}
\end{gathered} \tag{1.18}$$

Remark. In order to avoid possible confusion, we have chosen to write Fourier components X_n, $X = L, Y, \ldots$ with usual letters, and the currents themselves with calligraphic letters, $\mathcal{L}, \mathcal{Y}, \ldots$

---

[3]In conformal field theory (see Chap. 6), quantized fields $(X_m)_{m\in\mathbb{Z}}$ such that $[L_n, X_m] = ((\mu - 1)n - m)X_{n+m}$ are called *primary operators of weight* μ. In our language, the components of such a field generates the tensor-density module $\mathcal{F}_{\mu-1}$.

With the above definitions, one sees immediately that, under the action of $\langle \mathcal{L}_f \rangle_{f \in C^\infty(S^1)} \simeq \mathrm{Vect}(S^1)$, *the* $(Y_m^i)_{m \in \frac{1}{2}+\mathbb{Z}}$ *behave as elements of the module* $\mathcal{F}_{\frac{1}{2}}$, *while the* $(R_m^{ij})_{m \in \mathbb{Z}}$ *and the* $(M_m)_{m \in \mathbb{Z}}$ *define several copies of* $\mathcal{F}_0$. This remark turns out to be fundamental to understand the algebraic structure of the Schrödinger–Virasoro algebra, and will be taken up in Chap. 2 in more details.

The commutative Lie algebra generated by the Y_n^i has an infinite family of central extensions. If we want to leave unchanged the action of $\mathrm{Vect}(S^1)$ on the Y_n^i and to extend the action of $\mathrm{Vect}(S^1)$ to the central charges, though, the most natural possibility (originally discovered by M. Henkel, see [50], by extrapolating the relations (1.7) and (1.8) to integer or half-integer indices) containing $\mathfrak{sch}(d)$ as a Lie subalgebra, is the Lie algebra $\mathfrak{sv}(d)$ defined as follows.

Definition 1.9. We denote by $\mathfrak{sv}(d)$ the Lie algebra with generators $L_n, Y_m^i, M_n, R_n^{ij}$ $(n \in \mathbb{Z}, m \in \frac{1}{2}+\mathbb{Z})$ and following relations (where $n, p \in \mathbb{Z}, m, m' \in \frac{1}{2}+\mathbb{Z}$):

$$[L_n, L_p] = (n-p)L_{n+p}$$

$$[L_n, Y_m^i] = \left(\frac{n}{2} - m\right) Y_{n+m}^i,\ \left[L_n, R_p^{ij}\right] = -pR_{n+p}^{ij},\ [L_n, M_p] = -pM_{n+p}$$

$$\left[Y_m^i, Y_{m'}^j\right] = (m-m')M_{m+m'}, \left[R_p^{ij}, Y_m^k\right] = \delta_{j,k} Y_{m+p}^i - \delta_{i,k} Y_{m+p}^j$$

$$[Y_m^i, M_p] = 0,\ [R_n^{ij}, M_p] = 0,\ [M_n, M_p] = 0$$

$$\left[R_n^{ij}, R_p^{kl}\right] = \delta_{j,k} R_{n+p}^{il} + \delta_{i,l} R_{n+p}^{jk} - \delta_{j,l} R_{n+p}^{ik} - \delta_{i,k} R_{n+p}^{jl} \tag{1.19}$$

One sees immediately that $\mathfrak{sv}(d)$ has a semi-direct product structure $\mathfrak{sv}(d) \simeq \mathrm{Vect}(S^1) \ltimes \mathfrak{h}(d)$, with $\mathrm{Vect}(S^1) \simeq \langle L_n \rangle_{n \in \mathbb{Z}}$ and $\mathfrak{h}(d) = \langle Y_m^i \rangle_{m \in \mathbb{Z}, i \le d} \oplus \langle M_p \rangle_{p \in \mathbb{Z}} \oplus \langle R_m^{ij} \rangle_{m \in \mathbb{Z}, 1 \le i < j \le d}$.

Note that the Lie subalgebra $\langle L_{-1}, L_0, L_1, Y_{-\frac{1}{2}}^i, Y_{\frac{1}{2}}^i, R_0^{ij}, M_0 \rangle \subset \mathfrak{sv}(d)$ is isomorphic to $\mathfrak{sch}(d)$. The following Proposition shows that the above mass $\mathcal{M}$ representation of the Schrödinger algebra extends in a natural way to the Schrödinger–Virasoro algebra.

Proposition 1.10 (see [50]). *The realization* $d\tilde{\pi}^d$ *of* $\mathfrak{sch}(d)$ *(see (1.11)) extends to the following realization* $d\tilde{\pi}^d$ *of* $\mathfrak{sv}(d)$ *as vector fields on* $S^1 \times \mathbb{R}^{d+1}$*:*

$$\begin{aligned} d\tilde{\pi}^d(\mathcal{L}_f) &= -f(z)\partial_z - \frac{1}{2} f'(z) \left(\sum_{i=1}^d r_i \partial_{r_i} \right) - \frac{1}{4} f''(z) r^2 \partial_\zeta \\ d\tilde{\pi}^d(\mathcal{Y}_{g_i}^i) &= -g_i(z)\partial_{r_i} - g_i'(z) r_i \partial_\zeta \\ d\tilde{\pi}^d(\mathcal{R}_{k_{ij}}^{ij}) &= -k_{ij}(z)(r_i \partial_{r_j} - r_j \partial_{r_i}) \\ d\tilde{\pi}^d(\mathcal{M}_h) &= -h(z)\partial_\zeta \end{aligned} \tag{1.20}$$

Let us rewrite this action in Fourier components for completeness. In the following formulas, $n \in \mathbb{Z}$ while $m \in \frac{1}{2} + \mathbb{Z}$:

$$\begin{aligned} d\tilde{\pi}^d(L_n) &= -z^{n+1}\partial_z - \frac{1}{2}(n+1)z^n\left(\sum_{i=1}^{d} r_i \partial_{r_i}\right) - \frac{1}{4}(n+1)nz^{n-1}r^2\partial_\zeta \\ d\tilde{\pi}^d(Y_m) &= -z^{m+\frac{1}{2}}\partial_{r_i} - \left(m+\frac{1}{2}\right)z^{m-\frac{1}{2}} r_i \partial_\zeta \\ d\tilde{\pi}^d(R_n^{ij}) &= -z^n\left(r_i\partial_{r_j} - r_j\partial_{r_i}\right) \\ d\tilde{\pi}^d(M_n) &= -z^n\partial_\zeta \end{aligned} \tag{1.21}$$

Up to the extra scaling terms, $\lambda f'(z)$ or $\lambda(n+1)z^n$, for the Virasoro generators, this is the Fourier version of the *vector-field representation* originally introduced by M. Henkel (see formulas at the end of the Introduction in the case $d = 1$).

1.1.3 Our Object of Study: The Schrödinger–Virasoro Algebra in One Space Dimension

We shall restrict henceforth to the case $d = 1$ and write $\mathfrak{sv}$ for $\mathfrak{sv}(1)$, Sch for Sch(1), $d\pi$ for $d\pi^1$, $d\tilde{\pi}$ for $d\tilde{\pi}^1$, $\mathfrak{h}$ for $\mathfrak{h}(1)$, $\mathcal{Y}_f$ for $\mathcal{Y}_f^1$, Y_m for Y_m^1 to simplify notations.

Then (as one sees immediately) the *Schrödinger–Virasoro algebra* $\mathfrak{sv}$

$$\mathfrak{sv} \simeq \langle L_n \rangle_{n\in\mathbb{Z}} \ltimes \langle Y_m, M_p \rangle_{m\in\frac{1}{2}+\mathbb{Z}, p\in\mathbb{Z}} \tag{1.22}$$

is generated by three fields, L, Y and M, with commutators

$$\begin{aligned} &[L_n, L_p] = (n-p)L_{n+p},\ [L_n, Y_m] = \left(\frac{n}{2} - m\right)Y_{n+m},\ [L_n, M_p] = -pM_{n+p} \\ &[Y_m, Y_{m'}] = (m-m')M_{m+m'},\ [Y_m, M_p] = 0,\ [M_n, M_p] = 0 \end{aligned} \tag{1.23}$$

where $n, p \in \mathbb{Z}, m, m' \in \frac{1}{2} + \mathbb{Z}$, and $\mathfrak{h} = \langle Y_m, M_p \rangle_{m\in\frac{1}{2}+\mathbb{Z}, p\in\mathbb{Z}}$ is a two-step nilpotent infinite-dimensional Lie algebra.

Writing the currents $\mathcal{L}_f, \mathcal{Y}_g, \mathcal{M}_h$ with calligraphic letters as before, one obtains equivalently the following non-vanishing brackets:

$$\begin{aligned} &[\mathcal{L}_{f_1}, \mathcal{L}_{f_2}] = \mathcal{L}_{f_1' f_2 - f_1 f_2'}, \quad [\mathcal{L}_f, \mathcal{Y}_g] = \mathcal{L}_{\frac{1}{2}f'g - fg'} \\ &[\mathcal{L}_f, \mathcal{M}_h] = \mathcal{L}_{-fh'}, \quad [\mathcal{Y}_{g_1}, \mathcal{Y}_{g_2}] = \mathcal{M}_{g_1' g_2 - g_1 g_2'}. \end{aligned} \tag{1.24}$$

1.2 Integration of the Schrödinger–Virasoro Algebra to a Group

We let Diff(S^1) be the group of orientation-preserving C^∞-diffeomorphisms of the circle. This connected Lie group integrates the Lie algebra of vector fields on the circle, Vect(S^1). It comes naturally with a structure of Fréchet Lie group.[4] Orientation is important since we shall need to consider the square-root of the Jacobian of the diffeomorphism (see Proposition 1.11).

Theorem 1.1 (Definition of the Schrödinger–Virasoro group SV).

1. *Let $H = C^\infty(S^1) \times C^\infty(S^1)$ be the product of two copies of the space of infinitely differentiable functions on the circle, with its group structure modified as follows:*

$$(\alpha_2, \beta_2).(\alpha_1, \beta_1) = \left(\alpha_1 + \alpha_2, \beta_1 + \beta_2 + \frac{1}{2}\left(\alpha_1'\alpha_2 - \alpha_1\alpha_2'\right)\right). \qquad (1.25)$$

   Then H is a Fréchet-Lie group which integrates $\mathfrak{h}$.
2. *Let $SV = \mathrm{Diff}(S^1) \ltimes H$ be the group with semi-direct product given by*

$$(1; (\alpha, \beta)).(\phi; 0) = (\phi; (\alpha, \beta)) \qquad (1.26)$$

   and

$$(\phi; 0).(1; (\alpha, \beta)) = (\phi; ((\phi')^{\frac{1}{2}}(\alpha \circ \phi), \beta \circ \phi)). \qquad (1.27)$$

   Then SV is a Fréchet-Lie group which integrates $\mathfrak{sv}$.

Proof. 1. From classical arguments due to Hamilton (see [47]), one easily sees that H is a Fréchet-Lie group, its underlying manifold being the Fréchet space $C^\infty(S^1) \times C^\infty(S^1)$ itself.

One sees moreover that its group structure is unipotent.

By computing commutators

$$(\alpha_2, \beta_2)(\alpha_1, \beta_1)(\alpha_2, \beta_2)^{-1}(\alpha_1, \beta_1)^{-1} = (0, \alpha_1'\alpha_2 - \alpha_1\alpha_2') \qquad (1.28)$$

one recovers the formulas for the nilpotent Lie algebra $\mathfrak{h}$.

---

[4] A Fréchet Lie group is a group endowed with an infinite-dimensional manifold structure, locally modelled on a Fréchet space (i.e. a complete infinite-dimensional vector space equipped with a countable family of semi-norms), such that the multiplication and inverse mappings are differentiable with respect to this differentiable structure. Groups of diffeomorphisms of differentiable manifolds are typical examples of such structures (see [43], Chap. 4). Here the Fréchet-Lie group structure of Diff(S^1) is induced by its C^∞-topology, i.e by the topology of uniform convergence for the functions representing the vector fields in the Lie algebra and all their derivatives.

2. The group $\dot{H}$ is realized (as $\mathrm{Diff}(S^1)$-module) as a product of modules of densities $\mathscr{F}_{\frac{1}{2}} \times \mathscr{F}_0$, hence the semi-direct product $\mathrm{Diff}(S^1) \ltimes H$ integrates the semi-direct product $\mathrm{Vect}(S^1) \ltimes \mathfrak{h}$.

□

Let us denote quite generally by $\mathrm{Diff}(M)$ the Lie group of orientation-preserving diffeomorphisms of M if M is any orientable manifold, see again [47]. The representation $d\tilde{\pi}$, defined in Proposition 1.10, can be exponentiated into a representation of SV, given in the following proposition:

Proposition 1.11 (see [54]).

1. Define $\tilde{\pi} : SV \to \mathrm{Diff}(S^1 \times \mathbb{R}^2)$ *by*

$$\tilde{\pi}(\phi; (\alpha, \beta)) = \tilde{\pi}(1; (\alpha, \beta)).\tilde{\pi}(\phi; 0)$$

and

$$\tilde{\pi}(\phi; 0)(z, r, \zeta) = \left(\phi(z), r\sqrt{\phi'(z)}, \zeta - \frac{1}{4}\frac{\phi''(z)}{\phi'(z)}r^2\right).$$

Then $\tilde{\pi}$ *is a representation of* SV.

2. The infinitesimal representation of $\tilde{\pi}$ *is equal to* $d\tilde{\pi}$.

Proof. Point 1. may be checked by direct verification (note that the formulas were originally derived by exponentiating the vector fields in the realization $d\tilde{\pi}$).

For 2., it is clearly enough to show that, for any $f \in C^\infty(S^1)$ and $g, h \in C^\infty(\mathbb{R})$,

$$\frac{d}{du}|_{u=0}\tilde{\pi}(\exp u\mathscr{L}_f) = d\tilde{\pi}(\mathscr{L}_f), \quad \frac{d}{du}|_{u=0}\tilde{\pi}(\exp u\mathscr{Y}_g) = d\tilde{\pi}(\mathscr{Y}_g),$$

$$\frac{d}{du}|_{u=0}\tilde{\pi}(\exp u\mathscr{M}_h) = d\tilde{\pi}(\mathscr{M}_h).$$

Put $\phi_u = \exp u\mathscr{L}_f$, so that $\frac{d}{du}|_{u=0}\phi_u(z) = f(z)$. Then

$$\frac{d}{du}r(\phi_u')^{\frac{1}{2}} = \frac{1}{2}r(\phi_u')^{-\frac{1}{2}}\frac{d}{du}\phi_u' \to_{u\to 0} \frac{1}{2}rf'(z),$$

$$\frac{d}{du}\left(r^2\frac{\phi_u''}{\phi_u'}\right) = r^2\left(\frac{\frac{d}{du}\phi_u''}{\phi_u'} - \frac{\phi_u''}{(\phi_u')^2}\frac{d}{du}\phi_u'\right) \to_{u\to 0} r^2 f''(z)$$

so the equality $\frac{d}{du}|_{u=0}\tilde{\pi}(\exp u\mathscr{L}_f) = d\tilde{\pi}(\mathscr{L}_f)$ holds. The two other equalities can be proved in a similar way. □

Let us introduce another related representation, using the "triangular" structure of the representation $\tilde{\pi}$. The action $\tilde{\pi} : SV \to \mathrm{Diff}(S^1 \times \mathbb{R}^2)$ can be projected onto an action $\bar{\pi} : SV \to \mathrm{Diff}(S^1 \times \mathbb{R})$ by "forgetting" the coordinate ζ, since the way coordinates (t,r) are transformed does not depend on ζ. Note also that $\tilde{\pi}$ acts by (time- and space- dependent) translations on the coordinate ζ, so one may define a function $\Phi : SV \to C^\infty(\mathbb{R}^2)$ with coordinates (t,r) by

$$\tilde{\pi}(g)(t,r,\zeta) = (\bar{\pi}(g)(t,r,\zeta + \Phi_g(t,r))$$

(independently of $\zeta \in \mathbb{R}$). This action may be further projected onto $\bar{\pi}_{S^1} : SV \to \mathrm{Diff}(S^1)$ by "forgetting" the *second* coordinate r this time, so

$$\bar{\pi}_{S^1}(\phi; (\alpha,\beta)) = \phi.$$

Proposition 1.12. *1. One has the relation*

$$\Phi_{g_2 \circ g_1}(t,r) = \Phi_{g_1}(t,r) + \Phi_{g_2}(\bar{\pi}(g_1)(t,r)).$$

In other words, Φ is a trivial $\bar{\pi}$-cocycle: $\Phi \in Z^1(SV, C^\infty(\mathbb{R}^2))$.
2. The application $\pi_\lambda : SV \to Hom(C^\infty(S^1 \times \mathbb{R}), C^\infty(S^1 \times \mathbb{R}))$ defined by

$$\pi_\lambda(g)(\phi)(t,r) = \left(\bar{\pi}'_{S^1} \circ \bar{\pi}^{-1}_{S^1}(t)\right)^\lambda e^{\mathscr{M}\Phi_g(\bar{\pi}(g)^{-1}.(t,r))}\, \phi(\bar{\pi}(g)^{-1}.(t,r))$$

defines a representation of SV in $C^\infty(S^1 \times \mathbb{R})$.

Proof. Straightforward. □

Restricting $\pi_{d/4}$ to the Schrödinger group $\mathrm{Sch}(d)$, one obtains the original action of $\mathrm{Sch}(d)$ on the space of solutions of the free Schrödinger equation (see Preface). Perroud [98] obtains $\pi_{d/4}$ by classifying projective representations of the quotiented Lie group obtained by considering a zero mass (i.e. of the non-centrally extended Lie group).

Let us look at the associated infinitesimal representation. Introduce the function Φ' defined by $\Phi'(X) = \frac{d}{du}|_{u=0}\Phi(\exp uX)$, $X \in \mathfrak{sv}$. If now $g = \exp X$, $X \in \mathfrak{sv}$, then

$$\frac{d}{du}|_{u=0}\pi_\lambda(\exp uX)(\phi)(t,r) = \left(\mathscr{M}\Phi'(X) + \lambda(d\bar{\pi}_{S^1}(X))'(t) + d\bar{\pi}(X)\right)\phi(t,r) \tag{1.29}$$

so $d\pi_\lambda(X)$ may be represented as the differential operator of order one

$$d\bar{\pi}(X) + \mathscr{M}\Phi'(X) + \lambda(d\bar{\pi}_{S^1}(X))'(t).$$

So all this amounts to replacing formally ∂_ζ by $\mathscr{M}$ in the formulas of Proposition 1.10 in the case $\lambda = 0$. Then, for any λ,

$$d\pi_\lambda(\mathscr{L}_f) = d\pi_0(\mathscr{L}_f) - \lambda f', \tag{1.30}$$

while $d\pi_\lambda(\mathscr{Y}_g) = d\pi_0(\mathscr{Y}_g)$ and $d\pi_\lambda(\mathscr{M}_h) = d\pi_0(\mathscr{M}_h)$.

Let us write again explicitly the action of all generators, both for completeness and for future reference:

$$\begin{aligned} d\pi_\lambda(\mathcal{L}_f) &= -f(t)\partial_t - \frac{1}{2}f'(t)r\partial_r - \frac{1}{4}f''(t)\mathcal{M}r^2 - \lambda f'(t) \\ d\pi_\lambda(\mathcal{Y}_g) &= -g(t)\partial_r - \mathcal{M}g'(t)r \\ d\pi_\lambda(\mathcal{M}_h) &= -\mathcal{M}h(t) \end{aligned} \tag{1.31}$$

These formulas for the vector-field representations $d\pi_\lambda$ coincide with those of the Introduction.

Chapter 2
Basic Algebraic and Geometric Features

We gather in this chapter useful algebraic and geometric features of the Schrödinger–Virasoro algebra that have not been derived in the previous chapter because they are not directly related to Newton-Cartan structures. The unifying concept here is that of *graduations*. Recall that a *graduation* of a Lie algebra $\mathfrak{g}$ is a decomposition of $\mathfrak{g}$ into a direct sum $\mathfrak{g} = \oplus_{n\in\mathbb{Z}}\mathfrak{g}_n$ such that $[\mathfrak{g}_n, \mathfrak{g}_m] \subset \mathfrak{g}_{n+m}$. If $X \in \mathfrak{g}_n$ for some n, then X is said to be an element of degree n. Allowing for $\mathbb{Z}^k$-valued graduations, a particularly interesting case is that of the graduation given by the root system if $\mathfrak{g}$ is semi-simple.

Graduations – and more generally, root diagrams – play a prominent rôle for Lie algebras, both from a classification and a representation point of view; think for instance about highest-weight modules, or (in the infinite-dimensional case) about O. Mathieu's theorem [86] on the classification of simple graded Lie algebras with polynomial growth. In this respect, it is interesting to note the two following facts:

- The Schrödinger–Virasoro Lie algebra enjoys two independent graduations (both of which are equally interesting from a representation theoretic point of view, as we shall see in Chaps. 4–6), one of which is given by the indices of the generators L_n, Y_m, M_p (see Sect. 2.1). The Lie algebra $\mathfrak{sv}$ also appears naturally as a subalgebra of an *extended Poisson algebra* on the torus, quotiented out by the ideal generated by the elements with negative graduation (see Sect. 2.3). The latter idea is developed in full generality in Chap. 11 when considering supersymmetric extensions;
- The finite-dimensional Lie algebra $\mathfrak{sch} \subset \mathfrak{sv}$ may be embedded into the complexified *conformal Lie algebra* $\mathfrak{conf}(3)_{\mathbb{C}}$ in three dimensions (see Sect. 2.2). More generally, $\mathfrak{sch}(d)$ may be embedded into $\mathfrak{conf}(d+2)_{\mathbb{C}}$. This simple statement has already been made in the Preface (see 8.); it comes from the fact that a Laplace transformation $\mathscr{M} \rightsquigarrow \partial_\zeta$ formally intertwines the free Schrödinger operators in $(d+1)$-dimensions with the Klein-Gordon operator in $(d+2)$-dimensions. We give here explicit formulas for the embedding.

J. Unterberger and C. Roger, *The Schrödinger-Virasoro Algebra*, Theoretical and Mathematical Physics, DOI 10.1007/978-3-642-22717-2_2,

A natural question arises then: does there exist a Lie algebra containing both $\mathfrak{conf}(3)$ and $\mathfrak{sv}$, such that both natural embeddings $\mathfrak{sch} \subset \mathfrak{conf}(3)$, $\mathfrak{sch} \subset \mathfrak{sv}$ are preserved? It turns out that there is no natural answer to this problem. In analogy with the well-known result in gauge theory stating that there does not exist a non-trivial extension containing both the Poincaré group and the external gauge group, we shall call this a *'no-go' theorem* (see Sect. 2.3).

Finally, we recall in Sect. 2.4 classical results concerning Schrödinger and conformal *tensor invariants*. These tensor invariants are called *covariant n-point functions* in the context of statistical physics. The general idea is that symmetries constrain n-point functions, which in turn yields predictions for the correlators of physical quantities. This idea will be developed in Chap. 6, and also in a supersymmetric setting in Chap. 11.

The results of this chapter are taken from [106].

2.1 On Graduations and Some Deformations of the Lie Algebra $\mathfrak{sv}$

Let us emphasize the following statement from Sect. 1.1.2.

Proposition 2.1. *As a* Vect(S^1)*-module, the Schrödinger–Virasoro algebra is isomorphic to the semi-direct product* Vect$(S^1) \ltimes (\mathcal{F}_{\frac{1}{2}} \oplus \mathcal{F}_0)$.

Recall (see footnote in Sect. 1.1.2) that a field X in $\mathcal{F}_\lambda$ gives rise in the formalism of conformal field theory to a (quantum) *primary field* $\hat{X}$ with *conformal weight* $\lambda + 1$, i.e. such that $[L_n, \hat{X}_m] = (\lambda n - m)\hat{X}_{n+m}$ for the operator bracket, so that (depending on the terminology) the Y, resp. M field behaves as a $(-\frac{1}{2})$-density, resp. 0-density or, in other words, Y, resp. M has *conformal weight* $\frac{3}{2}$, resp. 1. We shall spend some time explaining the basics of conformal field theory in Chap. 6; for the moment, it is sufficient to know that it is based upon the explicit construction out of creation and annihilation operators of quantum fields acting on Fock spaces, whose Fourier components satisfy commutation relations of a certain type, including in particular those of the Virasoro algebra and those between the Virasoro field and a primary field. Conformal field theory is thus a very fruitful way of constructing infinite-dimensional algebras and representations thereof. Let us now remark that quantum fields with *half-integer weight* (such as Y) are usually *fermionic*. As a matter of fact, the celebrated spin-statistics theorem *in dimension* ≥ 3 [110] states that bosonic, resp. fermionic fields should have an integer-, resp. half-integer-valued spin. Fermionic fields produce *odd* generators in *superalgebras*. The most celebrated superalgebra containing the Virasoro field *and* a field with conformal weight $3/2$ is the so-called *Neveu-Schwarz superalgebra* $\mathfrak{ns}$ (see [66]). It is generated by the Virasoro field $(L_n)_{n\in\mathbb{Z}}$ and a fermionic current (G_m) with $m \in \mathbb{Z}$ or $\frac{1}{2} + \mathbb{Z}$, with supplementary anti-commutation relations $\{G_n, G_m\} = 2L_{n+m}$. Contrary to $\mathfrak{sv}$, it is a *simple* superalgebra. The liberty of choosing either integer of half-integer indices for the G-generators leas to subtle differences when considering its representations in conformal field theory.[1]

Turning back to the Schrödinger–Virasoro algebra, and trying to play with its definition, one may note the following facts:

- A theorem due to O. Mathieu [86], already alluded to in the introduction to this chapter, gives a classification of all *simple* $\mathbb{Z}$-graded algebras $\mathfrak{g} = \oplus_{n\in\mathbb{Z}}\mathfrak{g}_n$ with polynomial growth, i.e. such that $\dim \mathfrak{g}_n \le C|n|^k$ for some $C > 0$ and $k \in \mathbb{N}$. They come in four series: (1) *finite-dimensional simple Lie algebras*; (2) *Kac-Moody Lie algebras*, or in other words current Lie algebras on the circle; (3) the so-called *Cartan-type Lie algebras*, which are Lie algebras of formal vector fields in n variables;[2] (4) the two Virasoro-type algebras, $\mathrm{Vect}(S^1)$ and $\mathfrak{vir}$. From this theorem one sees clearly the impossibility of deforming the bracket of $\mathfrak{sv}$ so as to obtain a new *simple* algebra. On the other hand, the Neveu-Schwarz algebra is a counterexample to this statement in the category of *superalgebras*.
- The same ambiguity that exists in the range of indices for the generators of the Neveu-Schwarz algebra $\mathfrak{ns}$ is also found for $\mathfrak{sv}$, yielding a competition between two algebras, the Schrödinger–Virasoro algebra $\mathfrak{sv}$ and the *twisted Schrödinger–Virasoro algebra* $\mathfrak{sv}(0)$, to which we shall come back in Chaps. 7 and 9;
- Contrary to the case of the Neveu-Schwarz algebra, *changing continuously the conformal weight* of the Y and M generators yields two one-parameter families of algebras, $\mathfrak{sv}_\varepsilon$ and $\mathfrak{sv}_\varepsilon(0)$, $\varepsilon \in \mathbb{R}$, with $\mathfrak{sv}_0 = \mathfrak{sv}$ and $\mathfrak{sv}_0(0) = \mathfrak{sv}(0)$. We shall actually show by cohomological methods in Chap.7 that $\mathfrak{sv}$ and $\mathfrak{sv}(0)$ admit *three* independent families of deformations.

Definition 2.2. Let $\mathfrak{sv}_\varepsilon$, $\varepsilon \in \mathbb{R}$ (resp. $\mathfrak{sv}_\varepsilon(0)$) be the Lie algebra generated by L_n, Y_m, M_p, $n, p \in \mathbb{Z}, m \in \frac{1}{2} + \mathbb{Z}$ (resp. $m \in \mathbb{Z}$), with relations

$$[L_n, L_{n'}] = (n-n')L_{n+n'},\ [L_n, Y_m] = \left(\frac{(1+\varepsilon)n}{2} - m\right) Y_{n+m},\ [L_n, M_p] = (\varepsilon n - p)M_{n+p}$$

$$[Y_m, Y_{m'}] = (m-m')M_{m+m'},\ [Y_m, M_p] = 0,\ [M_p, M_{p'}] = 0, \qquad (2.1)$$

The Y, resp. M field is a $-(1+\varepsilon)n/2$, resp. $-\varepsilon$ density, or in other words, has primary weight $(3+\varepsilon)/2$, resp. $1+\varepsilon$.

In close connection to all this, let us now turn to graduations. Any graduation δ on a Lie algebra $\mathfrak{g}$ may be seen alternatively as a *derivation* $\bar{\delta} : \mathfrak{g} \to \mathfrak{g}$ by setting $\bar{\delta}(X_n) = nX_n$ if $\delta(X_n) = n$, i.e. if X_n has degree n. Recall that a *derivation* of $\mathfrak{g}$ is a

---

[1]Vertex representations with integer, resp. half-integer valued G-component indices correspond to the so-called *Ramond*, resp. *Neveu-Schwarz* sector.

[2]There are four series indexed by n: the Lie algebra $\mathrm{Vect}(n)$ of all formal vector fields, and the subalgebras of $\mathrm{Vect}(n)$ made up of symplectic, unimodular or contact vector fields.

linear map $\bar{\delta} : \mathfrak{g} \to \mathfrak{g}$ such that $\bar{\delta}[X,Y] = [\bar{\delta}(X),Y] - [\bar{\delta}(Y),X]$. Typical examples are *inner derivations* $\bar{\delta} := [X,.]$, given by bracketing against an element $X \in \mathfrak{g}$. *Outer derivations* are then simply non-inner derivations.

It turns out that there exist *two* linearly independent graduations on $\mathfrak{sv}_\varepsilon$ or $\mathfrak{sv}_\varepsilon(0)$.

Definition 2.3. Let δ_1, resp. δ_2, be the graduations on $\mathfrak{sv}_\varepsilon$ or $\mathfrak{sv}_\varepsilon(0)$ defined by

$$\delta_1(L_n) = n,\ \delta_1(Y_m) = m, \delta_1(M_p) = p \tag{2.2}$$

$$\delta_2(L_n) = n,\ \delta_2(Y_m) = m - \frac{1}{2},\ \delta_2(M_p) = p - 1 \tag{2.3}$$

with $n, p \in \mathbb{Z}$ and $m \in \mathbb{Z}$ or $\frac{1}{2} + \mathbb{Z}$.

One immediately checks that both δ_1 and δ_2 define graduations and that they are linearly independent.

Proposition 2.4. *The graduation δ_1, defined either on $\mathfrak{sv}_\varepsilon$ or on $\mathfrak{sv}_\varepsilon(0)$, is given by the inner derivation $\delta_1 = \mathrm{ad}(-L_0)$, while δ_2 is an outer derivation.*

Remark 2.5. As we shall see in Chap. 7, the space $H^1(\mathfrak{sv}, \mathfrak{sv})$ or $H^1(\mathfrak{sv}(0), \mathfrak{sv}(0))$ of outer derivations modulo inner derivations (see introduction to Chap. 7) is three-dimensional, but only δ_2 defines a graduation on the basis (L_n, Y_m, M_p).

Proof. The only non-trivial point is to prove that δ_2 is not an inner derivation. Suppose, to the contrary, that $\delta_2 = \mathrm{ad}X$, $X \in \mathfrak{sv}$ or $X \in \mathfrak{sv}(0)$ (we treat both cases simultaneously). Then $\delta_2(M_0) = 0$ since M_0 is central in $\mathfrak{sv}$ and in $\mathfrak{sv}(0)$. Hence the contradiction. Note that the graduation δ_2 is given by the Lie action of the Euler vector field $t\partial_t + r\partial_r + \zeta\partial_\zeta$ in the vector-field representation $d\tilde{\pi}$ (see Introduction). □

All these Lie algebras may be extended by using the natural extension of the Virasoro cocycle (1.14) of Sect. 1.1, yielding Lie algebras denoted by $\widetilde{\mathfrak{sv}}$, $\widetilde{\mathfrak{sv}}(0)$, $\widetilde{\mathfrak{sv}}_\varepsilon$, $\widetilde{\mathfrak{sv}}_\varepsilon(0)$.

2.2 The Conformal Embedding

Let $\mathfrak{conf}(d)$ be the *conformal algebra* in d dimensions ($d \geq 3$). By definition, it is the Lie algebra of the Lie group of global conformal transformations of $\mathbb{R}^d$ equipped with its standard Riemannian metric, i.e. the Lie group of C^1-diffeomorphisms $\phi : \mathbb{R}^d \to \mathbb{R}^d$ such that the differential $d\phi_x : T_x\mathbb{R}^d \to T_x\mathbb{R}^d$ at the point x preserves the metric tensor $g_{\mu\nu}dx^\mu dx^\nu$ up to a scalar multiplication.

Such a Lie group may also be defined on any Riemannian manifold M of dimension d. As shown for instance in [19], in dimension $d \geq 3$, the local constraints coming from the requirement that a one-parameter subgroup of local diffeomorphisms preserve the metric up to scalar multiplication is already strong enough to make the conformal group finite-dimensional. As mentioned in the Introduction, this is not true for $d = 2$, since local conformal transformations are then local holomorphic or anti-holomorphic transformations.

Computations show that $\mathfrak{conf}(d)$ is $\frac{(d+1)(d+2)}{2}$- dimensional and is isomorphic to $\mathfrak{so}(d+1,1)$. Clearly, rotations and translations preserve the metric, so the (Riemannian) Poincaré algebra $\mathfrak{so}(d) \ltimes \mathbb{R}^d$ is naturally embedded into $\mathfrak{so}(d+1,1)$. The dilatations $x \to \lambda x$ ($\lambda \in \mathbb{R}$) preserve the metric up to a constant. There also exist so-called *special conformal transformations* which preserve the metric only up to a non-trivial scalar (see below).

The idea of embedding $\mathfrak{sch}(d)$ into $\mathfrak{conf}(d+2)_{\mathbb{C}}$ comes naturally when considering the wave equation

$$(2\mathcal{M}\partial_t - \Delta_r)\psi(\mathcal{M};t,r) = 0 \tag{2.4}$$

where $\mathcal{M}$ is viewed no longer as a parameter, but as a coordinate. Then the formal Laplace transform of the wave function with respect to the mass

$$\tilde{\psi}(\zeta;t,r) = \int_{\mathbb{R}} \psi(\mathcal{M};t,r)e^{\mathcal{M}\zeta}\, d\mathcal{M} \tag{2.5}$$

satisfies the equation

$$(2\partial_\zeta\partial_t - \Delta_r)\psi(\zeta;t,r) = 0 \tag{2.6}$$

which is nothing but a zero mass Klein-Gordon equation $(\partial_x^2 - \partial_y^2 - \Delta_r)\psi = 0$ on $(d+2)$-dimensional space-time, put into light-cone coordinates $(\zeta,t) = (\frac{x+y}{\sqrt{2}}, \frac{x-y}{\sqrt{2}})$. In this sense, Schrödinger transformations may be seen as the *subgroup of conformal transformations which preserve the mass.*

This simple idea has been developed in a previous article (see [54]) for $d = 1$. Let us here give an explicit embedding for any dimension d.

We need first to fix some notations. Consider the conformal algebra in its standard representation as infinitesimal conformal transformations on $\mathbb{R}^{d+2}$ with coordinates $(\xi_1,\ldots,\xi_{d+2})$. Then there is a natural basis of $\mathfrak{conf}(d+2)$ made of $(d+2)$ translations P_μ, $\frac{1}{2}(d+1)(d+2)$ rotations $\mathcal{M}_{\mu,\nu}$, $(d+2)$ inversions K_μ and the Euler operator D: in coordinates, one has

$$P_\mu = \partial_{\xi_\mu} \tag{2.7}$$

$$\mathcal{M}_{\mu,\nu} = \xi_\mu\partial_\nu - \xi_\nu\partial_\mu \tag{2.8}$$

$$K_\mu = 2\xi_\mu\left(\sum_{\nu=1}^{d+2}\xi_\nu\partial_\nu\right) - \left(\sum_{\nu=1}^{d+2}\xi_\nu^2\right)\partial_\mu \tag{2.9}$$

$$D = \sum_{\nu=1}^{d+2}\xi_\nu\partial_\nu. \tag{2.10}$$

Proposition 2.6. *The formulas*

$$Y^j_{-\frac{1}{2}} = -2^{\frac{1}{2}} e^{-i\pi/4} P_j \tag{2.11}$$

$$Y^j_{\frac{1}{2}} = -2^{-\frac{1}{2}} e^{i\pi/4} \left(\mathscr{M}_{d+2,j} + i\mathscr{M}_{d+1,j}\right) \tag{2.12}$$

$$R_{j,k} = \mathscr{M}_{j,k} \tag{2.13}$$

$$L_{-1} = i(P_{d+2} - iP_{d+1}) \tag{2.14}$$

$$L_0 = -\frac{D}{2} + \frac{i}{2}\mathscr{M}_{d+2,d+1} \tag{2.15}$$

$$L_1 = -\frac{i}{4}(K_{d+2} + iK_{d+1}) \tag{2.16}$$

give an embedding of $\mathfrak{sch}(d)$ *into* $\mathfrak{conf}(d+2)_{\mathbb{C}}$.

Proof. Put

$$t = \frac{1}{2}(-\xi_{d+1} + i\xi_{d+2}), \zeta = \frac{1}{2}(\xi_{d+1} + i\xi_{d+2}), r_j = 2^{-\frac{1}{2}} e^{i\pi/4}\xi_j \ (j = 1, \ldots, d).$$

Then the previous definitions yield the representation $d\tilde{\pi}^d_{d/4}$ of $\mathfrak{sch}(d)$ (see (1.11)). □

When $d = 1$, this conformal embedding can be represented by the following root diagram of $\mathfrak{conf}(3)$, of type B_2, as drawn in [54] (see reproduction in Fig. 2.1). The *Schrödinger representation* $d\tilde{\pi}^d_{d/4}$ extends naturally to a representation of $\mathfrak{conf}(3)$. The Cartan subalgebra $\mathfrak{a}$ is two-dimensional, generated by L_0 and N_0, with $d\tilde{\pi}^d_{d/4}(N_0) = -r\partial_r - 2\zeta\partial_\zeta$ (see Sect. 6.1). It is convenient here to introduce instead the linear combinations $D = -L_0 - \frac{N_0}{2}$ and $N = L_0 - \frac{N_0}{2}$, realized in the Schrödinger representation as $t\partial_t + r\partial_r + \zeta\partial_\zeta$ and $-t\partial_t + \zeta\partial_\zeta$ respectively (see Chap. 11). Define the roots e_1, e_2 as $e_i(N) = \delta_{i,1}$, $e_i(D) = \delta_{i,2}$, yielding the coordinates along the horizontal and vertical axes respectively on Fig. 2.1. Then $\Pi := \{-e_2, e_1 + e_2\}$ defines a basis of the root system; the associated set of positive roots, i.e. of roots which are linear combinations with *positive* coefficients of $-e_2$ and $e_1 + e_2$, is $\Delta_+ = \{-e_2, e_1 + e_2, e_1, e_1 - e_2\}$. From Fig. 2.1 one reads directly that the positive Borel subalgebra (or *minimal parabolic subalgebra*) is

$$\mathfrak{a} \oplus \sum_{\alpha \in \Delta_+} \mathfrak{g}_\alpha = \langle L_0, N_0, L_1, Y_{\pm\frac{1}{2}}, M_0 \rangle, \tag{2.17}$$

where $\mathfrak{g}_\alpha = \{X \in \mathfrak{conf}(3) \mid \forall H \in \mathfrak{a}, [H, X] = \alpha(H)X\}$ is the one-dimensional root space associated to the root α. Up to the extra generator N_0, this is the *age subalgebra* of the Schrödinger subalgebra introduced in [54], which is a natural symmetry algebra for ageing systems (see Preface). The *extended Schrödinger*

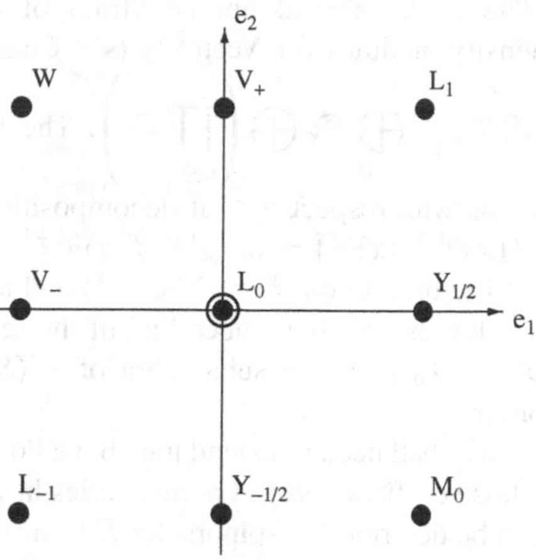

Fig. 2.1 Connected diagram

algebra $\overline{\mathfrak{sch}} = \langle N_0 \rangle \ltimes \mathfrak{sch}$, introduced in Chap. 6, may be understood algebraically as the *maximal parabolic subalgebra* associated to the subset $\Pi_{\overline{\mathfrak{sch}}} = \{e_1 + e_2\} \subset \Pi$; apart from the generators of the positive Borel subalgebra, it also contains $\mathfrak{g}_{-(e_1+e_2)} = \mathbb{C}L_{-1}$, realized as time-translations in the Schrödinger representation. All these notions are standard in the study of the structure and representations of semi-simple Lie groups and algebras [60, 72].

2.3 Relations Between $\mathfrak{sv}$ and the Poisson Algebra on T^*S^1 and 'no-go' Theorem

The relation between the Virasoro algebra and the Poisson algebra on T^*S^1 has been investigated in [93]. We shall consider more precisely the Lie algebra $\mathscr{A}(S^1)$ of smooth functions on $\dot{T}^*S^1 = T^*S^1 \setminus S^1$, the total space of the cotangent bundle with zero section removed, which are Laurent series on the fibers.[3] So $\mathscr{A}(S^1) = C^\infty(S^1) \otimes \mathbb{R}[\partial, \partial^{-1}]]$ and $F \in \mathscr{A}(S^1)$ is of the following form:

$$F(t, \partial) = \sum_{k \in \mathbb{Z}} f_k(t) \partial^k,$$

with $f_k = 0$ for large enough k. The Poisson bracket is defined as usual, following:

$$\{F, G\} = \frac{\partial F}{\partial \partial} \frac{\partial G}{\partial t} - \frac{\partial G}{\partial \partial} \frac{\partial F}{\partial t}.$$

[3]One might also consider the subalgebra of $\mathscr{A}(S^1)$ defined as $\mathbb{C}[z, z^{-1}] \otimes \mathbb{C}[\partial, \partial^{-1}]$; it gives the usual description of the Poisson algebra on the torus $\mathbb{T}^2$, sometimes denoted $SU(\infty)$ [59].

(The reader should not be afraid of the notation $\partial\partial$!). In terms of tensor-density modules for Vect(S^1) (see Chap. 1), one has the natural decomposition: $\mathscr{A}(S^1) = \bigoplus_{k>0} \mathscr{F}_k \bigoplus \left(\prod_{k\leq 0} \mathscr{F}_k \right)$. The Poisson bracket turns out to be homogeneous with respect to that decomposition: $\{\mathscr{F}_k, \mathscr{F}_l\} \subset \mathscr{F}_{k+l-1}$; more explicitly $\{f(x)\partial^k, g(x)\partial^l\} = (kfg' - lf'g)\partial^{k+l-1}$. One recovers the formulae of Chap. 1 for the Lie bracket on $\mathscr{F}_1 = \mathrm{Vect}(S^1)$ and its representations on modules of densities; one has as well the embedding of the semi-direct product $\mathrm{Vect}(S^1) \ltimes C^\infty(S^1) = \mathscr{F}_1 \ltimes \mathscr{F}_0$ as a Lie subalgebra of $\mathscr{A}(S^1)$, representing differential operators of order ≤ 1.

We shall need to extend the above Poisson algebra by considering *half-densities*, whose coefficients are Laurent series in $\sqrt{z}$. Geometrically speaking, half-densities can be described as spinors: let E be a vector bundle over S^1, square root of T^*S^1; in other words one has $E \otimes E = T^*S^1$. Then the space of Laurent polynomials on the fibers of E (with the zero-section removed) is exactly the Poisson algebra $\widetilde{\mathscr{A}}(S^1) = C^\infty(S^1) \otimes \mathbb{C}[\partial^{1/2}, \partial^{-1/2}]]$. We shall also need to consider half-integer power series or polynomials in $\sqrt{z}$ as coefficients of the Laurent series in ∂; one may obtain the corresponding algebra globally by using the pull-back though the application $S^1 \longrightarrow S^1$ defined as $z \mapsto z^2$.

Summarizing, one has obtained the subalgebra $\widehat{\mathscr{A}}(S^1) \supset \widetilde{\mathscr{A}}(S^1)$ generated by terms $z^m \partial^n$ where m and n are either integers or half-integers. One may represent such generators as the points with coordinates (m, n) in the plane $\mathbb{R}^2$. In particular, $\mathfrak{sv} = \mathrm{Vect}(S^1) \ltimes \mathfrak{h}$, with $\mathfrak{h} \simeq \mathscr{F}_{1/2} \ltimes \mathscr{F}_0$ as a Vect(S^1)-module, can be naturally embedded into $\widehat{\mathscr{A}}(S^1)$ as follows:

$$
\begin{array}{ccccccccccc}
\cdots & L_{-2} & & L_{-1} & & L_0 & & L_1 & & L_2 & \cdots \\
 & \cdots & Y_{-\frac{3}{2}} & & Y_{-\frac{1}{2}} & & Y_{\frac{1}{2}} & & Y_{\frac{3}{2}} & \cdots & \\
\cdots & M_{-2} & & M_{-1} & & M_0 & & M_1 & & M_2 & \cdots
\end{array}
\tag{2.18}
$$

while the twisted Schrödinger–Virasoro algebra $\mathfrak{sv}(0)$ (with integer-valued indices for the Y-field) may be represented as follows:

$$
\begin{array}{ccccccc}
\cdots & L_{-2} & L_{-1} & L_0 & L_1 & L_2 & \cdots \\
\cdots & Y_{-2} & Y_{-1} & Y_0 & Y_1 & Y_2 & \cdots \\
\cdots & M_{-2} & M_{-1} & M_0 & M_1 & M_2 & \cdots
\end{array}
$$

The graduation along the vertical axis is given by the outer derivation $\delta_2 - \delta_1$, and indicates the weights of the associated density modules. One can naturally ask whether this defines a Lie algebra embedding, just as in the case of $\mathrm{Vect}(S^1) \ltimes \mathscr{F}_0$. The answer is *no*:

Proposition 2.7. *The natural vector space embedding $\mathfrak{sv} \hookrightarrow \tilde{\mathscr{A}}(S^1)$ is not a Lie algebra homomorphism.*

Proof. One sees immediately that on the one hand $[Y_m, M_p] = 0$ and $[M_p, M_{p'}] = 0$, while on the other hand $\{\mathscr{F}_{1/2}, \mathscr{F}_0\} \subset \mathscr{F}_{-1/2}$ is non trivial. □

The vanishing of $\{\mathscr{F}_0, \mathscr{F}_0\}$ which makes the embedding of $\mathrm{Vect}(S^1) \ltimes \mathscr{F}_0$ as a Lie subalgebra possible was in some sense an accident. In fact, one can show that, starting from the image of the generators of $\mathfrak{sv}$ and computing successive Poisson brackets, one can generate all the $\mathscr{F}_\lambda$ with $\lambda \leq 0$. In other words,

$$\tilde{\mathscr{A}}(S^1)_{\leq 1} = \mathscr{F}_1 \oplus \mathscr{F}_{1/2} \oplus \mathscr{F}_0 \bigoplus_{\lambda \in \frac{\mathbb{Z}}{2},\ \lambda<0} \mathscr{F}_\lambda \tag{2.19}$$

defines the smallest possible Poisson subalgebra of $\tilde{\mathscr{A}}(S^1)$ which contains the image of $\mathfrak{sv}$. Now, let

$$\tilde{\mathscr{A}}(S^1)_{\leq -\frac{1}{2}} = \prod_{\lambda \in \frac{\mathbb{Z}}{2},\ \lambda<0} \mathscr{F}_\lambda \tag{2.20}$$

be the Poisson subalgebra of $\tilde{\mathscr{A}}(S^1)$ which contains only negative powers of ∂. One easily sees that it is an ideal of $\tilde{\mathscr{A}}(S^1)_{(1)}$. If one considers the quotient $\tilde{\mathscr{A}}(S^1)_{(1)}/\tilde{\mathscr{A}}(S^1)_{(0)}$, then the embedding becomes an isomorphism:

Proposition 2.8 ($\mathfrak{sv}$ as a Poisson subquotient). *There exists a natural Lie algebra isomorphism between $\mathfrak{sv}$ and quotient $\tilde{\mathscr{A}}(S^1)_{\leq 1}/\tilde{\mathscr{A}}(S^1)_{\leq -\frac{1}{2}}$. In other words, $\mathfrak{sv}$ may be seen as a subquotient of the Poisson algebra $\tilde{\mathscr{A}}(S^1)$.*

Now a natural question arises: the conformal embedding of Schrödinger algebra described in Sect. 2.2 yields $\mathfrak{sch} \subset \mathfrak{conf}(3)_{\mathbb{C}}$, so one would like to extend the construction of $\mathfrak{sv}$ as generalization of $\mathfrak{sch}$, in order that it contain $\mathfrak{conf}(3)$. In other words, we are looking for an hypothetic Lie algebra $\mathscr{G}$ making the following diagram of embeddings complete:

$$\begin{array}{ccc} \mathfrak{sch} & \hookrightarrow & \mathfrak{conf}(3) \\ \downarrow & & \downarrow \\ \mathfrak{sv} & \hookrightarrow & \mathscr{G} \end{array} \tag{2.21}$$

In the category of abstract Lie algebras, one has an obvious solution to this problem: simply take the amalgamated sum of $\mathfrak{sv}$ and $\mathfrak{conf}(3)$ over $\mathfrak{sch}$. Such a Lie algebra is defined though generators and relations, and is generally intractable. We are looking here for a natural, geometrically defined construction of such a $\mathscr{G}$; we shall give some evidence of its non-existence, a kind of *'no-go' theorem*, analogous to those well-known in gauge theory [4] (see e.g. [67]).

[4]Simply recall that this theorem states that there does not exist a common non-trivial extension containing both the Poincaré group and the external gauge group.

Let us consider once again the root diagram of $\mathfrak{conf}(3)$:

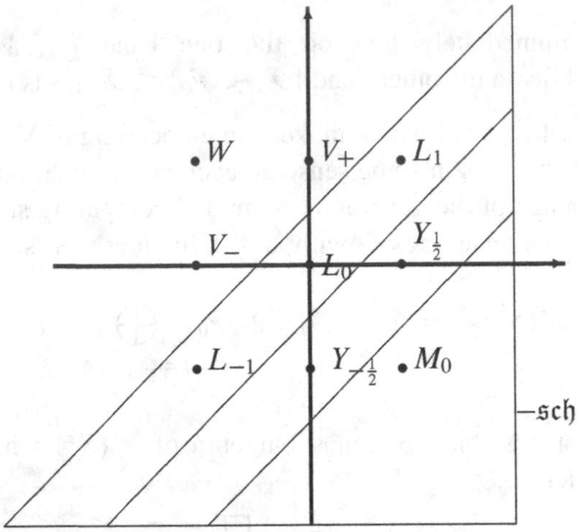

The graduation along the vertical (resp. horizontal) axis is given by the action of δ_2, resp. $2\delta_1 - \delta_2$. Comparing with (2.18), one sees that the successive diagonal strips are contained in $\mathcal{F}_1, \mathcal{F}_{1/2}, \mathcal{F}_0$ respectively. So the first idea might be to try to add $\mathcal{F}_{3/2}$ and $\mathcal{F}_2$, as an infinite-dimensional prolongation of $\langle V_-, V_+, W\rangle$, so that $V_- \longrightarrow t^{-1/2}\partial^{3/2}$, $V_+ \longrightarrow t^{1/2}\partial^{3/2}$, $W \longrightarrow \partial^2$.

Unfortunately, this construction fails at once for two reasons: first, one does not get the right brackets for $\mathfrak{conf}(3)$ with such a choice, and secondly the elements of $\mathcal{F}_\lambda$, $\lambda \in \{0, \frac{1}{2}, 1, \frac{3}{2}, 2\}$, taken together with their successive brackets generate the whole Poisson algebra $\tilde{\mathcal{A}}(S^1)$.

Another approach could be the following: take two copies of $\mathfrak{h}$, say $\mathfrak{h}^+$ and $\mathfrak{h}^-$ and consider the semi-direct product $\mathcal{G} = \mathrm{Vect}(S^1) \ltimes (\mathfrak{h}^+ \oplus \mathfrak{h}^-)$, so that $\mathfrak{h}^+$ extends $\langle Y_{-\frac{1}{2}}, Y_{\frac{1}{2}}, M_0\rangle$ as in $\mathfrak{sv}$ before, and so that $\mathfrak{h}^-$ extends $\langle V_-, V_+, W\rangle$. Then $\mathcal{G}$ is obtained from density modules. However, it does *not* extend $\mathfrak{conf}(3)$, but only a contraction of it: all the brackets between $\langle Y_{-\frac{1}{2}}, Y_{\frac{1}{2}}, M_0\rangle$ on the one hand and $\langle V_-, V_+, W\rangle$ on the other are vanishing. Now, we can try to deform $\mathcal{G}$ in order to obtain the right brackets for $\mathfrak{conf}(3)$. Let Y_m^+, M_m^+ and Y_m^-, M_m^- be the generators of $\mathfrak{h}^+$ and $\mathfrak{h}^-$; we want to find coefficients $a_{p,m}$ such that $[Y_m^+, Y_p^-] = a_{p,m} L_{m+p}$ defines a Lie bracket. So let us check Jacobi identity for (L_n, Y_m^+, Y_p^-). One obtains $(m - \frac{n}{2})a_{p,n+m} + (n - m - p)a_{p,m} + (p - \frac{n}{2})a_{p+n,m} = 0$. If one tries $a_{pm} = \lambda p + \mu m$, one deduces from this relation: $n\lambda(p - \frac{n}{2}) + n\mu(m - \frac{n}{2}) = 0$ for every $n \in \mathbb{Z}$, $p, m \in \frac{1}{2}\mathbb{Z}$, so obviously $\lambda = \mu = 0$.

So our computations show there does not exist a geometrically defined construction of $\mathcal{G}$ satisfying the conditions of diagram (2.16). The two possible extensions of $\mathfrak{sch}$, $\mathfrak{sv}$ and $\mathfrak{conf}(3)$ are shown to be incompatible, and this is our "no-go" theorem.

To finish, note that embeddings into Poisson algebras also appear naturally in a supersymmetric setting, yielding a large class of superizations of $\mathfrak{sv}$ (see Chap. 11, Sect. 11.3).

2.4 Conformal and Schrödinger Tensor Invariants

Consider quite generally a group representation $\rho : G \to Hom(\mathscr{H}, \mathscr{H})$ into some vector space $\mathscr{H}$, and its n-fold tensor product $\rho^{\otimes n} : G \to Hom(\mathscr{H}^{\otimes n}, \mathscr{H}^{\otimes n})$ defined as usual by $\rho^{\otimes n}(g)(v_1 \otimes \ldots v_n) = \rho(g)v_1 \otimes \ldots \otimes \rho(g)v_n$. Then an *n-tensor invariant* of ρ is an element $v \in \mathscr{H}^{\otimes n}$ such that $\rho^{\otimes n}(g) \cdot v = v$ for all $g \in G$. This notion is central in quantum field theory, and more specifically in conformal field theory, as we shall presently see. Let $\Psi_1(x), \ldots, \Psi_n(x)$ be (vector-valued) quantum fields on $\mathbb{R}^d$ (free fields in quantum field theory [115], or *quasi-primary fields* in conformal field theory [19]), acting on some Hilbert space $\mathscr{H}$, usually a Fock space. Assume that some symmetry group G (the Poincaré group for quantum field theory on $\mathbb{R}^{3+1}$, or the group of projective transformations for two-dimensional conformal field theory, resp. the conformal group Conf(d) in $d \geq 3$ dimensions) has been implemented as unitary operators $U(g) : \mathscr{H} \to \mathscr{H}$ such that

$$U(g)\Psi_i(x)U(g)^{-1} = \sum_i D_{i,j}(g^{-1}) \cdot \Psi_j(g \cdot x), \tag{2.22}$$

where D is a finite-dimensional matrix representation of G acting on the components of the vector Ψ_i, and $g \cdot x$ is the image of x by the natural coordinate transformation induced by $g \in G$.[5] Assume also that the *vacuum state* $|0\rangle \in \mathscr{H}$ is G-invariant. Then the *vacuum state expectation* (or *n-point function*) $\langle 0|\Psi_1(x_1) \ldots \Psi_n(x_n)|0\rangle$ is an n-tensor invariant of ρ.

We consider here two classical examples: conformal and Schrödinger tensor-invariants.

Example 1 (Conformal invariants). Consider the representation $\rho = \rho_\mu$ of the group of homographies, i.e. of holomorphic projective transformations of the sphere: it acts on a *quasi-primary field* of conformal weight μ as

$$\rho_\mu \begin{pmatrix} a & b \\ c & d \end{pmatrix} \Psi_\mu(z) = (ct + d)^{-2\mu} \Psi_\mu \left(\frac{dz - b}{-cz + a} \right). \tag{2.23}$$

In other words, $\rho_\mu(\phi)\Psi_\mu(z) = (\phi'(z))^\mu (\Psi_\mu \circ \phi^{-1})(z)$ for $g \in SL(2, \mathbb{R}) \subset \mathrm{Diff}(S^1)$. The associated classical action is the restriction to $SL(2, \mathbb{R})$ of the action of diffeomorphisms of the circle on μ-densities (see Chap. 1). A *primary field* is by definition transformed in the same way by *all* diffeomorphisms of the circle. However, the constraints induced by this much stronger property on n-point functions are much more subtle because the vacuum state $|0\rangle$ itself is *not* invariant under $\mathrm{Diff}(S^1)$.

---

[5] In many cases, by replacing the quantum fields $\Psi(x)$ by a function $\psi(x)$, one obtains simply an irreducible unitary representation of G on a one-particle state.

Tensor-invariants (or $SL(2,\mathbb{R})$-covariant n-point functions) are given as follows for $n = 2, 3, 4$:

$$\langle \Psi_{\mu_1}(z_1)\Psi_{\mu_2}(z_2)\rangle = C\delta_{\mu_1,\mu_2}(z_1 - z_2)^{-2\mu_1}; \tag{2.24}$$

$$\langle \Psi_{\mu_1}(z_1)\Psi_{\mu_2}(z_2)\Psi_{\mu_3}(z_3)\rangle = C(z_1 - z_2)^{\mu_3-\mu_1-\mu_2}(z_2 - z_3)^{\mu_1-\mu_2-\mu_3}(z_3 - z_1)^{\mu_2-\mu_3-\mu_1}; \tag{2.25}$$

$$\left\langle \prod_{i=1}^{4} \Psi_{\mu_i}(z_i) \right\rangle = C \prod_{1\le i<j\le 4} (z_i - z_j)^{-2\gamma_{ij}} \cdot f(F(z_1, z_2, z_3, z_4)) \tag{2.26}$$

where f is some arbitrary function (called: *scaling function*) and $F(z_1, z_2, z_3, z_4) := \frac{(z_1-z_3)/(z_1-z_4)}{(z_2-z_3)/(z_2-z_4)}$ is the *cross-ratio* of its arguments, and γ_{ij} are parameters fixed by the relations $\sum_{i\neq j} \gamma_{ij} = \mu_j$.

Example 2 (Schrödinger invariants). Consider the Schrödinger action π_λ on fields $\Psi_{\mathcal{M},\lambda}$ of mass $\mathcal{M}$ and scaling dimension 2λ as in the Preface. Then Sch(d)-covariant (or equivalently Sch(d)-*quasiprimary*) n-point functions are given as follows for $n = 2, 3$ (see Appendix B for references):

$$\langle \Psi_{\mathcal{M}_1,\lambda_1}(t_1, r_1)\Psi_{\mathcal{M}_2,\lambda_2}(t_2, r_2)\rangle = C\delta_{\mathcal{M}_1+\mathcal{M}_2,0}\delta_{\lambda_1,\lambda_2}(t_1-t_2)^{-2\lambda_1}e^{-\mathcal{M}_1\frac{(r_1-r_2)^2}{2(t_1-t_2)}}; \tag{2.27}$$

$$\begin{aligned}
&\langle \Psi_{\mathcal{M}_1,\lambda_1}(t_1, r_1)\Psi_{\mathcal{M}_2,\lambda_2}(t_2, r_2)\Psi_{\mathcal{M}_3\lambda_3}(t_3, r_3)\rangle \\
&\quad = C\delta_{\mathcal{M}_1+\mathcal{M}_2+\mathcal{M}_3,0}(t_1 - t_2)^{\lambda_3-\lambda_1-\lambda_2}(t_2 - t_3)^{\lambda_1-\lambda_2-\lambda_3}(t_3 - t_1)^{\lambda_2-\lambda_3-\lambda_1} \\
&\qquad \exp\left[-\frac{\mathcal{M}_1}{2}\frac{(r_1 - r_3)^2}{t_1 - t_3} - \frac{\mathcal{M}_2}{2}\frac{(r_2 - r_3)^2}{t_2 - t_3}\right] f(F(t_1, r_1; t_2, r_2; t_3, r_3)),
\end{aligned} \tag{2.28}$$

where f is an arbitrary scaling function, and $F(t_1, r_1; t_2, r_2; t_3, r_3) = \frac{((r_1-r_3)(t_2-t_3)-(r_2-r_3)(t_1-t_3))^2}{(t_1-t_2)(t_2-t_3)(t_1-t_3)}$.

The fact that (contrary to the conformally covariant case) Schrödinger-covariant three-point functions should only be determined up to a scaling function is of course very disturbing for physicists, since (1) it means that Schrödinger-covariant model are less constrained, (2) two-point functions alone give little evidence that a given model should be Schrödinger-covariant. Using the natural embedding $\mathfrak{sch} \subset \mathfrak{conf}(3)_{\mathbb{C}}$ exhibited earlier in this chapter, one can however consider three-point functions which are covariant under the action of the whole conformal group Conf(3). These are determined up to a constant; formulas (2.25) actually extend to arbitrary dimension. To be explicit,

$$\begin{aligned}
\langle \Psi_{\lambda_1}(\zeta_1, t_1, r_1)\Psi_{\lambda_2}(\zeta_2, t_2, r_2)\Psi_{\lambda_3}(\zeta_3, t_3, r_3)\rangle &= C|(\zeta_1 - \zeta_2, t_1 - t_2, r_1 - r_2)|^{\lambda_3-\lambda_1-\lambda_2} \\
&|(\zeta_2 - \zeta_3, t_2 - t_3, r_2 - r_3)|^{\lambda_1-\lambda_2-\lambda_3}|(\zeta_3 - \zeta_1, t_3 - t_1, r_3 - r_1)|^{\lambda_2-\lambda_3-\lambda_1},
\end{aligned} \tag{2.29}$$

where $|(\zeta,t,r)|^2 := 2\zeta t - r^2$ is the squared Minkowski norm of (ζ,t,r) in light-cone coordinates.

We shall find out in Chap. 6 (see Theorem 6.10) that the three-point functions of the *massive* $\mathfrak{sv}$*-primary field* ψ_{-1} look similar to these up to a time-dependent prefactor. Because of the "no-go" theorem described above, it would have been very surprising to find three-point functions which are conformally covariant *and* transform covariantly under the whole Schrödinger–Virasoro group.

The two appendices at the end of the book give variants and extensions of the above results (in a supersymmetric setting for Appendix B).

Chapter 3
Coadjoint Representation of the Schrödinger–Virasoro Group

Coadjoint representations have proven to be both a rather efficient tool in representation theory, and a source of nice and useful examples in Poisson and symplectic geometry (see [71] for a fascinating survey); we simply mention here the fundamental rôle of the coadjoint representation of the Virasoro algebra for the study of Hill operators (see Introduction and Chap. 9).

Let us recall some well-known definitions in group theory about adjoint and coadjoint representations. For any Lie group G, every $g \in G$ acts on G by the inner automorphism $x \mapsto gxg^{-1}$; the differential at e of this diffeomorphism gives an automorphism of the Lie algebra $\mathfrak{g}$ through

$$X \longrightarrow \mathrm{Ad}(g)(X) = \frac{d}{dt}\left(g \exp(tX) g^{-1}\right)_{|t=0}. \tag{3.1}$$

This is the *adjoint action* of G on $\mathfrak{g}$, which can also be seen as a group homomorphism $G \to \mathrm{GL}(\mathfrak{g})$. On the infinitesimal level, one obtains a Lie algebra homomorphism $\mathfrak{g} \to \mathfrak{gl}(\mathfrak{g})$ through $Y \to \mathrm{ad}(Y)(X) = [Y, X]$; this is the *adjoint representation* of $\mathfrak{g}$. The *coadjoint representation* of G is then the contragradient representation $\mathrm{Ad}^* : G \to \mathrm{GL}(\mathfrak{g}^*)$ defined as:

$$\forall \mu \in \mathfrak{g}^*, \mathrm{Ad}^*_g(\mu) := -^t \left[\mathrm{Ad}_{g^{-1}}\right](\mu). \tag{3.2}$$

The associated infinitesimal coadjoint action is then $\mathrm{ad}^* : \mathfrak{g} \to \mathfrak{gl}(\mathfrak{g}^*)$ defined as follows:

$$\left\langle \mathrm{ad}^*_\xi(\mu), \eta \right\rangle = -\langle \mu, \mathrm{ad}_\xi(\eta)\rangle = -\langle \mu, [\xi, \eta]\rangle. \tag{3.3}$$

This chapter is devoted to the study of the coadjoint representation of SV on its *regular dual* $\mathfrak{sv}^*$ defined below. In particular, we shall classify in Sect. 3.2 the orbits of the coadjoint action of the Schrödinger–Virasoro group SV, following an analogous derivation for the Virasoro group (see [43]). The results are somehow disappointingly simple. Recall $\mathfrak{sv} \simeq \mathrm{Vect}(S^1) \ltimes \mathfrak{h}$. The isotropy algebras become

J. Unterberger and C. Roger, *The Schrödinger-Virasoro Algebra*, Theoretical and Mathematical Physics, DOI 10.1007/978-3-642-22717-2_3,

simply $\mathfrak{g}_0' \ltimes \mathfrak{h}'$, $\mathfrak{g}_0' \subset \mathrm{Vect}(S^1)$, $\mathfrak{h}' \subset \mathfrak{h}$, with $\mathfrak{g}_0' = \{0\}$ or $\mathfrak{g}_0' \simeq \mathbb{R}$, and $\mathfrak{h}'$ commutative, $\dim \mathfrak{h}' = 1, 2$ or ∞, except in one trivial case where one is simply reduced to the study of the coadjoint orbits of the Virasoro group. In particular, one does not get any interesting isotropy algebra of the type $\mathfrak{sl}(2,\mathbb{R}) \ltimes \mathfrak{h}'$ as might be expected from the well-known structure of Virasoro coadjoint orbits of codimension 3. We shall encounter a rich variety of isotropy algebras of this kind in the study of the SV-orbits on the space of Schrödinger operators, see Chap. 9.

The results of this chapter have not been published elsewhere.

3.1 Coadjoint Action of $\mathfrak{sv}$

Let us recall some facts about coadjoint actions of centrally extended Lie groups and algebras, referring to [43], Chap. 6, for details. So let G be a C^∞ Lie group with Lie algebra $\mathfrak{g}$, and let us consider central extensions of them, in the categories of groups and algebras respectively:

$$(1) \longrightarrow \mathbb{R} \longrightarrow \tilde{G} \longrightarrow G \longrightarrow (1) \tag{3.4}$$

$$(0) \longrightarrow \mathbb{R} \longrightarrow \tilde{\mathfrak{g}} \longrightarrow \mathfrak{g} \longrightarrow (0) \tag{3.5}$$

with $\tilde{\mathfrak{g}}$ =Lie($\tilde{G}$), the extension (3.5) being the exact sequence of tangent spaces at the identity of the extension (3.4) (see [43], II 6.1.1. for explicit formulas). For $G = \mathrm{Diff}(S^1)$, one has the Bott-Thurston cocycle C in the second group cohomology space $H^2(\mathrm{Diff}(S^1),\mathbb{R})$, given by $C(\phi,\psi) = \int_{S^1} (\phi \circ \psi)'(z) \frac{\psi''(z)}{\psi'(z)}\, dz$. By differentiating, one obtains a cocycle on the Lie algebra level, $c(\ell_n,\ell_m) = \delta_{n+m,0}(n+1)n(n-1)$, which is the Virasoro cocycle already introduced in Chap. 1. We can consider the associated Souriau cocycle on the group level in $H^1(\mathrm{Diff}(S^1), \mathfrak{vir}^*)$; it is the famous *Schwarzian derivative* that will frequently appear in this book:

$$S(\varphi) := \frac{\varphi'''}{\varphi'} - \frac{3}{2}\left(\frac{\varphi''}{\varphi'}\right)^2. \tag{3.6}$$

The Schwarzian derivative has a long and extensive history: it appeared first in the works of Lagrange about geographic maps, so a long time before Hermann Schwarz (1843-1921) was born, and has a lot of implications in complex analysis, projective geometry, integrable systems, statistical physics [43, 95, 105].

Let us come back to the general case, and denote by C, resp. c the cocycles on the group, resp. Lie algebra level. We want to study the *coadjoint action* on the dual $\tilde{\mathfrak{g}}^* = \mathfrak{g}^* \times \mathbb{R}$. We shall denote by Ad^* and $\widetilde{\mathrm{Ad}}^*$ the coadjoint actions of G and $\tilde{G}$ respectively, and ad^* and $\widetilde{\mathrm{ad}}^*$ the coadjoint actions of $\mathfrak{g}$ and $\tilde{\mathfrak{g}}$. One then has the following formulas

$$\widetilde{\mathrm{Ad}}^*(g,\alpha)(u,\lambda) = (\mathrm{Ad}^*(g)u + \lambda\Theta(g), \lambda) \tag{3.7}$$

and

$$\widetilde{\mathrm{ad}}^*(\xi,\alpha)(u,\lambda) = (\mathrm{ad}^*(\xi)u + \lambda\theta(\xi), 0) \tag{3.8}$$

where $\Theta : G \to \hat{\mathfrak{g}}^*$ and $\theta : \mathfrak{g} \to \hat{\mathfrak{g}}^*$ are the Souriau cocycles for differentiable and Lie algebra cohomologies respectively; for θ one has the following formula : $\langle \theta(\xi), \eta \rangle = c(\xi,\eta)$. For details of the proof, as well as 'dictionaries' between the various cocycles, the reader is referred to [43], Chap. 6.

The formulas (3.7) and (3.8) define the *extended coadjoint action* of G and $\mathfrak{g}$ respectively; these are *affine representations*, different from the usual, *linear coadjoint actions* when $\lambda \neq 0$. The actions on hyperplanes $\mathfrak{g}_\lambda^* = \{(u,\lambda) \mid u \in \mathfrak{g}^*\} \subset \tilde{\mathfrak{g}}^*$ with fixed second coordinate will be denoted by ad_λ^* and Ad_λ^* respectively.

We shall consider here the central extension $\tilde{\mathfrak{sv}}$ of $\mathfrak{sv}$ obtained by the natural extension c of the Virasoro cocycle to $\mathfrak{sv}$, defined by

$$\begin{aligned} c(L_n, L_p) &= n(n+1)(n-1)\delta_{n+p,0} \\ c(L_n, Y_m) &= c(L_n, M_p) = c(Y_m, Y_{m'}) = 0 \end{aligned} \tag{3.9}$$

with $n, p \in \mathbb{Z}$ and $m, m' \in \frac{1}{2} + \mathbb{Z}$. As we shall prove later on (see Chap.7), this central extension is universal (a more 'pedestrian' proof was given in [50]).

As usual in the infinite-dimensional setting, the algebraic dual of $\tilde{\mathfrak{sv}}$ is intractable for cardinality reasons. So let us consider its *regular dual*; for a large class of Lie algebras consisting of generalizations of vector fields or functions on manifolds, an element of the regular dual is defined by the integration against a "density measure".[1] Here, specifically, the dual module $\mathcal{F}_\mu^*$ may be identified with $\mathcal{F}_{-1-\mu}$ through the pairing:

$$\langle u(dx)^{1+\mu}, f\,(dx)^{-\mu} \rangle = \int_{S^1} u(x) f(x)\, dx. \tag{3.10}$$

In particular, $\mathrm{Vect}(S^1) \simeq \mathcal{F}_1$, and, as already mentioned in Sect. 1.1.2, the components $(Y_m)_m$, resp. $(M_p)_p$ generate the tensor-density module $\mathcal{F}_{\frac{1}{2}}$, resp. $\mathcal{F}_0$, so that (as $\mathrm{Vect}(S^1)$-module)

$$\mathfrak{sv} = \mathcal{F}_1 \oplus \mathcal{F}_{\frac{1}{2}} \oplus \mathcal{F}_0 \tag{3.11}$$

and

$$\mathfrak{sv}^* = \mathcal{F}_{-2} \oplus \mathcal{F}_{-\frac{3}{2}} \oplus \mathcal{F}_{-1}. \tag{3.12}$$

[1] In our case, a measure with a C^∞ density with respect to the Lebesgue measure on S^1.

We shall identify the element $\gamma = \gamma_0 dx^2 + \gamma_1 dx^{\frac{3}{2}} + \gamma_2 dx \in \mathfrak{sv}^*$ with the triple $\begin{pmatrix} \gamma_0 \\ \gamma_1 \\ \gamma_2 \end{pmatrix} \in (C^\infty(S^1))^3$. In other words,

$$\left\langle \begin{pmatrix} \gamma_0 \\ \gamma_1 \\ \gamma_2 \end{pmatrix}, \mathcal{L}_{f_0} + \mathcal{Y}_{f_1} + \mathcal{M}_{f_2} \right\rangle = \sum_{i=0}^{2} \int_{S^1} (\gamma_i f_i)(z)\, dz. \tag{3.13}$$

The following general Lemma describes the coadjoint representation of a Lie algebra that can be written as a semi-direct product.

Lemma 3.1. *Let $\mathfrak{s} = \mathfrak{s}_0 \ltimes \mathfrak{s}_1$ be a semi-direct product of two Lie algebra $\mathfrak{s}_0$ and $\mathfrak{s}_1$. Then the coadjoint action of $\mathfrak{s}$ on $\mathfrak{s}^*$ is given by*

$$\mathrm{ad}^*_{\mathfrak{s}}(f_0, f_1).(\gamma_0, \gamma_1) = \left\langle \mathrm{ad}^*_{\mathfrak{s}_0}(f_0)\gamma_0 - \tilde{f}_1.\gamma_1, \tilde{f}_0^*(\gamma_1) + \mathrm{ad}^*_{\mathfrak{s}_1}(f_1)\gamma_1 \right\rangle$$

where by definition

$$\left\langle \tilde{f}_1.\gamma_1, X_0 \right\rangle_{\mathfrak{s}_0^* \times \mathfrak{s}_0} = \langle \gamma_1, [X_0, f_1] \rangle_{\mathfrak{s}_1^* \times \mathfrak{s}_1}$$

and

$$\left\langle \tilde{f}_0^*(\gamma_1), X_1 \right\rangle_{\mathfrak{s}_1^* \times \mathfrak{s}_1} = \langle \gamma_1, [f_0, X_1] \rangle_{\mathfrak{s}_1^* \times \mathfrak{s}_1}.$$

Proof. Straightforward. □

As a corollary, one gets:

Theorem 3.1 (extended coadjoint action). *The coadjoint action of $\mathfrak{sv}$ on the affine hyperplane $\mathfrak{sv}_c^*$ is given by the following formulas:*

$$\mathrm{ad}^*_c(\mathcal{L}_{f_0}) \begin{pmatrix} \gamma_0 \\ \gamma_1 \\ \gamma_2 \end{pmatrix} = \begin{pmatrix} c f_0''' + 2 f_0' \gamma_0 + f_0 \gamma_0' \\ f_0 \gamma_1' + \frac{3}{2} f_0' \gamma_1 \\ f_0 \gamma_2' + f_0' \gamma_2 \end{pmatrix} \tag{3.14}$$

$$\mathrm{ad}^*_c(\mathcal{Y}_{f_1}) \begin{pmatrix} \gamma_0 \\ \gamma_1 \\ \gamma_2 \end{pmatrix} = \begin{pmatrix} \frac{3}{2}\gamma_1 f_1' + \frac{1}{2}\gamma_1' f_1 \\ 2\gamma_2 f_1' + \gamma_2' f_1 \\ 0 \end{pmatrix} \tag{3.15}$$

$$\mathrm{ad}^*_c(\mathcal{M}_{f_2}) \begin{pmatrix} \gamma_0 \\ \gamma_1 \\ \gamma_2 \end{pmatrix} = \begin{pmatrix} -\gamma_2 f_2' \\ 0 \\ 0 \end{pmatrix}. \tag{3.16}$$

Proof. The action of $\mathrm{Vect}(S^1) \subset \mathfrak{sv}$ follows from the identification of $\mathfrak{sv}_c^*$ with $\mathfrak{vir}_c^* \oplus \mathcal{F}_{-\frac{3}{2}} \oplus \mathcal{F}_{-1}$.

Applying the preceding Lemma, one gets now

$$\left\langle \mathrm{ad}_c^*(\mathcal{Y}_{f_1}).\begin{pmatrix}\gamma_0\\ \gamma_1\\ \gamma_2\end{pmatrix}, \mathcal{L}_{h_0}\right\rangle = -\left\langle \tilde{\mathcal{Y}}_{f_1}.\begin{pmatrix}0\\ \gamma_1\\ \gamma_2\end{pmatrix}, \mathcal{L}_{h_0}\right\rangle$$

$$= \left\langle \begin{pmatrix}0\\ \gamma_1\\ \gamma_2\end{pmatrix}, \mathcal{Y}_{\frac{1}{2}h_0' f_1 - h_0 f_1'}\right\rangle$$

$$= \int_{S^1} \gamma_1 \left(\frac{1}{2}h_0' f_1 - h_0 f_1'\right) dz$$

$$= \int_{S^1} h_0 \left(-\frac{3}{2}\gamma_1 f_1' - \frac{1}{2}\gamma_1' f_1\right) dz;$$

$$\left\langle \mathrm{ad}_c^*(\mathcal{Y}_{f_1}).\begin{pmatrix}\gamma_0\\ \gamma_1\\ \gamma_2\end{pmatrix}, \mathcal{Y}_{h_1}\right\rangle = \left\langle \mathrm{ad}_{\mathfrak{h}}^*(\mathcal{Y}_{f_1}).\begin{pmatrix}0\\ \gamma_1\\ \gamma_2\end{pmatrix}, \mathcal{Y}_{h_1}\right\rangle$$

$$= -\left\langle \begin{pmatrix}0\\ \gamma_1\\ \gamma_2\end{pmatrix}, \mathcal{M}_{f_1' h_1 - f_1 h_1'}\right\rangle$$

$$= -\int_{S^1} \gamma_2 (f_1' h_1 - f_1 h_1')\, dz$$

$$= \int_{S^1} h_1 (-2\gamma_2 f_1' - \gamma_2' f_1)\, dz$$

and

$$\left\langle \mathrm{ad}_c^*(\mathcal{Y}_{f_1})\begin{pmatrix}\gamma_0\\ \gamma_1\\ \gamma_2\end{pmatrix}, \mathcal{M}_{h_2}\right\rangle = 0.$$

Hence the result for $\mathrm{ad}_c^*(\mathcal{Y}_{f_1})$.

For the action of $\mathrm{ad}_c^*(\mathcal{M}_{f_2})$, one gets similarly

$$\left\langle \mathrm{ad}_c^*(\mathcal{M}_{f_2}).\begin{pmatrix}\gamma_0\\ \gamma_1\\ \gamma_2\end{pmatrix}, \mathcal{L}_{h_0}\right\rangle = -\left\langle \begin{pmatrix}0\\ \gamma_1\\ \gamma_2\end{pmatrix}, \mathcal{M}_{f_2' h_0}\right\rangle$$

$$= -\int_{S^1} \gamma_2 f_2' h_0\, dz$$

and

$$\left\langle \mathrm{ad}_c^*(\mathcal{M}_{f_2}).\begin{pmatrix}\gamma_0\\ \gamma_1\\ \gamma_2\end{pmatrix}, \mathcal{Y}_{h_1}\right\rangle = \left\langle \mathrm{ad}_c^*(\mathcal{M}_{f_2}).\begin{pmatrix}\gamma_0\\ \gamma_1\\ \gamma_2\end{pmatrix}, \mathcal{M}_{h_2}\right\rangle = 0.$$

Hence the result for $\mathrm{ad}_c^*(\mathcal{M}_{f_2})$. □

One may now easily construct a coadjoint action of the group SV, which "integrates" the above defined coadjoint action of $\mathfrak{sv}$; as usual in infinite dimension, such an action should not be taken for granted, and one has to construct it explicitly case by case. The result is given by the following.

Theorem 3.2. *The coadjoint action of SV on the affine hyperplane $\mathfrak{sv}_c^*$ is given by the following formulas: letting $(\varphi;(\alpha,\beta)) \in SV$,*

$$\mathrm{Ad}_c^*(\varphi;(0,0))\begin{pmatrix}\gamma_0\\ \gamma_1\\ \gamma_2\end{pmatrix} = \begin{pmatrix} cS(\varphi) + (\gamma_0\circ\varphi)(\varphi')^2\\ (\gamma_1\circ\varphi)(\varphi')^{\frac{3}{2}}\\ (\gamma_2\circ\varphi)\varphi'\end{pmatrix} \tag{3.17}$$

$$Ad_c^*(\mathrm{Id};(\alpha,\beta))\begin{pmatrix}\gamma_0\\ \gamma_1\\ \gamma_2\end{pmatrix}$$

$$= \begin{pmatrix} \gamma_0 + \frac{3}{2}\gamma_1\alpha' + \frac{\gamma_1'}{2}\alpha + \gamma_2\beta' - \frac{\gamma_2}{2}(3\alpha'^2 + \alpha\alpha'') - \frac{3}{2}\gamma_2'\alpha\alpha' - \frac{\gamma_2''}{4}\alpha^2\\ \gamma_1 + 2\gamma_2\alpha' + \gamma_2'\alpha\\ \gamma_2\end{pmatrix} \tag{3.18}$$

Proof. The first part (3.17) is easily deduced from the natural action of $\mathrm{Diff}(S^1)$ on $\tilde{\mathfrak{sv}}_c^* = \mathfrak{vir}_c^* \oplus \mathcal{F}_{-3/2} \oplus \mathcal{F}_{-1}$. Here $S(\varphi) := \frac{\varphi'''}{\varphi'} - \frac{3}{2}\left(\frac{\varphi''}{\varphi'}\right)^2$ denotes the above-defined Schwarzian derivative of φ.

The problem of computing the coadjoint action of $(\alpha,\beta) \in H$ may be split into two pieces; the coadjoint action of H on $\mathfrak{h}^*$ is readily computed and one finds:

$$\mathrm{Ad}_c^*(\alpha,\beta)\begin{pmatrix}\gamma_1\\ \gamma_2\end{pmatrix} = \begin{pmatrix}\gamma_1 + 2\gamma_2\alpha' + \gamma_2'\alpha\\ \gamma_2\end{pmatrix}$$

The most delicate part is to compute the part of coadjoint action of $(\alpha,\beta) \in H$ coming from the adjoint action on $\mathrm{Vect}(S^1)$, by using:

$$\left\langle \mathrm{Ad}_c^*(\mathrm{Id};(\alpha,\beta))\begin{pmatrix}\gamma_0\\ \gamma_1\\ \gamma_2\end{pmatrix}, f\partial\right\rangle = \left\langle \begin{pmatrix}\gamma_0\\ \gamma_1\\ \gamma_2\end{pmatrix}, \mathrm{Ad}(\mathrm{Id};(\alpha,\beta))^{-1}(f\partial,0,0)\right\rangle.$$

Exponentiating the Lie bracket, one finds

$$\mathrm{Ad}(\mathrm{Id};(\alpha,\beta))^{-1}(f\partial,0,0)$$
$$=\left(f\partial, f\alpha' - \frac{1}{2}\alpha f', f\beta' + \frac{1}{2}\left(f\alpha''\alpha + \frac{f'}{2}\alpha\alpha' - \frac{\alpha^2}{2}f'' - f\alpha'^2 + \frac{f'}{2}\alpha\alpha''\right)\right).$$

Now, using an integration by parts, one easily finds the formula (3.18) above. □

3.2 Coadjoint Orbits of $\mathfrak{sv}$

Let $\begin{pmatrix}\gamma_0\\ \gamma_1\\ \gamma_2\end{pmatrix}$ be an element of $\mathfrak{sv}_c^*$: we shall determine its *isotropy algebra* $\mathcal{G}_{(\gamma_0,\gamma_1,\gamma_2)}$ – and in some cases its *isotropy group* $G_{(\gamma_0,\gamma_1,\gamma_2)}$ – under the coadjoint representation of $\mathfrak{sv}$. By definition,

$$G_{(\gamma_0,\gamma_1,\gamma_2)} := \left\{ g \in \mathrm{SV} \mid \mathrm{Ad}_c^* g.\begin{pmatrix}\gamma_0\\ \gamma_1\\ \gamma_2\end{pmatrix} = \begin{pmatrix}\gamma_0\\ \gamma_1\\ \gamma_2\end{pmatrix}\right\} \tag{3.19}$$

and $\mathcal{G}_{(\gamma_0,\gamma_1,\gamma_2)} = \mathrm{Lie}(G_{(\gamma_0,\gamma_1,\gamma_2)})$.

One has to solve the following equation

$$\mathrm{ad}_c^*(\mathcal{L}_{f_0})\begin{pmatrix}\gamma_0\\ \gamma_1\\ \gamma_2\end{pmatrix} = \mathrm{ad}_c^*(\mathcal{Y}_{f_1} + \mathcal{M}_{f_2})\begin{pmatrix}\gamma_0\\ \gamma_1\\ \gamma_2\end{pmatrix} \tag{3.20}$$

When (3.20) is satisfied, $\mathcal{L}_{f_0} - \mathcal{Y}_{f_1} - \mathcal{M}_{f_2}$ belongs to $\mathcal{G}_{(\gamma_0,\gamma_1,\gamma_2)}$.

The formulas in Theorem 3.1 above yield easily the following system of differential equations:

$$cf_0''' + 2f_0'\gamma_0 + f_0\gamma_0' = \frac{3}{2}\gamma_1 f_1' + \frac{1}{2}\gamma_1' f_1 - \gamma_2 f_2' \tag{3.21}$$

$$f_0\gamma_1' + \frac{3}{2}f_0'\gamma_1 = 2\gamma_2 f_1' + \gamma_2' f_1 \tag{3.22}$$

$$f_0\gamma_2' + f_0'\gamma_2 = 0 \tag{3.23}$$

Note that $\mathcal{G}_{(\gamma_0,\gamma_1,\gamma_2)}$ always contains $M_0 = \mathcal{M}_1$ since M_0 is in the center of $\mathfrak{sv}$.

We shall use techniques analogous to the case of the Virasoro group (see again [43], Chap. VI, cf. also [3, 92]). The discussion will be split into different cases, depending essentially on the set of zeros of γ_2.

Case a): γ_2 is everywhere non vanishing Equation (3.23) implies $(f_0\gamma_2)' = 0$ so $f_0 = \lambda\gamma_2^{-1}$ for some constant λ. One may then solve (3.22) using the classical method of variation of parameters. One finds:

$$f_1 = \frac{\lambda}{2}\gamma_2^{-2}\gamma_1 + \mu\gamma_2^{-\frac{1}{2}}$$

Then (3.21) may be solved by a mere integration:

$$f_2' = \gamma_2^{-1}\left(\frac{3}{2}\gamma_1 f_1' + \frac{1}{2}\gamma_1' f_1 - c f_0''' - 2f_0'\gamma_0 - f_0\gamma_0'\right)$$

which gives:

$$\begin{aligned} f_2' = {} & 6c\lambda\gamma_2^{-5}(\gamma_2')^3 - \lambda\left(6c\gamma_2^{-4}\gamma_2'\gamma_2'' + \frac{3}{2}\gamma_2^{-4}\gamma_2'\gamma_1^2\right) \\ & + \lambda\gamma_2^{-3}\left(c\gamma_2''' + 2\gamma_2'\gamma_0 + \gamma_1\gamma_1'\right) - \left(\frac{3\mu}{4}\gamma_2^{-\frac{5}{2}}\gamma_2'\gamma_1 - \lambda\gamma_2^{-2}\gamma_0' + \frac{\mu}{2}\gamma_2^{-\frac{3}{2}}\gamma_1'\right). \end{aligned} \tag{3.24}$$

So one obtains:

$$f_2 = -\frac{3}{2}\lambda c\gamma_2^{-4}(\gamma_2')^2 + \lambda\gamma_2^{-3}\left(\frac{\gamma_1^2}{2} + c\gamma_2''\right) - \lambda\gamma_2^{-2}\gamma_0 + \frac{\mu}{2}\gamma_2^{-\frac{3}{2}}\gamma_1 + \nu$$

We may now summarize our results as follows:

$$\mathcal{G}_{(\gamma_0,\gamma_1,\gamma_2)} = \{\mathcal{L}_{f_0} + \mathcal{Y}_{f_1} + \mathcal{M}_{f_2}, \lambda, \mu, \nu \in \mathbb{R}\} \tag{3.25}$$

where:

$$\left[\begin{aligned} & f_0 = f_0(\lambda) = \lambda\gamma_2^{-1} \\ & f_1 = f_1(\lambda,\mu) = -\frac{\lambda}{2}\gamma_2^{-1}\gamma_1 - \mu\gamma_2^{-\frac{1}{2}} \\ & f_2 = f_2(\lambda,\mu,\nu) = \frac{3}{2}c\lambda\gamma_2^{-4}(\gamma_2')^2 - \lambda\gamma_2^{-3}\left(\frac{\gamma_1^2}{2} + c\gamma_2''\right) + \lambda\gamma_2^{-2}\gamma_0 - \frac{\mu}{2}\gamma_2^{-\frac{3}{2}}\gamma_1 + \nu \end{aligned}\right. \tag{3.26}$$

So the integration constants determine an isomorphism of Lie algebras:

$$\begin{aligned} \mathbb{R}^3 & \longrightarrow \mathcal{G}_{(\gamma_0,\gamma_1,\gamma_2)} \\ (\lambda,\mu,\nu) & \longrightarrow \left(\mathcal{L}_{f_0} + \mathcal{Y}_{f_1} + \mathcal{M}_{f_2}\right) \end{aligned}$$

where (f_0, f_1, f_2) are given by formulas (3.26) above.

Subcase a'): $\gamma_2 \neq 0$ constant Under this stronger hypothesis, we shall now determine the isotropy group $G_{(\gamma_0,\gamma_1,\gamma_2)}$.

Following the formulas of Theorem 3.2, one has to find $\gamma \in \mathrm{Diff}(S^1)$ and $(\alpha,\beta) \in H$ such that the following system is satisfied

$$\lambda\Theta(\varphi) + (\gamma_0 \circ \varphi)\varphi'^2 = \gamma_0 + \frac{3}{2}\gamma_1\alpha' + \frac{\gamma_1'}{2}\alpha + \gamma_2\beta' - \frac{\gamma_2}{2}\left(3\alpha'^2 + \alpha\alpha''\right) \quad (3.27)$$

$$(\gamma_1 \circ \varphi)(\varphi')^{\frac{3}{2}} = \gamma_1 + 2\gamma_2\alpha' \quad (3.28)$$

$$(\gamma_2 \circ \varphi)\varphi' = \gamma_2 \quad (3.29)$$

Since γ_2 is constant, (3.29) implies that φ must be a rotation, so $\varphi = R_\theta$ and $\varphi' = 1$. In order to solve (3.28) we shall work on the universal covering of $\mathrm{Diff}(S^1)$ (see [43], Chap. IV)

$$0 \longrightarrow \mathbb{Z} \longrightarrow \mathrm{Diff}_{2\pi\mathbb{Z}}(\mathbb{R}) \longrightarrow \mathrm{Diff}\left(S^1\right) \longrightarrow 0$$

where $\mathrm{Diff}_{2\pi\mathbb{Z}}(\mathbb{R})$ denotes the group of diffeomorphisms of $\mathbb{R}$ such that $F(x + 2\pi n) = F(x) + 2\pi m$. So one has 2π-periodic functions $\Gamma_k : \mathbb{R} \longrightarrow \mathbb{R}$ for $k = 0,1,2$ such that $\gamma_k(e^{ix}) = e^{i\Gamma_k(x)}$, and analogously $R_\theta(e^{ix}) = e^{i(x+\theta)}$. Set $\Gamma_2(x) = \frac{a_2}{2}$ constant, then (3.28), when lifted to $\mathbb{R}$, gives:

$$\Gamma_1(x+\theta) - \Gamma_1(x) = a_2 A'(x)$$

where A is the lift of α.

Let us denote by I_θ the difference operator:

$$(I_\theta\Gamma_1)(x) = \Gamma_1(x+\theta) - \Gamma_1(x)$$

One finds $A(x) = \frac{1}{a_2}\int_0^x (I_\theta\Gamma_1)(s)ds + \mu$ (note that the condition $A(x+2\pi) = A(x)$ is satisfied).

One then obtains for (3.27) lifted to $\mathbb{R}$:

$$I_\theta\Gamma_0 = \frac{3}{2}\Gamma_1 A' + \frac{\Gamma_1'}{2}A + \Gamma_2 B' - \frac{\Gamma_2}{2}\left(3A'^2 + AA''\right)$$

or

$$\begin{aligned}\mathscr{B}' = &\frac{3I_\theta\Gamma_0}{a_2} + \frac{\mu}{2a_2}\left(I_\theta\Gamma_1 - 2\Gamma_1'\right)\\ &+\frac{1}{2a_2^2}\left(3(I_\theta\Gamma_1 - 2\Gamma_1)I_\theta\Gamma_1 + (I_\theta\Gamma_1 - 2\Gamma_1')\left(\int_0^x I_\theta\Gamma_1\right)\right)\end{aligned}$$

So $\mathscr{B}$ is uniquely determined up to an integration constant.

Summarizing our results, one finds the following expression for the elements of the isotropy group after lifting from S^1 to $\mathbb{R}$: let $G_{(\gamma_0,\gamma_1,\gamma_2)} \subset SV = \mathrm{Diff}(S^1) \ltimes H$ and $\widetilde{G}_{(\gamma_0,\gamma_1,\gamma_2)} \subset \widetilde{SV} = \mathrm{Diff}_{2\pi\mathbb{Z}}(\mathbb{R}) \ltimes \widetilde{H}$ be its lifting. Then

$$\widetilde{G}_{(\gamma_0,\gamma_1,\gamma_2)} = \{\tau_\theta, A, B\} \tag{3.30}$$

where

$$\tau_\theta(x) = x + \theta, \quad A(x) = \Phi_\theta(x) + \mu, B(x) = \psi_\theta(x) + \mu\xi_\theta(x) + \nu \tag{3.31}$$

and explicit expressions for Φ_θ, ψ_θ and ξ_θ may easily be deduced from the above formulas. It is an easy exercise to check that those mappings form a group for composition.

We have just proved:

Proposition 3.2. *If $\gamma_2 \neq 0$ is a constant, then the isotropy group $G_{(\gamma_0,\gamma_1,\gamma_2)}$ is isomorphic to a* direct *product $S^1 \times \mathbb{R} \times \mathbb{R}$, so the coadjoint orbit $O_{(\gamma_0,\gamma_1,\gamma_2)}$ is isomorphic to a product of a generic orbit* $\mathrm{Diff}(S^1)/S^1$ *of the Virasoro group by $H/\mathbb{R}^2$.*

Case b): γ_2 has only isolated zeros Relation (3.23) still implies $(f_0\gamma_2)' = 0$ so $f_0\gamma_2$ constant, but here, since γ_2 vanishes at some point, one has $f_0\gamma_2 \equiv 0$. So f_0 must vanish where γ_2 is non zero. Since the zeros of γ_2 are isolated, one has $\gamma_2 \neq 0$ almost everywhere so $f_0 \equiv 0$ almost everywhere, whence $f_0 \equiv 0$ by continuity.

From (3.22) one gets $2\gamma_2 f_1' + \gamma_2' f_1 = 0$, so $\gamma_2 f_1^2$ is constant and $\gamma_2^{-2} f_1^2$ vanishes by the same argument, hence $f_1 \equiv 0$. Finally (3.21) leads to $f_2' \equiv 0$, so f_2 is a constant. So

$$\mathcal{G}_{(\gamma_0,\gamma_1,\gamma_2)} = \langle M_0 \rangle = \{(0,0,\nu)\} \subset \mathfrak{sv} = \mathrm{Vect}\left(S^1\right) \ltimes \mathfrak{h} \tag{3.32}$$

is one-dimensional.

Note that the isotropy group may be non-trivial though: some discrete subgroups of $PSL(2,\mathbb{R}) \subset \mathrm{Diff}(S^1)$ may show up (such examples have been worked out by Laurent Guieu, see [43], Chap. VI).

Case c): γ_2 vanishes on some interval but $\gamma_2 \not\equiv 0$ In order to avoid pathologies (recall the famous theorem by Whitney which states that any closed subset of $\mathbb{R}^n$ can be the zero set of a smooth function), we shall assume there exists a finite union of closed intervals $I \subset S^1$ on which $\gamma_2 \equiv 0$, while on $S^1 \setminus I$, γ_2 has only isolated zeros. One still deduces $\gamma_2 f_0 = 0$ from (3.23) so $f_0 \equiv 0$ on $S^1 \setminus I$, while f_0 is a priori arbitrary on I. One gets from (3.22) $(f_0^3\gamma_1^2)' = 0$ so $f_0^3\gamma_1^2$ is a constant.

In all subcases considered below, we shall find the same conclusion, namely,

$$\mathcal{G}_{(\gamma_0,\gamma_1,\gamma_2)} = \langle M_0 \rangle \tag{3.33}$$

as in case b).

Assume first γ_1 has isolated zeros on I. One deduces $f_0 \equiv 0$ on I as above, so f_0 vanishes everywhere. By (3.21) one gets $(\gamma_1 f_1^3)' = 0$, so $f_1 \equiv 0$ on I.

If γ_2 has at least one zero on $S^1 \setminus I$, one deduces from $(\gamma_2 f_1^2)' = 0$ ((3.22)) that $f_1 \equiv 0$ on $S^1 \setminus I$, so both f_0 and f_1 vanish identically. One finds $f_2 =$ constant from (3.21). So $\mathcal{G}_{(\gamma_0,\gamma_1,\gamma_2)} \simeq \mathbb{R}$, exactly as in the previous case b) where the zeros of γ_2 were isolated.

Suppose now γ_2 is nowhere vanishing on $S^1 \setminus I$. Then $f_1 = k\gamma_2^{-\frac{1}{2}}$ should be a solution of (3.22), but since $| \gamma_2^{-\frac{1}{2}}(t) | \longrightarrow \infty$ when t tends to some point of ∂I, this solution cannot be globally defined, so one has once more $f_1 \equiv 0$.

Assume γ_1 vanishes identically on I or on some subinterval. Then one gets the same conclusion.

Assume now γ_1 is everywhere non-vanishing on I. Then (3.22) admits $f_0 = k\gamma_1^{-\frac{2}{3}}$ as a possible solution on I, but unfortunately, one easily sees that it is impossible to link it continuously with $f_0 \equiv 0$ on $S^1 \setminus I$, so necessarily $f_0 \equiv 0$ everywhere. In order to find f_1, one has to solve

$$\begin{bmatrix} 3\gamma_1 f_1' + \gamma_1' f_1 = 0 \quad \text{on } I \\ 2\gamma_2 f_1' + \gamma_2' f_1 = 0 \quad \text{on } S^1 \setminus I \end{bmatrix}$$

So: $\gamma_1 f_1$ is constant on I and $\gamma_2 f_1^2$ is constant on $S^1 \setminus I$. These two equations are incompatible for the same reasons as in the preceding arguments, so one necessarily gets $f_1 \equiv 0$, and the conclusion is the same as above.

Let us summarize the results for case b) and c):

Proposition 3.3. *If γ_2 has some zero without being identically zero, then the Lie algebra of the isotropy group $\mathcal{G}_{(\gamma_0,\gamma_1,\gamma_2)}$ is one-dimensional, $\mathcal{G}_{(\gamma_0,\gamma,\gamma_2)} = \{(0,0,a) \mid a \in \mathbb{R}\}$. One gets an analogous result for the isotropy group $G_{(\gamma_0,\gamma_1,\gamma_2)}$, save for a possible discrete subgroup of $PSL(2,\mathbb{R})$ on the* Diff(S^1)*-component.*

Case d): $\gamma_2 \equiv 0$. Equation (3.22) yields $(f_0^3 \gamma_1^2)' = 0$. We shall now distinguish three subcases.

Assume first that γ_1 is identically zero. Then one is led to (3.21) with right-hand side identically zero, $cf_0''' + 2f_0'\gamma_0 + f_0\gamma_0' = 0$, which is precisely the equation defining the isotropy algebra for the coadjoint orbits of the Virasoro group (see [43], Chap. VI), while f_1 and f_2 may be arbitrary.

Assume now γ_1 has isolated zeros. Then $f_0 \equiv 0$. Also, (3.21) yields $(f_1^3\gamma_1)' = 0$, hence for the same reason $f_1 \equiv 0$, while f_2 may be arbitrary. So $\mathcal{G}_{(\gamma_0,\gamma_1,\gamma_2)} = \langle M_n, n \in \mathbb{Z}\rangle$. The isotropy group may once more contain some discrete subgroup of $PSL(2,\mathbb{R})$, obviously impossible to detect on the Lie algebra level.

Assume finally that γ_1 is nowhere vanishing. Then one gets $f_0 = \lambda\gamma_1^{-\frac{2}{3}}$ for some coefficient λ. Equation (3.21) yields:

$$3\gamma_1 f_1' + \gamma_1' f_1 = 2\left(c f_0''' + 2 f_0' \gamma_0 + f_0 \gamma_0'\right) = F\left(\gamma_1, \gamma_1', \gamma_1'', \gamma_1''', \gamma_0, \gamma_0'\right)$$

So using the method of variation of parameters, one gets: $f_1 = \mu \gamma_1^{-\frac{1}{3}} + \lambda K(\gamma_1, \gamma_0)$ where K is some differential expression in γ_1 and γ_0. Once again that there is no condition on f_2, so finally:

$$G_{(\gamma_0, \gamma_1, 0)} = S^1 \ltimes \left(\mathbb{R} \times C^\infty(S^1)\right) . \tag{3.34}$$

Chapter 4
Induced Representations and Verma Modules

The results of this chapter may be found in an earlier electronic version of [106], see http://www.arxiv.org/abs/math-ph/0601050, but have not been published elsewhere. For related algebraically minded results, one may also refer to the recent papers by the Chinese school [81, 82, 112].

4.1 Introduction and Notations

There are a priori two different ways to define *induced representations* (or *Verma modules*) of $\mathfrak{sv}$, corresponding to the two natural graduations (see Definition 2.3): one of them (called *degree* and denote by deg in the following) corresponds to the adjoint action of L_0, so that $\deg(X_n) = -n$ for $X = L, Y$ or M; it is given by $-\delta_1$ in the notation of Definition 2.3. The other one, corresponding to the graduation of the Cartan prolongation (see Chap. 5), is given by the outer derivation δ_2. The action of both graduations is diagonal on the generators (L_n); the subalgebra of weight 0 is two-dimensional abelian, generated by L_0 and M_0, in the former case, and three-dimensional solvable, generated by $L_0, Y_{\frac{1}{2}}, M_1$ in the latter case.

Since Verma modules are usually defined by inducing a character of an abelian subalgebra to the whole Lie algebra (although this is by no means necessary), we shall forget altogether about the graduation given by δ_2 in this section and consider representations of $\mathfrak{sv}$ that are induced from $\langle L_0, M_0\rangle$.

Let $\mathfrak{sv}_{(n)} = \{Z \in \mathfrak{sv} \mid \mathrm{ad}L_0.Z = nZ\}$ $(n \in \frac{1}{2}\mathbb{Z})$, $\mathfrak{sv}_{>0} = \oplus_{n>0}\mathfrak{sv}_{(n)}$, $\mathfrak{sv}_{<0} = \oplus_{n<0}\mathfrak{sv}_{(n)}$, $\mathfrak{sv}_{\leq 0} = \mathfrak{sv}_{<0} \oplus \mathfrak{sv}_{(0)}$. Define $\mathbb{C}_{h,\mu} = \mathbb{C}\psi$ $(h, \mu \in \mathbb{C})$ to be the character of $\mathfrak{sv}_{(0)} = \langle L_0, M_0\rangle$ such that $L_0\psi = h\psi$, $M_0\psi = \mu\psi$. Following the usual definition of Verma modules (see [66] or [87]), we extend $\mathbb{C}_{h,\mu}$ trivially to $\mathfrak{sv}_{\leq 0}$ by putting $\mathfrak{sv}_{<0}.\psi = 0$ and call $\mathcal{V}_{h,\mu}$ the induction of the representation $\mathbb{C}_{h,\mu}$ to $\mathfrak{sv}$:

$$\mathcal{V}_{h,\mu} = \mathcal{U}(\mathfrak{sv}) \otimes_{\mathcal{U}(\mathfrak{sv}_{\leq 0})} \mathbb{C}_{h,\mu}. \tag{4.1}$$

J. Unterberger and C. Roger, *The Schrödinger-Virasoro Algebra*, Theoretical and Mathematical Physics, DOI 10.1007/978-3-642-22717-2_4,

In simpler words, $\mathscr{V}_{h,\mu}$ is generated by the vector ψ, and the elements X_n ($X = L, Y, M$, $n < 0$) act freely on $\mathscr{V}_{h,\mu}$, while the action of the elements $X_n, n \geq 0$ may be understood by commuting them to the right and applying the rules $X_n.\psi = 0$ ($n > 0$), $L_0.\psi = h\psi$, $M_0.\psi = \mu.\psi$. Notice that with this choice of signs, *negative* degree elements L_n, Y_n, M_n ($n > 0$) applied to ψ yield zero.

The Verma module $\mathscr{V}_{h,\mu}$ is positively graded through the natural extension of deg from $\mathfrak{sv}$ to $\mathscr{U}(\mathfrak{sv})$, namely, we put $(\mathscr{V}_{h,\mu})_{(n)} = \mathscr{V}_{(n)} = \mathscr{U}(\mathfrak{sv}_{>0})_{(n)} \otimes \mathbb{C}_{h,\mu}$.

There exists exactly one bilinear form $\langle \mid \rangle$ (called the *contravariant Hermitian form* or, as we shall also say, 'scalar product', although it is neither necessarily positive nor even necessarily non-degenerate) on $\mathscr{V}_{h,\mu}$ such that $\langle \psi \mid \psi \rangle = 1$ and $X_n^* = X_{-n}$, $n \in \frac{1}{2}\mathbb{Z}$ ($X = L, Y$ or M), where the star means taking the adjoint with respect to the bilinear form (see [66]). For the contravariant Hermitian form, $\langle \mathscr{V}_{(j)} \mid \mathscr{V}_{(k)} \rangle = 0$ if $j \neq k$. It is well-known that the module $\mathscr{V}_{h,\mu}$ is indecomposable and possesses a unique maximal proper sub-representation $\mathscr{K}_{h,\mu}$, which is actually the kernel of Hermitian form, and such that the quotient module $\mathscr{V}_{h,\mu}/\mathscr{K}_{h,\mu}$ is irreducible. When $\mathscr{K}_{h,\mu} \neq \{0\}$ is not trivial, the induced representation is called a *degenerate Verma module*, of which finite-dimensional representations of simple Lie algebras and *unitary minimal models* for the Virasoro Lie algebra (see Preface) are foremost examples.

Hence, in order to determine if $\mathscr{V}_{h,\mu}$ is irreducible and to find the irreducible quotient representation if it is not, one is naturally led to the computation of the *Kac determinants*, by which we mean the determinants of the Hermitian form restricted to $\mathscr{V}_{(n)} \times \mathscr{V}_{(n)}$ for each n.

Let us introduce some useful notations for *partitions*.

Definition 4.1 (partitions). A partition $A = (a^1, a^2, \ldots)$ of degree $n = \deg(A) \in \mathbb{N}^*$ is an ordered set $a^1, a^2, \ldots$ of non-negative integers such that $\sum_{i \geq 1} i a^i = n$.

A partition can be represented as a Young tableau: one associates to A a set of vertical stacks of boxes put side by side, with (from left to right) a^1 stacks of height 1, a^2 stacks of height 2, and so on.

The *width* $\mathrm{wid}(A)$ of the tableau is equal to $\sum_{i \geq 1} a^i$.

By convention, we shall say that there is exactly one partition of degree 0, denoted by $\emptyset$, and such that $\mathrm{wid}(\emptyset) = 0$.

Now any partition A defines elements of $\mathscr{U}(\mathfrak{sv})$, namely, let us put $X^{-A} = X_{-1}^{A_1} X_{-2}^{A_2} \ldots$ (X stands here for L or M) and $Y^{-A} = Y_{-\frac{1}{2}}^{A_1} Y_{-\frac{3}{2}}^{A_2} \ldots$, so that $\deg(X^{-A}) = \deg(A)$ and

$$\deg(Y^{-A}) = \sum_{i \geq 1} (i - \frac{1}{2}) A^i = \deg(A) - \frac{1}{2}\mathrm{wid}(A)$$

(we shall also call this expression the *shifted degree* of A, and write it $\widetilde{\deg}(A)$).

Definition 4.2. We denote by $\mathscr{P}(n)$ (resp. $\tilde{\mathscr{P}}(n)$) the set of partitions of degree (resp. shifted degree) n.

By Poincaré-Birkhoff-Witt's theorem (PBW for short, see for instance [16] or [60]), $\mathcal{V}_{(n)}$ is generated by the vectors $Z = L^{-A} Y^{-B} M^{-C} \psi$ where A, B, C range among all partitions such that $\deg(A) + \widetilde{\deg}(B) + \deg(C) = n$. On this basis of $\mathcal{V}_{(n)}$, that we shall call in the sequel the *PBW basis at degree n*, it is possible to define three partial graduations, namely, $\deg_L(Z) = \deg(A)$, $\widetilde{\deg_Y}(Z) = \widetilde{\deg}(B)$ and $\deg_M(Z) = \deg(C)$.

It is then of course easy to express the dimension of $\mathcal{V}_{(n)}$ as a (finite) sum of products of the partition function p of number theory, but we do not know how to simplify this (rather complicated) expression, so it is of practically no use. Let us rather write the set of above generators for degree $n = 0, \frac{1}{2}, 1, \frac{3}{2}$ and 2:

$$\begin{aligned}
\mathcal{V}_{(0)} &= \langle \psi \rangle \\
\mathcal{V}_{(\frac{1}{2})} &= \langle Y_{-\frac{1}{2}} \psi \rangle \\
\mathcal{V}_{(1)} &= \langle (M_{-1}\psi), (Y^2_{-\frac{1}{2}}\psi, L_{-1}\psi) \rangle \\
\mathcal{V}_{(\frac{3}{2})} &= \langle (M_{-1}Y_{-\frac{1}{2}}\psi), (Y^3_{-\frac{1}{2}}\psi, Y_{-\frac{3}{2}}\psi, L_{-1}Y_{-\frac{1}{2}}\psi) \rangle \\
\mathcal{V}_{(2)} &= \langle (M^2_{-1}\psi, M_{-2}\psi), (M_{-1}Y^2_{-\frac{1}{2}}\psi, M_{-1}L_{-1}\psi), \\
&\quad (Y^4_{-\frac{1}{2}}\psi, Y_{-\frac{1}{2}}Y_{-\frac{3}{2}}\psi, L_{-1}Y^2_{-\frac{1}{2}}\psi, L_{-2}\psi, L^2_{-1}\psi) \rangle. \qquad (4.2)
\end{aligned}$$

The elements of these Poincaré-Birkhoff-Witt bases have been written in the *M-order* and separated into *blocks* (see below, Sect. 4.2 for a definition of these terms).

The *Kac determinants* are quite easy to compute in the above bases at degree $0, \frac{1}{2}, 1$. If $\{x_1, \ldots, x_{\dim(\mathcal{V}_n)}\}$ is the PBW basis at degree n, put

$$\Delta_n^{\mathfrak{sv}} = \det(\langle x_i | x_j \rangle)_{i,j=1,\ldots,dim(\mathcal{V}_n)}.$$

Note that $\Delta_n^{\mathfrak{sv}}$ does not depend on the ordering of the elements of the basis $\{x_1, \ldots, x_{\dim(\mathcal{V}_n)}\}$.

Then

$$\begin{aligned}
\Delta_0^{\mathfrak{sv}} &= 1 \\
\Delta_{\frac{1}{2}}^{\mathfrak{sv}} &= \langle Y_{-\frac{1}{2}}\psi, Y_{-\frac{1}{2}}\psi \rangle = \mu \\
\Delta_1^{\mathfrak{sv}} &= \det \begin{pmatrix} 0 & 0 & \mu \\ 0 & 2\mu^2 & \mu \\ \mu & \mu & 2h \end{pmatrix} = -2\mu^4.
\end{aligned}$$

For higher degrees, straightforward computations become quickly laborious: even at degree 2, one gets a 9×9 determinant, to be compared with the simple 2×2-determinant that one gets when computing the Kac determinant of the Virasoro algebra at level 2.

The essential idea for calculating this determinant at degree n is to find two permutations σ, τ of the set of elements of the basis, $x_1, \ldots, x_{\dim(\mathcal{V}_n)}$, in such a way that the matrix $(\langle x_{\sigma_i} | x_{\tau_j} \rangle)_{i,j=1,\ldots,p}$ be upper-triangular. Then the Kac determinant $\Delta_n^{\mathfrak{sv}}$, as computed in the basis $\{x_1, \ldots, x_{\dim(\mathcal{V}_n)}\}$, is equal (up to a sign) to the product of diagonal elements of that matrix, which leads finally to the following theorem.

Theorem 4.1. *The Kac determinant $\Delta_n^{\mathfrak{sv}}$ is given (up to a non-zero constant) by*

$$\Delta_n^{\mathfrak{sv}} = \mu^{\sum_{0\le j\le n}\sum_{B\in\tilde{\mathscr{P}}(j)}\left(wid(B)+2\sum_{0\le i\le n-j} p(n-j-i)(\sum_{A\in\mathscr{P}(i)} wid(A))\right)} \tag{4.3}$$

where $p(k) := \#\mathscr{P}(k)$ is the usual partition function.

The same formula holds for the central extension of $\mathfrak{sv}$.

The proof is technical but conceptually easy, depending essentially on the fact that $\mathfrak{h}$ contains a central subalgebra, namely $\langle M_n \rangle_{n\in\mathbb{Z}}$, that is a module of tensor densities for the action of the Virasoro algebra. This fact implies that, in the course of the computations, the (L, M)-generators decouple from the other generators (see Lemma 4.10).

As a matter of fact, we shall need on our way to compute the Kac determinants for the subalgebra $\langle L_n, M_n \rangle_{n\in\mathbb{Z}} \simeq \mathrm{Vect}(S^1) \ltimes \mathscr{F}_0$ or $\mathfrak{vir} \ltimes \mathscr{F}_0$. The result is very similar and encaptures, so we think, the main characteristics of the Kac determinants of Lie algebras $\mathrm{Vect}(S^1) \ltimes \mathfrak{k}$ or $\mathrm{Vect}(S^1) \ltimes \mathfrak{k}$ such that $\mathfrak{k}$ contains in its center a module of tensor densities. To the contrary, the very first computations for the deformations and central extensions of $\mathrm{Vect}(S^1) \ltimes \mathscr{F}_0$ obtained through the cohomology spaces $H^2(\mathrm{Vect}(S^1), \mathscr{F}_0)$ and $H^2(\mathrm{Vect}(S^1)\tilde{\ltimes}\mathscr{F}_0, \mathbb{C})$ (denoting by $\tilde{\ltimes}$ any deformed product) show that the Kac determinants look completely different as soon as the image of $\langle M_n \rangle_{n\in\mathbb{Z}}$ is not central anymore in $\mathfrak{k}$.

We state the result for $\mathfrak{vir} \ltimes \mathscr{F}_0$ as follows.

Theorem 4.2. *Let $\mathfrak{vir}_c$ be the Virasoro algebra with central charge $c \in \mathbb{R}$ and let $\mathcal{V}' = \mathcal{V}'_{h,\mu} = \mathscr{U}((\mathfrak{vir}_c \ltimes \mathscr{F}_0)_0) \otimes_{\mathscr{U}((\mathfrak{vir}_c\ltimes\mathscr{F}_0)_{\le 0})} \mathbb{C}_{h,\mu} \subset \mathcal{V}_{h,\mu}$ be the Verma module representation of $\mathfrak{vir}_c \ltimes \mathscr{F}_0$ induced from the character $\mathbb{C}_{h,\mu}$, with the graduation naturally inherited from that of $\mathcal{V}_{h,\mu}$.*

Then the Kac determinant (computed in the PBW basis) $\Delta_n^{\mathfrak{vir}_c\ltimes\mathscr{F}_0}$ of $\mathcal{V}'_{h,\mu}$ at degree n is equal (up to a positive constant) to

$$\Delta_n^{\mathfrak{vir}_c\ltimes\mathscr{F}_0} = (-1)^{\dim(\mathcal{V}'_\mu)}\mu^{2\sum_{0\le i\le n} p(n-i)(\sum_{A\in\mathscr{P}(i)} wid(A))}. \tag{4.4}$$

It does not depend on c.

4.2 Kac Determinant Formula for $\mathrm{Vect}(S^1) \ltimes \mathscr{F}_0$

We first need to introduce a few notations and define two different orderings for the PBW basis of $\mathrm{Vect}(S^1) \otimes \mathscr{F}_0$.

Definition 4.3. Let A, B be two partitions of $n \in \mathbb{N}^*$. One says that A is *finer* than B, and write $A \preceq B$, if A can be obtained from B by a finite number of transformations $B \to \cdots \to D \to D' \to \cdots A$ where

$$D'^i = D^i + C^i \quad (i \neq p), \quad D'^p = D^i - 1,$$

$C = (C^i)$ being a non-trivial partition of degree p.

Graphically, this means that the Young tableau of A is obtained from the Young tableau of B by splitting some of the stacks of boxes into several stacks.

The relation $\preceq$ gives a *partial* order on the set of partitions of fixed degree n, with smallest element $(1, 1, \ldots, 1)$ and largest element the trivial partition (n). One *chooses* arbitrarily, for every degree n, a total ordering $\leq$ of $\mathscr{P}(n)$ compatible with $\preceq$, i.e. such that $(A \preceq B) \Rightarrow (A \leq B)$.

Definition 4.4. If A is a partition, then $X^A := (X^{-A})^*$ is given by $X^A = \ldots X_2^{A_2} X_1^{A_1}$ (where X stands for L or M).

Let $n \in \mathbb{N}$. By the Poincaré-Birkhoff-Witt theorem, the $L^{-A} M^{-C}$ (A partition of degree p, C partition of degree q, $p, q \geq 0$, $p + q = n$) form a basis $\mathscr{B}'$ of $\mathscr{V}'_n$, the subspace of $\mathscr{V}'$ made up of the vectors of degree n.

We now give two different orderings of the set $\mathscr{B}'$, that we call *horizontal ordering* (or *M-ordering*) and *vertical ordering* (or *L-ordering*). For the *horizontal ordering*, we proceed as follows:

- we split $\mathscr{B}'$ into $(n + 1)$ *blocks* $\mathscr{B}'_0, \ldots, \mathscr{B}'_n$ such that

$$\mathscr{B}'_j = \{L^{-A} M^{-C} \psi \mid \deg(A) = j, \deg(C) = n - j\}; \tag{4.5}$$

- we split each block $\mathscr{B}'_j$ into *sub-blocks* $\mathscr{B}'_{j,C}$ (also called j-sub-blocks if one wants to be more explicit) such that C runs among the set of partitions of $n - j$ in the *increasing* order chosen above, and

$$\mathscr{B}'_{j,C} = \{L^{-A} M^{-C} \psi \mid \deg(A) = j\}.$$

- finally, inside each sub-block $\mathscr{B}'_{j,C}$, we take the elements $L^{-A} M^{-C}$, $A \in \mathscr{P}_j$, in *decreasing order*.

For the *vertical ordering*, we split $\mathscr{B}'$ into blocks

$$\widetilde{\mathscr{B}}'_j = \{L^{-A} M^{-C} \psi \mid \deg(A) = n - j, \deg(C) = j\},$$

each of these blocks into sub-blocks $\widetilde{\mathscr{B}}'_{j,A}$ where A runs among all partitions of $n - j$ in the increasing order, and take the elements $L^{-A} M^{-C} \in \widetilde{\mathscr{B}}'_{j,A}$ according to the decreasing order of the partitions C of degree j.

As one easily checks, $L^{-C} M^{-A} \psi$ is at the same place in the vertical ordering as $L^{-A} M^{-C}$ in the horizontal ordering. Note that the vertical ordering can also be obtained by reversing the horizontal ordering. Yet we maintain the separate definitions both for clarity and because these definitions will be extended in the next paragraph to the case of $\mathfrak{sv}$, where there is no simple relation between the two orderings.

Roughly speaking, one may say that, for the horizontal ordering, the degree in M decreases from one block to the next one, the $(M^C)_{C\in\mathscr{P}(n-j)}$ are chosen in the increasing order inside each block, and then the L^A are chosen in the decreasing order inside each sub-block; the vertical ordering is defined in exactly the same way, except that L and M are exchanged.

We shall compute the Kac determinant $\Delta'_n := \Delta_n^{\mathfrak{vir}_c\ltimes\mathscr{F}_0}$ relative to $\mathscr{B}'$ by representing it as

$$\Delta'_n = \pm\det\mathscr{A}'_n, \quad \mathscr{A}'_n = (\langle H_j|V_i\rangle)_{i,j} \tag{4.6}$$

where the $(H_j)_j \in \mathscr{B}'$ are chosen in the horizontal order and the $(V_i)_i \in \mathscr{B}'$ in the vertical order.

The following facts are clear from the above definitions:

- the horizontal and vertical blocks $\mathscr{B}'_j, \widetilde{\mathscr{B}}'_j$ (j fixed) and sub-blocks $\mathscr{B}'_{j,C}, \widetilde{\mathscr{B}}'_{j,C}$ (j, C fixed) have same size, so one has matrix diagonal blocks and sub-blocks;
- diagonal elements are of the form $\langle L^{-A}M^{-C}\psi \mid L^{-C}M^{-A}\psi\rangle$;
- define *sub-diagonal* elements to be the $\langle H|V\rangle$, $H \in \mathscr{B}'_j, V \in \mathscr{B}'_i$ such that $i > j$; then $\deg_L(V) < \deg_M(H)$;
- define *j-sub-sub-diagonal* elements (or simply sub-sub-diagonal elements if one doesn't need to be very definite) to be the $\langle H|V\rangle$, $H \in \mathscr{B}'_{j,C_1}, V \in \mathscr{B}'_{j,C_2}$ with $C_2 > C_1$. Then $\deg_L(V) = \deg_M(H) = n - j$ and $H = L^{-A_1}M^{-C_1}, V = L^{-C_2}M^{-A_2}$ for certain partitions A_1, A_2 of degree j;
- define *(j, C)-sub^3*-diagonal elements to be the $\langle H|V\rangle$, $H, V \in \mathscr{B}'_{j,C}$, such that $H = L^{-A_1}M^{-C}, V = L^{-C}M^{-A_2}$ with $A_1 > A_2$.

Then the set of sub-diagonal elements is the union of the matrix blocks situated under the diagonal; the set of j-sub-sub-diagonal elements is the union of the matrix sub-blocks situated under the diagonal of the j-th matrix diagonal blocks; the set of (j, C)-sub^3-diagonal elements is the union of the elements situated under the diagonal of the (j, C)-diagonal sub-block. All these elements together form the set of lower-diagonal elements of the matrix $\mathscr{A}'_n$.

Elementary computations show that

$$\mathscr{A}'_1 = \begin{pmatrix} \mu & 2h \\ 0 & \mu \end{pmatrix}$$

with horizontal ordering $(M^{-1}\psi, L^{-1}\psi)$;

$$\mathscr{A}'_2 = \begin{pmatrix} 2\mu^2 & 2\mu & 2\mu(1+2h) & 6h & 4h(2h+1) \\ 0 & 2\mu & 3\mu & 4h+c/2 & 6h \\ 0 & 0 & \mu^2 & 3\mu & 2\mu(1+2h) \\ 0 & 0 & 0 & 2\mu & 2\mu \\ 0 & 0 & 0 & 0 & 2\mu^2 \end{pmatrix}$$

with horizontal ordering $((M_{-1}^2\psi, M_{-2}\psi), (L_{-1}M_{-1}\psi), (L_{-2}\psi, L_{-1}^2\psi))$ (the blocks being separated by parentheses). Note that the Kac determinant of $\mathfrak{vir}_c$ at level 2 appears as the top rightmost 2×2 upper-diagonal block of $\mathscr{A}_2'$, and hence does not play any role in the computation of Δ_n'.

So $\mathscr{A}_1'$, $\mathscr{A}_2'$ are upper-diagonal matrices, with diagonal elements that are (up to a coefficient) simply powers of μ. More specifically, $\Delta_1' = -\det \mathscr{A}_1' = -\mu^2$ and $\Delta_2' = \det \mathscr{A}_2' = 16\mu^8$.

The essential technical lemmas for the proof of Theorem 4.2 and Theorem 4.1 are Lemma 4.5 and Lemma 4.6, which show, roughly speaking, how to move the M's through the L's.

Lemma 4.5. *Let A, C be two partitions of degree n. Then:*

(i) If $A \not\preceq C$, then $\langle L^{-A}\psi \mid M^{-C}\psi\rangle = 0$.
(ii) If $A \preceq C$, then

$$\langle L^{-A}\psi \mid M^{-C}\psi\rangle = a.\, \mu^{\mathrm{wid}(C)} \tag{4.7}$$

for a certain positive constant a depending only on A and C.

Proof of Lemma 4.5. We use induction on n. Take $A = (a_i), C = (c_i)$ of degree n, then

$$\langle L^{-A}\psi | M^{-C}\psi\rangle = \left\langle \psi \mid \left(\prod_{i=\infty}^{1} L_i^{a_i}\right)\left(\prod_{i=1}^{\infty} M_{-i}^{c_i}\right)\psi\right\rangle.$$

We shall compute $\langle L^{-A}\psi | M^{-C}\psi\rangle$ by moving successively to the left the M_{-1}'s, then the M_{-2}'s and so on, and taking care of the commutators that show up in the process.

- Suppose $c_1 > 0$. By commuting M_{-1} with the L's, there appear terms of two types, modulo positive constants:
  - either of type

$$\left\langle \psi \mid \left(\prod_{i=\infty}^{k+1} L_i^{a_i}\right)\left(L_k^{a_k'} M_{k-1} L_k^{a_k''}\right)\left(\prod_{i=k-1}^{1} L_i^{a_i}\right) M_{-1}^{c_1-1}\left(\prod_{i=2}^{\infty} M_{-i}^{c_i}\right)\psi\right\rangle,$$

    with $a_k' + a_k'' = a_k - 1$ (by commuting with L_k, $k \geq 2$). But this is zero since commuting M_{k-1} (of negative degree $1-k$) to the right past the L's can only lower its degree.
  - or of type

$$\left\langle \psi \mid \left(\prod_{i=\infty}^{2} L_i^{a_i}\right)\left(L_1^{a_1'} M_0 L_1^{a_1''}\right) M_{-1}^{c_1-1}\left(\prod_{i=2}^{\infty} M_{-i}^{c_i}\right)\psi\right\rangle,$$

    with $a_1' + a_1'' = a_1 - 1$ (by commuting with L_1). Since the central element $M_0 \equiv \mu$ can be taken out of the brackets, we may compute this as μ times a scalar product between two elements of degree $n-1$. Removing one M_{-1} and one L_{-1} means removing the leftmost column of the Young tableaux C and A. Call C', A' the new tableaux: it is clear that $A \npreceq C \Rightarrow A' \npreceq C'$. So we may conclude the proof by induction.

- Suppose that all $c_1 = \ldots = c_{j-1} = 0$ and $c_j > 0$. Then, by similar arguments, one sees that all potentially non-zero terms appearing on the way (while moving M_{-j} to the left) are of the form

$$\alpha\mu \,.\, \left\langle \psi \mid \left(\prod_{i=\infty}^{1} L_i^{a_i'}\right) M_{-j}^{c_j-1} \left(\prod_{ij} M_{-i}^{c_i}\right) \psi \right\rangle,$$

$\sum_{i=\infty}^{1} i\,(a_i - a_i') = j$, with α defined by

$$\left(\prod_{i=\infty}^{1} (adL_i)^{a_i - a_i'}\right) . \; M_{-j} = \alpha M_0.$$

Without computing α explicitly, it is clear that $\alpha > 0$. On the Young tableaux, this corresponds to removing one stack of height j from C and $(a_i - a_i')$ stacks of height i $(i = \infty, \ldots, 1)$ from A. Once again, $A \npreceq C \Rightarrow A' \npreceq C'$. So one may conclude by induction. □

Lemma 4.6. *Let* A_1, A_2, C_1, C_2 *be partitions such that* $\deg A_1 + \deg C_1 = \deg A_2 + \deg C_2$. *Then:*

(i)

$$\langle L^{-C_2} M^{-A_2} \psi \mid L^{-A_1} M^{-C_1} \psi \rangle = 0$$

if $\deg A_1 < \deg A_2$.

(ii) If $\deg A_1 = \deg A_2$, *then*

$$\langle L^{-C_2} M^{-A_2} \psi \mid L^{-A_1} M^{-C_1} \psi \rangle = \langle L^{-C_2} \psi | M^{-C_1} \psi \rangle \langle M^{-A_2} \psi | L^{-A_1} \psi \rangle. \tag{4.8}$$

Remark. The central argument in this Lemma can be trivially generalized (just by using the fact that the M's are central in $\mathfrak{h}$) in a form that will be used again and again in the next section: namely,

$$\langle \psi \mid U L^{C_2} V L^{-A} W M^{-C_1} \psi \rangle = 0 \quad (U, V, W \in \mathcal{U}(\mathfrak{h}))$$

if $\deg(C_1) > \deg(C_2)$.

Proof. Putting all generators on one side, one gets

$$\langle L^{-C_2} M^{-A_2} \psi \mid L^{-A_1} M^{-C_1} \psi \rangle = \langle \psi \mid M^{A_2} L^{C_2} L^{-A_1} M^{-C_1} \psi \rangle.$$

Assume that $\deg(A_1) \le \deg(A_2)$ (so that $\deg(C_2) \le \deg(C_1)$). Let us move M^{-C_1} to the left and consider the successive commutators with the L's: it is easy to see that one gets (apart from the trivially commuted term $M^{-C_1}L^{-A_1}$) a sum of terms of the type $M^{-C_1'}L^{-A_1'}$ with

$$\deg(C_1')+\deg(A_1')=\deg(C_1)+\deg(A_1), \quad \deg(C_1')>\deg(C_1), \deg(A_1')<\deg(A_1).$$

Now comes the central argument : let us commute $M^{-C_1'}$ through L^{C_2}. It yields terms of the type $M^{A_2}M^{-C_1''}M^{C_1'''}L^{C_2'}$ with $\deg(C_2) - \deg(C_1') = \deg(C_2') + \deg(C_1''') - \deg(C_1'')$.

If $C_1'' \neq \emptyset$, then $M^{-C_1''}$ commutes with M^{A_2} and gives 0 when set against $\langle\psi$; so one may assume $C_1'' = \emptyset$. But this is impossible, since it would imply

$$\begin{aligned} \deg(C_1''') &= \deg(C_2) - \deg(C_1') - \deg(C_2') && (4.9)\\ &\le \deg(C_1) - \deg(C_1') \quad (\text{even } < \text{ if } \deg(C_2) < \deg(C_1)) && (4.10)\\ &< 0. && (4.11) \end{aligned}$$

So

$$\langle L^{-C_2}M^{-A_2}\psi \mid L^{-A_1}M^{-C_1}\psi\rangle = \langle \psi \mid M^{A_2}L^{C_2}M^{-C_1}L^{-A_1}\psi\rangle.$$

Moreover, the above argument applied to M^{-C_1} instead of $M^{-C_1'}$, shows that

$$\langle \psi \mid M^{A_2}(L^{C_2}M^{-C_1})L^{-A_1}\psi\rangle = \langle \psi \mid M^{A_2}(M^{-C_1}L^{C_2})L^{-A_1}\psi\rangle = 0$$

if $\deg(C_2) < \deg(C_1)$. So (i) holds and one may assume that $\deg(A_1) = \deg(A_2)$, $\deg(C_1) = \deg(C_2)$ in the sequel.

Now move M^{A_2} to the right: the same argument proves that

$$\langle \psi \mid M^{A_2}L^{C_2}M^{-C_1}L^{-A_1}\psi\rangle = \langle \psi \mid (L^{C_2}M^{-C_1})(M^{A_2}L^{-A_1})\psi\rangle.$$

But $M^{A_2}L^{-A_1}$ has degree 0, so

$$\begin{aligned} \langle \psi \mid (L^{C_2}M^{-C_1})(M^{A_2}L^{-A_1})\psi\rangle &= \langle \psi \mid L^{C_2}M^{-C_1}\psi\rangle\langle \psi \mid M^{A_2}L^{-A_1}\psi\rangle && (4.12)\\ &= \langle L^{-C_2}\psi \mid M^{-C_1}\psi\rangle\langle M^{-A_2}\psi \mid L^{-A_1}\psi\rangle. && (4.13) \end{aligned}$$

□

Corollary 4.7. *The matrix $\mathscr{A}_n'$ is upper-diagonal.*

Proof. Lower-diagonal elements come in three classes: let us give an argument for each class.

By Lemma 4.6, (i), sub-diagonal elements are zero.

Consider a j-sub-sub-diagonal element $\langle L^{-C_2}M^{-A_2}\psi \mid L^{-A_1}M^{-C_1}\psi\rangle$ with $\deg(C_1) = \deg(C_2) = n - j$, $\deg(A_1) = \deg(A_2)$, $C_2 > C_1$. Then, by Lemma 4.6 (ii), and Lemma 4.5 (i),

$$\langle L^{-C_2}M^{-A_2}\psi \mid L^{-A_1}M^{-C_1}\psi\rangle = \langle L^{-C_2}\psi \mid M^{-C_1}\psi\rangle\langle M^{-A_2}\psi \mid L^{-A_1}\psi\rangle = 0.$$

Finally, consider a (C, j)-sub^3-diagonal element $\langle L^{-C}M^{-A_2}\psi \mid L^{-A_1}M^{-C}\psi\rangle$ with $A_1 > A_2$. By the same arguments, this is zero. □

Proof of Theorem 4.2. By Corollary 4.7, one has

$$\Delta'_n = (-1)^{\dim(\mathcal{V}'_n)} \prod_{i=1}^{\dim(\mathcal{V}'_n)} (\mathcal{A}'_n)_{ii}.$$

By Lemma 4.6 (ii) and Lemma 4.5 (ii), the diagonal elements of $\mathcal{A}'_n$ are equal (up to a positive constant) to μ to a certain power. Now the total power of μ is equal to

$$\sum_{0\le j\le n} \sum_{A\in\mathcal{P}(j)} \sum_{C\in\mathcal{P}(n-j)} (\mathrm{wid}(A) + \mathrm{wid}(C)) = 2 \sum_{0\le j\le n} p(n-j) \left(\sum_{A\in\mathcal{P}(j)} \mathrm{wid}(A) \right). \tag{4.14}$$

□

4.3 Kac Determinant Formula for $\mathfrak{sv}$

Let $n \in \frac{1}{2}\mathbb{N}$. We shall define in this case also a *horizontal ordering* (also called *M-ordering*) and a *vertical ordering* (also called *L-ordering*) of the Poincaré-Birkhoff-Witt basis

$$\mathcal{B} = \{L^{-A}Y^{-B}M^{-C}\psi \mid \deg(A) + \widetilde{\deg}(B) + \deg(C) = n\}$$

of $\mathcal{V}_{(n)}$.

The *M-ordering* is defined as follows (note that blocks and sub-blocks are defined more or less as in the preceding sub-section):

- split $\mathcal{B}$ into $(n + 1)$-blocks $\mathcal{B}_0, \ldots, \mathcal{B}_n$ such that $\mathcal{B}_j = \{L^{-A}Y^{-B}M^{-C}\psi \in \mathcal{B} \mid \deg(C) = n - j\}$;
- split each block $\mathcal{B}_j$ into sub-blocks $\mathcal{B}_{jC}$ such that $\mathcal{B}_{jC} = \{L^{-A}Y^{-B}M^{-C}\psi \in \mathcal{B}\}$, with C running among the set of partitions of $n - j$ in the *increasing* order;
- split each sub-block $\mathcal{B}_{jC}$ into sub^2-blocks $\mathcal{B}_{jC,\kappa}$ (for *decreasing* κ) with $L^{-A}Y^{-B}M^{-C} \in \mathcal{B}_{jC,\kappa} \Leftrightarrow L^{-A}Y^{-B}M^{-C} \in \mathcal{B}_{jC}$ and $\widetilde{\deg}(B) = \kappa$;
- split each sub^2-block $\mathcal{B}_{jC,\kappa}$ into sub^3-blocks $\mathcal{B}_{jC,\kappa B}$ where B runs among all partitions of shifted degree κ (in any randomly chosen order);

– finally, order the elements $L^{-A}Y^{-B}M^{-C}\psi$ of $\mathscr{B}_{jC,\kappa B}$ ($A \in \mathscr{P}(n - \deg(C) - \widetilde{\deg}(B)) = \mathscr{P}(j - \kappa)$) so that the A's appear in the *decreasing* order.

Then the *L-ordering* is chosen in such a way that $L^{-C}Y^{-B}M^{-A}$ appears vertically in the same place as $L^{-A}Y^{-B}M^{-C}$ in the M-ordering. From formula (4.2) giving the M-ordering of $\mathscr{V}_{(2)}$, it is clear that the L-ordering is *not* the opposite of the M-ordering.

These two orderings define a block matrix $\mathscr{A}_n$ whose determinant is equal to $\pm\Delta_n^{\mathfrak{sv}}$.

We shall need three preliminary lemmas.

Lemma 4.8. *Let $B = (b_j)$, $B' = (b'_j)$ be two partitions with same shifted degree: then*

$$\langle Y^{-B}\psi \mid Y^{-B'}\psi\rangle = \delta_{B,B'}\left[\prod_{j\geq 1}(b_j)!(2j-1)^{b_j}\right] . \mu^{\mathrm{wid}(B)}. \tag{4.15}$$

Proof. Consider the sub-module $\mathscr{W}\subset\mathscr{V}$ generated by the $Y^{-P}M^{-Q}\psi$ (P, Q partitions). Then, inside this module, and as long as vacuum expectation values $\langle\psi \mid Y_{\pm(j_1-\frac{1}{2})}\cdots Y_{\pm(j_k-\frac{1}{2})}\psi\rangle$ are concerned, the $(Y_{-(j-\frac{1}{2})}, Y_{j-\frac{1}{2}})$ can be considered as independent pairs of creation/annihilation operators with $[Y_{(j-\frac{1}{2})}, Y_{-(j-\frac{1}{2})}] = (2j-1)M_0$. Namely, other commutators $[Y_k, Y_l]$ with $k+l \neq 0$ yield $(k-l)M_{k+l}$ which commutes with all other generators Y's and M's and gives 0 when applied to $\psi\rangle$ or $\psi\langle$ according to the sign of $k+l$. The result now follows for instance by an easy application of Wick's theorem, or by induction. □

Lemma 4.9. *Let B_1, B_2, C_1, C_2 be partitions such that $C_1 \neq \emptyset$ or $C_2 \neq \emptyset$. Then*

$$\langle Y^{-B_1}M^{-C_1}\psi \mid Y^{-B_2}M^{-C_2}\psi\rangle = 0.$$

Proof. Obvious (the $M^{\pm C}$'s are central in the subalgebra $\mathfrak{h} \subset \mathfrak{sv}$ and can thus be commuted freely with the Y's and the M's; set against $\langle\psi$ or $\psi\rangle$ according to the sign of their degree, they give zero). □

Lemma 4.10. *Let $A_1, A_2, B_1, B_2, C_1, C_2$ be partitions such that* $\deg(A_1) + \deg(B_1) + \deg(C_1) = \deg(A_2) + \deg(B_2) + \deg(C_2)$. *Then*

(i) If $\deg(A_1) < \deg(A_2)$ *or* $\deg(C_2) < \deg(C_1)$, *then*

$$\langle L^{-C_2}Y^{-B_2}M^{-A_2}\psi \mid L^{-A_1}Y^{-B_1}M^{-C_1}\psi\rangle = 0.$$

(ii) If $\deg(A_1) = \deg(A_2)$ *and* $\deg(C_1) = \deg(C_2)$, *then*

$$\langle L^{-C_2} Y^{-B_2} M^{-A_2} \psi \mid L^{-A_1} Y^{-B_1} M^{-C_1} \psi \rangle$$
$$= \langle Y^{-B_2} \psi \mid Y^{-B_1} \psi \rangle \langle M^{-A_2} \psi \mid L^{-A_1} \psi \rangle \langle L^{-C_2} \psi \mid M^{-C_1} \psi \rangle. \quad (4.16)$$

Proof of Lemma 4.10. (i) is a direct consequence of the remark following Lemma 4.6. So we may assume that $\deg(A_1) = \deg(A_2)$ and $\deg(C_1) = \deg(C_2)$. Using the hypothesis $\deg(A_1) = \deg(A_2)$, we may choose this time to move L^{-A_1} to the left in the expression $\langle \psi \mid M^{A_2} Y^{B_2} L^{C_2} L^{-A_1} Y^{-B_1} M^{-C_1} \psi \rangle$. Commuting L^{-A_1} through L^{C_2} and Y^{B_2}, one obtains terms of the type

$$\langle \psi \mid M^{A_2} (L^{-A'_1} Y^{B'_2} L^{C'_2}) Y^{-B_1} M^{-C_1} \psi \rangle,$$

with $\deg(A'_1) \leq \deg(A_1)$, and $(\deg(A'_1) = \deg(A_1)) \Rightarrow (A'_1 = A_1, B'_2 = B_2, C'_2 = C_2)$. So, by the Remark following Lemma 4.6,

$$\langle L^{-C_2} Y^{-B_2} M^{-A_2} \psi \mid L^{-A_1} Y^{-B_1} M^{-C_1} \psi \rangle$$
$$= \langle \psi \mid M^{A_2} Y^{B_2} L^{C_2} L^{-A_1} Y^{-B_1} M^{-C_1} \psi \rangle$$
$$= \langle \psi \mid Y^{B_2} L^{C_2} M^{A_2} L^{-A_1} Y^{-B_1} M^{-C_1} \psi \rangle$$
$$= \langle \psi \mid M^{A_2} L^{-A_1} \psi \rangle \langle \psi \mid Y^{B_2} L^{C_2} Y^{-B_1} M^{-C_1} \psi \rangle. \quad (4.17)$$

Moving L^{C_2} to the right in the same way leads to (ii), thanks to the hypothesis $\deg(C_1) = \deg(C_2)$ this time. □

Corollary 4.11. *The matrix $\mathscr{A}_n$ is upper-diagonal.*

Proof. Lower-diagonal elements $f = \langle L^{-C_2} Y^{-B_2} M^{-A_2} \psi \mid L^{-A_1} Y^{-B_1} M^{-C_1} \psi \rangle$ come this time in five classes. Let us treat each class separately.

By Lemma 4.10 (i), sub-diagonal elements (characterized by $\deg(C_2) < \deg(C_1)$) are zero.

Next, j-sub^2-diagonal elements (characterized by $\deg(C_2) = \deg(C_1) = n - j$, $C_2 > C_1$) are also zero: namely, the hypothesis $\deg(C_1) = \deg(C_2)$ gives as in the proof of Lemma 4.10

$$f = \langle \psi \mid M^{A_2} Y^{B_2} L^{C_2} L^{-A_1} Y^{-B_1} M^{-C_1} \psi \rangle \quad (4.18)$$
$$= \langle \psi \mid L^{C_2} M^{-C_1} \psi \rangle \langle \psi \mid M^{A_2} Y^{B_2} L^{-A_1} Y^{-B_1} \psi \rangle \quad (4.19)$$
$$= 0 \text{ (by Lemma 3.1.1)}. \quad (4.20)$$

Now (j, C)-sub^3-diagonal elements (characterized by $C_1 = C_2 := C$, $\widetilde{\deg}(B_2) < \widetilde{\deg}(B_1)$) are again zero because, as we have just proved,

$$f = \langle \psi \mid L^{C_2} M^{-C_1} \psi \rangle \langle \psi \mid M^{A_2} Y^{B_2} L^{-A_1} Y^{-B_1} \psi \rangle$$

and this time $\langle \psi \mid M^{A_2} Y^{B_2} L^{-A_1} Y^{-B_1} \psi \rangle = 0$ since

$$\deg(A_1) = j - \widetilde{\deg}(B_1) < j - \widetilde{\deg}(B_2) = \deg(A_2).$$

Finally, (jC, κ)-sub^4-diagonal elements (with $C_1 = C_2 = C$, $\deg B_1 = \deg B_2$, $\deg A_1 = \deg A_2$, $B_1 \neq B_2$) are 0 by Lemma 4.10 (ii) and Lemma 4.8 since $\langle Y^{-B_2}\psi \mid Y^{-B_1}\psi\rangle = 0$, and $(jC, \kappa B)$-sub^5-diagonal elements (such that $B_1 = B_2$ but $A_2 < A_1$) are 0 by Lemma 4.10 (ii) and Lemma 4.5 (i) since $\langle M^{-A_2}\psi \mid L^{-A_1}\psi\rangle = 0$. □

Proof of Theorem 4.1. By Corollary 4.11, $\Delta_n^{\mathfrak{sv}} = \pm\prod_{i=1}^{\dim(\mathcal{V}_{(n)})}(\mathcal{A}_n)_{ii}$. By Lemma 4.10 (ii), Lemma 4.5 (ii) and Lemma 4.8, the diagonal element $\langle L^{-C}Y^{-B}M^{-A}\psi \mid L^{-A}Y^{-B}M^{-C}\psi\rangle$ is equal (up to a positive constant) to $\mu^{\mathrm{wid}(A)+\mathrm{wid}(B)+\mathrm{wid}(C)}$, so (see proof of Theorem 4.2) $\Delta_n^{\mathfrak{sv}} = c_n\mu^{a_n}$ (c_n non-zero constant, $a_n \in \mathbb{N}$) with

$$a_n = \sum_{0\leq j\leq n}\sum_{B\in\tilde{\mathcal{P}}(j)}(\mathrm{wid}(B) + a'_{n-j}),$$

where a'_{n-j} is the power of μ appearing in $\Delta_{n-j}^{\mathrm{Vect}(S^1)\ltimes\mathcal{F}_0}$. Hence the final result. □

Chapter 5
Coinduced Representations

In order to classify 'reasonable' representations of the Virasoro algebra, V.G. Kac made the following conjecture: the *Harish-Chandra representations*, those for which L_0 (see Sect. 1.1.2) acts semi-simply with finite-dimensional eigenspaces, are either highest- (or lowest-) weight modules, or tensor density modules. As proved in [85] and [86], one has essentially two types of Harish-Chandra representations of the Virasoro algebra:

- Verma modules which are *induced* to $\mathrm{Vect}(S^1)$ from a character of $\mathrm{Vect}(S^1)_+ = \langle L_0, L_1, \ldots \rangle$, zero on the subalgebra $\mathrm{Vect}(S^1)_{\geq 1} = \langle L_1, \ldots \rangle$, and quotients of degenerate Verma modules.
- Tensor modules of formal densities which are *coinduced* to the subalgebra of formal or polynomial vector fields $\mathrm{Vect}(S^1)_{\geq -1} = \langle L_{-1}, L_0, \ldots \rangle$ from a character of $\mathrm{Vect}(S^1)_{\geq 0}$ that is zero on the subalgebra $\mathrm{Vect}(S^1)_{\geq 1}$. These modules extend naturally to representations of $\mathrm{Vect}(S^1)$.

Verma modules for the Schrödinger–Virasoro algebra have been constructed in the previous chapter. The purpose of the present chapter is to introduce a natural generalization of the tensor density $\mathrm{Vect}(S^1)$-modules to the Schrödinger–Virasoro case. Contrary to the previous chapter, the natural graduation here is given by δ_2, which is associated to the polynomial degree of the coefficients of the vector fields in the vector-field representation $d\tilde{\pi}$ defined in the Introduction. The Lie algebra $\mathfrak{sv}$ appears actually as a natural prolongation (called: *Cartan prolongation*) of its finite-dimensional Lie subalgebra made up of vector fields with affine coefficients (see Sect. 5.1). This supplementary structure makes it possible to obtain explicit formulas for the coinduced representations in Sect. 5.3. We give examples in Sects. 5.2 and 5.4, including the vector-field representation and the linearization of an action on an affine space of Dirac-Lévy-Leblond operators.

Contrary to the case of the Virasoro algebra, it appears that there are representations which are neither of highest-weight type nor are coinduced: for instance, the coadjoint representation. All explicit realizations of $\mathfrak{sv}$ arising from symmetries of physical equations are of coinduced type, while the coadjoint representation is of

J. Unterberger and C. Roger, *The Schrödinger-Virasoro Algebra*, Theoretical and Mathematical Physics, DOI 10.1007/978-3-642-22717-2_5,

a totally different nature. The reason is that coinduced representations are instead closely related to the *adjoint* representation.

Section 5.2 may be skipped in a first approach. The main results are Theorem 5.1 in Sect. 5.1 (yielding the Cartan prolongation structure of $\mathfrak{sv}$) and Theorem 5.3 (explicit formulas for the coinduced representations).

The contents of this chapter are adapted from [106].

5.1 The Lie Algebra $\mathfrak{sv}$ as a Cartan Prolongation

As in the case of vector fields on the circle, it is natural, starting from the representation $d\tilde{\pi}$ of $\mathfrak{sv}$ given by formula (1.20), to consider the subalgebra $\mathfrak{fsv} \subset \mathfrak{sv}$ made up of the vector fields with polynomial coefficients, also called formal vector fields. Recall from Definition 2.3 that the outer derivation δ_2 of $\mathfrak{sv}$ is defined by

$$\delta_2(L_n) = n,\ \delta_2(Y_m) = m - \frac{1}{2},\ \delta_2(M_n) = n - 1 \quad \left(n \in \mathbb{Z}, m \in \frac{1}{2} + \mathbb{Z}\right). \tag{5.1}$$

and that δ_2 is simply obtained from the Lie action of the Euler operator $t\partial_t + r\partial_r + \zeta\partial_\zeta$ in the representation $d\tilde{\pi}$. The Lie subalgebra $\mathfrak{fsv}$ is given more abstractly, using δ_2, as

$$\mathfrak{fsv} = \oplus_{k=-1}^{+\infty} \mathfrak{sv}_k \tag{5.2}$$

where $\mathfrak{sv}_k = \{X \in \mathfrak{sv} \mid \delta_2(X) = kX\} = \langle L_k, Y_{k+\frac{1}{2}}, M_{k+1}\rangle$ is the eigenspace of δ_2 corresponding to the eigenvalue $k \in \mathbb{Z}$.

Note in particular that $\mathfrak{sv}_{-1} = \langle L_{-1}, Y_{-\frac{1}{2}}, M_0\rangle$ is commutative, generated by the infinitesimal translations $\partial_t, \partial_r, \partial_\zeta$ in the vector field representation $d\tilde{\pi}$ (see Introduction), and that $\mathfrak{sv}_0 = \langle L_0, Y_{\frac{1}{2}}, M_1\rangle = \langle L_0\rangle \ltimes \langle Y_{\frac{1}{2}}, M_1\rangle$ is solvable.

We shall now proceed to define the notion of *Cartan prolongation*. For that we need first some algebraical and geometrical preliminaries. For basic constructions and properties of Cartan prolongations, see [108].

Let E be a finite-dimensional real vector space with dual E^*, and $\mathfrak{g} \subset \mathfrak{gl}(E)$ be a subset of $\mathfrak{gl}(E)$. Define the Lie algebra of *formal vector fields* on E as follows:

$$\mathrm{Vect}(E) := \oplus_{p=0}^{\infty} S^p E^* \otimes E, \tag{5.3}$$

where $S^p E^*$ stands for the space of symmetric covariant p-tensors on E. Its naturally defined Lie bracket is graded of weight (-1) with respect to polynomial degree. Then $\mathfrak{gl}(E) \simeq E^* \otimes E$ is naturally embedded as a subalgebra of $\mathrm{Vect}(E)$. For any $z \in E$ one has a naturally defined inner product $i(z) : S^p E^* \otimes E \to S^{p-1} E^* \otimes E$ (in terms of the Lie bracket on $\mathrm{Vect}(E)$ one has simply $i(z)u = [z, u]$). One may then define the p-prolongation of $\mathfrak{g}$ inductively as:

$$\mathfrak{g}^{(0)} = E, \quad \mathfrak{g}^{(1)} = \mathfrak{g}, \quad \mathfrak{g}^{(p)} = \{u \in S^p E^* \otimes E \mid \forall z \in E,\ i(z)u \in \mathfrak{g}^{(p-1)}\}. \tag{5.4}$$

One easily checks that $\mathfrak{g}^{(\infty)} := \oplus_{p=0}^{\infty} \mathfrak{g}^{(p)}$ is a Lie subalgebra of Vect(E).

Definition 5.1 (Cartan prolongation). The Lie algebra $\mathfrak{g}^{(\infty)}$ is called the Cartan prolongation of $\mathfrak{g}$.

The classical families of examples exhibited by E. Cartan are Vect(n), the Lie algebra of polynomial vector fields on $\mathbb{R}^n$, and its following subalgebras: SVect(n) (divergence-free vector fields, exponentiating into volume-preserving diffeormorphisms); Sympl($2n$) (Hamiltonian vector fields, exponentiating into symplectomorphisms, i.e. diffeomorphisms preserving the canonical symplectic form ω of $\mathbb{R}^{2n}$); ConfSympl($2n$) (vector fields X such that $X^*\omega$ is proportional to ω, exponentiating into "conformal symplectic" diffeomorphisms, i.e. diffeomorphisms preserving the symplectic form up to a constant); and finally, Cont(n) (contact vector fields, exponentiating into diffeomorphisms preserving the kernel of a contact form).

Now $\mathfrak{fsv}$ may be obtained by this construction.

Theorem 5.1. *The Lie algebra $\mathfrak{fsv}$ is isomorphic to the Cartan prolongation of $\mathfrak{sv}_{-1} \oplus \mathfrak{sv}_0$ where $\mathfrak{sv}_{-1} \simeq \langle L_{-1}, Y_{-\frac{1}{2}}, M_0\rangle$ and $\mathfrak{sv}_0 \simeq \langle L_0, Y_{\frac{1}{2}}, M_1\rangle$.*

Proof. Let $\mathfrak{sv}_n$ ($n = 1, 2, \ldots$) be the n-th level vector space obtained from Cartan's construction, so that the Cartan prolongation of $\mathfrak{sv}_{-1} \oplus \mathfrak{sv}_0$ is equal to the Lie algebra $\mathfrak{sv}_{-1} \oplus \mathfrak{sv}_0 \oplus \oplus_{n\geq 1}\mathfrak{sv}_n$. It will be enough, to establish the required isomorphism, to prove the following. Consider the representation $d\tilde{\pi}$ of $\mathfrak{sv}$. Then the space $\mathfrak{h}_n$ defined through induction on n by

$$\mathfrak{h}_{-1} = d\tilde{\pi}(\mathfrak{sv}_{-1}) = \langle \partial_t, \partial_r, \partial_\zeta \rangle \tag{5.5}$$

$$\mathfrak{h}_0 = d\tilde{\pi}(\mathfrak{sv}_0) = \langle t\partial_t + \frac{1}{2} r\partial_r, t\partial_r + r\partial_\zeta, t\partial_\zeta \rangle \tag{5.6}$$

$$\mathfrak{h}_{k+1} = \{X \in \mathscr{X}_{k+1} \mid [X, \mathfrak{h}_{-1}] \subset \mathfrak{h}_k\}, \ (k \geq 0) \tag{5.7}$$

(where $\mathscr{X}_k$ is the space of vector fields with polynomial coefficients of degree k) is equal to $d\tilde{\pi}(\mathfrak{sv}_n)$ for any $n \geq 1$.

So assume that $X = f(t,r,\zeta)\partial_t + g(t,r,\zeta)\partial_r + h(t,r,\zeta)\partial_\zeta$ satisfies

$$[X, \mathfrak{h}_{-1}] \subset \pi(\mathfrak{sv}_n) = \Big\langle t^{n+1}\partial_t + \frac{1}{2}(n+1)t^n r\partial_r + \frac{1}{4}(n+1)nt^{n-1}r^2\partial_\zeta, t^{n+1}\partial_r + (n+1)t^n r\partial_\zeta, t^{n+1}\partial_\zeta \Big\rangle. \tag{5.8}$$

In the following lines, C_1, C_2, C_3 are undetermined constants. Then (by comparing the coefficients of ∂_t)

$$f(t,r,\zeta) = C_1 t^{n+2}.$$

By inspection of the coefficients of ∂_r, one gets then

$$\partial_t g(t,r,\zeta) = \frac{C_1}{2}(n+1)(n+2)t^n r + C_2(n+2)t^{n+1}$$

so

$$g(t,r,\zeta) = \frac{C_1}{2}(n+2)t^{n+1}r + C_2 t^{n+2} + G(r,\zeta)$$

with an unknown polynomial $G(r,\zeta)$. But

$$[X, Y_{-\frac{1}{2}}] = \left(\frac{C_1}{2}(n+2)t^{n+1} + \partial_r G(r,\zeta)\right)\partial_r \bmod \partial_\zeta$$

so $\partial_r G(r,\zeta) = 0$.

Finally, by comparing the coefficients of ∂_ζ, one gets

$$[X, L_{-1}] = (n+2)C_1[t^{n+1}\partial_t + \frac{1}{2}(n+1)t^n r\partial_r] + C_2(n+2)t^{n+1}\partial_r + \partial_t h\, \partial_\zeta$$

so

$$\partial_t h(t,r,\zeta) = \frac{C_1}{4}(n+2)(n+1)nt^{n-1}r^2 + C_2(n+2)(n+1)t^n r + C_3(n+2)t^{n+1}$$

whence

$$h(t,r,\zeta) = \frac{C_1}{4}(n+2)(n+1)t^n r^2 + C_2(n+2)t^{n+1}r + C_3 t^{n+2} + H(r,\zeta)$$

where $H(r,\zeta)$ is an unknown polynomial. Also

$$[X, Y_{-\frac{1}{2}}] = \frac{C_1}{2}(n+2)t^{n+1}\partial_r + \frac{C_1}{2}(n+2)(n+1)t^n r\partial_\zeta + C_2(n+2)t^{n+1}\partial_\zeta + \partial_r H(r,\zeta)\partial_\zeta,$$

so $H = H(\zeta)$ does not depend on r; finally

$$[X, M_0] = \frac{dG(\zeta)}{d\zeta}\partial_r + \frac{dH(\zeta)}{d\zeta}\partial_\zeta$$

so $G = H = 0$. □

Remark. by modifying slightly the definition of $\mathfrak{sv}_0$, one gets related Lie algebras. For instance, substituting $L_0^\varepsilon := -t\partial_t - (1+\varepsilon)r\partial_r - (1+2\varepsilon)\zeta\partial_\zeta$ for L_0 leads to the 'polynomial part' of $\mathfrak{sv}_{1+2\varepsilon}$ (see Theorems 5.2 and 5.4 for explicit realizations of $\mathfrak{sv}_{1+2\varepsilon}$).

5.2 About Multi-Diagonal Differential Operators and Some Virasoro-Solvable Lie Algebras

This section may be skipped in a first approach. We introduce new wave equations and related realizations of $\mathfrak{sv}$ – and, more generally, of its family of deformations $\mathfrak{sv}_\varepsilon$, see Chap. 2 – that will also appear as a particular case of the general construction of Sect. 5.3.

The original remark that prompted the introduction of multi-diagonal differential operators in our context (see below for a definition) was that a certain 'constrained' Dirac equation is $\mathfrak{sv}$-invariant. Namely, consider the space $\mathbb{R}^3$ with coordinates r, t, ζ as in Proposition 1.10 or 1.11. We introduce the two-dimensional Dirac operator

$$\widetilde{\mathscr{D}}_0 = \begin{pmatrix} \partial_r & -2\partial_\zeta \\ \partial_t & -\partial_r \end{pmatrix}, \tag{5.9}$$

acting on spinors $\begin{pmatrix} \phi_1 \\ \phi_2 \end{pmatrix} \in (C^\infty(\mathbb{R}^3))^2$. It can be obtained from the Dirac-Lévy-Leblond operator $\mathscr{D}_0$ (see Sect. 8.3) by taking a formal Laplace transform with respect to the mass, so that $\mathscr{M}$ goes to ∂_ζ. The kernel of $\widetilde{\mathscr{D}}_0$ is given by the equations of motion obtained from the Lagrangian density

$$\left(\bar{\phi}_2(\partial_r\phi_1 - 2\partial_\zeta\phi_2) - \bar{\phi}_1(\partial_t\phi_1 - \partial_r\phi_2)\right)\, dt\, dr\, d\zeta.$$

Let $d\tilde{\pi}^\sigma_{\frac{1}{2}}$ be the Laplace transform with respect to $\mathscr{M}$ of the infinitesimal representation $d\pi^\sigma_{\frac{1}{2}}$ of $\mathfrak{sv}$ given in Sect. 8.3. Then $d\tilde{\pi}^\sigma_{\frac{1}{2}}(\mathfrak{sch})$ preserves the space of solutions of the equation $\widetilde{\mathscr{D}}_0 \begin{pmatrix} \phi_1 \\ \phi_2 \end{pmatrix} = 0$, $\phi_1, \phi_2 \in C^\infty(\mathbb{R}^3)$. Now, by computing $\widetilde{\mathscr{D}}_0 \left(d\tilde{\pi}^\sigma_{\frac{1}{2}}(X) \begin{pmatrix} \phi_1 \\ \phi_2 \end{pmatrix} \right)$ for $X \in \mathfrak{sv}$ and $\begin{pmatrix} \phi_1 \\ \phi_2 \end{pmatrix}$ in the kernel of $\widetilde{\mathscr{D}}_0$, it clearly appears (do check it!) that if one adds the constraint $\partial_\zeta\phi_1 = 0$, then $\widetilde{\mathscr{D}}_0 \left(d\tilde{\pi}^\sigma_{\frac{1}{2}}(X) \begin{pmatrix} \phi_1 \\ \phi_2 \end{pmatrix} \right) = 0$ for every $X \in \mathfrak{sv}$, and, what is more, the transformed spinor $\begin{pmatrix} \psi_1 \\ \psi_2 \end{pmatrix} = d\tilde{\pi}_{\frac{1}{2}}(X) \begin{pmatrix} \phi_1 \\ \phi_2 \end{pmatrix}$ also satisfies the same constraint $\partial_\zeta\psi_1 = 0$. (See also Definition 6.11 and Theorem 6.4 for a quantized version of this spinor field). One may realize this constraint by adding to the Lagrangian density the Lagrange multiplier term $(h\partial_\zeta\bar{\phi}_1 - \bar{h}\partial_\zeta\phi_1)\, dt\, dr\, d\zeta$. The new equations of motion read then $\nabla \begin{pmatrix} -h/2 \\ \phi_2 \\ -\phi_1 \end{pmatrix} = 0$, with

$$\nabla = \begin{pmatrix} 2\partial_\zeta & \partial_r & \partial_t \\ & 2\partial_\zeta & \partial_r \\ & & 2\partial_\zeta \end{pmatrix}. \tag{5.10}$$

This is our main example of a multi-diagonal differential operator. Quite generally, we shall say that a function is *multi-diagonal* if it is a function- or operator-valued matrix $M = (M_{i,j})_{0 \le i,j \le d-1}$ such that $M_{i,j} = M_{i+k,j+k}$ for every admissible triple of indices i, j, k. So M is defined for instance by the d independent coefficients $M_{0,0}, \ldots, M_{0,j}, \ldots, M_{0,d-1}$, with $M_{0,j}$ located on the j*-shifted diagonal.*

An obvious generalization in d dimensions leads to the following definition.

Definition 5.2. Let ∇^d be the $d \times d$ matrix differential operator of order one, acting on d-uples of functions $H = \begin{pmatrix} h_0 \\ \vdots \\ h_{d-1} \end{pmatrix}$ on $\mathbb{R}^d$ with coordinates $t = (t_0, \ldots, t_{d-1})$, given by

$$\nabla^d = \begin{pmatrix} \partial_{t_{d-1}} & \partial_{t_{d-2}} & \cdots & \partial_{t_1} & \partial_{t_0} \\ 0 & \ddots & & & \partial_{t_1} \\ \vdots & & \ddots & & \vdots \\ & & & & \partial_{t_{d-2}} \\ 0 & \cdots & & 0 & \partial_{t_{d-1}} \end{pmatrix} \tag{5.11}$$

So ∇^d is upper-triangular, with coefficients $\nabla^d_{i,j} = \partial_{t_{i-j+d-1}}$, $i \le j$. The kernel of ∇^d is defined by a system of equations linking $h_0, \ldots, h_{d-1}$. The set of differential operators of order one is of the form

$$X = X_1 + \Lambda = \left(\sum_{i=0}^{d-1} f_i(t) \partial_{t_i} \right) \otimes \mathrm{Id} + \Lambda, \quad \Lambda = (\lambda_{i,j}) \in Mat_{d \times d}(C^\infty(\mathbb{R}^d)) \tag{5.12}$$

preserves the equation $\nabla^d H = 0$, forms a Lie algebra, that is much too large for our purpose.

Suppose now (this is a very restrictive condition) that $\Lambda = \mathrm{diag}(\lambda_0, \ldots, \lambda_{d-1})$ is diagonal. Since ∇^d is an operator with constant coefficients, $[X_1, \nabla^d]$ has no term of zero order, whereas $[\Lambda, \nabla^d]_{i,j} = \lambda_i \partial_{t_{i-j+d-1}} - \partial_{t_{i-j+d-1}} \lambda_j$ $(i \le j)$ does have terms of zero order in general. One possibility to solve this constraint, motivated by the preceding examples (see for instance the representation $d\tilde{\pi}_\lambda$ in Chap. 1), but also by the theory of scaling in statistical physics (see commentary following Proposition 1.5), is to impose $\lambda_i = \lambda_i(t_0)$, $i = 0, \ldots, d-2$, and $\lambda_{d-1} = 0$. Since $[X, \nabla^d]$ is of first order, preserving $\mathrm{Ker}\nabla^d$ is equivalent to a relation of the type $[X, \nabla^d] = A\nabla^d$, with $A = A(X) \in Mat_{d \times d}(C^\infty(\mathbb{R}^d))$. Then the matrix operator $[X_1, \nabla^d]$ is upper-triangular, and multi-diagonal, so this must also hold for $A\nabla^d - [\Lambda, \nabla^d]$. By looking successively at the coefficients of $\partial_{t_{d-1-l}}$ on the l-shifted diagonals, $l = 0, \ldots, d-1$, one sees easily that A must also be upper-triangular and multi-diagonal, and that one must have $\lambda_i(t_0) - \lambda_{i+1}(t_0) = \lambda(t_0)$ for a certain function λ independent of i, so $\Lambda = \begin{pmatrix} (d-1)\lambda & & & \\ & \ddots & & \\ & & \lambda & \\ & & & 0 \end{pmatrix}$. Also, denoting by $a_0 = A_{0,0}, \ldots, a_{d-1} = A_{0,d-1}$ the coefficients of the first line of the matrix A, one obtains:

$$\partial_{t_i} f_j = 0 \quad (i > j); \tag{5.13}$$

$$a_0 = \partial_{t_0} f_0 + (d-1)\lambda = \partial_{t_1} f_1 + (d-2)\lambda = \ldots = \partial_{t_{d-1}} f_{d-1}; \tag{5.14}$$

$$a_i = \partial_{t_0} f_i = \partial_{t_1} f_{i+1} = \ldots = \partial_{t_{d-i-1}} f_{d-1} \quad (i = 1, \ldots, d-1). \tag{5.15}$$

In particular, f_0 depends only on t_0.

From all these considerations follows quite naturally the following definition. We let $\Lambda_0 \in Mat_{d\times d}(\mathbb{R})$ be the diagonal matrix $\Lambda_0 = \begin{pmatrix} d-1 & & \\ & \ddots & \\ & & 0 \end{pmatrix}$.

Lemma 5.3. *Let* $\mathfrak{md}^d_\varepsilon$ $(\varepsilon \in \mathbb{R})$ *be the set of differential operators of order one of the type*

$$X = \left(f_0(t_0)\partial_{t_0} + \sum_{i=1}^{d-1} f_i(t)\partial_{t_i} \right) \otimes \mathrm{Id} - \varepsilon f_0'(t_0) \otimes \Lambda_0 \tag{5.16}$$

preserving $\mathrm{Ker}\nabla^d$.

Then $\mathfrak{md}_\varepsilon$ *forms a Lie algebra.*

Proof. Let $\mathscr{X}$ be the Lie algebra of vector fields X of the form

$$X = X_1 + X_0 = \left(f_0(t_0)\partial_{t_0} + \sum_{i=1}^{d-1} f_i(t)\partial_{t_i} \right) \otimes \mathrm{Id} + \Lambda,$$

$$\Lambda = \mathrm{diag}(\lambda_i)_{i=0,\ldots,d-1} \in Mat_{d\times d}(C^\infty(\mathbb{R}^d))$$

preserving $\mathrm{Ker}\nabla^d$. Then the set $\{Y = \sum_{i=0}^{d-1} f_i(t)\partial_{t_i} \mid \exists\lambda \in Mat_{d\times d}(C^\infty(\mathbb{R}^d)), Y + \Lambda \in \mathscr{X}\}$ of the differential parts of order one of the elements of $\mathscr{X}$ forms a Lie algebra, say $\mathscr{X}_1$. Define

$$\mathscr{X}_1^\varepsilon := \left\{ \sum_{i=0}^{d-1} f_i(t)\partial_{t_i} - \varepsilon f_0'(t) \otimes \Lambda_0 \mid \sum_{i=0}^{d-1} f_i(t)\partial_{t_i} \in \mathscr{X}_1 \right\}.$$

Let $Y = (\sum f_i(t)\partial_{t_i}) \otimes \mathrm{Id} - \varepsilon f_0'(t) \otimes \Lambda_0, Z = (\sum g_i(t)\partial_{t_i}) \otimes \mathrm{Id} - \varepsilon g_0'(t) \otimes \Lambda_0$ be two elements of $\mathscr{X}_1^{\varepsilon}$: then

$$[Y, Z] = ((f_0 g_0' - f_0' g_0)(t_0)\partial_{t_0} + \dots) \otimes \mathrm{Id} - \varepsilon (f_0 g_0' - f_0' g_0)'(t_0) \otimes \Lambda_0$$

belongs to $\mathscr{X}_1^{\varepsilon}$, so $\mathscr{X}_1^{\varepsilon}$ forms a Lie algebra. Finally, $\mathfrak{mo}_{\varepsilon}^d$ is the Lie subalgebra of $\mathscr{X}_1^{\varepsilon}$ consisting of all differential operators preserving $\mathrm{Ker}\nabla^d$. □

It is quite possible to give a family of generators and relations for $\mathfrak{mo}_{\varepsilon}^d$. The surprising fact, though, is the following: for $d \geq 4$, one finds by solving the equations that f_0'' is necessarily zero if $\varepsilon \neq 0$ (see proof of Theorem 5.2). So in any case, the only Lie algebra that deserves to be considered for $d \geq 4$ is $\mathfrak{mo}_0^d$.

The algebras $\mathfrak{mo}_{\varepsilon}^d$ $(d = 2, 3)$, $\mathfrak{mo}_0^d$ $(d \geq 4)$ are semi-direct products of a Lie subalgebra isomorphic to $\mathrm{Vect}(S^1)$, with generators

$$\mathscr{L}_{f_0}^{(0)} = (-f_0(t_0)\partial_{t_0} + \dots) \otimes \mathrm{Id} - \varepsilon f_0'(t_0) \otimes \Lambda_0$$

and commutators $[\mathscr{L}_{f_0}^{(0)}, \mathscr{L}_{g_0}^{(0)}] = \mathscr{L}_{f_0' g_0 - f_0 g_0'}^{(0)}$, with a nilpotent Lie algebra consisting of all generators with coefficient of ∂_{t_0} vanishing. When $d = 2, 3$, one recovers realizations of the familiar Lie algebras $\mathrm{Vect}(S^1) \ltimes \mathscr{F}_{1+\varepsilon}$ and $\mathfrak{sv}_{\varepsilon}$ (the latter being defined in Definition 2.2).

Theorem 5.2 (structure of $\mathfrak{mo}_{\varepsilon}^d$).

1. *(case $d = 2$). Put $t = t_0, r = t_1$: then* $\mathfrak{mo}_{\varepsilon}^2 = \langle \mathscr{L}_f^{(0)}, \mathscr{L}_g^{(1)} \rangle_{f,g \in C^{\infty}(S^1)}$ *with*

$$\mathscr{L}_f^{(0)} = \left(-f(t)\partial_t - (1+\varepsilon)f'(t)r\partial_r\right) \otimes \mathrm{Id} + \varepsilon f'(t) \otimes \begin{pmatrix} 1 & \\ & 0 \end{pmatrix}, \tag{5.17}$$

$$\mathscr{L}_g^{(1)} = -g(t)\partial_r. \tag{5.18}$$

It is isomorphic to $\mathrm{Vect}(S^1) \ltimes \mathscr{F}_{1+\varepsilon}$.

2. *(case d=3). Put $t = t_0, r = t_1, \zeta = t_2$: then* $\mathfrak{mo}_{\varepsilon}^3 = \langle \mathscr{L}_f^{(0)}, \mathscr{L}_g^{(1)}, \mathscr{L}_h^{(2)} \rangle_{f,g,h \in C^{\infty}(S^1)}$ *with*

$$\mathscr{L}_f^{(0)} = \left(-f(t)\partial_t - (1+\varepsilon)f'(t)r\partial_r - \left[(1+2\varepsilon)f'(t)\zeta + \frac{1+\varepsilon}{2} f''(t)r^2\right]\partial_{\zeta}\right) \otimes \mathrm{Id}$$
$$+ \varepsilon f'(t) \otimes \begin{pmatrix} 2 & & \\ & 1 & \\ & & 0 \end{pmatrix}, \tag{5.19}$$

$$\mathscr{L}_g^{(1)} = -g(t)\partial_r - g'(t)r\partial_{\zeta}, \tag{5.20}$$

$$\mathscr{L}_h^{(2)} = -h(t)\partial_{\zeta}. \tag{5.21}$$

The Lie algebra obtained by taking the modes

$$L_n = \mathscr{L}^{(0)}_{t^{n+1}},\ Y_m = \mathscr{L}^{(1)}_{t^{m+1+\varepsilon}},\ M_p = \mathscr{L}^{(2)}_{t^{p+1+2\varepsilon}} \tag{5.22}$$

is isomorphic to $\mathfrak{sv}_{1+2\varepsilon}$. In particular, the differential parts give three independent copies of the vector field representation $d\tilde{\pi}$ of $\mathfrak{sv}$ when $\varepsilon = -\frac{1}{2}$.

3. *(case $\varepsilon = 0, d \geq 2$) Then $\mathfrak{md}^d_\varepsilon \simeq \mathrm{Vect}(S^1) \otimes \mathbb{R}[\eta]/\eta^d$ is generated by the*

$$\mathscr{L}^{(k)}_g = -g(t_0)\partial_{t_k} - \sum_{i=1}^{d-1-k} g^{(i)}(t_0) t_1^{i-1}$$

$$\times \left(\frac{1}{i!} t_1 \partial_{t_{i+k}} + \frac{1}{(i-1)!} \sum_{j=2}^{d-i-k} t_j \partial_{t_{i+j+k-1}} \right), \quad g \in C^\infty(S^1) \tag{5.23}$$

$k = 0, \ldots, d-1$, with commutators $[\mathscr{L}^{(i)}_g, \mathscr{L}^{(j)}_h] = \mathscr{L}^{(i+j)}_{g'h-gh'}$ if $i + j \leq d - 1$, 0 else.

Proof. Let $X = -\left(f_0(t_0)\partial_{t_0} + \sum_{i=1}^{d-1} f_i(t)\partial_{t_i} \right) \otimes \mathrm{Id} + \varepsilon f_0'(t) \otimes \Lambda_0$: a set of necessary and sufficient conditions for X to be in $\mathfrak{md}^d_\varepsilon$ has been given before Lemma 5.3, namely

$$\partial_{t_i} f_j = 0 \text{ if } i > j,$$

$$(1+\varepsilon(d-1)) f_0'(t_0) = \partial_{t_1} f_1(t_0, t_1) + \varepsilon(d-2) f_0'(t_0) = \ldots = \partial_{t_{d-1}} f_{d-1}(t_0, \ldots, t_{d-1})$$

and

$$\partial_{t_0} f_i = \partial_{t_1} f_{i+1} = \ldots = \partial_{t_{d-i-1}} f_{d-1} \quad (i = 1, \ldots, d-1).$$

Solving successively these equations yields

$$f_i(t_0, \ldots, t_i) = (1 + \varepsilon i) f_0'(t_0) \,.\, t_i + f_i^{[1]}(t_0, \ldots, t_{i-1}), \quad i \geq 1; \tag{5.24}$$

$$f_i^{[1]}(t_0, \ldots, t_{i-1}) = \partial_{t_0} f_1^{[1]}(t_0) \,.\, t_{i-1} + (1+\varepsilon) f_0''(t_0) \int_0^{t_{i-1}} t_1 dt_{i-1} + f_i^{[2]}(t_0, \ldots, t_{i-2}); \tag{5.25}$$

At the next step, the relation $\partial_{t_0} f_2 = \partial_{t_1} f_3$ yields the equation

$$(1 + 2\varepsilon) f_0''(t_0) \,.\, t_2 + (f_1^{[1]})''(t_0) \,.\, t_1 + 2(1+\varepsilon) f_0'''(t_0) \,.\, t_1 + (f_2^{[2]})'(t_0)$$

$$= (1+\varepsilon) f_0''(t_0) \,.\, t_2 + \partial_{t_1} f_3^{[2]}(t_0, t_1)$$

which has no solution as soon as $\varepsilon \neq 0$ and $f_0'' \neq 0$. So, as we mentioned without proof before the theorem, the most interesting case is $\varepsilon = 0$ when $d \geq 4$.

The previous computations completely solve the cases $d = 2$ and $d = 3$. So let us suppose that $d \geq 4$ and $\varepsilon = 0$.

Then, by solving the next equations, one sees by induction that $f_0, \ldots, f_{d-1}$ may be expressed in terms of d arbitrary functions of t_0, namely, $f_0 = f_0^{[0]}, f_1^{[1]}, f_2^{[2]}, \ldots, f_{d-1}^{[d-1]}$, and that generators satisfying $f_i^{[i]} = 0$ for every $i \neq k$, k fixed, are necessarily of the form

$$f_k^{[k]}(t_0)\partial_{t_k} + \sum_{j=1}^{d-1-k} g_{k+j}(t_0, \ldots, t_j)\partial_{t_{k+j}}$$

for functions g_{k+j} that may be expressed in terms of $f_k^{[k]}$ and its derivatives.

One may then easily check that $\mathscr{L}^{(k)}_{-f_k^{[k]}}$ is of this form and satisfies the conditions for being in $\mathfrak{mod}^d_\varepsilon$, so we have proved that the $\mathscr{L}_f^{(k)}, k = 0, \ldots, d-1, f \in C^\infty(S^1)$, generate $\mathfrak{mod}^d_\varepsilon$.

All there remains to be done is to check for commutators. Since $\mathscr{L}_f^{(i)}$ is homogeneous of degree $-i$ for the Euler-type operator $\sum_{k=0}^{d-1} k t_k \partial_{t_k}$, one necessarily has $[\mathscr{L}_f^{(i)}, \mathscr{L}_g^{(j)}] = \mathscr{L}^{(i+j)}_{C(f,g)}$ for a certain function C (depending on f and g) of the time-coordinate t_0. One gets immediately $[\mathscr{L}_f^{(0)}, \mathscr{L}_g^{(0)}] = \mathscr{L}^{(0)}_{f'g-fg'}$. Next (supposing $l > 0$), since

$$\mathscr{L}_g^{(0)} = -\sum_{i=0}^{l-1} E_i^0(g)\partial_{t_i} - (g'(t_0)t_l + F_l^0(t_0, \ldots, t_{l-1}))\partial_{t_l} + \ldots$$

where $E_i^0(g), i = 0, \ldots, l-1$ do not depend on t_l, and

$$\mathscr{L}_h^{(l)} = -h(t_0)\partial_{t_l} + \ldots,$$

one gets $[\mathscr{L}_g^{(0)}, \mathscr{L}_h^{(l)}] = (gh' - g'h)(t_0)\partial_{t_l} + \ldots$, so $[\mathscr{L}_g^{(0)}, \mathscr{L}_h^{(l)}] = \mathscr{L}^{(l)}_{g'h-gh'}$. Considering now $k, l > 0$, then one has

$$\mathscr{L}_g^{(k)} = -\sum_{i=0}^{l-1} E_i^k(g)\partial_{t_{i+k}} - (h'(t_0)t_l + F_l^k(t_0, \ldots, t_{l-1}))\partial_{t_{l+k}}$$

where $E_i^k(g), i = 0, \ldots, l-1$, do not depend on t_l, and a similar formula for $\mathscr{L}_h^{(l)}$, which give together the right formula for $[\mathscr{L}_g^{(k)}, \mathscr{L}_h^{(l)}]$. □

Let us come back to the original motivation, that is, finding new representations of $\mathfrak{sv}$ arising in a geometric context. Denote by $d\pi^{(3,0)}$ the realization of $\mathfrak{sv}$ given in Theorem 5.2.

Definition 5.4. Let $d\pi^\nabla$ be the infinitesimal representation of $\mathfrak{sv}$ on the space $\mathscr{H}^\nabla \simeq (C^\infty(\mathbb{R}^2))^3$ with coordinates t, r, defined by

$$d\tilde{\rho}(\mathcal{L}_f) = \left(-f(t)\partial_t - \frac{1}{2}f'(t)r\partial_r\right) \otimes \mathrm{Id} + f'(t) \otimes \begin{pmatrix} -1 & & \\ & -\frac{1}{2} & \\ & & 0 \end{pmatrix} \tag{5.26}$$

$$+\frac{1}{2}f''(t)r \otimes \begin{pmatrix} 0 & 1 & 0 \\ & 0 & 1 \\ & & 0 \end{pmatrix} + \frac{1}{4}f'''(t)r^2 \otimes \begin{pmatrix} 0 & 0 & 1 \\ & 0 & 0 \\ & & 0 \end{pmatrix}; \tag{5.27}$$

$$d\tilde{\rho}(\mathcal{Y}_f) = -f(t)\partial_r \otimes \mathrm{Id} + f'(t) \otimes \begin{pmatrix} 0 & 1 & 0 \\ & 0 & 1 \\ & & 0 \end{pmatrix} + f''(t)r \otimes \begin{pmatrix} 0 & 0 & 1 \\ & 0 & 0 \\ & & 0 \end{pmatrix}; \tag{5.28}$$

$$d\tilde{\rho}(\mathcal{M}_f) = f'(t) \otimes \begin{pmatrix} 0 & 0 & 1 \\ & 0 & 0 \\ & & 0 \end{pmatrix}. \tag{5.29}$$

Proposition 5.5. *For every $X \in \mathfrak{sv}$, $d\pi^{\nabla}(X) \circ \nabla - \nabla \circ d\pi^{(3,0)}(X) = 0$.*

Proof. Let $X \in \mathfrak{sv}$; put $d\pi^{(3,0)}(X) = -(f_0(t)\partial_t + f_1(t,r)\partial_r + f_2(t,r,\zeta)\partial_\zeta) \otimes \mathrm{Id} - f_0'(t) \otimes \begin{pmatrix} 1 & & \\ & \frac{1}{2} & \\ & & 0 \end{pmatrix}$.

The computations preceding Lemma 5.3 prove that $[d\pi^{(3,0)}(X), \nabla^d] = A(X)\nabla^d$, $A(X)$ being the upper-triangular, multi-diagonal matrix defined by

$$A(X)_{0,0} = \partial_\zeta f_2,\ A(X)_{0,1} = \partial_r f_2,\ A(X)_{0,2} = \partial_t f_2.$$

Hence one has

$$A(\mathcal{L}_f) = \frac{r}{2}f''(t) \begin{pmatrix} 0 & 1 & 0 \\ & 0 & 1 \\ & & 0 \end{pmatrix} + \frac{r^2}{4}f'''(t) \begin{pmatrix} 0 & 0 & 1 \\ & 0 & 0 \\ & & 0 \end{pmatrix},$$

$$A(\mathcal{Y}_g) = g'(t) \begin{pmatrix} 0 & 1 & 0 \\ & 0 & 1 \\ & & 0 \end{pmatrix} + rg''(t) \begin{pmatrix} 0 & 0 & 1 \\ & 0 & 0 \\ & & 0 \end{pmatrix},$$

and

$$A(\mathcal{M}_h) = f'(t) \otimes \begin{pmatrix} 0 & 0 & 1 \\ & 0 & 0 \\ & & 0 \end{pmatrix}.$$

Hence the result. □

Remark. Consider the affine space

$$\mathscr{H}_\nabla^{aff} = \{\nabla + \begin{pmatrix} g_0 & g_1 & g_2 \\ & g_0 & g_1 \\ & & g_0 \end{pmatrix} \mid g_0, g_1, g_2 \in C^\infty(S^1 \times \mathbb{R}^2)\}.$$

Then one may define an infinitesimal left-and-right action $d\sigma$ of $\mathfrak{sv}$ on $\mathscr{H}_\nabla^{aff}$ by putting

$$d\sigma(X)(\nabla + V) = d\pi^\nabla(X) \circ (\nabla + V) - (\nabla + V) \circ d\pi^{(3,0)}(X),$$

but the action is simply linear this time, since $d\pi^\nabla \circ \nabla = \nabla \circ d\pi^{(3,0)}(X)$. So this action is not very interesting and doesn't give anything new.

5.3 Coinduced Representations of $\mathfrak{sv}$

We shall generalize in this paragraph this second type of representations to the case of $\mathfrak{sv}$. Note that although we have two natural graduations on $\mathfrak{sv}$, the one given by the structure of Cartan prolongation is most useful here since $\mathfrak{sv}_{-1}$ is commutative (see [2]).

Let $d\rho$ be a representation of $\mathfrak{sv}_0 = \langle L_0, Y_{\frac{1}{2}}, M_1 \rangle$ into a vector space $\mathscr{H}_\rho$. Then $d\rho$ can be trivially extended to $\mathfrak{sv}_+ = \oplus_{i \geq 0} \mathfrak{sv}_i$ by setting $d\rho(\sum_{i>0} \mathfrak{sv}_i) = 0$. Let $\mathfrak{fsv} = \oplus_{i \geq -1} \mathfrak{sv}_i \subset \mathfrak{sv}$ be the subalgebra of 'formal' vector fields: in the vector field representation $d\tilde{\pi}$, the image of $\mathfrak{fsv}$ is the subset of vector fields that are polynomial in the time coordinate.

Let us now define the representation of $\mathfrak{fsv}$ coinduced from $d\rho$.

Definition 5.6 (coinduced representations). The *ρ-formal density module* $(\tilde{\mathscr{H}}_\rho, d\tilde{\rho})$ is the coinduced module

$$\tilde{\mathscr{H}}_\rho = \mathrm{Hom}_{\mathscr{U}(\mathfrak{sv}_+)}(\mathscr{U}(\mathfrak{fsv}), \mathscr{H}_\rho) \tag{5.30}$$

$$= \{\phi : \mathscr{U}(\mathfrak{fsv}) \to \mathscr{H}_\rho \text{ linear} \mid \phi(U_0 V) = d\rho(U_0).\phi(V),$$

$$U_0 \in |\mathscr{U}(\mathfrak{sv}_+)| V \in \mathscr{U}(\mathfrak{fsv})\}\} \tag{5.31}$$

with the natural action of $\mathscr{U}(\mathfrak{fsv})$ on the right

$$(d\tilde{\rho}(U).\phi)(V) = \phi(VU), \quad U, V \in \mathscr{U}(\mathfrak{fsv}). \tag{5.32}$$

By Poincaré-Birkhoff-Witt's theorem (see for instance [16] or [60]), this space can be identified with

$$\mathrm{Hom}(\mathscr{U}(\mathfrak{sv}_+) \setminus \mathscr{U}(\mathfrak{fsv}), \mathscr{H}_\rho) \simeq \mathrm{Hom}(\mathrm{Sym}(\mathfrak{sv}_{-1}), \mathscr{H}_\rho)$$

(linear maps from the symmetric algebra on $\mathfrak{sv}_{-1}$ into $\mathscr{H}_\rho$), and this last space is in turn isomorphic with the space $\mathscr{H}_\rho \otimes \mathbb{R}[[t,r,\zeta]]$ of $\mathscr{H}_\rho$-valued functions of t,r,ζ, through the application

$$\mathscr{H}_\rho \otimes \mathbb{R}[[t,r,\zeta]] \longrightarrow \mathrm{Hom}(\mathrm{Sym}(\mathfrak{sv}_{-1}), \mathscr{H}_\rho) \tag{5.33}$$

$$F(t,r,\zeta) \longrightarrow \phi_F : (U \to \partial_U F|_{t=0,r=0,\zeta=0}) \tag{5.34}$$

where ∂_U stands for the product derivative $\partial_{L_{-1}^j Y_{-\frac{1}{2}}^k M_0^l} = (-\partial_t)^j(-\partial_r)^k(-\partial_\zeta)^l$ (note our choice of signs!).

We shall really be interested in the action of $\mathfrak{fsv}$ on functions $F(t,r,\zeta)$ that we shall denote by $d\sigma_\rho$, or $d\sigma$ for short.

The above morphisms allow one to compute the action of $\mathfrak{fsv}$ on monomials through the equality

$$\left(\frac{\partial_t^j}{j!}\frac{\partial_r^k}{k!}\frac{\partial_\zeta^l}{l!}\right)|_{t=0,r=0,\zeta=0}(d\sigma(X).F) = \frac{(-1)^{j+k+l}}{j!k!l!}(d\tilde{\rho}(X).\phi_F)(L_{-1}^j Y_{-\frac{1}{2}}^k M_0^l) \tag{5.35}$$

$$= \frac{(-1)^{j+k+l}}{j!k!l!}\phi_F(L_{-1}^j Y_{-\frac{1}{2}}^k M_0^l X), \quad X \in \mathfrak{fsv}. \tag{5.36}$$

In particular,

$$\partial_t^j \partial_r^k \partial_\zeta^l|_{t=0,r=0,\zeta=0}(d\sigma(L_{-1}).F) = -\partial_t^{j+1}\partial_r^k\partial_\zeta^l|_{t=0,r=0,\zeta=0}F$$

so $d\sigma(L_{-1}).F = -\partial_t F$; similarly, $d\sigma(Y_{-\frac{1}{2}}).F = -\partial_r F$ and $d\sigma(M_0).F = -\partial_\zeta F$.

So one may assume that $X \in \mathfrak{sv}_+$: by the Poincaré-Birkhoff-Witt theorem, $L_{-1}^j Y_{-\frac{1}{2}}^k M_0^l X$ can be rewritten as $U + V$ with

$$U \in \mathfrak{sv}_{>0}\mathscr{U}(\mathfrak{fsv})$$

and

$$V = V_1 V_2, \quad V_1 \in \mathscr{U}(\mathfrak{sv}_0), V_2 \in \mathscr{U}(\mathfrak{sv}_{-1}).$$

Then $\phi_F(U) = 0$ by definition of $\tilde{\mathscr{H}_\rho}$, and $\phi_F(V)$ may easily be computed as $\phi_F(V) = d\rho(V_1) \otimes \partial_{V_2}|_{t=0,r=0,\zeta=0}F$.

Theorem 5.3. *Let $f \in \mathbb{R}[t]$, the coinduced representation $d\tilde{\rho}$ of $\mathfrak{sv}$ is given by the action of the following matrix differential operators on functions:*

$$d\tilde{\rho}(\mathscr{L}_f) = \left(-f(t)\partial_t - \frac{1}{2}f'(t)r\partial_r - \frac{1}{4}f''(t)r^2\partial_\zeta\right)\otimes \mathrm{Id}_{\mathscr{H}_\rho} + f'(t)d\rho(L_0)$$
$$+\frac{1}{2}f''(t)rd\rho(Y_{\frac{1}{2}}) + \frac{1}{4}f'''(t)r^2 d\rho(M_1); \tag{5.37}$$

$$d\tilde{\rho}(\mathscr{Y}_f) = (-f(t)\partial_r - f'(t)r\partial_\zeta)\otimes\mathrm{Id}_{\mathscr{H}_\rho} + f'(t)d\rho(Y_{\frac{1}{2}}) + f''(t)r\, d\rho(M_1); \tag{5.38}$$

$$d\tilde{\rho}(\mathscr{M}_f) = -f(t)\partial_\zeta \otimes \mathrm{Id}_{\mathscr{H}_\rho} + f'(t)\, d\rho(M_1). \tag{5.39}$$

Proof. One easily checks that these formulas define a representation of $\mathfrak{sv}$. Since $(L_{-1}, Y_{-\frac{1}{2}}, M_0, L_0, L_1, L_2)$ generated $\mathfrak{sv}$ as a Lie algebra, it is sufficient to check the above formulas for L_0, L_1, L_2 (they are obviously correct for $L_{-1}, Y_{-\frac{1}{2}}, M_0$).

Note first that M_0 is central in $\mathfrak{sv}$, so

$$\partial_\zeta^l(d\sigma(X).F) = d\sigma(X).(\partial_\zeta^l F).$$

Hence it will be enough to compute the action on monomials of the form $t^j r^l \otimes v$, $v \in \mathscr{H}_\rho$.

We shall give a detailed proof since the computations in $\mathscr{U}(\mathfrak{sv})$ are rather involved.

Let us first compute $d\sigma(L_0)$: one has

$$\begin{aligned}
&(-\partial_t)^j(-\partial_r)^k|_{t=0,r=0}(d\sigma(L_0).F)\\
&= \phi_F(L_{-1}^j Y_{-\frac{1}{2}}^k L_0)\\
&= \phi_F(L_{-1}^j L_0 Y_{-\frac{1}{2}}^k - \frac{k}{2}L_{-1}^j Y_{-\frac{1}{2}}^k)\\
&= \phi_F(L_0 L_{-1}^j Y_{-\frac{1}{2}}^k - (j+\frac{k}{2})L_{-1}^j Y_{-\frac{1}{2}}^k)\\
&= \left[d\rho(L_0)(-\partial_t)^j(-\partial_r)^k - (j+\frac{k}{2})(-\partial_t)^j(-\partial_r)^k\right]F(0)
\end{aligned}$$

so

$$d\sigma(L_0) = -t\partial_t - \frac{1}{2}r\partial_r + d\rho(L_0).$$

Next,

$$\begin{aligned}
\phi(L_{-1}^j Y_{-\frac{1}{2}}^k L_1) &= \phi\left(L_{-1}^j L_1 Y_{-\frac{1}{2}}^k - kL_{-1}^j Y_{\frac{1}{2}} Y_{-\frac{1}{2}}^{k-1} + \frac{k(k-1)}{2}L_{-1}^j Y_{-\frac{1}{2}}^{k-2}M_0\right)\\
&= \phi\left(\left(-2jL_0L_{-1}^{j-1} + j(j-1)L_{-1}^{j-1}\right)Y_{-\frac{1}{2}}^k\right) - k\phi\left(Y_{\frac{1}{2}}L_{-1}^j Y_{-\frac{1}{2}}^{k-1}\right)
\end{aligned}$$

$$
\begin{aligned}
&+ jk\phi\left(L_{-1}^{j-1}Y_{-\frac{1}{2}}^{k}\right) + \frac{k(k-1)}{2}\phi\left(L_{-1}^{j}Y_{-\frac{1}{2}}^{k-2}M_0\right) \\
&= \Big[(-2j)(-\partial_r)^k(-\partial_t)^{j-1}d\rho(L_0) + j(j-1)(-\partial_r)^k(-\partial_t)^{j-1} \\
&\quad - k(-\partial_t)^j(-\partial_r)^{k-1}d\rho(Y_{\frac{1}{2}}) + jk(-\partial_t)^{j-1}(-\partial_r)^k \\
&\quad - \frac{k(k-1)}{2}\partial_\zeta(-\partial_t)^j(-\partial_r)^{k-2}\Big]F(0)
\end{aligned}
$$

hence the result for $d\sigma(L_1)$.

Finally,

$$
\begin{aligned}
\phi\left(L_{-1}^{j}Y_{-\frac{1}{2}}^{k}L_2\right) &= \phi\left(L_{-1}^{j}L_2Y_{-\frac{1}{2}}^{k} - \frac{3}{2}kL_{-1}^{j}Y_{\frac{3}{2}}Y_{-\frac{1}{2}}^{k-1} + 3\frac{k(k-1)}{2}L_{-1}^{j}M_1Y_{-\frac{1}{2}}^{k-2}\right) \\
&= \phi\left(\left(-3jL_1L_{-1}^{j-1} + 3j(j-1)L_0L_{-1}^{j-2} - j(j-1)(j-2)L_{-1}^{j-3}\right)Y_{-\frac{1}{2}}^{k}\right) \\
&\quad - \frac{3}{2}k\phi\left(-2jY_{\frac{1}{2}}L_{-1}^{j-1}Y_{-\frac{1}{2}}^{k-1} + j(j-1)L_{-1}^{j-2}Y_{-\frac{1}{2}}^{k}\right) \\
&\quad + \frac{3}{2}k(k-1)\phi\left(M_1L_{-1}^{j}Y_{-\frac{1}{2}}^{k-2} - jM_0L_{-1}^{j-1}Y_{-\frac{1}{2}}^{k-2}\right) \\
&= \Big[\left(3j(j-1)d\rho(L_0)(-\partial_t)^{j-2} - j(j-1)(j-2)(-\partial_t)^{j-2}\right)(-\partial_r)^k \\
&\quad - \frac{3}{2}k\left(-2jd\rho(Y_{\frac{1}{2}})(-\partial_t)^{j-1}(-\partial_r)^{k-1} + j(j-1)(-\partial_t)^{j-2}(-\partial_r)^k\right) \\
&\quad + \frac{3}{2}k(k-1)\left(d\rho(M_1)(-\partial_t)^j(-\partial_r)^{k-2} + j\partial_\zeta(-\partial_t)^{j-1}(-\partial_r)^{k-2}\right)\Big] \\
&\quad \times F(0).
\end{aligned}
$$

Hence

$$
d\sigma(L_2) = -t^3\partial_t - \frac{3}{2}t^2r\partial_r - \frac{3}{2}tr^2\partial_\zeta + 3t^2d\rho(L_0) + \frac{3}{2}t^2r\, d\rho(Y_{\frac{1}{2}}) + \frac{3}{2}r^2\, d\rho(M_1).
$$

□

Remark. The formulas for the coinduced representation may be generalized to the family of deformed Lie algebras $\mathfrak{sv}_\varepsilon$ (see Sect. 2.1). Let us simply state the results:

Theorem 5.4. *Let $f \in \mathbb{R}[t]$, the coinduced representation $d\tilde{\rho}_\varepsilon$ of $\mathfrak{sv}_\varepsilon$ is given by the action of the following matrix differential operators on functions:*

$$d\tilde{\rho}(\mathcal{L}_f) = \left(-f(t)\partial_t - \frac{1+\varepsilon}{2} f'(t) r\partial_r - \varepsilon f'(t)\zeta\partial_\zeta - \frac{1+\varepsilon}{4} f''(t) r^2 \partial_\zeta\right) \otimes \mathrm{Id}_{\mathcal{H}_\rho}$$
$$+ f'(t) d\rho(L_0) + \frac{1+\varepsilon}{2} f''(t) r d\rho(Y_{\frac{1}{2}})$$
$$+ \left(\frac{1+\varepsilon}{4} f'''(t) r^2 + \varepsilon\zeta f''(t)\right) d\rho(M_1); \tag{5.40}$$

$$d\tilde{\rho}(\mathcal{Y}_f) = (-f(t)\partial_r - f'(t) r\partial_\zeta) \otimes \mathrm{Id}_{\mathcal{H}_\rho} + f'(t) d\rho(Y_{\frac{1}{2}}) + f''(t) r\, d\rho(M_1); \tag{5.41}$$

$$d\tilde{\rho}(\mathcal{M}_f) = -f(t)\partial_\zeta \otimes \mathrm{Id}_{\mathcal{H}_\rho} + f'(t)\, d\rho(M_1). \tag{5.42}$$

5.4 Examples of Coinduced Representations

The Schrödinger–Virasoro Lie algebra is shown in various places in the present work to act on the space of solutions of several partial differential equations: multi-diagonal differential operators (see Sect. 5.2); more interestingly, Schrödinger or Dirac operators (see Chap. 8). The action on Schrödinger operators is actually derived in a very straightforward way from that of the vector field representation $d\tilde{\pi}_\lambda$ (see Introduction) on functions.

All these representations are instances of coinduced representations. This is also the case for the Harish-Chandra modules of the intermediate series studied in [81]. On the other hand, looking for representations which are *not* coinduced representations (leaving aside of course the Verma modules of Chap. 4, see introduction to Sect. 5.3), one encounters the *coadjoint representation* of Chap. 3. This is actually the only example we know.

Let us show how all these examples fit (or do *not* fit) inside the general theory.

Example 3. Take $\mathcal{H}_{\rho_\lambda} = \mathbb{R}$, $d\rho_\lambda(L_0) = -\lambda$, $d\rho_\lambda(Y_{\frac{1}{2}}) = d\rho_\lambda(M_1) = 0$ $(\lambda \in \mathbb{R})$. Then $d\tilde{\rho}_\lambda = d\tilde{\pi}_\lambda$ (see Introduction for a definition of the vector field representation $d\tilde{\pi}_\lambda$).

Example 4. The linear part of the infinitesimal action on the affine space of Schrödinger operators (see Chap. 8) is given by the restriction of $d\tilde{\rho}_{-1}$ to functions of the type $g_0(t)r^2 + g_1(t)r + g_2(t)$.

Example 5. Take $\mathcal{H}_{\rho_\lambda} = \mathbb{R}^2$, $d\rho_\lambda(L_0) = \begin{pmatrix} 1/4 & \\ & -1/4 \end{pmatrix} - \lambda\mathrm{Id}$, $d\rho_\lambda(Y_{\frac{1}{2}}) = -\begin{pmatrix} 0 & 0 \\ 1 & 0 \end{pmatrix}$, $d\rho_\lambda(M_1) = 0$. Then the infinitesimal action $d\pi_\lambda^\sigma$ on the affine space of Dirac operators, see Sect. 8.3, is equal to $d\tilde{\rho}_\lambda$ (up to a Laplace transform in the mass).

Example 6. (action on multi-diagonal matrix differential operators) Take $\mathscr{H}_\rho = \mathbb{R}^3$,
$d\rho(L_0) = \begin{pmatrix} -1 & & \\ & -\frac{1}{2} & \\ & & 0 \end{pmatrix}$, $d\rho(Y_{\frac{1}{2}}) = d\rho(M_1) = 0$ on the one hand;

$$\mathscr{H}_\sigma = \mathbb{R}^3,\ d\rho(L_0) = \begin{pmatrix} -1 & & \\ & -\frac{1}{2} & \\ & & 0 \end{pmatrix},\ d\rho(Y_{\frac{1}{2}}) = \begin{pmatrix} 0 & 1 & 0 \\ & 0 & 1 \\ & & 0 \end{pmatrix}$$

and $d\rho(M_1) = \begin{pmatrix} 0 & 0 & 1 \\ & 0 & 0 \\ & & 0 \end{pmatrix}$ on the other. Then $d\pi^{(3,0)} = d\tilde{\rho}$ and $d\pi^{\nabla} = d\tilde{\sigma}$ (see Sect. 5.2).

Counterexample. The coadjoint representation is *not* a coinduced representation, as follows easily by comparing the formulas for the action of the Y and M generators of Theorem 3.1 and Theorem 5.3: the second derivative f'' does not appear in $\mathrm{ad}^*(Y_f)$, while it does in $d\tilde{\rho}(Y_f)$ for any representation ρ such that $d\rho(M_1) \neq 0$; if $d\rho(M_1) = 0$, then, on the contrary, there's no way to account for the first derivative f' in $\mathrm{ad}^*(M_f)$.

Remark. The problem of classifying all coinduced representations is hence reduced to the problem of classifying the representations $d\rho$ of the Lie algebra $\langle L_0, Y_{\frac{1}{2}}, M_1\rangle$. This is *a priori* an intractable problem (due to the non-semi-simplicity of this Lie algebra), even if one is satisfied with finite-dimensional representations. An interesting class of examples (to which examples 1 through 4 belong) is provided by extending a (finite-dimensional, say) representation $d\rho$ of the $(ax+b)$-type Lie algebra $\langle L_0, Y_{\frac{1}{2}}\rangle$ to $\langle L_0, Y_{\frac{1}{2}}, M_1\rangle$ by putting $d\rho(M_1) = d\rho(Y_{\frac{1}{2}})^2$. In particular, one may consider the spin s-representation $d\sigma$ of $\mathfrak{sl}(2,\mathbb{R})$, restrict it to the Borel subalgebra considered as $\langle L_0, Y_{\frac{1}{2}}\rangle$, 'twist' it by putting $d\sigma^\lambda := d\sigma + \lambda \mathrm{Id}$ and extend it to $\langle L_0, Y_{\frac{1}{2}}, M_1\rangle$ as we just explained.

Chapter 6
Vertex Representations

Two-dimensional conformal field theory – as explained in the Introduction – is a language introduced by A.A. Belavin, A.M. Polyakov and A.B. Zamolodchikov in the 1980s to describe the interactions between the physically relevant quantities of models at equilibrium and at the critical temperature. It rests fundamentally on the assumption that these interactions are invariant under conformal transformations. In two dimensions, this is translated into the following axioms for a *vertex algebra*, extending the *Wightman axioms* [110] (see [64] for a thorough introduction to vertex algebras):

1. Following the usual framework of quantum statistical physics, correlations are evaluated in terms of *vacuum state expectations*,
$$\langle X_1(x_1)\ldots X_n(x_n)\rangle = \langle 0|\hat{X}_1(x_1)\ldots\hat{X}_n(x_n)|0\rangle, \tag{6.1}$$
where $\hat{X}_1(x_1),\ldots,\hat{X}_n(x_n)$ are quantum operators, and $|0\rangle$ is a *vacuum state* belonging to a Fock space.
2. The physically relevant quantities may be rewritten in terms of tensor products $\hat{X}(z)\otimes\hat{Y}(\bar{z})$, where $\hat{X}(z) = \sum_{n\in\mathbb{Z}} X_n z^{-n}$, resp. $\hat{Y}(z) = \sum_{n\in\mathbb{Z}} Y_n \bar{z}^{-n}$ are *chiral operators*, resp. *anti-chiral operators*, depending only on z, resp. $\bar{z}$. Then the *causality property* for *chiral operators* translates into a commutation relation $[\hat{X}_1(z),\hat{X}_2(z')] = 0$ for $z\neq z'$, while chiral operators commute with anti-chiral operators.
3. The above support hypothesis for the Lie bracket of the operator-valued distributions $\hat{X}_1,\hat{X}_2$ may be rewritten as a *finite sum*
$$[\hat{X}_1(z),\hat{X}_2(z')] = \sum_{n=1}^{N}\hat{\Phi}_n(z')\delta^{(n-1)}(z-z'), \tag{6.2}$$
in terms of derivatives of the δ-function. Equivalently, the following *operator product expansion* (*OPE* for short) holds:

J. Unterberger and C. Roger, *The Schrödinger-Virasoro Algebra*, Theoretical and Mathematical Physics, DOI 10.1007/978-3-642-22717-2_6,

$$\hat{X}_1(z)\hat{X}_2(z') \sim \sum_{n=1}^{N} \frac{\hat{\Phi}_n(z')}{(z-z')^n}, \tag{6.3}$$

where the symbol "$\sim$" means: up to $O_{|z-z'|\to 0}(1)$.

4. The two Virasoro algebras, $\mathfrak{vir} = \langle L_n, n \in \mathbb{Z}\rangle$ and $\overline{\mathfrak{vir}} = \langle \overline{L_n}, n \in \mathbb{Z}\rangle$ act on the Fock space; the central generator acts as a constant, c, called the *central charge*. The generators L_{-1}, L_0, L_1, corresponding to the finite subalgebra $\mathfrak{sl}(2,\mathbb{R})$ which generates projective conformal transformations, leave invariant the vacuum state $|0\rangle$. Hence n-point functions $\langle X_1(x_1)\ldots X_n(x_n)\rangle$ are invariant under projective conformal transformations. This fixes already (up to a constant) the value of two- and three-point functions, as explained in Sect. 2.4.
5. Chiral operators belong to the algebra generated by *primary operators*, i.e. chiral operators whose commutation relations with $\mathfrak{vir}$ define a *tensor-density representation* as in Chap. 1. To be precise, a primary operator $\hat{X} = \sum_n \hat{X}_n z^{-n-\mu}$ has *conformal weight* μ if its components satisfy the commutation relations $[L_n, \hat{X}_m] = ((\mu-1)n - m)\hat{X}_{n+m}$, or equivalently, if they generate the tensor-density module $\mathcal{F}_{\mu-1}$. Note that the Virasoro field $L(z) = \sum_n L_n z^{-n-2}$ is *not* a primary field unless $c = 0$.

Primary fields in conformal field theory are usually constructed by means of creation and annihilation operators, or by means of *vertex operators* $e^{\alpha \int a}$, where the primary operator a is the derivative of the *free boson*,[1] with OPE $a(z)a(w) \sim \frac{1}{(z-w)^2}$. This holds true in particular – after suitable generalizations though – for the primary operators corresponding to the famous *unitary minimal models* which describe a number of physical systems at criticality (see Preface).

Let us consider now the case of a physical theory which would be invariant under the action of the Schrödinger–Virasoro group. Then the previous constructions may be generalized, with the following modifications:

1. The couple of variables $(z,\bar{z})$ disappear in favor of a single variable t representing the time coordinate, playing the same rôle as z in conformal field theory, so fields are automatically chiral. OPEs are written by considering Laurent series in the time coordinate.
2. Primary operators are assumed to behave covariantly under a *coinduced representation* of $\mathfrak{sv}$, or more precisely $\widetilde{\mathfrak{sv}} \supset \mathfrak{sv}$ (see below).
3. The vacuum state must be invariant under the Schrödinger group. The associated covariance of correlators still fixes two-point functions, but leaves one undetermined scaling function for three-point functions, as explained in Sect. 2.4.

It turns out to be difficult to find primary operators X with fixed mass $\mathcal{M}$, i.e. such that $[M_0, X(t)] = -\mathcal{M} X(t)$. On the other hand, using a Laplace transform

[1] Mind that some authors call free boson the field $\int a$ – whose logarithmic covariance is a Green kernel for the two-dimensional Laplacian –, others its derivative a, because it is primary, contrary to $\int a$.

with respect to the mass (see Preface and Introduction) so that formally $\mathcal{M} \rightsquigarrow \partial/\partial_\zeta$, one finds primary operators that are polynomial in ζ. However, as noted in Appendix A, considering ζ as an independent coordinate releases constraints on n-point functions due to covariance under the finite-dimensional Schrödinger group, which is bothersome. Hence it is natural to assume covariance under the generator N_0 in the Cartan subalgebra of $\mathfrak{conf}(3)$, which writes in the representation by conformal transformations $-r\partial_r - 2\zeta\partial_\zeta$ (see Sect. 2.2). Considering the brackets of this new vector field with the generators of $\mathfrak{sv}$ in their Fourier vector-field realization (see formulas at the end of the Introduction) leads to an extended Lie algebra $\overline{\mathfrak{sv}}$ which is introduced below.

Here is a summary of the most important results of the chapter.

We introduce the *extended Schrödinger–Virasoro algebra* in Sect. 6.1. It contains a new current $N(z)$ which has conformal weight 1, i.e. whose components behave as functions in $\mathcal{F}_0$. This new algebra has *three* independent central extensions, including the natural extension of the Virasoro cocycle, and the natural extension of the usual cocycle of $\mathcal{F}_0$, which turns $N(z)$ into the derivative of a free boson.

We construct a Schrödinger–Virasoro field and Schrödinger–Virasoro primary fields (defined in Sect. 6.2) in Sect. 6.3, out of the derivative of a *free boson* a, and a pair of charged *symplectic bosons* $(\bar{b}^+, \bar{b}^-)$. For this reason we call this construction the *ab-model*. These primary fields $\Phi_{j,k}$, polynomial in a and $\bar{b}$, and are called *polynomial fields*; the *generalized polynomial fields* ${}_\alpha\Phi_{j,k}$ also include a vertex operator $e^{\alpha \int a}$. We compute their two- and three-point functions in Sect. 6.4. The correlators are polynomial in the dual coordinate ζ. Hence, after an inverse Fourier transform, they define singular non-massive fields, which is unsatisfactory from a physical point of view.

Finally, we write out in Sect. 6.5 a conjectural construction of *massive fields* as an analytic continuation of series of polynomial or generalized polynomial fields, whose two-point and (in one case at least) three-point functions are shown to be those of a Schrödinger-covariant massive field.

Some tensor-invariant computations (see Sect. 2.4) are collected in Appendix A. This chapter is adapted from [114].

6.1 On the Extended Schrödinger–Virasoro Lie Algebra $\tilde{\mathfrak{sv}}$ and its Coinduced Representations

As is known from Chap. 2 (the 'no-go' theorem), there is no 'reasonable' Lie algebra including both the extension of the Schrödinger algebra $\mathfrak{sch}$ by the conformal algebra $\mathfrak{conf}(3)$ *and* the extension of $\mathfrak{sch}$ to the Schrödinger–Virasoro algebra $\mathfrak{sv}$. Yet $\mathfrak{sv}$ may be extended by adding to the currents $(\mathcal{Y}, \mathcal{M})$ another current $\mathcal{N}$ whose component N_0 (see Sect. 2.2) comes from $\mathfrak{conf}(3)$. This extra generator is known to fix the two-point function for Schrödinger-covariant fields in the Laplace-transformed coordinates (t, r, ζ), see Appendix to Chap. 11. It turns out that the covariance under the N-current for (quantized) primary fields is a reasonable additional requirement (in our construction, N_0 simply counts the charge of the states).

The definition of the *extended Schrödinger–Virasoro Lie algebra* $\tilde{\mathfrak{sv}}$ is rather straightforward.

Definition 6.1 (extended Schrödinger algebra).
Let $\overline{\mathfrak{sch}} \supset \mathfrak{sch}$ be the semi-direct product Lie algebra $\overline{\mathfrak{sch}} \simeq \langle N_0 \rangle \ltimes \mathfrak{sch}$, where N_0 acts as a derivation on $\mathfrak{sch}$, namely,

$$[N_0, L_{0,\pm 1}] = 0, \quad [N_0, Y_{\pm\frac{1}{2}}] = Y_{\pm\frac{1}{2}}, \quad [N_0, M_0] = 2M_0. \tag{6.4}$$

One obtains thus a 7-dimensional maximal parabolic Lie subalgebra of $\mathfrak{conf}_3$ (see [54]).

Definition 6.2 (extended Schrödinger–Virasoro Lie algebra).
Let $\overline{\mathfrak{sv}} \supset \mathfrak{sv}$ be the (abstract) Lie algebra generated by L_n, M_n, N_n $(n \in \mathbb{Z})$ and Y_m $(m \in \frac{1}{2} + \mathbb{Z})$ with the following additional brackets:

$$[L_n, N_p] = -pN_{n+p}, [N_n, N_p] = 0, \ [N_n, Y_p] = Y_{n+p}, \ [N_n, M_p] = 2M_{n+p} \tag{6.5}$$

Note that the $N_n, n \in \mathbb{Z}$, may be interpreted as a second L-conformal current with conformal weight 1.

Lemma 6.3. *1. Let*

$$\mathfrak{h} = \left\langle Y_m \mid m \in \frac{1}{2} + \mathbb{Z} \right\rangle \oplus \langle M_p \mid p \in \mathbb{Z} \rangle \tag{6.6}$$

and

$$\tilde{\mathfrak{h}} = \langle N_n \mid n \in \mathbb{Z} \rangle \oplus \mathfrak{h}. \tag{6.7}$$

Then $\mathfrak{h}$ and $\tilde{\mathfrak{h}}$ are Lie subalgebras of $\overline{\mathfrak{sv}}$ and one has the following double semi-direct product structure:

$$\tilde{\mathfrak{h}} = \langle N_n \mid n \in \mathbb{Z} \rangle \ltimes \mathfrak{h}, \quad \overline{\mathfrak{sv}} = \mathrm{Vect}(S^1) \ltimes \tilde{\mathfrak{h}}. \tag{6.8}$$

The Lie algebra $\tilde{\mathfrak{h}}$ is solvable.

2. *The Lie algebra $\overline{\mathfrak{sch}} = \langle N_0 \rangle \ltimes \mathfrak{sch}$ is a maximal Lie subalgebra of $\overline{\mathfrak{sv}}$.*
3. *The Lie algebra $\overline{\mathfrak{sv}}$ has three independent classes of central extensions given by the cocycles*

$$c_1(L_n, L_m) = \frac{1}{12} n(n^2 - 1)\delta_{n+m,0}; \tag{6.9}$$

$$c_2(N_n, N_m) = n\delta_{n+m,0}; \tag{6.10}$$

$$c_3(L_n, N_m) = n^2\delta_{n+m,0} \tag{6.11}$$

(the zero components of the cocycles have been omitted).

Proof. Points 1 and 2 are straightforward. Let us turn to the proof of point 3 (see introduction to Chap. 7 for prerequisites on cohomology if needed).

The Lie subalgebra $\mathfrak{sv}$ is known (see [50] or Chap. 7) to have only one class of central extensions given by the multiples of the Virasoro cocycle c_1; it extends straightforwardly by zero to $\overline{\mathfrak{sv}}$. Then any central cocycle c of $\overline{\mathfrak{sv}}$ which is non-trivial on the N-generators may be decomposed by L_0-homogeneity (see [43]) into the following components

$$c(N_m, N_p)=a_m\delta_{m+p,0}, c(N_m, M_p) = b_m\delta_{m+p,0}, c(L_m, N_p)=c_m\delta_{m+p,0} \quad (6.12)$$

The b_m are easily seen to vanish by applying the Jacobi relation to $[N_n, [Y_m, Y_p]]$ where $n + m + p = 0$. The same relation applied to $[L_n, [N_m, N_p]]$, respectively $[L_n, [L_m, N_p]]$, yields $pa_m = ma_p$, viz. $(n + m)(c_n - c_m) = (n - m)c_{n+m}$, hence $a_m = \kappa m$ and $c_m = \alpha m^2 + \beta m$ for some coefficients κ, α, β. The coefficient β may be set to zero by adding a constant to N_0. Finally, the two remaining cocycles are easily seen to be non-trivial and independent. □

Definition 6.4. Let $\overline{\mathfrak{sv}}_{c,\kappa,\alpha}$ be the central extension of $\mathfrak{sv}$ corresponding to the cocycle $cc_1 + \kappa c_2 + \alpha c_3$, i.e. such that

$$[L_n, L_m] = (n - m)L_{n+m} + \frac{1}{12}cn(n^2 - 1)\delta_{n+m,0}; \quad (6.13)$$

$$[N_n, N_m] = \kappa n\delta_{n+m,0}; \quad [L_n, N_m] = -mN_{n+m} + \alpha n^2\delta_{n+m,0}. \quad (6.14)$$

The Lie algebra $\overline{\mathfrak{sv}}$ is provided with a graduation δ (extending δ_2 for $\mathfrak{sv}$, see Definition 2.3) defined by

$$\delta(L_n) = nL_n,\ \delta(N_n) = nN_n,\ \delta(Y_m) = \left(m - \frac{1}{2}\right)Y_m,$$

$$\delta(M_n) = (n - 1)M_n \quad \left(n \in \mathbb{Z}, m \in \frac{1}{2} + \mathbb{Z}\right) \quad (6.15)$$

Note that $\delta = \mathrm{ad}(-\frac{1}{2}N_0 - L_0) = -\frac{1}{2}[N_0, .] - [L_0, .]$.

Set $\overline{\mathfrak{sv}}_n = \{X \in \overline{\mathfrak{sv}} \mid \delta(X) = nX\} = \langle L_n, N_n, Y_{n+\frac{1}{2}}, M_{n+1}\rangle$ for $n = 0, 1, 2, \ldots$ and $\overline{\mathfrak{sv}}_{-1} = \langle L_{-1}, Y_{-\frac{1}{2}}, M_0\rangle$. Note that we choose to exclude N_{-1} from $\overline{\mathfrak{sv}}_{-1}$ although $\delta(N_{-1}) = -N_{-1}$. Then $\widetilde{\mathfrak{fsv}} := \oplus_{n\geq -1}\overline{\mathfrak{sv}}_n$ is a Lie subalgebra of $\overline{\mathfrak{sv}}$. The subspace $\overline{\mathfrak{sv}}_{-1}$ is commutative and the Lie subalgebra $\overline{\mathfrak{sv}}_0 := \{X \in \overline{\mathfrak{sv}} \mid \delta(X) = 0\}$ is a double extension of the commutative Lie algebra $\langle Y_{\frac{1}{2}}, M_1\rangle \cong \mathbb{R}^2$ by L_0 and N_0 as follows:

$$\overline{\mathfrak{sv}}_0 = (\langle L_0\rangle \oplus \langle N_0\rangle) \ltimes \langle Y_{\frac{1}{2}}, M_1\rangle = \langle N_0\rangle \ltimes \mathfrak{sv}_0 \quad (6.16)$$

Namely, one has

$$[L_0, Y_{\frac{1}{2}}] = -\frac{1}{2}Y_{\frac{1}{2}},\ [L_0, M_1] = -M_1;$$

$$[N_0, L_0] = 0,\ [N_0, Y_{\frac{1}{2}}] = Y_{\frac{1}{2}},\ [N_0, M_1] = 2M_1. \quad (6.17)$$

Note that N_0 acts by conjugation as $-2L_0$ on $\overline{\mathfrak{sv}}_0$. Also, the adjoint action of $\overline{\mathfrak{sv}}_0$ preserves $\overline{\mathfrak{sv}}_{-1}$, so that $\overline{\mathfrak{sv}}_0 \oplus \overline{\mathfrak{sv}}_{-1} = \overline{\mathfrak{sv}}_0 \ltimes \overline{\mathfrak{sv}}_{-1}$ is a Lie algebra too. Actually, $\widetilde{\mathfrak{fsv}}$ appears to be the Cartan prolongation of $\overline{\mathfrak{sv}}_0 \ltimes \overline{\mathfrak{sv}}_{-1}$ (see Sect. 5.1): if one realizes $\overline{\mathfrak{sv}}_0 \ltimes \overline{\mathfrak{sv}}_{-1}$ as the following polynomial vector fields [2]

$$L_{-1} = -\partial_t, \; Y_{-\frac{1}{2}} = -\partial_r, \; M_0 = -\partial_\zeta \tag{6.18}$$

$$L_0 = -t\partial_t - \frac{1}{2} r\partial_r, \; N_0 = -r\partial_r - 2\zeta\partial_\zeta, \; Y_{\frac{1}{2}} = -t\partial_r - r\partial_\zeta, \; M_1 = -t\partial_\zeta \tag{6.19}$$

then the Lie algebra $\overline{\mathfrak{sv}}_{-1} \oplus \overline{\mathfrak{sv}}_0 \oplus \overline{\mathfrak{sv}}_1 \oplus \ldots$ defined inductively by

$$\overline{\mathfrak{sv}}_n := \{\mathscr{X} \in \mathscr{P}_n \mid [\mathscr{X}, \overline{\mathfrak{sv}}_{-1}] \subset \overline{\mathfrak{sv}}_{n-1}\}, \quad n \geq 1 \tag{6.20}$$

(where $\mathscr{P}_n$ stands for the vector space of homogeneous polynomial vector fields on $\mathbb{R}^3$ of degree $n+1$) defines a vector field realization of $\widetilde{\mathfrak{fsv}}$ which extends straightforwardly into a representation of $\overline{\mathfrak{sv}}$. Namely, let $f \in \mathbb{C}[t, t^{-1}]$: then

$$\mathscr{L}_f = -f(t)\partial_t - \frac{1}{2} f'(t) r\partial_r - \frac{1}{4} f''(t) r^2 \partial_\zeta \tag{6.21}$$

$$\mathscr{N}_f = -f(t)(r\partial_r + 2\zeta\partial_\zeta) - \frac{1}{2} f'(t) r^2 \partial_\zeta \tag{6.22}$$

$$\mathscr{Y}_f = -f(t)\partial_r - f'(t) r\partial_\zeta \tag{6.23}$$

$$\mathscr{M}_f = -f(t)\partial_\zeta \tag{6.24}$$

One recognizes the extension to $\overline{\mathfrak{sv}}$ of the vector field representation $d\tilde{\pi}$ of Chap. 1.

Let us now give the *coinduced representations* of $\widetilde{\mathfrak{fsv}}$. The work was done in Chap. 5 for the Lie algebra $\mathfrak{sv}$. The generalization to $\overline{\mathfrak{sv}}$ is only a matter of easy computations. Hence we merely recall the definition and give the results.

Let ρ be a representation of $\overline{\mathfrak{sv}}_0 = (\langle L_0\rangle \oplus \langle N_0\rangle) \ltimes \langle Y_{\frac{1}{2}}, M_1\rangle$ into a vector space $\mathscr{H}_\rho$. Then ρ can be trivially extended to $\overline{\mathfrak{sv}}_+ = \oplus_{i\geq 0}\overline{\mathfrak{sv}}_i$ by setting $\rho(\sum_{i>0}\overline{\mathfrak{sv}}_i) = 0$. Standard examples are provided:

(i) either by choosing a representation ρ of the $(ax+b)$-Lie algebra $\langle L_0, Y_{\frac{1}{2}}\rangle$ and extending it to $\overline{\mathfrak{sv}}_0$ by setting

$$\rho(N_0) = -2\rho(L_0) + \mu \mathrm{Id} \; (\mu \in \mathbb{R}), \quad \rho(M_1) = C\rho(Y_{\frac{1}{2}})^2 \; (C \in \mathbb{R}); \tag{6.25}$$

[2] Note that this realization was originally obtained in [54], where the generator denoted by N (reintroduced in Chap. 11) coincides with $L_0 - \frac{N_0}{2} = -t\partial_t + \zeta\partial_\zeta$.

(ii) or by choosing a representation ρ of the $(ax+b)$-Lie algebra $\langle L_0, M_1\rangle$ and extending it to $\overline{\mathfrak{sv}}_0$ by setting

$$\rho(N_0) = -2\rho(L_0) + \mu \mathrm{Id}\ (\mu \in \mathbb{R}), \quad \rho(Y_{\frac{1}{2}}) = 0. \tag{6.26}$$

Actually, one may show easily that finite-dimensional indecomposable representations of $\langle L_0, Y_{\frac{1}{2}}\rangle$ or $\langle L_0, M_1\rangle$ are given (up to the addition of a constant to L_0) by restricting any finite-dimensional representation of $\mathfrak{sl}(2,\mathbb{R})$ to its Borel subalgebra (i.e. to strictly upper-triangular matrices). (On the other hand, the classification of all indecomposable finite-dimensional representations of $\overline{\mathfrak{sv}}_0$ is probably a very difficult task).

It happens so that all examples considered in this monograph are obtained as in (i) or (ii).

One may now realized the coinduced module $(\tilde{\mathscr{H}}_\rho, \tilde\rho) = \mathrm{Hom}_{\mathscr{U}(\overline{\mathfrak{sv}}_+)}(\mathscr{U}(\widetilde{\mathfrak{fsv}}), \mathscr{H}_\rho)$ (see Theorem 5.3 for more details) in the following explicit way.

Theorem 6.1. *The $\tilde\rho$-action of $\widetilde{\mathfrak{fsv}}$ on the coinduced module $\tilde{\mathscr{H}}_\rho$ is isomorphic to the action of the following matrix differential operators on functions:*

$$\begin{aligned}
\tilde\rho(\mathscr{L}_f) &= \left(-f(t)\partial_t - \frac{1}{2}f'(t)r\partial_r - \frac{1}{4}f''(t)r^2\partial_\zeta\right) \otimes \mathrm{Id}_{\mathscr{H}_\rho} \\
&\quad + f'(t)\rho(L_0) + \frac{1}{2}f''(t)r\rho(Y_{\frac{1}{2}}) + \frac{1}{4}f'''(t)r^2\rho(M_1); \\
\tilde\rho(\mathscr{N}_f) &= \left(-f(t)(r\partial_r + 2\zeta\partial_\zeta) - \frac{1}{2}f'(t)r^2\partial_\zeta\right) \otimes \mathrm{Id}_{\mathscr{H}_\rho} + f(t)\rho(N_0) \\
&\quad + f'(t)r\rho(Y_{\frac{1}{2}}) + \left(\frac{1}{2}f''(t)r^2 + 2\zeta f'(t)\right)\rho(M_1); \\
\tilde\rho(\mathscr{Y}_f) &= (-f(t)\partial_r - f'(t)r\partial_\zeta) \otimes \mathrm{Id}_{\mathscr{H}_\rho} + f'(t)\rho(Y_{\frac{1}{2}}) + f''(t)r\,\rho(M_1); \\
\tilde\rho(\mathscr{M}_f) &= -f(t)\partial_\zeta \otimes \mathrm{Id}_{\mathscr{H}_\rho} + f'(t)\,\rho(M_1).
\end{aligned} \tag{6.27}$$

It may be extended into a representation of $\overline{\mathfrak{sv}}$ by simply extrapolating the above formulas to $f \in \mathbb{R}[t,t^{-1}]$.

The representations of $\overline{\mathfrak{sv}}$ thus obtained will be called *coinduced representations*.

6.2 The Schrödinger–Virasoro Primary Fields and the Superfield Interpretation of $\mathfrak{sv}$

Just as conformal fields are given by quantizing density modules in the Virasoro representation theory, we shall define in this section $\overline{\mathfrak{sv}}$-primary fields by quantizing the coinduced representations $\tilde\rho$ introduced in the previous section.

6.2.1 Definition of the Schrödinger–Virasoro Primary Fields

Our fundamental hypothesis is that correlators of $\overline{\mathfrak{sv}}$-primary fields $\langle \Phi_1(t_1, r_1, \zeta_1) \ldots \Phi_n(t_n, r_n, \zeta_n)\rangle$ should be singular only when some of the time coordinates coincide; this is confirmed by the computations of two- and three-point functions for scalar massive Schrödinger-covariant fields (see [50] or [54], or also Appendix, Chap. 12). Hence one is led to the following assumption:

A $\overline{\mathfrak{sv}}$-primary field $\Phi(t, r, \zeta)$ may be written as

$$\Phi(t, r, \zeta) = \sum_{\mu} \Phi^{(\mu)}(t, r, \zeta) e_{\mu}, \tag{6.28}$$

where $(e_{\mu})_{\mu=1,\ldots,\dim \mathscr{H}_\rho}$ is a basis of the representation space $\mathscr{H}_\rho$ (see Sect. 6.1) and

$$\Phi^{(\mu)}(t, r, \zeta) := \sum_{\xi} \Phi^{(\mu),\xi}(t, \zeta) r^{\xi} \tag{6.29}$$

where ξ varies in a denumerable set of real values which is bounded below (so that it is possible to multiply two such formal series) and stable with respect to translations by positive integers. It may have been more logical to decompose further $\Phi^{(\mu),\xi}(t, \zeta)$ as $\sum_{\sigma} \Phi^{(\mu),\xi,\sigma}(t)\zeta^{\sigma}$, as we shall occasionally do (see Sect. 6.3.2), but this leads to unnecessarily complicated notations and turns out to be mostly counter-productive. In any case, $\Phi^{(\mu),\xi}(t, \zeta)$ is to be seen as a ζ-indexed quantum field in the variable t, the latter playing the same role as the complex variable z of conformal field theory, implying the possibility of defining normal ordering, operator product expansions and so on. Note that the $\mathscr{H}_\rho$-components of the field Φ are written systematically *inside parentheses* in order to avoid any possible confusion with other indices.

Suppose now that $\overline{\mathfrak{sv}}$ (or any of its central extensions) acts on Φ by the coinduced representation $\tilde{\rho}$ of Theorem 6.1. This action decomposes naturally as an action on each field component $\Phi^{(\mu),\xi}$ as follows (where Einstein's summation convention is implied):

$$\begin{aligned}
[L_m, \Phi^{(\mu),\xi}(t, \zeta)] = &- t^{m+1}\partial_t \Phi^{(\mu),\xi}(t, \zeta) - \frac{\xi}{2}(m+1)t^m \Phi^{(\mu),\xi}(t, \zeta) \\
&- \frac{1}{4}(m+1)m t^{m-1} \partial_\zeta \Phi^{(\mu),\xi-2} \\
&+ (m+1)t^m \rho(L_0)^{\mu}_{\nu} \Phi^{(\nu),\xi}(t, \zeta) \\
&+ \frac{1}{2}(m+1)m t^{m-1} \rho(Y_{\frac{1}{2}})^{\mu}_{\nu} \Phi^{(\nu),\xi-1}(t, \zeta) \\
&+ \frac{1}{4}(m+1)m(m-1)t^{m-2} \rho(M_1)^{\mu}_{\nu} \Phi^{(\nu),\xi-2}(t, \zeta);
\end{aligned} \tag{6.30}$$

$$
\begin{aligned}
[N_m, \Phi^{(\mu),\xi}(t,\zeta)] = & - t^m(\xi + 2\zeta\partial_\zeta)\Phi^{(\mu),\xi}(t,\zeta) - \frac{m}{2} t^{m-1}\partial_\zeta \Phi^{(\mu),\xi-2}(t,\zeta) \\
& + t^m \rho(N_0)^\mu_\nu \Phi^{(\nu),\xi}(t,\zeta) + m t^{m-1}\rho(Y_{\frac{1}{2}})^\mu_\nu \Phi^{(\nu),\xi-1}(t,\zeta) \\
& + \frac{m(m-1)}{2} t^{m-2}\rho(M_1)^\mu_\nu \Phi^{(\nu),\xi-2}(t,\zeta) \\
& + 2m t^{m-1}\zeta\rho(M_1)^\mu_\nu \Phi^{(\nu),\xi}(t,\zeta);
\end{aligned} \tag{6.31}
$$

$$
\begin{aligned}
[Y_m, \Phi^{(\mu),\xi}(t,\zeta)] = & -t^{m+\frac{1}{2}}(\xi+1)\Phi^{(\mu),\xi+1}(t,\zeta) - \left(m+\frac{1}{2}\right) t^{m-\frac{1}{2}}\partial_\zeta \Phi^{(\mu),\xi-1}(t,\zeta) \\
& + \left(m+\frac{1}{2}\right) t^{m-\frac{1}{2}} \rho(Y_{\frac{1}{2}})^\mu_\nu \Phi^{(\nu),\xi}(t,\zeta) \\
& + \left(m+\frac{1}{2}\right)\left(m-\frac{1}{2}\right) t^{m-3/2}\rho(M_1)^\mu_\nu \Phi^{(\nu),\xi-1}(t,\zeta);
\end{aligned} \tag{6.32}
$$

$$
[M_m, \Phi^{(\mu),\xi}(t,\zeta)] = -t^m \partial_\zeta \Phi^{(\mu),\xi}(t,\zeta) + m t^{m-1}\rho(M_1)^\mu_\nu \Phi^{(\nu),\xi}(t,\zeta). \tag{6.33}
$$

In order to define $\overline{\mathfrak{sv}}_{c,\kappa,\alpha}$-primary fields, one needs first the following assumption: there exist four mutually local fields

$$
L(t) = \sum_{n\in\mathbb{Z}} L_n t^{-n-2}, \quad Y(t) = \sum_{n\in\mathbb{Z}+\frac{1}{2}} Y_n t^{-n-3/2}, \quad M(t) = \sum_{n\in\mathbb{Z}} M_n t^{-n-1},
$$

$$
N(t) = \sum_{n\in\mathbb{Z}} N_n t^{-n-1}
$$

with the following OPE's:

$$
L(t_1)L(t_2) \sim \frac{\partial L(t_1)}{t_1 - t_2} + \frac{2L(t_2)}{(t_1-t_2)^2} + \frac{c/2}{(t_1-t_2)^4}, \quad c \in \mathbb{R} \tag{6.34}
$$

so that L is a Virasoro field with central charge c;

$$
L(t_1)Y(t_2) \sim \frac{\partial Y(t_2)}{t_1-t_2} + \frac{\frac{3}{2}Y(t_2)}{(t_1-t_2)^2}, \quad L(t_1)M(t_2) \sim \frac{\partial M(t_2)}{t_1-t_2} + \frac{M(t_2)}{(t_1-t_2)^2} \tag{6.35}
$$

and

$$
L(t_1)N(t_2) \sim \frac{\partial M(t_2)}{t_1-t_2} + \frac{M(t_2)}{(t_1-t_2)^2} + \frac{\alpha}{(t_1-t_2)^3} \tag{6.36}
$$

so that Y (resp. M) is an L-primary field with conformal weight $\frac{3}{2}$ (resp. 1) and N is primary with conformal weight 1 up to the term $\frac{\alpha}{(t_1-t_2)^3}$ due to the central extension;

$$Y(t_1)Y(t_2) \sim \frac{\partial M}{t_1 - t_2} + \frac{2M(t_2)}{(t_1 - t_2)^2}, \quad Y(t_1)M(t_2) \sim 0, \quad M(t_1)M(t_2) \sim 0 \quad (6.37)$$

and

$$N(t_1)M(t_2) \sim \frac{2M(t_2)}{t_1 - t_2}, \quad N(t_1)Y(t_2) \sim \frac{Y(t_2)}{t_1 - t_2}, \quad N(t_1)N(t_2) \sim \frac{\kappa}{(t_1 - t_2)^2} \quad (6.38)$$

which all together yield in mode decomposition the centrally extended Lie algebra $\overline{\mathfrak{sv}}_{c,\kappa,\alpha}$.

We may now give the definition for a *ρ-$\overline{\mathfrak{sv}}$-primary field*. Note that we leave aside for the time being the essential condition which states that the values of the index ξ should be bounded from below; we shall actually see in Sect. 6.3.2 that our free field construction works only for fields $\Phi^{(\mu)} = \sum_\xi \Phi^{(\mu),\xi} r^\xi$ such that $\Phi^{(\mu),\xi} = 0$ for all negative indices ξ. For technical reasons that will be explained below, we shall also define $\mathfrak{sv}$-primary fields and $\langle N_0 \rangle \ltimes \mathfrak{sv}$-primary fields.

In the following definition, we call (following [64]) *mutually local fields* a set $X_1, \ldots, X_n$ of operator-valued formal series in t whose commutators $[X_i(t_1), X_j(t_2)]$ are distributions of finite order supported on the diagonal $t_1 = t_2$. In other words, the fields $X_1, \ldots, X_n$ have meromorphic operator-product expansions (OPE).

Definition 6.5 (primary fields).

1. ($\mathfrak{sv}$-primary fields)
   Let $\rho : \mathfrak{sv}_0 \to \mathscr{L}(\mathscr{H}_\rho)$ be a finite-dimensional representation of $\mathfrak{sv}_0 = \langle L_0 \rangle \ltimes \langle Y_{\frac{1}{2}}, M_1 \rangle$. A *$\rho$-$\mathfrak{sv}$-primary field* $\Phi(t, r, \zeta) = \sum_\mu \Phi^{(\mu)}(t, r, \zeta) e_\mu$ is given (at least in a formal sense) as an infinite series

$$\Phi^{(\mu)}(t, r, \zeta) = \sum_\xi \Phi^{(\mu),\xi}(t, \zeta) r^\xi$$

   where ξ varies in a denumerable set of real values which is stable with respect to integer translations, and the $\Phi^{(\mu),\xi}(t, \zeta)$ are mutually local fields with respect to the time variable t– which are also mutually local with the $\mathfrak{sv}$-fields $L(t), Y(t), M(t)$ – with the following OPE:

$$L(t_1)\Phi^{(\mu),\xi}(t_2, \zeta) \sim \frac{\partial \Phi^{(\mu),\xi}(t_2, \zeta)}{t_1 - t_2} + \frac{(\frac{1}{2}\xi)\Phi^{(\mu),\xi}(t_2, \zeta) - \rho(L_0)^\mu_\nu \Phi^{(\nu),\xi}(t_2, \zeta)}{(t_1 - t_2)^2}$$
$$+ \frac{\frac{1}{2}\partial_\zeta \Phi^{(\mu),\xi-2}(t_2, \zeta) - \rho(Y_{\frac{1}{2}})^\mu_\nu \Phi^{(\nu),\xi-1}(t_2, \zeta)}{(t_1 - t_2)^3}$$
$$- \frac{\frac{3}{2}\rho(M_1)^\mu_\nu \Phi^{(\nu),\xi-2}(t_2)}{(t_1 - t_2)^4} \quad (6.39)$$

$$Y(t_1)\Phi^{(\mu),\xi}(t_2,\zeta) \sim \frac{(1+\xi)\Phi^{(\mu),\xi+1}(t_2,\zeta)}{t_1-t_2} + \frac{\partial_\zeta \Phi^{(\mu),\xi-1}(t_2,\zeta) - \rho(Y_{\frac{1}{2}})^\mu_\nu \Phi^{(\nu),\xi}(t_2,\zeta)}{(t_1-t_2)^2} - \frac{2\rho(M_1)^\mu_\nu \Phi^{(\nu),\xi-1}(t_2,\zeta)}{(t_1-t_2)^3} \tag{6.40}$$

$$M(t_1)\Phi^{(\mu),\xi}(t_2,\zeta) \sim \frac{\partial_\zeta \Phi^{(\mu),\xi}(t_2,\zeta)}{t_1-t_2} - \frac{\rho(M_1)^\mu_\nu \Phi^{(\nu),\xi}(t_2,\zeta)}{(t_1-t_2)^2} \tag{6.41}$$

2. ($(\langle N_0\rangle \ltimes \mathfrak{sv}$-primary fields) Let $\bar\rho : \overline{\mathfrak{sv}}_0 = \langle N_0\rangle \ltimes \mathfrak{sv}_0 \to \mathscr{L}(\mathscr{H}_\rho)$ be a finite-dimensional representation of $\overline{\mathfrak{sv}}_0$, and ρ be the restriction of $\bar\rho$ to $\mathfrak{sv}_0$. A $\bar\rho$-$\langle N_0\rangle \ltimes$ $\mathfrak{sv}$*-primary field* $\Phi(t,r,\zeta)$ is a ρ-$\mathfrak{sv}$-primary field such that

$$[N_0, \Phi^{(\mu),\xi}(t,\zeta)] = (\xi + 2\zeta\partial_\zeta)\Phi^{(\mu),\xi}(t,\zeta) - \bar\rho(N_0)^\mu_\nu \Phi^{(\nu),\xi}(t,\zeta). \tag{6.42}$$

3. ($\overline{\mathfrak{sv}}$-primary fields) Let $\bar\rho : \overline{\mathfrak{sv}}_0 = \langle N_0\rangle \ltimes \mathfrak{sv}_0 \to \mathscr{L}(\mathscr{H}_\rho)$ be a finite-dimensional representation of $\overline{\mathfrak{sv}}_0$ and $\Omega : \mathscr{H}_\rho \to \mathscr{H}_\rho$ be a linear operator such that $[\rho(L_0),\Omega] = \Omega$, $[\rho(Y_{\frac{1}{2}}),\Omega] = [\rho(M_1),\Omega] = [\rho(N_0),\Omega] = 0$. Then a $(\bar\rho,\Omega)$-$\overline{\mathfrak{sv}}$*-primary field* is a $\bar\rho|_{\mathfrak{sv}_0}$-$\mathfrak{sv}$-primary field $\Phi(t,r,\zeta)$, local with N, such that

$$N(t_1)\Phi^{(\mu),\xi}(t_2,\zeta) \sim \frac{(\xi+2\zeta\partial_\zeta)\Phi^{(\mu),\xi}(t_2,\zeta) - \rho(N_0)^\mu_\nu \Phi^{(\nu),\xi}(t_2,\zeta)}{t_1-t_2} + \frac{\frac{1}{2}\partial_\zeta \Phi^{(\mu),\xi-2}(t_2,\zeta) - \rho(Y_{\frac{1}{2}})^\mu_\nu \Phi^{(\nu),\xi-1}(t_2,\zeta) - 2\zeta\rho(M_1)^\mu_\nu \Phi^{(\nu),\xi}(t_2,\zeta) - \Omega^\mu_\nu \Phi^{(\nu),\xi}(t_2,\zeta)}{(t_1-t_2)^2} - \frac{\rho(M_1)^\mu_\nu \Phi^{(\nu),\xi-2}(t_2,\zeta)}{(t_1-t_2)^3} \tag{6.43}$$

In the case $\Omega = 0$, we shall simply say that Φ is $\bar\rho$-$\overline{\mathfrak{sv}}$-primary.

Remark. Bear in mind that in these OPE and in all the following ones, ζ is considered only as a parameter, as we mentioned earlier.

The operator Ω for $\overline{\mathfrak{sv}}$-primary fields does not follow from the coinduction method. However, it appears in all our examples, including for the superfield $\mathscr{L}$ with components L, Y, M, N with the adjoint action of $\overline{\mathfrak{sv}}$ on itself (see Sect. 6.2.2 below).

Proposition 6.6. *Suppose Φ is a (ρ,Ω)-$\overline{\mathfrak{sv}}$-primary field. Then the adjoint action of $\overline{\mathfrak{sv}}$ on Φ is given by the formulas of Theorem 6.1 except for the action of the N-generators which are twisted as follows:*

$$\tilde{\rho}(N_f) = \left(-f(t)(r\partial_r + 2\zeta\partial_\zeta) - \frac{1}{2}f'(t)r^2\partial_\zeta\right) \otimes \mathrm{Id}_{\mathscr{H}_\rho} + f(t)\rho(N_0)$$
$$+ f'(t)r\rho(Y_{\frac{1}{2}}) + \left(\frac{1}{2}f''(t)r^2 + 2\zeta f'(t)\right)\rho(M_1) + f'(t)\Omega; \quad (6.44)$$

Proof. Straightforward computations. One may in particular check that the twisted representation is indeed a representation of $\overline{\mathfrak{sv}}$. □

Note that the usual conformal fields of weight λ are a particular case of this construction: they correspond to ρ-$\mathfrak{sv}$-conformal fields Φ with only one component $\Phi = \Phi^{(0)}(t)$, commuting with $\mathcal{N}, \mathcal{Y}, \mathcal{M}$, such that ρ is the one-dimensional character given by $\rho(L_0) = -\lambda$, $\rho(N_0) = \rho(Y_{\frac{1}{2}}) = \rho(M_1) = 0$.

6.2.2 A Superfield Interpretation

Similarly to the case of superconformal field theory (see [64], Sect. 5.9), one may consider the fields $L(t), Y(t), M(t), N(t)$ as four components of the same *superfield* $Z(t)$. To construct $Z(t)$, we first need to go over to the 'Heisenberg' point of view by setting

$$\bar{L}(t, r, \zeta) := e^{\zeta M_0} e^{rY_{-\frac{1}{2}}} L(t) e^{-rY_{-\frac{1}{2}}} e^{-\zeta M_0} \quad (6.45)$$

and similarly for $\bar{Y}, \bar{M}, \bar{N}$, the quantum generators $Y_{-\frac{1}{2}}$, resp. M_0 corresponding to the infinitesimal generators of space, resp. ζ-translations.

In the following, the sign ∂ alone always indicates a derivative with respect to time. Differences of coordinates are abbreviated as $t_{12} = t_1 - t_2$, $r_{12} = r_1 - r_2$, $\zeta_{12} = \zeta_1 - \zeta_2$.

Lemma 6.7. *1. The Heisenberg fields $\bar{L}, \bar{Y}, \bar{N}, \bar{M}$ read*

$$\bar{L}(t, r) = L(t) + \frac{1}{2}r\partial Y(t) + \frac{r^2}{4}\partial^2 M(t);$$
$$\bar{Y}(t, r) = Y(t) + r\partial M(t);$$
$$\bar{M}(t) = M(t);$$
$$\bar{N}(t, r, \zeta) = N(t) - rY(t) - \frac{r^2}{2}\partial M(t) - 2\zeta M(t). \quad (6.46)$$

2. Operator product expansions are given by the following formulas:

$$\bar{L}(t_1, r_1)\bar{L}(t_2, r_2) \sim \frac{\partial_{t_2}\bar{L}(t_2, r_2)}{t_{12}} + \frac{2\bar{L}(t_2, r_2) - \frac{1}{4}r_{12}\partial_{t_2}\bar{Y}(t_2, r_2)}{t_{12}^2} \quad (6.47)$$
$$-\frac{3}{2}\frac{r_{12}}{t_{12}^3}\bar{Y}(t_2, r_2) + \frac{\frac{c}{2} + \frac{3}{2}r_{12}^2\bar{M}(t_2)}{t_{12}^4};$$

$$\bar{L}(t_1,r_1)\bar{Y}(t_2,r_2) \sim \frac{\partial \bar{Y}(t_2,r_2)}{t_{12}} + \frac{\frac{3}{2}\bar{Y}(t_2,r_2) - \frac{1}{2}r_{12}\partial_{r_2}\bar{Y}(t_2,r_2)}{t_{12}^2} - 2\frac{r_{12}}{t_{12}^3}\bar{M}(t_2);$$

$$\bar{Y}(t_1,r_1)\bar{L}(t_2,r_2) \sim \frac{\partial_{r_2}\bar{L}(t_2,r_2)}{t_{12}} + \frac{3}{2}\frac{\bar{Y}(t_2,r_2)}{t_{12}^2} - \frac{2r_{12}\bar{M}(t_2)}{t_{12}^3};$$

$$\bar{L}(t_1,r_1)\bar{M}(t_2) \sim \frac{\partial \bar{M}(t_2)}{t_{12}} + \frac{\bar{M}(t_2)}{t_{12}^2}, \quad \bar{M}(t_1)\bar{L}(t_2,r_2) \sim \frac{\bar{M}(t_2)}{t_{12}^2};$$

$$\bar{L}(t_1,r_1)\bar{N}(t_2,r_2,\zeta_2) \sim \frac{\partial \bar{N}(t_2,r_2,\zeta_2)}{t_{12}} + \frac{\bar{N}(t_2,r_2,\zeta_2) - \frac{1}{2}r_{12}\partial_{r_2}\bar{N}(t_2,r_2,\zeta_2)}{t_{12}^2}$$
$$+\frac{1}{2}\frac{r_{12}^2\partial_\zeta \bar{N}(t_2,r_2,\zeta_2)}{t_{12}^3} + \frac{\alpha}{t_{12}^3};$$

$$\bar{N}(t_1,r_1,\zeta_1)\bar{L}(t_2,r_2) \sim \left(-\frac{r_{12}\partial_{r_2}\bar{L}(t_2,r_2)}{t_{12}} + \frac{-\frac{3}{2}r_{12}\bar{Y}(t_2,r_2) - 2\zeta_{12}\bar{M}(t_2)}{t_{12}^2}\right.$$
$$\left. + \frac{r_{12}^2\bar{M}(t_2)}{t_{12}^3}\right) + \frac{\bar{N}(t_2,r_2,\zeta_2)}{t_{12}^2};$$

$$\bar{Y}(t_1,r_1)\bar{Y}(t_2,r_2) \sim \frac{\partial \bar{M}(t_2)}{t_{12}} + \frac{2\bar{M}(t_2)}{t_{12}^2}, \quad \bar{Y}(t_1,r_1)\bar{M}(t_2) \sim \bar{M}(t_1)\bar{M}(t_2) \sim 0;$$

$$\bar{N}(t_1,r_1,\zeta_1)\bar{Y}(t_2,r_2) \sim \frac{-r_{12}\partial_{r_2}\bar{Y}(t_2,r_2) + \bar{Y}(t_2,r_2)}{t_{12}} - \frac{2r_{12}\bar{M}(t_2)}{t_{12}^2};$$

$$\bar{Y}(t_1,r_1)\bar{N}(t_2,r_2,\zeta_2) \sim \frac{-\bar{Y}(t_2,r_2)}{t_{12}} + \frac{2r_{12}\bar{M}(t_2)}{t_{12}^2};$$

$$\bar{N}(t_1,r_1,\zeta_1)\bar{M}(t_2) \sim \frac{2\bar{M}(t_2)}{t_{12}}, \quad \bar{M}(t_1)\bar{N}(t_2,r_2,\zeta_2) \sim \frac{\partial_{\zeta_2}\bar{N}(t_2,r_2,\zeta_2)}{t_{12}};$$

$$\bar{N}(t_1,r_1,\zeta_1)\bar{N}(t_2,r_2,\zeta_2) \sim \left(\frac{-(2\zeta_{12}\partial_{\zeta_2} + r_{12}\partial_{r_2})\bar{N}(t_2,r_2,\zeta_2)}{t_{12}}\right.$$
$$\left. +\frac{1}{2}\frac{r_{12}^2\partial_{\zeta_2}\bar{N}(t_2,r_2,\zeta_2)}{t_{12}^2}\right) + \frac{\kappa}{t_{12}^2}. \tag{6.48}$$

3. *A field $\Phi = \sum_\mu \Phi^{(\mu)} e_\mu$ is a (ρ, Ω)-$\overline{\mathfrak{sv}}$-primary field if and only if the following relations hold (we omit the argument (t_2, r_2, ζ_2) of the field Φ in the right-hand side of the equations):*

$$\bar{L}(t_1, r_1)\Phi^{(\mu)}(t_2, r_2, \zeta_2) \sim \frac{\partial_{t_2}\Phi^{(\mu)}}{t_{12}} - \frac{1}{2}\frac{r_{12}\partial_{r_2}\Phi^{(\mu)}}{t_{12}^2} - \frac{\rho(L_0)^\mu_\nu \Phi^{(\nu)}}{t_{12}^2}$$
$$+ \frac{\frac{1}{2}r_{12}^2\partial_\zeta \Phi^{(\mu)} + r_{12}\rho(Y_{\frac{1}{2}})^\mu_\nu \Phi^{(\nu)}}{t_{12}^3} - \frac{3}{2}\frac{r_{12}^2\rho(M_1)^\mu_\nu \Phi^{(\nu)}}{t_{12}^4}; \tag{6.49}$$

$$\bar{Y}(t_1, r_1)\Phi^{(\mu)}(t_2, r_2, \zeta_2) \sim \frac{\partial_{r_2}\Phi^{(\mu)}}{t_{12}} - \frac{r_{12}\partial_\zeta \Phi^{(\mu)}}{t_{12}^2} - \frac{\rho(Y_{\frac{1}{2}})^\mu_\nu \Phi^{(\nu)}}{t_{12}^2}$$
$$+ \frac{2r_{12}\rho(M_1)^\mu_\nu \Phi^{(\nu)}}{t_{12}^3}; \tag{6.50}$$

$$\bar{M}(t_1)\Phi^{(\mu)}(t_2, r_2, \zeta_2) \sim \frac{\partial_{\zeta_2}\Phi^{(\mu)}}{t_{12}} - \frac{\rho(M_1)^\mu_\nu \Phi^{(\nu)}}{t_{12}^2}; \tag{6.51}$$

$$\bar{N}(t_1, r_1, \zeta_1)\Phi^{(\mu)}(t_2, r_2, \zeta_2) \sim \frac{-(r_{12}\partial_{r_2} + 2\zeta_{12}\partial_{\zeta_2})\Phi^{(\mu)} - \rho(N_0)^\mu_\nu \Phi^{(\nu)}}{t_{12}}$$
$$+ \frac{\frac{1}{2}r_{12}^2\partial_{\zeta_2}\Phi^{(\mu)} + r_{12}\rho(Y_{\frac{1}{2}})^\mu_\nu \Phi^{(\nu)} + 2\zeta_{12}\rho(M_1)^\mu_\nu \Phi^{(\nu)} + \Omega^\mu_\nu \Phi^{(\nu)}}{t_{12}^2} - \frac{r_{12}^2\rho(M_1)^\mu_\nu \Phi^{(\nu)}}{t_{12}^3}. \tag{6.52}$$

Putting all this together, one gets:

Theorem 6.2. *Set $c = \kappa = \alpha = 0$. Then:*

(i) The four-dimensional field

$$Z(t, r, \zeta) = \begin{pmatrix} \bar{L} \\ \bar{Y} \\ \bar{M} \\ \bar{N} \end{pmatrix} (t, r, \zeta) \tag{6.53}$$

is ρ-$\mathfrak{sv}$-primary for the representation ρ defined by:

$$\rho(L_0) = \begin{pmatrix} -2 & & & \\ & -\frac{3}{2} & & \\ & & -1 & \\ & & & -1 \end{pmatrix}, \ \rho(Y_{\frac{1}{2}}) = \begin{pmatrix} 0 & -\frac{3}{2} & & \\ & 0 & -2 & \\ & & 0 & \\ & & & 0 \end{pmatrix}, \ \rho(M_1) = \begin{pmatrix} 0 & 0 & -1 & 0 \\ & 0 & & \\ & & 0 & \\ & & & 0 \end{pmatrix}. \tag{6.54}$$

(ii) It is not $\rho - \overline{\mathfrak{sv}}$-primary.

Proof. Straightforward computations. Note that $\rho(M_1)$ is proportional to $\rho(Y_{\frac{1}{2}})^2$, see the remarks preceding Theorem 6.1.

So what happened? Setting $\rho(N_0) = \begin{pmatrix} 0 & & & \\ & -1 & & \\ & & -2 & \\ & & & 0 \end{pmatrix}$, one gets a representation of $\overline{\mathfrak{sv}}_0 = (\langle L_0 \rangle \oplus \langle N_0 \rangle) \ltimes \langle Y_{\frac{1}{2}}, M_1 \rangle$ and Z looks $\rho - \widetilde{\mathfrak{sv}}$-primary, except for the last term $\frac{\bar{N}}{t_{12}^2}$ in the above OPE $\bar{N}.\bar{L}$. Fortunately, a supplementary matrix Ω as in Definition 6.5 (3) allows to take into account this term:

Theorem 6.3. *Set* $c = \kappa = \alpha = 0$. *Then* Z *is* (ρ, Ω)-$\overline{\mathfrak{sv}}$*-primary if one sets* $\Omega = \begin{pmatrix} 0 & 0 & 0 & -1 \\ & 0 & & \\ & & 0 & \\ & & & 0 \end{pmatrix}$.

Proof. Straightforward computations.

6.3 Construction by $U(1)$-Currents or $a\bar{b}$-Theory

Now that the definition of what is intended by $\overline{\mathfrak{sv}}$*-primary* has been completed, we proceed to give explicit examples. The rest of the article is devoted to the detailed analysis of a vertex algebra constructed out of two bosons (called : *$a\bar{b}$-model*) containing a representation of $\overline{\mathfrak{sv}}$ and $\overline{\mathfrak{sv}}$-primary fields of any L_0-weight.

6.3.1 Definition of the $\overline{\mathfrak{sv}}$-Fields

We shall use here a classical construction of current algebras given in all generality in [64]. Let $V = V_{\bar{0}} \oplus V_{\bar{1}}$ be a (finite-dimensional) super-vector space, with even generators $a^i, i = 1, \ldots, N$ for $V_{\bar{0}}$ and odd generators $b^{+,i}, b^{-,i}, i = 1, \ldots, M$ for $V_{\bar{1}}$ (supposed to be even-dimensional).

Definition 6.8. 1. The *bosonic supercurrents* associated with V (see [64], Sect. 3.5) are the mutually local N bosonic fields $a^i(z) = \sum_{n\in\mathbb{Z}} a_n^i z^{-n-1}$ and the $2M$ fermionic fields $b^{\pm,i}(z) = \sum_{n\in\mathbb{Z}} b_n^{\pm,i} z^{-n-1}$ with the following non-trivial OPE's:

$$a^i(z)a^j(w) \sim \frac{g^{i,j}}{(z-w)^2} \tag{6.55}$$

$$b^{\pm,i}(z)b^{\mp,j}(w) \sim \pm\frac{g^{i,j}}{(z-w)^2} \tag{6.56}$$

or equivalently, with the following non-trivial Lie brackets in mode decomposition

$$[a_n^i, a_m^j]_- = n\delta^{i,j}\delta_{n+m,0} \tag{6.57}$$

$$[b_n^{+,i}, b_m^{-,j}]_+ = n\delta^{i,j}\delta_{n+m,0}. \tag{6.58}$$

2. The *fermionic supercurrents* associated with V (see [64], Sects. 2.5 and 3.6) are the mutually local N fermionic fields $\bar{a}^i(z) = \sum_{n\in\mathbb{Z}} \bar{a}_n^i z^{-n-\frac{1}{2}}$ and the $(2M)$ bosonic fields $\bar{b}^{\pm,i}(z) = \sum_{n\in\mathbb{Z}} \bar{b}_n^{\pm,i} z^{-n-\frac{1}{2}}$ with the following non-trivial OPE's:

$$\bar{a}^i(z)\bar{a}^j(w) \sim \frac{\delta^{i,j}}{z-w} \tag{6.59}$$

$$\bar{b}^{\pm,i}(z)\bar{b}^{\mp,j}(w) \sim \pm\frac{\delta^{i,j}}{z-w} \tag{6.60}$$

or equivalently, with the following non-trivial Lie brackets in mode decomposition

$$[\bar{a}_n^i, \bar{a}_m^j]_+ = \delta^{i,j}\delta_{n+m,0} \tag{6.61}$$

$$[\bar{b}_n^{+,i}, \bar{b}_m^{-,j}]_- = \delta^{i,j}\delta_{n+m,0}. \tag{6.62}$$

Remark. The bosonic supercurrents $\bar{b}^{\pm,i}$ (with unusual parity considering their half-integer weight) are sometimes called *symplectic bosons* in the physical literature, see for instance [12, 30].

Proposition 6.9. (see [64], Sects. 3.5 and 3.6) *Consider the canonical Fock realization of the superalgebra generated by $a^i, b^{i,\pm}$ (obtained by requiring that $a^i, b^{i,\pm}$, $i \geq 0$, vanish on the vacuum vector $|0\rangle$). Then*

$$\langle 0 \mid a^i(z)a^j(w) \mid 0\rangle = \delta^{i,j}(z-w)^{-2}, \quad \langle 0 \mid b^{\pm,i}(z)b^{\mp,j}(w) \mid 0\rangle = \pm\delta^{i,j}(z-w)^{-2} \tag{6.63}$$

and

$$\langle 0 \mid \bar{a}^i(z)\bar{a}^j(w) \mid 0\rangle = \delta^{i,j}(z-w)^{-1}, \quad \langle 0 \mid \bar{b}^{\pm,i}(z)\bar{b}^{\mp,j}(w) \mid 0\rangle = \pm\delta^{i,j}(z-w)^{-1}. \tag{6.64}$$

One may build Virasoro fields out of these supercurrents, one for each type of current:

$$L_a(t) = \frac{1}{2} : a^2 : (t), \quad L_b(t) =: b^+b^- : (t) \tag{6.65}$$

with central charge 1, viz. -2;

$$L_{\bar{a}}(t) = -\frac{1}{2} : \bar{a}(\partial\bar{a}) : (t), \quad L_{\bar{b}}(t) = \frac{1}{2}\left(: \bar{b}^+\partial\bar{b}^- : (t) - : \bar{b}^-\partial\bar{b}^+ : (t)\right) \tag{6.66}$$

with central charge $\frac{1}{2}$, viz. -1.

For the appropriate Virasoro field, the bosonic supercurrents a^i, b^i are primary with conformal weight 1, while the fermionic supercurrents $\bar{a}^i, \bar{b}^{\pm,i}$ are primary with conformal weight $\frac{1}{2}$. The simplest way to construct a Lie algebra isomorphic to an appropriately centrally extended $\overline{\mathfrak{sv}}$ with these generating fields is the following:

Definition 6.10 ($a\bar{b}$-theory).
Let $V = V_{\bar{0}} \oplus V_{\bar{1}}$ with $V_{\bar{0}} = \mathbb{R}a$ and $V_{\bar{1}} = \mathbb{R}b^+ \oplus \mathbb{R}b^-$. Then $\overline{\mathfrak{sv}}_{(0,-1,0)}$-fields L, N, Y, M may be defined as follows:

$$L = L_a + L_{\bar{b}} \quad \text{with zero central charge;} \tag{6.67}$$

$$N = -:\bar{b}^+\bar{b}^-: \quad \text{with central charge } -1; \tag{6.68}$$

$$Y = :a\bar{b}^+:; \tag{6.69}$$

$$M = \frac{1}{2} :(\bar{b}^+)^2:. \tag{6.70}$$

Let us first check explicitly that one retrieves the OPE (6.34), (6.35), (6.36), (6.37), (6.38) with this definition:

$$:a^2:(t_1):a\bar{b}^+:(t_2) \sim \frac{2:a(t_1)\bar{b}^+(t_2):}{(t_1-t_2)^2} \sim \frac{2:a\bar{b}^+:(t_2)}{(t_1-t_2)^2} + \frac{2:\partial a\bar{b}^+:(t_2)}{t_1-t_2};$$

$$:\bar{b}^+\partial\bar{b}^-:(t_1)\ :a\bar{b}^+:(t_2) \sim \frac{:\bar{b}^+(t_1)a(t_2):}{(t_1-t_2)^2} \sim \frac{:a\bar{b}^+:(t_2)}{(t_1-t_2)^2} + \frac{:a\partial\bar{b}^+:(t_2)}{t_1-t_2};$$

$$:\bar{b}^-\partial\bar{b}^+:(t_1)\ :a\bar{b}^+:(t_2) \sim -\frac{:\partial\bar{b}^+(t_1)a(t_2):}{t_1-t_2} \sim -\frac{:\partial\bar{b}^+a:(t_2)}{t_1-t_2}$$

so $L(t_1)Y(t_2) \sim \frac{\partial Y(t_2)}{t_1-t_2} + \frac{\frac{3}{2}Y(t_2)}{(t_1-t_2)^2}$.
Similarly,

$$:a^2:(t_1)\ :(\bar{b}^+)^2:(t_2) \sim 0;$$

$$:\bar{b}^+\partial\bar{b}^-:(t_1):(\bar{b}^+)^2:(t_2) \sim \frac{2:\bar{b}^+(t_1)\bar{b}^+(t_2):}{(t_1-t_2)^2} \sim \frac{2:(\bar{b}^+)^2(t_2)}{(t_1-t_2)^2} + 2\frac{:\partial\bar{b}^+\bar{b}^+:(t_2)}{t_1-t_2};$$

$$:\bar{b}^-\partial\bar{b}^+:(t_1):(\bar{b}^+)^2:(t_2) \sim -2\frac{:\partial\bar{b}^+\bar{b}^+:(t_2)}{t_1-t_2}$$

so $L(t_1)M(t_2) \sim \frac{\partial M(t_2)}{t_1-t_2} + \frac{M(t_2)}{(t_1-t_2)^2}$.

Finally,

$$Y(t_1)Y(t_2) = :a\bar{b}^+:(t_1):a\bar{b}^+:(t_2) \sim \frac{:\bar{b}^+(t_1)\bar{b}^+(t_2):}{(t_1-t_2)^2}$$

$$\sim \frac{:(\bar{b}^+)^2:(t_2)}{(t_1-t_2)^2} + \frac{:\bar{b}^+\partial\bar{b}^+:(t_2)}{t_1-t_2} = \frac{2M(t_2)}{(t_1-t_2)^2} + \frac{\partial M(t_2)}{t_1-t_2} \quad (6.71)$$

and $Y(t_1)M(t_2) \sim 0$, $M(t_1)M(t_2) \sim 0$, so one is done for the $\mathfrak{sv}$-fields L, Y, M. Then

$$N(t_1)N(t_2) =: \bar{b}^+\bar{b}^-:(t_1):\bar{b}^+\bar{b}^-:(t_2) \sim -\frac{1}{(t_1-t_2)^2}$$

(the terms of order one cancel each other);

$$:\bar{b}^+\partial\bar{b}^-:(t_1):\bar{b}^+\bar{b}^-:(t_2) \sim \frac{:\partial(\bar{b}^+\bar{b}^-):(t_2)}{t_1-t_2} + \frac{:\bar{b}^+\bar{b}^-:(t_2)}{(t_1-t_2)^2}$$

$$:\bar{b}^-\partial\bar{b}^+:(t_1):\bar{b}^+\bar{b}^-:(t_2) \sim -\frac{:\partial(\bar{b}^+\bar{b}^-):(t_2)}{t_1-t_2} - \frac{:\bar{b}^+\bar{b}^-:(t_2)}{(t_1-t_2)^2}$$

hence $L(t_1)N(t_2) \sim \frac{\partial N(t_2)}{t_1-t_2} + \frac{N(t_2)}{t_1-t_2}$; finally,

$$N(t_1)Y(t_2) = -:\bar{b}^+\bar{b}^-:(t_1):a\bar{b}^+:(t_2) \sim \frac{Y(t_2)}{t_1-t_2}$$

and

$$N(t_1)M(t_2) = -\frac{1}{2}:\bar{b}^+\bar{b}^-:(t_1):(\bar{b}^+)^2:(t_2) \sim \frac{2M(t_2)}{t_1-t_2}.$$

Definition 6.11. The *constrained 3D-Dirac equation* (or: constrained Dirac equation for short) is the set of following equations for a spinor field $(\phi_1, \phi_2) = (\phi_1(t,r,\zeta), \phi_2(t,r,\zeta))$ on $\mathbb{R}^3$:

$$\partial_r\phi_0 = \partial_t\phi_1 \quad (6.72)$$

$$\partial_r\phi_1 = \partial_\zeta\phi_0 \quad (6.73)$$

$$\partial_\zeta\phi_1 = 0. \quad (6.74)$$

It is the same constrained Dirac equation as the one defined in Sect. 5.2. We shall now quantize this $\mathfrak{sv}$-invariant classical equation, and show that quantized spinors define a $\tilde{\mathfrak{sv}}$-primary field.

Theorem 6.4. *1. The space of solutions of the constrained 3D-Dirac equation is in one-to-one correspondence with the space of triples (h_0^-, h_0^+, h_1) of functions of t only: a natural bijection may be obtained by setting*

$$\phi_0(t,r,\zeta) = (h_0^-(t) + \zeta h_0^+(t)) + r h_1(t) + \frac{r^2}{2}\partial h_0^+(t) \quad (6.75)$$

$$\phi_1(t,r,\zeta) = \int_0^t h_1(u)\,du + r h_0^+(t) \quad (6.76)$$

2. *Put*

$$\Phi^{(0)}(t,r,\zeta) = (\bar{b}^-(t) + \zeta\bar{b}^+(t)) + r a(t) + \frac{r^2}{2}\partial\bar{b}^+(t) \quad (6.77)$$

and

$$\Phi^{(1)}(t,r,\zeta) = \left(\int a\right)(t) + r\bar{b}^+(t) \quad (6.78)$$

where

$$\int a = -\sum_{n\neq 0} a_n \frac{t^{-n}}{n} + a_0 \log t + \pi_0, \quad [a_0,\pi_0] = 1 \quad (6.79)$$

is the logarithmic bosonic field defined for instance in [19]. Then $\Phi := \begin{pmatrix} \Phi^{(0)} \\ \Phi^{(1)} \end{pmatrix}$ *is a* ρ-$\overline{\mathfrak{sv}}$-*primary field, where* ρ *is the two-dimensional character defined by*

$$\rho(L_0) = \begin{pmatrix} -\frac{1}{2} & \\ & 0 \end{pmatrix}, \quad \rho(N_0) = \begin{pmatrix} 1 & \\ & 0 \end{pmatrix}, \quad \rho(Y_{\frac{1}{2}}) = \rho(M_1) = 0. \quad (6.80)$$

3. *The two-point functions* $\mathscr{C}^{\mu,\nu}(t_1,r_1,\zeta_1;t_2,r_2,\zeta_2) := \langle 0|\ \Phi^{(\mu)}(t_1,r_1,\zeta_1)\Phi^{(\nu)}(t_2,r_2,\zeta_2)\ |0\rangle$, $\mu,\nu = 0,1$, *are given by*

$$\mathscr{C}^{0,0} = \frac{1}{t}\left(\zeta - \frac{r^2}{2t}\right), \ \mathscr{C}^{0,1} = \mathscr{C}^{1,0} = r, \ \mathscr{C}^{1,1} = \ln t \quad (6.81)$$

where $t = t_1 - t_2, r = r_1 - r_2, \zeta = \zeta_1 - \zeta_2$.

Remark. The free boson $\int a$ is not conformal in the usual sense since it contains a logarithmic term, contrary to the vertex operators built as exponentials of $\int a$ that we shall use in the following sections. In this very particular case, one needs to consider a_0, π_0 as a pair of usual annihilation/creation operators in order for the scalar product $\langle 0|\ (\int a)(t_1)(\int a)(t_2)\ |0\rangle$ to make sense, so that a_0 and π_0 are adjoint to each other. The usual definition of the void state $|0\rangle$ is different.

Proof. 1. Let (ϕ,ψ) be a solution of the constrained Dirac equation. Then $\partial_r^2\psi = \partial_t\partial_\zeta\psi = 0$ so

$$\psi(t,r,\zeta) = \psi_0(t) + r\psi_1(t). \quad (6.82)$$

On the other hand, $\partial_\zeta^2\phi = \partial_\zeta\partial_r\psi = 0$, $\partial_\zeta\phi = \psi_1$ and $\partial_r\phi = \partial_t\psi_0 + r\partial_t\psi_1$, hence, by putting together everything,

$$\phi(t,r,\zeta) = \phi^{00}(t) + \psi_1(t)\zeta + \psi_0'(t)r + \psi_1'(t)\frac{r^2}{2}. \quad (6.83)$$

Now one just needs to set $h_0^- := \phi^{00}, h_0^+ = \psi_1$ and $h_1 = \psi_0'$.

2. This follows directly from Definition 6.5 once one has established the following easy relations

$$L(t_1)\partial\bar{b}^+(t_2) \sim \frac{\partial^2\bar{b}^+(t_2)}{t_1-t_2} + \frac{3}{2}\frac{\partial\bar{b}^+(t_2)}{(t_1-t_2)^2} + \frac{\bar{b}^+(t_2)}{(t_1-t_2)^3} \tag{6.84}$$

$$N(t_1)(\bar{b}^-(t_2)+\zeta\bar{b}^+(t_2)) \sim \frac{-\bar{b}^-(t_2)+\zeta\bar{b}^+(t_2)}{t_1-t_2} \tag{6.85}$$

$$N(t_1)a(t_2) \sim 0 \tag{6.86}$$

$$N(t_1)\partial\bar{b}^+(t_2) \sim \partial_{t_2}\left(\frac{\bar{b}^+(t_2)}{t_1-t_2}\right) = \frac{\partial\bar{b}^+(t_2)}{t_1-t_2} + \frac{\bar{b}^+(t_2)}{(t_1-t_2)^2} \tag{6.87}$$

$$Y(t_1)(\bar{b}^-(t_2)+\zeta\bar{b}^+(t_2)) \sim \frac{a(t_2)}{t_1-t_2} \tag{6.88}$$

$$Y(t_1)a(t_2) \sim \frac{\partial\bar{b}^+(t_2)}{t_1-t_2} + \frac{\bar{b}^+(t_2)}{(t_1-t_2)^2} \tag{6.89}$$

$$Y(t_1)\bar{b}^+(t_2) \sim 0 \tag{6.90}$$

together with the fact that $\bar{b}^\pm$, resp. a, are L-conformal with conformal weight $\frac{1}{2}$ (resp. 1).

3. Straightforward.

In particular, the classical constrained Dirac equation is $\mathfrak{sv}$-invariant, as already pointed out in Sect. 5.2. Unfortunately, one can hardly say that this is an interesting physical equation.

We give thereafter two other examples. They exhaust all possibilities of $\overline{\mathfrak{sv}}$-primary linear fields of this model and are only given for the sake of completeness.

Lemma 6.12. *1. The trivial field $\bar{b}^+(t)$ is a ρ-Schrödinger-conformal field, where ρ is the one-dimensional character defined by*

$$\rho(L_0) = -\frac{1}{2}, \quad \rho(N_0) = -1, \quad \rho(Y_{\frac{1}{2}}) = \rho(M_1) = 0. \tag{6.91}$$

The associated two-point function vanishes.

2. Put $\Phi^{(0)}(t,r,\zeta) = a(t) + r\partial\bar{b}^+$ and $\Phi^{(1)}(t,r,\zeta) = -\bar{b}^+$. Then $\Phi = \begin{pmatrix}\Phi^{(0)}\\ \Phi^{(1)}\end{pmatrix}$ is a ρ-$\overline{\mathfrak{sv}}$-primary field, where ρ is the two-dimensional representation defined by

$$\rho(L_0) = \begin{pmatrix}-\frac{1}{2} & \\ & -1\end{pmatrix}, \quad \rho(N_0) = \begin{pmatrix}-1 & \\ & 0\end{pmatrix}, \quad \rho(Y_{\frac{1}{2}}) = \begin{pmatrix}0 & 1\\ 0 & 0\end{pmatrix}, \quad \rho(M_1) = 0. \tag{6.92}$$

The two-point functions of the field Φ are given by

$$\mathscr{C}^{0,0} = t^{-2},\ \mathscr{C}^{0,1} = \mathscr{C}^{1,0} = \mathscr{C}^{1,1} = 0. \tag{6.93}$$

Proof. 1. Straightforward.
2. Follows from preceding computations. □

6.3.2 Construction of the Generalized Polynomial Fields ${}_\alpha\Phi_{j,k}$

We shall introduce in this paragraph more general fields. Take any polynomial $P = P(\bar{b}^-, \bar{b}^+, \partial\bar{b}^+, a, \int a)$ where

$$\left(\int a\right)(t) := -\sum_{n\neq 0} a_n \frac{t^{-n}}{n} + a_0 \log t + \pi_0, \quad [a_0, \pi_0] = 1 \tag{6.94}$$

is the usual logarithmic bosonic field of conformal field theory from which vertex operators are built. Since $[\bar{b}^+_n, \bar{b}^-_m] = 0$ if $nm > 0$ and similarly for the commutators of any of the fields $\bar{b}^-, \bar{b}^+, \partial\bar{b}^+, a, \int a$, the normal ordering is *commutative* and the field $: P :$ is well defined.

Let us introduce first for convenience the following notation for the coefficients of OPE of two mutually local fields.

Definition 6.13 (operator product expansions (OPE)). Let A, B be two mutually local fields: their OPE is given as

$$A(t_1)B(t_2) \sim \sum_{k=0}^{\infty} \frac{C_k(t_2)}{t_{12}^{k+1}} \tag{6.95}$$

for some fields $C_0(t), C_1(t), \ldots, C_p(t), \ldots$ which vanish for p large enough.

We shall denote by $A_{(k)}B$, $k = 0, 1, \ldots$ the field C_k.

Theorem 6.5. *Let P be any polynomial in the fields $\bar{b}^\pm$, $\partial\bar{b}^+$, a and $\int a$. Then*

$$\begin{aligned} L(t_1) : P : (t_2) &\sim \frac{: \partial P : (t_2)}{t_{12}} \\ &+ \frac{: \left(\frac{1}{2}(\bar{b}^-\partial_{\bar{b}^-} + \bar{b}^+\partial_{\bar{b}^+}) + \frac{3}{2}\partial\bar{b}^+\partial_{\partial\bar{b}^+} + a\partial_a + \frac{1}{2}\partial^2_{\int a}\right) P : (t_2)}{t_{12}^2} \\ &+ \frac{: (\bar{b}^+\partial_{\partial\bar{b}^+} + \partial_{\int a}\partial_a)\, P : (t_2)}{t_{12}^3} + \frac{1}{2}\frac{: (\partial_a^2 + \partial_{\bar{b}^-}\partial_{\partial\bar{b}^+}) P : (t_2)}{t_{12}^4} \end{aligned} \tag{6.96}$$

$$N(t_1):P(t_2):\sim \frac{:(\bar{b}^+\partial_{\bar{b}^+}+\partial\bar{b}^+\partial_{\partial\bar{b}^+}-\bar{b}^-\partial_{\bar{b}^-})\,P:(t_2)}{t_{12}} \\ +\frac{:(\bar{b}^+\partial_{\partial\bar{b}^+}+\partial_{\bar{b}^-}\partial_{\bar{b}^+})\,P:(t_2)}{t_{12}^2} \\ +\frac{:(\partial_{\bar{b}^-}\partial_{\partial\bar{b}^+})P:(t_2)}{t_{12}^3} \tag{6.97}$$

$$Y(t_1):P:(t_2)\sim \frac{:(a\partial_{\bar{b}^-}+\partial\bar{b}^+\partial_a+\bar{b}^+\partial_{\int a})\,P:(t_2)}{t_{12}}+\frac{:(\bar{b}^+\partial_a+\partial_{\bar{b}^-}\partial_{\int a})P:(t_2)}{t_{12}^2} \\ +\frac{:\partial_a\partial_{\bar{b}^-}P:(t_2)}{t_{12}^3} \tag{6.98}$$

$$M(t_1):P:(t_2)\sim \frac{:\bar{b}^+\partial_{\bar{b}^-}P:(t_2)}{t_{12}}+\frac{1}{2}\frac{:\partial_{\bar{b}^-}^2 P:(t_2)}{t_{12}^2} \tag{6.99}$$

Proof. Consider the monomial $P=P_{jklmn}=(\bar{b}^-)^j(\bar{b}^+)^k(\partial\bar{b}^+)^l a^m(\int a)^n$. Let us compute $L_{(n)}:P:, n\geq 0$ first. Apart from the contribution of the logarithmic field $\int a$ which has special properties, one may deduce the coefficient of the terms of order t_{12}^{-1} and t_{12}^{-2} directly from general considerations (see [64]): the field $(\bar{b}^-)^j(\bar{b}^+)^k(\partial\bar{b}^+)^l a^m$ is quasiprimary with conformal weight $\frac{j+k}{2}+\frac{3l}{2}+m$. The contribution from the field $\int a$ reads

$$L(t_1):P:(t_2)\sim \frac{1}{2}:a^2:(t_1):P_{jkl00}a^m\left(\int a\right)^n:(t_2)+\dots \\ \sim n\frac{:P_{jkl00}a^{m+1}(\int a)^{n-1}:(t_2)}{t_{12}}+\frac{n(n-1)}{2}\frac{:P_{jkl00}a^m(\int a)^{n-1}:(t_2)}{t_{12}^2} \\ +mn\frac{:P_{jkl00}a^{m-1}(\int a)^{n-1}:(t_2)}{t_{12}^3}+\dots \tag{6.100}$$

in accordance with the Theorem. So, if we prove that $L_{(n)}:P_{jklm0}:$ agree with (6.96) for $n\geq 2$, we are done. One gets (leaving aside the poles of order 1 or 2)

$$\frac{1}{2}:a^2:(t_1)a^m(t_2)\sim\frac{1}{2}m(m-1)\frac{(\bar{b}^-)^j(\bar{b}^+)^k(\partial\bar{b}^+)^l a^{m-2}:(t_2)}{t_{12}^4} \\ +\dots\text{ (double contraction)};$$

$$\frac{1}{2} : \bar{b}^+ \partial \bar{b}^- : (t_1) : (\partial \bar{b}^+)^l : (t_2) \sim l \frac{: (\bar{b}^-)^j (\bar{b}^+)^k (\partial \bar{b}^+)^{l-1} a^m : (t_2)}{t_{12}^3} + \ldots \text{ (simple contraction);}$$

$$\frac{1}{2} : \bar{b}^+ \partial \bar{b}^- : (t_1) : (\bar{b}^-)^j (\bar{b}^+)^k : (t_2) \sim \frac{jk}{2} \frac{(\bar{b}^-)^{j-1} (\bar{b}^+)^{k-1} (\partial \bar{b}^+)^l a^m : (t_2)}{t_{12}^3}$$

(double contraction);

$$\frac{1}{2} : \bar{b}^+ \partial \bar{b}^- : (t_1) : (\bar{b}^-)^j (\partial \bar{b}^+)^l : (t_2) \sim jl \frac{: (\bar{b}^-)^{j-1} (\bar{b}^+)^k (\partial \bar{b}^+)^{l-1} a^m : (t_2)}{t_{12}^4}$$

(double contraction);

$$-\frac{1}{2} : \bar{b}^- \partial \bar{b}^+ : (t_1) : (\bar{b}^-)^j (\bar{b}^+)^k : (t_2) \sim -\frac{jk}{2} \frac{(\bar{b}^-)^{j-1} (\bar{b}^+)^{k-1} (\partial \bar{b}^+)^l a^m : (t_2)}{t_{12}^3}$$

(double contraction);

$$-\frac{1}{2} : \bar{b}^- \partial \bar{b}^+ : (t_1) : (\bar{b}^-)^j (\partial \bar{b}^+)^l : (t_2) \sim -\frac{jl}{2} \frac{: (\bar{b}^-)^{j-1} (\bar{b}^+)^k (\partial \bar{b}^+)^{l-1} a^m : (t_2)}{t_{12}^4}$$

(double contraction)

hence the result.

Let us consider now the OPE of N with P. The fields a and $\int a$ giving no contribution, one may just as well assume that $m = n = 0$. Then

$$- : \bar{b}^+ \bar{b}^- : (t_1) : (\bar{b}^-)^j (\bar{b}^+)^k (\partial \bar{b}^+)^l : (t_2) \sim \left(\frac{(k-j)(\bar{b}^-)^j (\bar{b}^+)^k (\partial \bar{b}^+)^l}{t_{12}} + l \frac{: \bar{b}^+ (t_1) (: (\bar{b}^-)^j (\bar{b}^+)^k (\partial \bar{b}^+)^{l-1} :)(t_2) :}{t_{12}^2} \right)$$
$$+ \left(\frac{jk : (\bar{b}^-)^{j-1} (\bar{b}^+)^{k-1} (\partial \bar{b}^+)^l : (t_2)}{t_{12}^2} + jl \frac{: (\bar{b}^-)^{j-1} (\bar{b}^+)^{k-1} (\partial \bar{b}^+)^{l-1} : (t_2)}{t_{12}^3} \right)$$

adding the terms coming from a single contraction to the terms coming from a double contraction. Hence the result. The OPE of M with P follows easily from the same rules. Finally,

$$Y(t_1):P:(t_2) =: a\bar{b}^+ : (t_1) : (\bar{b}^-)^j(\bar{b}^+)^k(\partial\bar{b}^+)^l a^m \left(\int a\right)^n : (t_2) \quad (6.101)$$

$$\sim \left(j\frac{:(\bar{b}^-)^{j-1}(\bar{b}^+)^k(\partial\bar{b}^+)^l a^{m+1}(\int a)^n : (t_2)}{t_{12}} \right.$$

$$+n\frac{:(\bar{b}^-)^j(\bar{b}^+)^{k+1}(\partial\bar{b}^+)^l a^m(\int a)^{n-1} : (t_2)}{t_{12}}$$

$$\left. +m\frac{:\bar{b}^+(t_1)(:(\bar{b}^-)^j(\bar{b}^+)^k(\partial\bar{b}^+)^l a^{m-1}(\int a)^n : (t_2):}{t_{12}^2} \right)$$

$$\left(nj\frac{:(\bar{b}^-)^{j-1}(\bar{b}^+)^k(\partial\bar{b}^+)^l a^m(\int a)^{n-1} : (t_2)}{t_{12}^2} \right.$$

$$\left. +mj\frac{:(\bar{b}^-)^{j-1}(\bar{b}^+)^k(\partial\bar{b}^+)^l a^{m-1}(\int a)^n : (t_2)}{t_{12}^3} \right) \quad (6.102)$$

(separating once more the terms coming from a single contraction from the terms with a double contraction) hence (6.98). □

In any case, the $\overline{\mathfrak{sv}}$-fields preserve this space of polynomial fields. The reason why we chose not to include powers of $\partial\bar{b}^-$ or ∂a for instance, or higher derivatives of the field $\bar{b}^+$, will become clear in a moment. Take a ρ-Schrödinger-conformal field $\Phi = (\Phi^{(\mu)})_\mu$ and suppose it has a formal expansion of the type $\sum_{\xi,\nu} \Phi^{(\mu),\xi,\sigma}(t) r^\xi \zeta^\sigma$ as in Sect. 2.1, with σ varying in a set of real values of the same type as for ξ, while the $\Phi^{(\mu),\xi,\sigma}$ are polynomials in the variable $\int a$, $\bar{b}^\pm$ and their derivatives of any order. Suppose $\Phi^{(\mu),\xi,\sigma} \neq 0$ for a negative value of ξ. Then

$$\Phi^{(\mu),\xi,\sigma} = Y_{(0)}\frac{\Phi^{(\mu),\xi-1,\sigma}}{\xi} = \frac{1}{\xi}(a\partial_{\bar{b}^-} + \ldots)\Phi^{(\mu),\xi-1,\sigma} \quad (6.103)$$

hence $\Phi^{(\mu),\xi-1,\sigma}$ must include a monomial P_{jklmn} with m strictly less than for all the monomials in $\Phi^{(\mu),\xi,\sigma}$. But this argument can be repeated indefinitely, going down one step $\xi \to \xi - 1$ at a time, and one ends with a contradiction if negative powers of a are not allowed. The same goes for σ since

$$\Phi^{(\mu),\xi,\sigma} = M_{(0)}\frac{\Phi^{(\mu),\xi,\sigma-1}}{\sigma} = \frac{1}{\sigma}\bar{b}^+\partial\bar{b}^-\Phi^{(\mu),\xi,\sigma-1}. \quad (6.104)$$

A moment's thought proves then that if the $\Phi^{(\mu),\xi,\sigma}$ are to be polynomials, then the indices ξ and σ should be positive integers and all the terms $\Phi^{(\mu),\xi,\sigma}$ may be obtained from the *lowest degree component fields* $\Phi^{(\mu),0,0}$ by using Definition 6.5; in particular, $Y_{(0)}\Phi^{(\mu)} = \partial_r\Phi^{(\mu)}$ and $M_{(0)}\Phi^{(\mu)} = \partial_\zeta\Phi^{(\mu)}$: by applying the operators $Y_{(0)}$ and $M_{(0)}$ to $\Phi^{(\mu),0,0}$, one retrieves the whole series $\Phi^{(\mu)} = \sum\sum_{\xi,\sigma=0}^{\infty} \Phi^{(\mu),\xi,\sigma}(t) r^\xi \zeta^\sigma$.

Now $\Phi^{(\mu),0,0}$ may contain neither powers of $\partial\bar{b}^{\pm}$ (otherwise Theorem 6.5 gives $L_{(2)}\Phi^{(\mu),0,0} \neq 0$ and formula (6.39) proves that this is impossible) nor powers of a, except, possibly, for fields of the type $(\bar{b}^{+})^{k}a$ (otherwise Theorem 6.5 shows that $Y_{(2)}\Phi^{(\mu),0,0} \neq 0$ or $L_{(2)}\Phi^{(\mu),0,0} \neq 0$ or $L_{(3)}\Phi^{(\mu),0,0} \neq 0$, and this is contradictory with formula (6.39) or (6.40)). Higher derivatives of the previous fields would yield higher order singularities in the OPE with L for instance. Note also that powers of $\int a$ may be freely included under the previous conditions and entail no supplementary constraint.

Hence (discarding fields such that $\Phi^{(\mu),0,0}$ is linear in a, which are not very interesting, as one sees by considering the rather trivial action of the $\overline{\mathfrak{sv}}$-fields on them and their disappointingly simple n-point functions), one is led to consider the following family of fields, where we make use of the *vertex operator* $V_\alpha := \exp \alpha \int a$ $(\alpha \in \mathbb{R})$, see [19] for instance. Vertex operators are known to be primary; with our normalization, V_α is L-primary with conformal weight $\frac{\alpha^2}{2}$.

Definition 6.14 (polynomial fields).
Set for $\alpha \in \mathbb{C}$, $j, k = 0, 1, \ldots$

$$_{\alpha}\phi_{j,k}(t) =: (\bar{b}^{-})^{j}(\bar{b}^{+})^{k}V_{\alpha} : (t) \tag{6.105}$$

and

$$\phi_{j,k}(t) = {}_{0}\phi_{j,k}(t) =: (\bar{b}^{-})^{j}(\bar{b}^{+})^{k} : (t). \tag{6.106}$$

All these fields appear to be the lowest-degree component fields of ρ-$\overline{\mathfrak{sv}}$-primary fields. The operator $\rho(Y_{\frac{1}{2}})$ is trivial if $\alpha = 0$; in the contrary case, $\rho(M_1)$ may be expressed as a coefficient times $(\rho(Y_{\frac{1}{2}}))^2$, in accordance with the discussion preceding Theorem 6.1. Since ρ is quite different according to whether $\alpha \neq 0$ or $\alpha = 0$, and also for the sake of clarity, we will state two different theorems.

Theorem 6.6. (construction of the polynomial fields $\Phi_{j,k}$)

1. Set

$$\Phi^{(0),0}_{j,k}(t,\zeta) = \sum_{m=0}^{j} \binom{m}{j} \zeta^{m} : (\bar{b}^{-})^{j-m}(\bar{b}^{+})^{k+m} : (t) \tag{6.107}$$

and define inductively a series of fields $\Phi^{(\mu),\xi}_{j,k}$ $(\mu, \xi = 0, 1, 2, \ldots)$ by setting

$$\Phi^{(\mu+1),\xi}_{j,k}(t,\zeta) = -\frac{1}{2} : \partial^{2}_{\bar{b}^{-}} \Phi^{(\mu),\xi}_{j,k} : (t,\zeta) \tag{6.108}$$

and

$$\Phi^{(\mu),\xi+1}_{j,k}(t,\zeta) = \frac{1}{1+\xi} : (a\partial_{\bar{b}^{-}} + \partial\bar{b}^{+}\partial_{a})\Phi^{(\mu),\xi}_{j,k} : (t,\zeta) \tag{6.109}$$

Then $\Phi^{(\mu)}_{j,k} = 0$ for $\mu > \lfloor j/2 \rfloor$ ($\lfloor\,.\,\rfloor$ =entire part), and

$$\Phi_{j,k} := (\Phi_{j,k}^{(\mu)})_{0\le\mu\le[j/2]}, \quad \Phi_{j,k}^{(\mu)}(t,r,\zeta) := \sum_{\xi\ge 0} \Phi_{j,k}^{(\mu),\xi}(t,\zeta) r^{\xi} \tag{6.110}$$

defines a ρ-$\langle N_0\rangle \ltimes \mathfrak{sv}$-primary field, ρ being the representation of $\overline{\mathfrak{sv}}_0$ defined by

$$\rho(L_0) = -\left[\frac{j+k}{2}\mathrm{Id} - \sum_{\mu=0}^{[j/2]} \mu E_{\mu}^{\mu}\right] \tag{6.111}$$

$$\rho(N_0) = -\left[(k-j)\mathrm{Id} + 2\sum_{\mu=0}^{[j/2]} \mu E_{\mu}^{\mu}\right] \tag{6.112}$$

$$\rho(Y_{\frac{1}{2}}) = 0 \tag{6.113}$$

$$\rho(M_1) = \sum_{\mu=0}^{[j/2]-1} E_{\mu+1}^{\mu} \tag{6.114}$$

where E_{ν}^{μ} is the $(\lfloor j/2\rfloor + 1) \times (\lfloor j/2\rfloor + 1)$ elementary matrix, with a single coefficient 1 at the intersection of the μ-th line and the ν-th row.

2. *Set $\Phi = (\Phi_{j,k}^{(0)})_{j,k=0,1,\ldots}$. Then Φ is a (ρ,Ω)-$\overline{\mathfrak{sv}}$-primary field if ρ, Ω are defined as follows:*

$$\begin{aligned}
\rho(L_0)\Phi_{j,k}^{(0)} &= -\frac{j+k}{2}\Phi_{j,k}^{(0)}; \\
\rho(Y_{\frac{1}{2}})\Phi_{j,k}^{(0)} &= 0; \quad \rho(M_1)\Phi_{j,k}^{(0)} = -\frac{1}{2}j(j-1)\Phi_{j-2,k}^{(0)}; \\
\rho(N_0)\Phi_{j,k}^{(0)} &= (j-k)\Phi_{j,k}^{(0)}; \\
\Omega\Phi_{j,k}^{(0)} &= jk\Phi_{j-1,k-1}^{(0)}.
\end{aligned} \tag{6.115}$$

Remark. 1. Both representations ρ are of course the same; the passage from the first action on the $\Phi_{j,k}^{(\mu)}$ to the action on Φ is given by the relation

$$\Phi_{j,k}^{(\mu)} = \left(-\frac{1}{2}\right)^{\mu} j(j-1)\ldots(j-2\mu+1)\Phi_{j-2\mu,k}^{(0)}. \tag{6.116}$$

The second case in the Theorem is an extension of the first one when one wants to consider covariance under all N-generators (not only under N_0), which makes things more complicated.

2. Formally, one has

$$\Phi_{j,k}^{(\mu)} =: \exp r(a\partial_{\bar{b}^-} + \partial\bar{b}^+\partial_a).\Phi_{j,k}^{(\mu),0} : \tag{6.117}$$

since $Y_{(0)} \equiv \partial_r \equiv a\partial_{\bar{b}^-} + \partial\bar{b}^+\partial_a$ when applied to a polynomial $\mathfrak{sv}$-primary field of the form $P(\bar{b}^{\pm}, \partial\bar{b}^+, a)$. Hence, by the Campbell-Hausdorff formula

$$\exp(A + B) = \exp\frac{1}{2}[B, A]\exp A \exp B, \tag{6.118}$$

valid if $[A, [A, B]] = [B, [A, B]] = 0$, one may also write

$$\Phi_{j,k}^{(\mu)} =: \exp ra\partial_{\bar{b}^-} \exp\frac{r^2}{2}\partial\bar{b}^+\partial_{\bar{b}^-} . \Phi_{j,k}^{(\mu),0}. \tag{6.119}$$

Proof. 1. First of all, $\Phi^{(\mu),\xi}$ is well-defined only because the operators $\partial^2_{\bar{b}^-}$ and $a\partial_{\bar{b}^-} + \partial\bar{b}^+\partial_a$ (giving the shifts $\mu \to \mu + 1$ and $\xi \to \xi + 1$) commute. Let us check successively the covariance under the action of M, Y, N_0, L.

- One finds from (6.99)

$$\begin{aligned} M_{(0)}\Phi^{(0),0}(t,\zeta) = \bar{b}^+\partial_{\bar{b}^-}\Phi^{(0),0}(t,\zeta) &= \sum_{m=0}^{j}\binom{j}{m}(j-m) \\ &: (\bar{b}^-)^{j-m-1}(\bar{b}^+)^{k+m+1} : (t)\zeta^m \\ &= \partial_\zeta\Phi^{(0),0}(t,\zeta); \end{aligned} \tag{6.120}$$

$$M_{(1)}\Phi^{(0),0}(t,\zeta) = \frac{1}{2}\partial^2_{\bar{b}^-}\Phi^{(0),0}(t,\zeta) = -\Phi^{(1),0}(t,\zeta) \tag{6.121}$$

  which is coherent with formula (6.41) and the definition (6.114) of $\rho(M_1)$. The field $\Phi_{j,k}$ is M-covariant if $\bar{b}^+\partial_{\bar{b}^-}\Phi^{(\mu),\xi}(t,\zeta) = \partial_\zeta\Phi^{(\mu),\xi}(t,\zeta)$ for every $\mu, \xi \geq 0$. But this is true for $\mu, \xi = 0$ and $[\bar{b}^+\partial_{\bar{b}^-}, \partial_{\bar{b}^-}] = [\bar{b}^+\partial_{\bar{b}^-}, a\partial_{\bar{b}^-} + \partial\bar{b}^+\partial_a] = 0$. Hence this is true for all values of μ, ξ by induction.
- The action of $Y_{(0)}$ on $\Phi^{(\mu),\xi}$ is correct by definition - compare with formulas (6.98) and (6.109). One has $Y_{(1)}\Phi^{(\mu),0} = 0$ because $\partial_a\Phi^{(\mu),0} = 0$, which is in accord with (6.40) if one sets $\rho(Y_{\frac{1}{2}}) = 0$. To prove that $Y_{(1)}\Phi^{(\mu),\xi} = \bar{b}^+\partial_a\Phi^{(\mu),\xi}$ coincides with $\partial_\zeta\Phi^{(\mu),\xi-1} = \bar{b}^+\partial_{\bar{b}^-}\Phi^{(\mu),\xi-1}$, one uses induction on ξ and the commutator relation $[\bar{b}^+\partial_a, a\partial_{\bar{b}^-} + (\partial\bar{b}^+)\partial_a] = \bar{b}^+\partial_{\bar{b}^-}$. If this holds for some $\xi \geq 0$, then

$$\begin{aligned} \bar{b}^+\partial_a\Phi^{(\mu),\xi+1} &= \frac{1}{1+\xi}(\bar{b}^+\partial_a)(a\partial_{\bar{b}^-} + (\partial\bar{b}^+)\partial_a)\Phi^{(\mu),\xi} \\ &= \frac{1}{1+\xi}\left[(a\partial_{\bar{b}^-} + (\partial\bar{b}^+)\partial_a)(\bar{b}^+\partial_{\bar{b}^-})\Phi^{(\mu),\xi-1} + \bar{b}^+\partial_{\bar{b}^-}\Phi^{(\mu),\xi}\right] \\ &= \frac{1}{1+\xi}\left[(\bar{b}^+\partial_{\bar{b}^-})(a\partial_{\bar{b}^-} + (\partial\bar{b}^+)\partial_a)\Phi^{(\mu),\xi-1} + \bar{b}^+\partial_{\bar{b}^-}\Phi^{(\mu),\xi}\right] \end{aligned}$$

$$= \frac{1}{1+\xi}\left[\xi \bar{b}^+ \partial_{\bar{b}^-} \Phi^{(\mu),\xi} + \bar{b}^+ \partial_{\bar{b}^-} \Phi^{(\mu),\xi}\right]$$

$$= \bar{b}^+ \partial_{\bar{b}^-} \Phi^{(\mu),\xi} \tag{6.122}$$

by (6.109).

- One has $Y_{(2)}\Phi^{(\mu),0} = 0$ by (6.98) and, supposing that $Y_{(2)}\Phi^{(\mu),\xi} = \partial_{\bar{b}^-}\partial_a \Phi^{(\mu),\xi}$ coincides with $-2\Phi^{(\mu+1),\xi-1} = -2\partial_\zeta \Phi^{(\mu),\xi-1} = \partial^2_{\bar{b}^-}\Phi^{(\mu),\xi-1}$ for some $\xi \geq 0$, then

$$\partial_{\bar{b}^-}\partial_a \Phi^{(\mu),\xi+1} = \frac{1}{1+\xi}(\partial_{\bar{b}^-}\partial_a)(a\partial_{\bar{b}^-} + \partial\bar{b}^+\partial_a)\Phi^{(\mu),\xi} = \partial^2_{\bar{b}^-}\Phi^{(\mu),\xi} \tag{6.123}$$

by a proof along the same lines, since $[\partial_{\bar{b}^-}\partial_a, a\partial_{\bar{b}^-} + \partial_{\bar{b}^+}\partial_a] = \partial^2_{\bar{b}^-}$.

- Since $N_{(0)}$ acts as $\bar{b}^+\partial_{\bar{b}^+} - \bar{b}^-\partial_{\bar{b}^-}$ on $\Phi^{(\mu),0}$, it simply measures the difference of degrees in $\bar{b}^+$ and $\bar{b}^-$ (for polynomial fields which depends only on $\bar{b}^\pm$ and not on their derivatives). Hence one sees easily that $N_{(0)}\Phi^{(\mu),0} = (2\zeta\partial_\zeta - j + k + 2\mu)\Phi^{(\mu),0}$, which is formula (6.43). Then $Y_{(0)} \equiv a\partial_{\bar{b}^-} + (\partial\bar{b}^+)\partial_a$ increases by 1 the eigenvalue of $N_{(0)}$, see (6.97), which is also in accord with (6.43).
- There remains to check for the action of $L_{(i)}$, $i = 2, 3$. Supposing that $L_{(2)}\Phi^{(\mu),\xi} = \bar{b}^+\partial_{\partial\bar{b}^+}\Phi^{(\mu),\xi}$ coincides with $\frac{1}{2}\partial_\zeta\Phi^{(\mu),\xi-2} = \frac{1}{2}\bar{b}^+\partial_{\bar{b}^-}\Phi^{(\mu),\xi-2}$ for some $\xi \geq 0$, then

$$\bar{b}^+\partial_{\partial\bar{b}^+}\Phi^{(\mu),\xi+1} = \frac{1}{1+\xi}\left[\frac{1}{2}(a\partial_{\bar{b}^-} + \partial\bar{b}^+\partial_a)(\bar{b}^+\partial_{\bar{b}^-})\Phi^{(\mu),\xi-2} + \bar{b}^+\partial_a\Phi^{(\mu),\xi}\right]; \tag{6.124}$$

since (as we just proved) $Y_{(1)}\Phi^{(\mu),\xi} = \bar{b}^+\partial_a\Phi^{(\mu),\xi} = \bar{b}^+\partial_{\bar{b}^-}\Phi^{(\mu),\xi-1}$, one gets $L_{(2)}\Phi^{(\mu),\xi+1} = \frac{1}{2}\partial_\zeta\Phi^{(\mu),\xi-1}$.

Finally, supposing that $2L_{(3)}\Phi^{(\mu),\xi} = (\partial_a^2 + \partial_{\bar{b}^-}\partial_{\partial\bar{b}^+})\Phi^{(\mu),\xi}$ coincides with $-3\Phi^{(\mu)+1,\xi-2} = \frac{3}{2}\partial^2_{\bar{b}^-}\Phi^{(\mu),\xi-2}$, then, using the commutator relation $[\partial_a^2 + \partial_{\bar{b}^-}\partial_{\partial\bar{b}^+}, a\partial_{\bar{b}^-} + (\partial\bar{b}^+)\partial_a] = 3\partial_{\bar{b}^-}\partial_a$ and the above equality $Y_{(2)}\Phi^{(\mu),\xi} = \partial_{\bar{b}^-}\partial_a\Phi^{(\mu),\xi} = \partial^2_{\bar{b}^-}\Phi^{(\mu),\xi-1}$, one finds

$$(\partial_a^2 + \partial_{\bar{b}^-}\partial_{\partial\bar{b}^+})\Phi^{(\mu),\xi+1} = \frac{1}{\xi+1}\left[\frac{3}{2}(a\partial_{\bar{b}^+} + \partial\bar{b}^+\partial_a)\partial^2_{\bar{b}^-}\Phi^{(\mu),\xi-2} + 3\partial_{\bar{b}^-}\partial_a\Phi^{(\mu),\xi}\right]$$
$$= \frac{3}{2}\partial^2_{\bar{b}^-}\Phi^{(\mu),\xi-1}. \tag{6.125}$$

2. First note that

$$\partial_{\bar{b}^+}\partial_{\bar{b}^-}\Phi^{(0)}_{j,k} = \sum_{m\geq 0}(j-m)(k+m)\binom{j}{m}(\bar{b}^-)^{j-m-1}(\bar{b}^+)^{k+m-1}$$

$$= jk\sum_{m\geq 0}\binom{j-1}{m}(\bar{b}^-)^{j-m-1}(\bar{b}^+)^{k+m-1}$$

$$+j(j-1)\sum_{m\geq 1}\binom{j-2}{m-1}(\bar{b}^-)^{j-m-1}(\bar{b}^+)^{k+m-1}$$
$$= jk\Phi^{(0)}_{j-1,k-1}+\zeta j(j-1)\Phi^{(0)}_{j-2,k}$$
$$= -\Omega\Phi^{(0)}_{j,k}-2\zeta\rho(M_1)\Phi^{(0)}_{j,k} \tag{6.126}$$

by Remark 1. following Theorem 6.6.

Hence one has identified the action of $N_{(1)}$ on $\Phi^{(0),0}_{j,k}$ as the correct one. Suppose now that $N_{(1)}\Phi^{(0),\xi} = (\bar{b}^+\partial_{\partial\bar{b}^+}+\partial_{\bar{b}^+}\partial_{\bar{b}^-})\Phi^{(0),\xi}$ coincides with $\frac{1}{2}\partial_\zeta\Phi^{(0),\xi-2}+\zeta j(j-1)\Phi^{(0),\xi}_{j-2,k}+jk\Phi^{(0),\xi}_{j-1,k-1}$ for some ξ. Then, by commuting $\bar{b}^+\partial_{\partial\bar{b}^+}+\partial_{\bar{b}^+}\partial_{\bar{b}^-}$ through $Y_{(0)} = a\partial_{\bar{b}^-}+\partial\bar{b}^+\partial_a$, one gets

$$(\bar{b}^+\partial_{\partial_{\bar{b}^+}}+\partial_{\bar{b}^+}\partial_{\bar{b}^-})\Phi^{(0),\xi+1}_{j,k} = \frac{1}{1+\xi}\Big[(a\partial_{\bar{b}^-}+\partial\bar{b}^+\partial_a)\Big(\frac{1}{2}\partial_\zeta\Phi^{(0),\xi-2}$$
$$+\zeta j(j-1)\Phi^{(0),\xi}_{j-2,k}+jk\Phi^{(0),\xi}_{j-1,k-1}\Big)+\bar{b}^+\partial_a\Phi^{(0),\xi}\Big] \tag{6.127}$$

and $Y_{(1)}\Phi^{(0),\xi} = \bar{b}^+\partial_a\Phi^{(0),\xi}_{j,k} = \partial_\zeta\Phi^{(0),\xi-1}_{j,k}$ as we have just proved, hence $N_{(1)}\Phi^{(0),\xi+1}$ is given by the correct formula.

Finally, $N_{(2)}\Phi^{(0),\xi}_{j,k} = \partial_{\bar{b}^-}\partial_{\partial\bar{b}^+}\Phi^{(0),\xi}$ must be identified with $-\partial_\zeta\Phi^{(0),\xi-2}_{j,k}$ (which is certainly true for $\xi=0$). Supposing this holds for some ξ,

$$\partial_{\bar{b}^-}\partial_{\partial\bar{b}^+}\Phi^{(0),\xi+1} = \frac{1}{1+\xi}(\partial_{\bar{b}^-}\partial_{\partial\bar{b}^+})(a\partial_{\bar{b}^-}+\partial\bar{b}^+\partial_a)\Phi^{(0),\xi}$$
$$= \frac{1}{1+\xi}\left[-(a\partial_{\bar{b}^-}+\partial\bar{b}^+\partial_a)\partial_\zeta\Phi^{(0),\xi-2}+\partial_{\bar{b}^-}\partial_a\Phi^{(0),\xi}\right] \tag{6.128}$$

since $[\partial_{\bar{b}^-}\partial_{\partial\bar{b}^+}, a\partial_{\bar{b}^-}+\partial\bar{b}^+\partial_a] = \partial_{\bar{b}^-}\partial_a$; we now use the previous result $Y_{(2)}\Phi^{(\mu),\xi} = \partial_{\bar{b}^-}\partial_a\Phi^{(\mu),\xi} = -2\partial_\zeta\Phi^{(0),\xi-1}$ and conclude by induction.

Theorem 6.7. (construction of the generalized polynomial fields ${}_\alpha\Phi_{j,k}$)

1. Set

$${}_\alpha\Phi^{(0),0}_{j,k}(w,\zeta) = \sum_{m=0}^{j}\binom{j}{m}\zeta^m : (\bar{b}^-)^{j-m}(\bar{b}^+)^{k+m}V_\alpha : \tag{6.129}$$

and define inductively a series of fields ${}_\alpha\Phi^{(\mu),\xi} = {}_\alpha\Phi^{(\mu),\xi}_{j,k}$ $(\mu,\xi = 0,1,2,\dots)$ *by setting*

$${}_\alpha\Phi^{(\mu+1),\xi}(w,\zeta) = \frac{i}{\sqrt{2}} : \partial_{\bar{b}^-}\,{}_\alpha\Phi^{(\mu),\xi} : (w,\zeta) \tag{6.130}$$

and

$$ {}_\alpha\Phi^{(\mu),\xi+1}(w,\zeta) = \frac{1}{1+\xi} : (a\partial_{\bar{b}}^{-} + \partial\bar{b}^{+}\partial_a + \alpha\bar{b}^{+})\,{}_\alpha\Phi^{(\mu),\xi} : (w,\zeta) \qquad (6.131) $$

Then

$$ {}_\alpha\Phi_{j,k} := ({}_\alpha\Phi^{(\mu)})_{0\le i\le j}, \quad {}_\alpha\Phi^{(\mu)}(t,r,\zeta) = \sum_{\xi\ge 0} {}_\alpha\Phi^{(\mu),\xi}(w,\zeta) r^{\xi} \qquad (6.132) $$

defines a ρ*-*$\langle N_0\rangle \ltimes \mathfrak{sv}$*-primary field,* ρ *being the representation of* $\overline{\mathfrak{sv}}_0$ *defined by*

$$ \rho(L_0) = -\left[\frac{j+k+\alpha^2}{2}\mathrm{Id} - \frac{1}{2}\sum_{\mu=0}^{j} \mu E_{\mu,\mu}\right] \qquad (6.133) $$

$$ \rho(N_0) = -\left[(k-j)\mathrm{Id} + \sum_{\mu=0}^{j} \mu E_{\mu,\mu}\right] \qquad (6.134) $$

$$ \rho(Y_{\frac{1}{2}}) = \mathrm{i}\alpha\sqrt{2}\sum_{\mu=0}^{j-1} E_{\mu,\mu+1} \qquad (6.135) $$

$$ \rho(M_1) = -\frac{1}{2}\left(\frac{1}{\alpha}\rho(Y_{\frac{1}{2}})\right)^2 = \sum_{\mu=0}^{j-2} E_{\mu,\mu+2} \qquad (6.136) $$

2. *Set* ${}_\alpha\Phi = ({}_\alpha\Phi^{(0)}_{j,k})_{j,k=0,1,\dots}$*. Then* ${}_\alpha\Phi$ *is a* (ρ,Ω)*-*$\overline{\mathfrak{sv}}$*-primary field if* ρ,Ω *are defined as follows:*

$$ \begin{aligned} \rho(L_0)\,{}_\alpha\Phi^{(0)}_{j,k} &= -\frac{j+k}{2}\,{}_\alpha\Phi^{(0)}_{j-2,k}; \\ \rho(Y_{\frac{1}{2}})\Phi^{(0)}_{j,k} &= -\alpha j\Phi^{(0)}_{j-1,k}; \quad \rho(M_1)\,{}_\alpha\Phi^{(0)}_{j,k} = -\frac{1}{2}j(j-1)\,{}_\alpha\Phi^{(0)}_{j-2,k}; \\ \rho(N_0)\,{}_\alpha\Phi^{(0)}_{j,k} &= (j-k)\,{}_\alpha\Phi^{(0)}_{j,k}; \\ \Omega\,{}_\alpha\Phi^{(0)}_{j,k} &= jk\,{}_\alpha\Phi^{(0)}_{j-1,k-1}. \end{aligned} \qquad (6.137) $$

Remark. The coherence between the two representations is given this time by:

$$ {}_\alpha\Phi^{(\mu)}_{j,k} = \left(\frac{\mathrm{i}}{\sqrt{2}}\right)^k j(j-1)\dots(j-k+1)\Phi^{(0)}_{j-\mu,k}. \qquad (6.138) $$

- One may write formally

$$ {}_{\alpha}\Phi^{(\mu)}_{j,k} = :\exp r\left(a\partial_{\bar{b}^-} + \partial\bar{b}^+\partial_a + \alpha\bar{b}^+\right).\Phi^{(\mu),0}_{j,k}: $$
$$ = :\exp\alpha r\bar{b}^+ \,.\, \exp ra\partial_{\bar{b}^-} \,.\, \exp\frac{r^2}{2}\partial\bar{b}^+\partial_{\bar{b}^-}\Phi^{(\mu),0}_{j,k} : \quad (6.139) $$

Proof. The proof is almost the same, with just a few modifications. We shall follow the proof of Theorem 6.6 line by line and rewrite only what has to be changed.

- One has $Y_{(1)}\,{}_{\alpha}\Phi^{(\mu),0}(\zeta) = \alpha\partial_{\bar{b}^-}\,{}_{\alpha}\Phi^{(\mu),0}(\zeta)$, to be identified with $-\rho(Y_{\frac{1}{2}})^{\mu}_{\nu}\,{}_{\alpha}\Phi^{(\nu),0}$. Hence one must set, in accordance with (6.135)

$$ {}_{\alpha}\Phi^{(\mu)+1,0} = \frac{i}{\sqrt{2}}\partial_{\bar{b}^-}\,{}_{\alpha}\Phi^{(\mu),0} \quad (6.140) $$

so ${}_{\alpha}\Phi^{(\mu)+2,0} = -\frac{1}{2}\partial^2_{\bar{b}^-}\,{}_{\alpha}\Phi^{(\mu),0}$ as in Theorem 6.6, with a double shift instead in the indices i.

Suppose now $Y_{(1)}\,{}_{\alpha}\Phi^{(\mu),\xi} = (\bar{b}^+\partial_a + \alpha\partial_{\bar{b}^-})\,{}_{\alpha}\Phi^{(\mu),\xi}$ coincides with

$$ \partial_{\zeta}\,{}_{\alpha}\Phi^{(\mu),\xi-1} - i\alpha\sqrt{2}\,{}_{\alpha}\Phi^{(\mu)+1,\xi} = \bar{b}^+\partial_{\bar{b}^-}\,{}_{\alpha}\Phi^{(\mu),\xi-1} - i\alpha\sqrt{2}\,{}_{\alpha}\Phi^{(\mu)+1,\xi} ; \quad (6.141) $$

then the commutator relation $[\bar{b}^+\partial_a + \alpha\partial_{\bar{b}^-}, a\partial_{\bar{b}^-} + \partial\bar{b}^+\partial_a + \alpha\bar{b}^+] = \bar{b}^+\partial_{\bar{b}^-}$ yields

$$ Y_{(1)}\,{}_{\alpha}\Phi^{(\mu),\xi+1} = \frac{1}{\xi+1}\Big\{(a\partial_{\bar{b}^-} + (\partial\bar{b}^+)\partial_a + \alpha\bar{b}^+) \quad (6.142) $$
$$ (\bar{b}^+\partial_{\bar{b}^-}\,{}_{\alpha}\Phi^{(\mu),\xi-1} - i\alpha\sqrt{2}\,{}_{\alpha}\Phi^{(\mu)+1,\xi}) + \bar{b}^+\partial_{\bar{b}^-}\,{}_{\alpha}\Phi^{(\mu),\xi}\Big\} $$
$$ = \bar{b}^+\partial_{\bar{b}^-}\,{}_{\alpha}\Phi^{(\mu),\xi} - i\alpha\sqrt{2}\Phi^{(\mu)+1,\xi+1} \quad (6.143) $$

Then $Y_{(2)} = \partial_{\bar{b}^-}\partial_a$ and $[\partial_{\bar{b}^-}\partial_a, Y_{(0)}] = \partial^2_{\bar{b}^-}$ as in Theorem 6.6, so covariance under $Y_{(2)}$ holds true.

- The action of $N_{(0)}, N_{(1)}, N_{(2)}$ on ${}_{\alpha}\Phi^{(\mu)}$ or ${}_{\alpha}\Phi^{(0)}_{j,k}$ is exactly as in Theorem 6.6 since $N = -:\bar{b}^+\bar{b}^-:$ does not involve neither the free boson a nor its integral.
- One must still check for $L_{(2)}$ (nothing changes for $L_{(3)}$). Suppose that $L_{(2)}\,{}_{\alpha}\Phi^{(\mu),\xi} = \bar{b}^+\partial_{\partial\bar{b}^+} + \alpha\partial_a$ coincides with

$$ \frac{1}{2}\partial_{\zeta}\,{}_{\alpha}\Phi^{(\mu),\xi-2} - \mathrm{i}\alpha\sqrt{2}\,{}_{\alpha}\Phi^{(\mu+1),\xi-1} = \frac{1}{2}\bar{b}^+\partial_{\bar{b}^-}\,{}_{\alpha}\Phi^{(\mu),\xi-2} - \mathrm{i}\alpha\sqrt{2}\,{}_{\alpha}\Phi^{(\mu)+1,\xi-1}. $$

Then, using

$$ [\bar{b}^+\partial_{\partial\bar{b}^+} + \alpha\partial_a, a\partial_{\bar{b}^-} + (\partial\bar{b}^+)\partial_a + \alpha\bar{b}^+]\,{}_{\alpha}\Phi^{(\mu),\xi} $$
$$ = (\bar{b}^+\partial_a + \alpha\partial_{\bar{b}^-})\,{}_{\alpha}\Phi^{(\mu),\xi} = \partial_{\zeta}\,{}_{\alpha}\Phi^{(\mu),\xi-1} - \mathrm{i}\alpha\sqrt{2}\Phi^{(\mu+1),\xi} \quad (6.144) $$

(see computation of $Y_{(1)}\,{}_\alpha\Phi^{(\mu),\xi}$ above) one gets

$$(\bar{b}^+\partial_{\partial\bar{b}^+} + \alpha\partial_a)\,{}_\alpha\Phi^{(\mu),\xi+1} = \frac{1}{1+\xi}\Bigg[(a\partial_{\bar{b}^-} + (\partial\bar{b}^+)\partial_a + \alpha\bar{b}^+)$$

$$\left(\frac{1}{2}\bar{b}^+\partial_{\bar{b}^-}\,{}_\alpha\Phi^{(\mu),\xi-2} - \mathrm{i}\alpha\sqrt{2}\,{}_\alpha\Phi^{(\mu)+1,\xi-1}\right)$$

$$+\partial_\zeta\Phi^{(\mu),\xi-1} - \mathrm{i}\alpha\sqrt{2}\,{}_\alpha\Phi^{(\mu)+1,\xi}\Bigg]$$

$$= \frac{1}{2}\bar{b}^+\partial_{\bar{b}^-}\,{}_\alpha\Phi^{(\mu),\xi-1} - \mathrm{i}\alpha\sqrt{2}\,{}_\alpha\Phi^{(\mu)+1,\xi}. \tag{6.145}$$

□

We shall now start computing explicitly the simplest n-point functions of the $\overline{\mathfrak{sv}}$-primary fields we have just defined.

6.4 Correlators of the Polynomial and Generalized Polynomial Fields

We obtain below the two-point functions of the generalized polynomial fields ${}_\alpha\Phi_{j,k}$ (see Propositions 6.15 and 6.16) and the three-point functions in the case $\alpha = 0$, see Proposition 6.17 (computations are much more involved in the case $\alpha \neq 0$).

Proposition 6.15. (computation of the two-point functions when $\alpha = 0$) *Set $t = t_1 - t_2, r = r_1 - r_2, \zeta = \zeta_1 - \zeta_2$ for the differences of coordinates. Then the two-point function*

$$\mathscr{C}(t_1, r_1, \zeta_1; t_2, r_2, \zeta_2) := \langle 0 \mid \Phi^{(0)}_{j_1,k_1}(t_1, r_1, \zeta_1)\Phi^{(0)}_{j_2,k_2}(t_2, r_2, \zeta_2) \mid 0\rangle$$

is equal to

$$\mathscr{C}(t, r, \zeta) = \delta_{j_1+k_1, j_2+k_2}(-1)^{k_2} j_1!\, k_1! \binom{j_2}{k_1} t^{-(j_1+k_1)} \left(\zeta - \frac{r^2}{2t}\right)^{j_1-k_2} \tag{6.146}$$

if $j_1 \geq k_2$ and 0 else.

Proof. We use the covariance of $\mathscr{C}$ under the finite subalgebra $\rho_{j,k}(L_{\pm1,0})$, $\rho_{j,k}(Y_{\pm\frac{1}{2}})$, $\rho_{j,k}(N_0)$. In particular, $\mathscr{C}$ is a function of the differences of coordinates t, r, ζ only. Covariance under

$$\rho(L_0) = -\sum_{i=1}^{2}\left(t_i\partial_{t_i} + \frac{1}{2}r_i\partial_{r_i}\right) - \frac{1}{2}\sum_{i=1}^{2}(j_i + k_i), \quad \rho(Y_{\frac{1}{2}}) = -\sum_{i=1}^{2} t_i\partial_{r_i} + r_i\partial_{\zeta_i}, \tag{6.147}$$

$$\rho(L_1) = -\sum_{i=1}^{2} t_i^2 \partial_{t_i} + t_i r_i \partial_{r_i} + \frac{r_i^2}{2}\partial_{\zeta_i} - \sum_{i=1}^{2} t_i(j_i + k_i) \tag{6.148}$$

yields quite generally (see [50], [54])

$$\mathscr{C} = C.\delta_{j_1+k_1,j_2+k_2} t^{-(j_1+k_1)} f\left(\zeta - \frac{r^2}{2t}\right) \tag{6.149}$$

for some function f.

Suppose $j_1 + k_1 = j_2 + k_2$. Assuming the extra covariance under $\tilde{\rho}_k(N_0) = -\sum_{i=1}^{2}(r_i\partial_{r_i} + 2\zeta_i\partial_{\zeta_i}) + (j_1-k_1) + (j_2-k_2) \equiv -(r\partial_r + 2\zeta\partial_\zeta) + 2(j_1-k_2)$, one gets $vf'(v) = (j_1-k_2)f$, hence $f(v) = v^{j_1-k_2}$ up to a constant (see also Proposition A.2 in Appendix A). The coefficient $(-1)^{k_2} j_1! k_1! \binom{j_2}{k_1}$ may be obtained from the coefficient C of the term of highest degree in t (i.e. the least singular term in t) in the formal series in $r_{1,2}, \zeta_{1,2}$. Since $Y_{(0)} \equiv a\partial_{\bar{b}^-} + \partial\bar{b}^+\partial_a$ maps an L-quasiprimary field of weight, say, λ into an L-quasiprimary field of weight $\lambda + \frac{1}{2}$, it is clear that C can be read from

$$\begin{aligned}\mathscr{C}_0 &= \langle 0 \mid \Phi^{(0),0}(t_1, r_1, \zeta_1)\Phi^{(0),0}(t_2, r_2, \zeta_2) \mid 0\rangle \\ &= \sum_{m_1,m_2} \zeta^{m_1+m_2} \binom{j_1}{m_1}\binom{j_2}{m_2} \langle 0| : (\bar{b}^-)^{j_1-m_1}(\bar{b}^+)^{k_1+m_1} : (t_1) \\ &\quad : (\bar{b}^-)^{j_2-m_2}(\bar{b}^+)^{k_2+m_2} : (t_2)|0\rangle\end{aligned} \tag{6.150}$$

For the same reason, $\mathscr{C}_0$ must be equal to $Ct^{-(j_1+k_1)}(\zeta_1 - \zeta_2)^{j_1-k_2}$. One gets immediately $C = 0$ for $j_1 < k_2$. In the contrary case, one gets C by looking for the coefficient of $\zeta_2^{j_1-k_2}$, which is given by

$$\begin{aligned}&(-1)^{j_1-k_2}\langle 0 \mid \Big(: (\bar{b}^-)^{j_1}(\bar{b}^+)^{k_1} : (t_1)\Big) \Bigg(\binom{j_2}{j_1-k_2} \\ &\quad : (\bar{b}^-)^{j_2-(j_1-k_2)}(\bar{b}^+)^{k_2+(j_1-k_2)} : (t_2)\Bigg) \mid 0\rangle \\ &= (-1)^{k_2} t^{-(j_1+k_1)} \binom{j_2}{k_1} j_1! k_1!.\end{aligned} \tag{6.151}$$

□

Proposition 6.16. *(computation of the two-point functions when $\alpha \neq 0$)*
Set $t = t_1 - t_2$, $r = r_1 - r_2$, $\zeta = \zeta_1 - \zeta_2$ for the differences of coordinates. Write

$$\mathscr{C}^{\mu_1,\mu_2}_{(\alpha_1,j_1,k_1),(\alpha_2,j_2,k_2)} := \langle 0 \mid {}_{\alpha_1}\Phi^{(\mu_1)}_{j_1,k_1}(t_1, r_1, \zeta_1)\, {}_{\alpha_2}\Phi^{(\mu_2)}_{j_2,k_2}(t_2, r_2, \zeta_2) \mid 0\rangle. \tag{6.152}$$

Then:

(i) the two-point functions vanish unless $\alpha_1 = -\alpha_2$ *and* $j_1 \geq k_2$ *and* $j_2 \geq k_1$;
(ii) suppose that $j_1 = j_2 := j$, $k_1 = k_2 = 0$ *and* $\alpha := \alpha_1 = -\alpha_2$. *Then*

$$\mathscr{C}^{\mu_1,\mu_2}_{(\alpha,j,0),(-\alpha,j,0)} = t^{-j-\alpha^2+\frac{\mu_1+\mu_2}{2}} \sum_{\delta=\max(\mu_1,\mu_2)}^{j} c_j^{\delta} \frac{(i\alpha\sqrt{2})^{\delta-\mu_1}(-i\alpha\sqrt{2})^{\delta-\mu_2}}{(\delta-\mu_1)!(\delta-\mu_2)!}$$

$$\left(\frac{r^2}{t}\right)^{\delta-\frac{\mu_1+\mu_2}{2}} \left(\zeta - \frac{r^2}{2t}\right)^{j-\delta} \tag{6.153}$$

where $c_j^{\delta} = (-1)^{\delta} \frac{(j!)^2}{2^{\delta}(j-\delta)!}$.

Remark. All the other cases may be deduced easily from formula (6.153) since, if $j_1 \geq k_2$ and $j_2 \geq k_1$ and (without loss of generality) $(j_2+k_2)-(j_1+k_1) = \Delta \geq 0$,

$$\mathscr{C}^{\mu_1,\mu_2}_{(\alpha,j_1,k_1),(-\alpha,j_2,k_2)}$$

$$= \left(k_1!\binom{j_1+k_1}{k_1} . k_2!\binom{j_2+k_2}{k_2}\right)^{-1} \langle 0 \mid \left[(\bar{b}^+\partial_{\bar{b}^-})^{k_1}\ {}_{\alpha}\Phi^{(\mu_1)}_{j_1+k_1,0}\right]$$

$$\left[(\bar{b}^+\partial_{\bar{b}^-})^{k_2}\ {}_{-\alpha}\Phi^{(\mu_2)}_{j_2+k_2,0}\right] \mid 0\rangle$$

$$= \left(k_1!\binom{j_1+k_1}{k_1} . k_2!\binom{j_2+k_2}{k_2} . \Delta!\binom{j_1+k_1+\Delta}{\Delta}\right)^{-1}$$

$$\langle 0 \mid \left[\partial^{\Delta}_{\bar{b}^-}(\bar{b}^+\partial_{\bar{b}^-})^{k_1}\ {}_{\alpha}\Phi^{(\mu_1)}_{j_2+k_2,0}\right]\left[(\bar{b}^+\partial_{\bar{b}^-})^{k_2}\ {}_{-\alpha}\Phi^{(\mu_2)}_{j_2+k_2,0}\right] \mid 0\rangle$$

$$= \frac{(i\sqrt{2})^{\Delta}}{k_1!\binom{j_1+k_1}{k_1}\cdot k_2!\binom{j_2+k_2}{k_2}\cdot \Delta!\binom{j_1+k_1+\Delta}{\Delta}} \partial^{k_1}_{\zeta_1}\partial^{k_2}_{\zeta_2}\mathscr{C}^{\mu_1+\Delta,\mu_2}_{(\alpha,j_2+k_2,0),(-\alpha,j_2+k_2,0)} \tag{6.154}$$

thanks to the fact that $\partial_{\zeta}\ {}_{\alpha}\Phi^{(\mu)}_{j,k} = \bar{b}^+\partial_{\bar{b}^-}\ {}_{\alpha}\Phi^{(\mu)}_{j,k}$ and ${}_{\alpha}\Phi^{(\mu+1)}_{j,k} = -\frac{i}{\sqrt{2}}\partial_{\bar{b}^-}\ {}_{\alpha}\Phi^{(\mu)}_{j,k}$.

Proof. We only prove (ii) since (i) is clear from the preceding computations. Applying Theorem A.1 from Appendix A, with $d = j+1$, $\lambda_{1,2} = \frac{\alpha^2+j}{2}$, $\alpha_{1,2} = \pm i\alpha\sqrt{2}$ and $\lambda'_{1,2} = -j$, one gets (6.153). There remains to find the coefficients c_{δ}^j. Let us first explain how to find c_j^j. One has

$$\mathscr{C}^{j,j} = \langle 0 \mid {}_{\alpha}\Phi^{(j)}_{j,0}\ {}_{-\alpha}\Phi^{(j)}_{j,0} \mid 0\rangle = c_j^j t^{-\alpha^2}$$

$$= \left(\frac{i}{\sqrt{2}}\right)^{2j} \langle 0 \mid \left((\partial_{\bar{b}^-})^j\ {}_{\alpha}\Phi^{(0)}_{j,0}\right)\left((\partial_{\bar{b}^-})^j\ {}_{-\alpha}\Phi^{(0)}_{j,0}\right) \mid 0\rangle \text{ by Theorem 6.7}$$

$$
\begin{aligned}
&= (-1)^j 2^{-j} (j!)^2 \langle 0 \mid {}_{\alpha}\Phi^{(0)}_{0,0} \, {}_{-\alpha}\Phi^{(0)}_{0,0} \mid 0 \rangle \\
&= (-1)^j 2^{-j} (j!)^2 \langle 0 \mid : \exp \alpha r_1 \bar{b}^+ V_\alpha(t_1) : : \exp -\alpha r_2 \bar{b}^+ V_{-\alpha}(t_2) : \mid 0 \rangle \\
&= (-1)^j 2^{-j} (j!)^2 t^{-\alpha^2}. \qquad (6.155)
\end{aligned}
$$

By the same trick, one gets (by deriving $j - \varepsilon$ times with respect to $\bar{b}^-$)

$$
\begin{aligned}
\mathscr{C}^{j-\varepsilon, j-\varepsilon} &= \left(c_j^{j-\varepsilon} \left(\zeta - \frac{r^2}{2t}\right)^{\varepsilon} + O(r)\right) t^{-\varepsilon-\alpha^2} \\
&= (-1)^{j-\varepsilon} 2^{-(j-\varepsilon)} (j(j-1)\dots(\varepsilon+1))^2 \langle 0 \mid {}_{\alpha}\Phi^{(0)}_{\varepsilon,0} \, {}_{-\alpha}\Phi^{(0)}_{\varepsilon,0} \mid 0 \rangle \quad (6.156)
\end{aligned}
$$

and one may identify the lowest degree component in r – which does *not* depend on α, up to a multiplication by the factor $t^{-\alpha^2}$ – by setting $r_1 = r_2$,

$$
\begin{aligned}
\mathscr{C}^{j-\varepsilon, j-\varepsilon}(r_1 = r_2) &= c_j^{j-\varepsilon} \zeta^{\varepsilon} t^{-\varepsilon-\alpha^2} \\
&= (-1)^{j-\varepsilon} 2^{-(j-\varepsilon)} \left(\frac{j!}{\varepsilon!}\right)^2 t^{-\alpha^2} \langle 0 \mid \Phi^{(0)}_{\varepsilon,0} \Phi^{(0)}_{\varepsilon,0} \mid 0 \rangle \\
&= (-1)^{j-\varepsilon} 2^{-(j-\varepsilon)} \left(\frac{(j!)^2}{\varepsilon!}\right) \zeta^{\varepsilon} t^{-\varepsilon-\alpha^2} \qquad (6.157)
\end{aligned}
$$

by Proposition 6.15. □

Proposition 6.17. *(computation of the three-point functions when $\alpha = 0$)*
The following formula holds:

$$
\begin{aligned}
&\langle \Phi^{(0)}_{j_1,0}(t_1, r_1, \zeta_1) \Phi^{(0)}_{j_2,0}(t_2, r_2, \zeta_2) \Phi^{(0)}_{j_3,0}(t_3, r_3, \zeta_3) \rangle \\
&= \frac{j_1! j_2! j_3!}{(\frac{1}{2}(j_1 + j_3 - j_2))! (\frac{1}{2}(j_2 + j_3 - j_1))! (\frac{1}{2}(j_1 + j_2 - j_3))!} \\
&\quad \left(\frac{\xi_{12}}{t_{12}}\right)^{\frac{1}{2}(j_1+j_2-j_3)} \left(\frac{\xi_{13}}{t_{13}}\right)^{\frac{1}{2}(j_1+j_3-j_2)} \left(\frac{\xi_{23}}{t_{23}}\right)^{\frac{1}{2}(j_2+j_3-j_1)} \qquad (6.158)
\end{aligned}
$$

where $\xi_{ij} := \zeta_{ij} - \frac{r_{ij}^2}{2t_{ij}}$.

Remark. All three-point correlators for the case $\alpha = 0$ can be obtained easily from these results by applying a number of times the operator $\bar{b}^+ \partial_{\bar{b}^-}$ or equivalently ∂_ζ.

Proof. Denote by $\mathscr{C}(t_i, r_i, \zeta_i) = \langle \Phi_{j_1}(t_1, r_1, \zeta_1) \Phi_{j_2}(t_2, r_2, \zeta_2) \Phi_{j_3}(t_3, r_3, \zeta_3) \rangle$ the three-point function. By Theorem A.2 in Appendix A,

$$
\mathscr{C} = C t_{12}^{-\alpha} t_{23}^{-\beta} t_{13}^{-\gamma} \left(\xi_{12}^{\alpha} \xi_{23}^{\beta} \xi_{31}^{\gamma} + \Gamma(\xi_{12}, \xi_{13}, \xi_{23})\right) \qquad (6.159)
$$

where C is a constant, $\alpha = \frac{j_1+j_2-j_3}{2}, \beta = \frac{j_1+j_3-j_2}{2}, \gamma = \frac{j_2+j_3-j_1}{2}$, and $\Gamma = \Gamma(\xi_{12}, \xi_{23}, \xi_{13})$ is any linear combination (with constant coefficients) of monomials $\xi_{12}^{\alpha'}\xi_{23}^{\beta'}\xi_{31}^{\gamma'}$ with $(\alpha', \beta', \gamma') \neq (\alpha, \beta, \gamma)$ and $\alpha' + \beta' + \gamma' = \frac{J}{2}$. Suppose $t_3 \neq t_1, t_2$ and look at the degree of the pole in $\frac{1}{t_{12}}$ in $\mathscr{C}$ considered as a function of t_1, t_2, t_3 and $\zeta_1, \zeta_2, \zeta_3$. Each term in the asymptotic expansion of Φ_{j_i} in powers of r_i, ζ_i is a polynomial of degree j_i in the fields $a, \bar{b}^-, b^+, \partial_{\bar{b}^+}$. The covariance $\mathscr{C} = \langle \Phi_1 \Phi_2 \Phi_3 \rangle$ may be computed as any polynomial of Gaussian variables by using Wick's theorem; calling a_{ij} the number of couplings of Φ_i with Φ_j, an easy argument yields $j_1 = a_{12} + a_{13}$, $j_2 = a_{12} + a_{23}$, $j_3 = a_{13} + a_{23}$, hence in particular $a_{12} = \alpha$. Hence $\mathscr{C}$ has a pole in $\frac{1}{t_{12}}$ of degree at most 2α and Γ may not contain any term of the type $\xi_{12}^{\alpha'}\xi_{23}^{\beta}\xi_{31}^{\gamma'}$ with $\alpha' > \alpha$. By taking into account the poles in $\frac{1}{t_{23}}$ and $\frac{1}{t_{13}}$, one sees that $\Gamma = 0$.

There remains to compute the coefficient C. By rewriting $\mathscr{C}$ as

$$\mathscr{C} = \sum_{\alpha'+\beta'+\gamma'=J/2} C_{\alpha',\beta',\gamma'} \xi_{12}^{\alpha'}\xi_{13}^{\beta'} (\xi_{12} + \xi_{23} + \xi_{31})^{\gamma'}, \tag{6.160}$$

and using $\xi_{31} = (\xi_{12} + \xi_{23} + \xi_{31}) - \xi_{12} - \xi_{23}$, one sees that $C = C_{\alpha,\beta,\gamma}$. Now a minute's thought shows that the coefficient of

$$(\zeta_1^0 \zeta_2^\beta \zeta_3^\gamma)(r_1^0 r_2^0 r_2^{2\gamma}) t_{23}^{-2\gamma} t_{12}^{-\alpha} t_{13}^{-\beta} \tag{6.161}$$

in $\mathscr{C}$ is equal to $(-1)^{J/2} 2^{-\gamma} C_{\alpha,\beta,\gamma}$. hence (using the asymptotic expansion of Φ_1, Φ_2 and Φ_3 in powers of ζ_i, r_i) $C_{\alpha,\beta,\gamma}$ may be computing by extracting the coefficient of $t_{23}^{-2\gamma} t_{12}^{-\alpha} t_{13}^{-\beta}$ in

$$\begin{aligned} \langle 0 | \ &: (\bar{b}^-)^{j_1} : (t_1) : \binom{j_2}{\alpha} (\bar{b}^-)^{j_2-\alpha} (\bar{b}^+)^\alpha : (t_2) \\ &: \frac{(a\partial_{\bar{b}^-} + \partial \bar{b}^+ \partial_a)^{2\gamma}}{(2\gamma)!} \binom{j_3}{\beta} (\bar{b}^-)^{j_3-\beta} (\bar{b}^+)^\beta : (t_3) \, |0\rangle, \end{aligned}$$

and multiplying by $(-1)^{J/2} 2^{-\gamma}$. Now the coefficient of $r^{2\gamma}$ in $\exp r Y_{(0)} \cdot \left((\bar{b}^-)^{j_3-\beta} (\bar{b}^+)^\beta \right)$ is equal to

$$\begin{aligned} &\sum_{i+2j=2\gamma} r^{2\gamma} \frac{(a\partial_{\bar{b}^-})^i}{i!} \left[2^{-j} \binom{j_3-\beta}{j} (\bar{b}^-)^{j_3-\beta-j} (\partial \bar{b}^+)^j (\bar{b}^+)^\beta \right] \\ &= \sum_{i+2j=2\gamma} r^{2\gamma} 2^{-j} \frac{(j_3-\beta)!}{j!\,i!\,(j_3-\beta-i-j)!} (\partial \bar{b}^+)^j a^i (\bar{b}^+)^\beta. \end{aligned} \tag{6.162}$$

The terms with $i > 0$ do not contribute to $C_{\alpha,\beta,\gamma}$ since a can only be found in the field with the variable t_3 and does not couple to the other fields. Hence

$$C_{\alpha,\beta,\gamma} t_{23}^{-2\gamma} t_{12}^{-\alpha} t_{13}^{-\beta} = (-1)^{J/2} \binom{j_3}{\beta} \frac{(j_3-\beta)!}{\gamma!(j_3-\beta-\gamma)!} \binom{j_2}{j_2-\alpha} \langle 0 \mid \, : (\bar{b}^-)^{j_1} : (t_1)$$

$$:(\bar{b}^-)^{j_2-\alpha} (\bar{b}^+)^{\alpha} : (t_2) : (\partial \bar{b}^+)^{J/2-\alpha-\beta} (\bar{b}^+)^{\beta} : (t_3) \mid 0 \rangle. \quad (6.163)$$

The first field $(\bar{b}^-)^{j_1}$ couples α times (resp. β times) with the second (resp. third) fields, yielding $(t_{12})^{-\alpha}(t_{13})^{-\beta}$ times

$$\binom{j_1}{\alpha} \alpha!(-1)^{\alpha} \beta!(-1)^{\beta}. \quad (6.164)$$

There remains the coupling of the second and third fields, namely,

$$\langle 0 \mid \, : (\bar{b}^-)^{j_2-\alpha}(t_2) : (\partial \bar{b}^+)^{J/2-\alpha-\beta} : (t_3) \quad (6.165)$$

which yields $t_{23}^{-2\gamma}$ times $\binom{j_2-\alpha}{J/2-\alpha-\beta} (J/2-\alpha-\beta)!(-1)^{j_2-\alpha}$.

All together one gets

$$C_{\alpha,\beta,\gamma} = \frac{j_1! j_2! j_3!}{(\frac{1}{2}(j_1+j_3-j_2))!(\frac{1}{2}(j_2+j_3-j_1))!(\frac{1}{2}(j_1+j_2-j_3))!}. \quad (6.166)$$

Hence the result. □

6.5 Construction of the Massive Fields

All the fields constructed until now involve only *polynomials* in the unphysical variable ζ. Inverting the (formal) Laplace transform $\mathcal{L} : f_{\mathcal{M}} \to \mathcal{L} f(\zeta) = \int_0^{\infty} f_{\mathcal{M}} e^{\mathcal{M}\zeta} \, d\mathcal{M}$ is a priori impossible since polynomials in ζ only give derivatives of the delta-function $\delta_{\mathcal{M}}$; one may say that these fields represent singular zero-mass fields, which are a priori irrelevant from a physical point of view.

However, we believe it is possible to construct *massive fields* by combining the above polynomial fields into a formal series depending on a parameter Ξ and taking an analytic continuation. The status of this construction is yet unclear. Let us formalize this as a conjecture:

Conjecture. *Massive fields may be obtained as an analytic continuation for* $\Xi \to 0$ *of series in* $\Phi_{j,k}$, ${}_{\alpha}\Phi_{j,k}$ *of the form*

$$\Xi^{\lambda} \sum_{j,k \geq 0} a_{j,k} \Xi^{-\frac{j+k}{2}} \Phi_{j,k}(t,r,\zeta) \quad \text{or} \quad \Xi^{\lambda} \sum_{j,k \geq 0} a_{j,k} \Xi^{-\frac{j+k}{2}} {}_{\alpha}\Phi_{j,k}(t,r,\zeta) \quad (6.167)$$

for some exponent λ, *with a non-zero radius of convergence in* Ξ^{-1}.

The idea lying behind this is that the discrepancy in the scaling behaviours in t of the fields $\Phi_{j,k}$ (namely, $\Xi^{-\frac{j+k}{2}}\Phi_{j,k}(t)$ behaves as $(\Xi t)^{-\frac{j+k}{2}}$ when $t \to \infty$ since $\Phi_{j,k}$ has L_0-weight $\frac{j+k}{2}$) disappears in the above sums in the limit $\Xi \to 0$ (see details of the proofs for an explanation). As for the appearance of a *massive* behaviour in the limit $\Xi \to 0$, it is reminiscent of the construction of the coherent state $e^{\mathscr{M}a^\dagger}|0\rangle$, an eigenvector of the annihilation operator a in the theory of the harmonic oscillator. We hope to make this analogy more precise in the future.

We introduce in Theorem 6.8 and Theorem 6.9 below good potential candidates for massive fields. Theorems 6.8, 6.9 and 6.10 show that all two-point functions and (at least) some three-point functions may indeed be analytically extended, and give explicit expressions for the corresponding n-point functions of the *conjectured* massive field. The missing part in the picture is a formal proof that all n-point functions have an analytic extension to $\Xi \to 0$. An encouraging fact is that the limit for $\Xi \to 0$ seems to be universal in some sense, i.e. it does not seem to depend (up to a physically irrelevant overall coefficient depending only on the mass) on the precise choice of the asymptotic series (compare e.g. Theorem 6.8, points 2. and 4.).

We made some attempts to prove the existence of the desired analytic extension by constructing the n-point functions as solutions of differential equations coming from the symmetries (for instance, the two-point function $\langle \psi_{-1}^{\Xi}\psi_{-1}^{\Xi}\rangle$, see below, may be computed – up to a constant – by using the covariance under $\mathfrak{sch}$ and under N_1, and it should be possible to compute more generally $\langle \psi_{d_1}^{\Xi}\psi_{d_2}^{\Xi}\rangle$ in the same way by induction in d_1, d_2). This scheme may work, at least for the three-point functions, but it looks like a difficult task in general, involving a precise analysis of the singularities at $\Xi = 0$ of differential operators with regular singularities.

In the case of the polynomial fields $\Phi_{j,k}$, one obtains (up to an irrelevant function of $\mathscr{M}$) the heat kernel in any even dimension, which occurs of course in many physical models (this is impossible for odd dimensions because the heat kernel then involves a square root of $t_1 - t_2$ and one should use instead non-local conformal fields in the first place instead of the bosons). In the case of the generalized polynomial fields ${}_\alpha\Phi_{j,k}$, the two-point function is non-standard, which is not surprising since the ${}_\alpha\Phi_{j,k}$ are themselves non-scalar. The exact form is new and involves a Bessel function. There are (to the best of our knowledge) no known examples at the moment of a physical model with a two-point function of this form.

Theorem 6.8 (massive fields).

1. Let $d = -1, 0, 1, \ldots$ *and* $\Xi > 0$. *Set*

$$\phi_d^{\Xi} := \sum_{j=0}^{\infty} \frac{\mathrm{i}^{j+d}}{\Xi^{\frac{j+1}{2}}} \frac{\sqrt{j!}}{(j+d+1)!} \Phi^{(0)}_{j+d+1,d+1}. \tag{6.168}$$

Then the inverse Laplace transform of the two-point function

$$\mathscr{C}^{\Xi}(t,r,\zeta) = \langle 0 \mid \phi_d^{\Xi}(t_1,r_1,\zeta_1)\phi_d^{\Xi}(t_2,r_2,\zeta_2) \mid 0\rangle,$$

defined a priori for $\Xi \gg 1$, may be analytically extended to the following function:

$$(\mathscr{L}^{-1}\mathscr{C}^{\Xi})(\mathscr{M};t,r) = e^{\mathscr{M}\Xi t} t^{-2d-1} e^{-\mathscr{M}\frac{r^2}{2t}}. \tag{6.169}$$

When $\Xi \to 0$, this goes to the standard heat kernel $K_{4d+2}(t,r) = t^{-2d-1} e^{-\mathscr{M}\frac{r^2}{2t}}$.

2. *Let $d = 0, 1, \ldots$ and $\Xi > 0$. Set*

$$\tilde{\phi}_d^{\Xi} := \sum_{j=1}^{\infty} \frac{\mathrm{i}^{j+d}}{\Xi^{\frac{j+1}{2}}} \frac{\sqrt{j!}}{(j+d)!} \Phi^{(0)}_{j+d,d+1}. \tag{6.170}$$

Then the inverse Laplace transform of the two-point function $\langle \tilde{\phi}_d^{\Xi} \tilde{\phi}_d^{\Xi} \rangle$ may be analytically extended into the function $\mathscr{M} e^{\mathscr{M}\Xi t} t^{-2d} e^{-\mathscr{M}\frac{r^2}{2t}}$. When $\Xi \to 0$, this goes to $\mathscr{M}$ times the standard heat kernel $K_{4d}(t,r) = t^{-2d} e^{-\mathscr{M}\frac{r^2}{2t}}$.

3. *(same hypotheses) Set*

$$\psi_{2d}^{\Xi} := \sum_{j=0}^{\infty} \mathrm{i}^j \frac{\Xi^{-j-d-\frac{3}{2}}}{j!} \Phi^{(0)}_{2j+2d+1,2d+1}. \tag{6.171}$$

Then the two-point function $\langle \psi_d^{\Xi} \psi_d^{\Xi} \rangle$ has an analytic continuation to small Ξ. The inverse Laplace transform of its value for $\Xi = 0$ is equal (up to a constant) to $\mathscr{M}^{2d+2} K_{4d-2}(t,r)$.

4. *(same hypotheses) Set*

$$\tilde{\psi}_{2d}^{\Xi} := \sum_{j=0}^{\infty} \mathrm{i}^j \frac{\Xi^{-j-d-\frac{1}{2}}}{j!} \Phi^{(0)}_{2j+2d,2d+1}. \tag{6.172}$$

Then the two-point function $\langle \tilde{\psi}_d^{\Xi} \tilde{\psi}_d^{\Xi} \rangle$ has an analytic continuation to small Ξ. The inverse Laplace transform of its value for $\Xi = 0$ is equal (up to a constant) to $\mathscr{M}^{2d+1} K_{4d}(t,r)$.

Remark. One may also define

$$\psi_d^{\Xi} := \sum_{j=0}^{\infty} \mathrm{i}^j \frac{\Xi^{-j-\frac{d}{2}-1}}{j!} \Phi^{(0)}_{2j+d+1,d+1} \tag{6.173}$$

for d odd, but similar computations (using a different connection formula for the hypergeometric function though, see proof below) show that its two-point function is equal (up to a constant) to that of ψ_{d+1}^{Ξ}, i.e. (up to a polynomial in $\mathscr{M}$) to K_{2d}. (Note however the strange-looking but necessary shift by $\frac{1}{2}$ in the powers of Ξ in the expression of the ψ_d^{Ξ} with odd index d with respect to those with an even index). Hence the need for introducing $\tilde{\psi}_{2d}^{\Xi}$.

Proof. Note first quite generally that, if $K_d(\mathcal{M};t,r) := \frac{e^{-\mathcal{M}\frac{r^2}{2t}}}{t^{d/2}}$ is the standard heat kernel in d dimensions, then

$$\mathcal{L}(\mathcal{M}^n K_d(\mathcal{M};t,r)) = \partial_\zeta^n \left(t^{-d/2} \left(\frac{r^2}{2t} - \zeta \right)^{-1} \right) = (-1)^{n+1} n! t^{-d/2} \left(\zeta - \frac{r^2}{2t} \right)^{-n-1}. \tag{6.174}$$

We shall use the notation $\xi := \zeta - r^2/2t$ in the proof.

1. The Laplace transform of the function $g_\Xi^{(d)}(\mathcal{M};t,r) := e^{\mathcal{M}\Xi t} t^{-2d-1} e^{-\mathcal{M}\frac{r^2}{2t}}$ is equal to

$$(\mathcal{L}g^{(d)})(t,r,\zeta) = -t^{-2d-1} \frac{1}{\Xi t + (\zeta - \frac{r^2}{2t})}$$
$$= -\sum_{j=0}^{\infty} (-1)^j \Xi^{-j-1} t^{-2(d+1)-j} \left(\zeta - \frac{r^2}{2t} \right)^j$$

(provided that the series converges, or taken in a formal sense). Then Proposition 6.15 shows that the two-point function of the field ϕ_d^Ξ defined above is equal to this series.

2. Set $\tilde{g}_\Xi^{(d)}(\mathcal{M};t,r) := \mathcal{M} e^{\mathcal{M}\Xi t} t^{-2d} e^{-\mathcal{M}\frac{r^2}{2t}}$: then

$$(\mathcal{L}\tilde{g}_\Xi^{(d)})(t,r,\zeta) = \partial_\zeta (\mathcal{L} g_\Xi^{(d-\frac{1}{2})})(t,r,\zeta) = -\sum_{j=1}^{\infty} j(-1)^j t^{-2d-j-1} \left(\zeta - \frac{r^2}{2t} \right)^{j-1}$$

is easily checked to be equal to the two-point function $\langle \tilde{\phi}_d^\Xi \tilde{\phi}_d^\Xi \rangle$.

3. Set
$$I^\Xi := \Xi^{2d+3} \partial_t^{-(2d+1)} (t^{2d} \langle \psi_d^\Xi \psi_d^\Xi \rangle)$$
where $\partial_t^{-1} = \int_0^t dt$ is the integration operator from 0 to t. Then Proposition 6.15, together with the duplication formula for the Gamma function, yield

$$I^\Xi = \sum_{j\geq 0} (2j+2d+1)!\, t^{-2j-1} \frac{(-\xi^2/\Xi^2)^j}{(j!)^2}$$

$$= \frac{1}{t} \cdot \frac{2^{2j+2d+1}}{\sqrt{\pi}} \sum_{j\geq 0} \frac{\Gamma(j+d+1)\Gamma(j+d+\frac{3}{2})}{\Gamma(j+1)} \frac{\left(-\left(\frac{\xi}{\Xi t}\right)^2 \right)^j}{j!}$$

$$= \frac{1}{t} \frac{2^{2d+1}}{\sqrt{\pi}} \Gamma(d+1)\Gamma(d+\frac{3}{2}) {}_2F_1 \left(d+1, d+\frac{3}{2}; 1; -\left(\frac{2\xi}{\Xi t} \right)^2 \right) \tag{6.175}$$

which is defined for $\Xi \gg 1$. The connection formula (see [1], 15.3.3) for the Gauss hypergeometric function ${}_2F_1$

$$ {}_2F_1(a,b,c;z) = (1-z)^{c-a-b}\, {}_2F_1(c-a,c-b,c;z) \tag{6.176} $$

yields

$$ \begin{aligned} {}_2F_1\left(d+1, d+\frac{3}{2}, 1; -\left(\frac{2\xi}{\Xi t}\right)^2\right) &= \left[1+\left(\frac{2\xi}{\Xi t}\right)^2\right]^{-2d-\frac{3}{2}} \\ & {}_2F_1\left(-d, -d-\frac{1}{2}, 1; -\left(\frac{2\xi}{\Xi t}\right)^2\right). \end{aligned} \tag{6.177} $$

The hypergeometric function on the preceding line is simply a polynomial in Ξ^{-1} since $-d$ is a negative integer. By extracting the most singular term in Ξ^{-1}, one sees that

$$ I^{\Xi} \sim_{\Xi\to 0} (-1)^d \frac{d!}{4\pi} \Gamma\left(d+\frac{3}{2}\right)^2 \Xi^{2d+3} \xi^{-2d-3} t^{2d+2}. $$

Hence

$$ \begin{aligned} \langle \psi_d^{\Xi} \psi_d^{\Xi} \rangle &\to_{\Xi\to 0} (-1)^d \frac{(2d+2)!d!}{4\pi} \Gamma\left(d+\frac{3}{2}\right)^2 \xi^{-2d-3} t^{1-2d} \\ &= (-1)^{d+1} \frac{d!}{4\pi} \Gamma\left(d+\frac{3}{2}\right)^2 \mathscr{L}(\mathscr{M}^{2d+2} K_{4d-2}(\mathscr{M}; t, r)). \end{aligned} $$

4. Same method. □

Theorem 6.9 (*generalized massive fields*).
Let $\alpha \in \mathbb{R}$ and $\Xi > 0$.

1. Set

$$ {}_\alpha\phi^{\Xi} := \sum_{j=0}^{\infty} \frac{\mathrm{i}^j}{\Xi^{\frac{j+1}{2}}} \frac{1}{\sqrt{j!}}\, {}_\alpha\Phi_{j,0}^{(0)}. \tag{6.178} $$

Then the two-point function

$$ \mathscr{C}^{\Xi}(t,r,\zeta) = \langle 0 \mid {}_\alpha\phi^{\Xi}(t_1,r_1,\zeta_1)\, {}_{-\alpha}\phi^{\Xi}(t_2,r_2,\zeta_2) \mid 0 \rangle $$

has an analytic continuation to small Ξ, and its inverse Laplace transform at $\Xi = 0$ is equal to

$$ \widetilde{\mathscr{C}}_{\mathscr{M}}(t,r) = -t^{1-\alpha^2} e^{-\mathscr{M}r^2/2t} I_0(2|\alpha|\sqrt{\mathscr{M}r^2/t}) \tag{6.179} $$

where I_0 is the modified Bessel function of order 0.

2. *Set*

$$ {}_{\alpha}\psi^{\Xi} := \sum_{j=0}^{\infty} (-1)^j \frac{\Xi^{-j-\frac{1}{2}}}{j!} \, {}_{\alpha}\Phi^{(0)}_{2j,0}. \tag{6.180} $$

Then the same results hold for the two-point function $\langle {}_{\alpha}\psi^{\Xi} \; {}_{-\alpha}\psi^{\Xi} \rangle$ (up to an overall multiplicative constant).

Remark. if one replaces α with iα, then the two-point function involves this time the Bessel function J_0 (which decreases for long distances).

Proof. 1. By applying Proposition 6.16, one gets

$$ \begin{aligned} \mathscr{C}(t,r,\zeta) &= \sum_{j\geq 0} \frac{(-1)^j}{\Xi^{j+1}} t^{-j-\alpha^2} \sum_{\delta=0}^{j} (-1)^{\delta} \frac{j!}{2^{\delta}(j-\delta)!} \frac{(2\alpha^2)^{\delta}}{(\delta!)^2} \left(\frac{r^2}{t}\right)^{\delta} \left(\zeta - \frac{r^2}{2t}\right)^{j-\delta} \\ &= \frac{t^{-\alpha^2}}{\Xi} \sum_{n=0}^{\infty} \left(-\frac{\zeta - r^2/2t}{\Xi t}\right)^n \sum_{\delta=0}^{\infty} \frac{(n+\delta)!}{n!(\delta!)^2} \left(\frac{\alpha^2 r^2}{\Xi t^2}\right)^{\delta}. \end{aligned} \tag{6.181} $$

The function

$$ f(y) = \sum_{\delta=0}^{\infty} \frac{\binom{n+\delta}{n}}{\delta!} y^{\delta} = \sum_{\delta=0}^{\infty} \frac{(n+\delta)!}{n!(\delta!)^2} y^{\delta} $$

is entire and admits a Laplace transform

$$ \begin{aligned} h(\lambda) = \mathscr{L} f(\lambda) &= \int_0^{\infty} f(y) e^{-\lambda y} \, dy = \sum_{\delta=0}^{\infty} \binom{n+\delta}{n} \lambda^{-\delta-1} \\ &= \frac{1}{\lambda}(1 - 1/\lambda)^{-n-1} = \lambda^n (\lambda - 1)^{-n-1} \end{aligned} $$

which is given by a converging series for $\lambda > 1$; by inverting the Laplace transform, one gets

$$ f(y) = \partial_y^n \left(\frac{(-1)^n y^n}{n!} e^y \right). $$

An application of Leibniz formula gives

$$ \partial_y^n \left(\frac{y^n}{n!} e^y \right) = \left(\sum_{k=0}^{n} \binom{n}{k} \frac{y^k}{k!} \right) e^y. $$

By putting everything together and setting $y = \frac{\alpha^2 r^2}{\Xi t^2}$, one gets

$$\mathscr{C}(t,r,\zeta) = \frac{t^{-\alpha^2}}{\Xi} \sum_{n=0}^{\infty} \left(\frac{\zeta - r^2/2t}{\Xi t} \right)^n \left(\sum_{k=0}^{n} \frac{\binom{n}{k}}{k!} \left(\frac{\alpha^2 r^2}{\Xi t^2} \right)^k \right) e^{\frac{\alpha^2 r^2}{\Xi t^2}}$$

$$= \frac{t^{-\alpha^2}}{\Xi} \sum_{k=0}^{\infty} \frac{1}{k!} \left[\sum_{n=k}^{\infty} \binom{n}{k} \left(\frac{\zeta - r^2/2t}{\Xi t} \right)^n \left(\frac{\alpha^2 r^2}{\Xi t^2} \right)^k \right] e^{\frac{\alpha^2 r^2}{\Xi t^2}}. \quad (6.182)$$

By comparing with the generating series

$$\sum_{k=0}^{\infty} \left(\sum_{n=k}^{\infty} \binom{n}{k} a^n \right) x^k = \sum_{n=0}^{\infty} a^n \sum_{k=0}^{n} \binom{n}{k} x^k = \sum_{n=0}^{\infty} [a(1+x)]^n$$

$$= \frac{1}{1-a} \sum_{k=0}^{\infty} \left(\frac{a}{1-a} \right)^k x^k,$$

one gets

$$\mathscr{C}(t,r,\zeta) = -\frac{t^{1-\alpha^2}}{\zeta - r^2/2t - \Xi t} \exp\left[-\alpha^2 \frac{r^2}{t} \left(\frac{1}{\zeta - r^2/2t - \Xi t} \right) \right]. \quad (6.183)$$

One finds in [27] $\mathscr{L}^{-1}(\lambda^{-1} e^{a/\lambda})(t) = I_0(2\sqrt{at})$, $a > 0$ (mind our unusual convention for the Laplace transform with respect to the mass $\mathscr{M}$!), where I_0 is the modified Bessel function of order 0. Hence

$$\mathscr{C}_{\mathscr{M}}(t,r) = -t^{1-\alpha^2} e^{\mathscr{M}\Xi t} e^{-\mathscr{M}r^2/2t} I_0(2|\alpha| \sqrt{\mathscr{M}r^2/t}).$$

2. The method is the same but computations are considerably more involved. Set $y := \frac{\alpha^2 r^2}{\Xi t^2}$ and $x = -\frac{4\xi}{\Xi t}$. Applying Proposition 6.16 yields this time

$$\mathscr{C}(t,r,\zeta) = \frac{t^{-\alpha^2}}{\Xi} \sum_{n=0}^{\infty} \left(-\frac{\zeta - r^2/2t}{\Xi t} \right)^n \sum_{\delta \geq 0, \delta + n \equiv 0[2]} (-1)^{\frac{n+\delta}{2}}$$

$$\frac{((n+\delta)!)^2}{n![(\frac{1}{2}(n+\delta))!]^2 (\delta!)^2} \left(\frac{\alpha^2 r^2}{\Xi t^2} \right)^\delta.$$

Let $h(\lambda)$ be the Laplace transform of $\mathscr{C}$ with respect to y. Formally, this is equivalent to replacing $y^\delta/\delta!$ by $\lambda^{-\delta-1}$. Separating the cases n, δ even, resp. odd, and using the duplication formula for the Gamma function, one gets

$$h(\lambda) = \frac{1}{\Xi \sqrt{\pi} \lambda t^{\alpha^2}} \left[\sum_{n=0}^{\infty} (-1)^n \frac{x^{2n}}{(2n)!} \sum_{m=0}^{\infty} (-1)^m \frac{\Gamma(n+m+\frac{1}{2})^2}{\Gamma(m+\frac{1}{2}) m!} \left(\frac{2}{\lambda} \right)^{2m} \right.$$

$$- \sum_{n=0}^{\infty}(-1)^n \frac{x^{2n+1}}{(2n+1)!} \sum_{m=0}^{\infty}(-1)^m \frac{\Gamma(n+m+\frac{3}{2})^2}{\Gamma(m+1)\Gamma(m+\frac{3}{2})}\left(\frac{2}{\lambda}\right)^{2m+1}\Bigg]$$

$$= \frac{1}{\Xi\sqrt{\pi}\lambda t^{\alpha^2}}\left[\sum_{n=0}^{\infty}(-1)^n \frac{x^{2n}}{(2n)!}\,{}_2F_1\left(n+\frac{1}{2}, n+\frac{1}{2}; \frac{1}{2}; -\frac{4}{\lambda^2}\right)\frac{\Gamma(n+\frac{1}{2})^2}{\Gamma(\frac{1}{2})}\right.$$

$$\left. - \sum_{n=0}^{\infty}(-1)^n \frac{x^{2n+1}}{(2n+1)!}\,{}_2F_1\left(n+\frac{3}{2}, n+\frac{3}{2}; \frac{3}{2}; -\frac{4}{\lambda^2}\right)\frac{2}{\lambda}\frac{\Gamma(n+\frac{3}{2})^2}{\Gamma(\frac{3}{2})}\right]. \tag{6.184}$$

Hence $h(\lambda) = \frac{1}{\Xi\sqrt{\pi}\lambda t^{\alpha^2}}(T_1(\lambda) + T_2(\lambda))$ where (using once more the duplication formula and connection formulas for the hypergeometric function)

$$T_1(\lambda) = \sqrt{\pi}\left(1+\frac{4}{\lambda^2}\right)^{-\frac{1}{2}} \sum_{n=0}^{\infty}(-1)^n \frac{(\frac{1}{2})_n}{n!}\left(\frac{x}{2}\right)^{2n}\left(1+\frac{4}{\lambda^2}\right)^{-2n}$$
$$\times {}_2F_1\left(-n, -n; \frac{1}{2}; -\frac{4}{\lambda^2}\right)$$

and

$$T_2(\lambda) = -\sqrt{\pi}\frac{x}{\lambda}\left(1+\frac{4}{\lambda^2}\right)^{-\frac{3}{2}} \cdot \sum_{n=0}^{\infty}(-1)^n \frac{(\frac{3}{2})_n}{n!}\left(\frac{x}{2}\right)^{2n}\left(1+\frac{4}{\lambda^2}\right)^{-2n}$$
$$\times {}_2F_1\left(-n, -n; \frac{3}{2}; -\frac{4}{\lambda^2}\right).$$

These hypergeometric functions are simple polynomials since they have negative integer arguments; however, the sum obtained by expanding these polynomials looks hopelessly complicated. We use instead the following formula

$$\sum_{n=0}^{\infty}\frac{(c)_{2n}}{n!(a)_n}(-v^2)^n\,{}_2F_1(-n, 1-a-n; b; u^2)$$
$$= 2^{a+b-c-2}u^{1-b}v^{-c}\frac{\Gamma(a)\Gamma(b)}{\Gamma(c)}\int_0^{\infty} J_{a-1}(w)J_{b-1}(uw)\exp(-\frac{w}{2v})w^{c-a-b+1}dw, \tag{6.185}$$

(see [48], formula (65.3.11)), valid if Re $(c) > 0$, Re $(v) > 0$, Re $(\frac{1}{2v} \pm iu) > 0$, Re $(a+b+c) > 0$.

Hence

$$T_1(\lambda) = -2x^{-1}\sqrt{\pi}\left(1+\frac{4}{\lambda^2}\right)^{\frac{1}{2}}\int_0^\infty dw J_0(w)\cosh\left(\frac{2w}{\lambda}\right)$$
$$\times\exp\left(-2wx^{-1}\left(1+\frac{4}{\lambda^2}\right)\right) \tag{6.186}$$

and

$$T_2(\lambda) = -2x^{-1}\sqrt{\pi}\left(1+\frac{4}{\lambda^2}\right)^{\frac{1}{2}}\int_0^\infty dw J_0(w)\sinh\left(\frac{2w}{\lambda}\right)$$
$$\times\exp\left(-2wx^{-1}\left(1-\frac{4}{\lambda^2}\right)\right). \tag{6.187}$$

By applying the following formula [38]

$$\int_0^\infty dw\, e^{-\beta w}J_0(w) = \frac{1}{\sqrt{1+\beta^2}}$$

(Laplace transform of the Bessel function) and expanding the cosh and sinh functions into exponentials, one gets

$$T_1 = -\sqrt{\pi}x^{-1}\left(1+\frac{4}{\lambda^2}\right)^{\frac{1}{2}}\left[\left(1+\left(-\frac{2}{\lambda}+\frac{2}{x}(1+\frac{4}{\lambda^2})\right)^2\right)^{-\frac{1}{2}}\right.$$
$$\left.+\left(1+\left(\frac{2}{\lambda}+\frac{2}{x}(1+\frac{4}{\lambda^2})\right)^2\right)^{-\frac{1}{2}}\right] \tag{6.188}$$

and

$$T_2 = -\sqrt{\pi}x^{-1}\left(1+\frac{4}{\lambda^2}\right)^{\frac{1}{2}}\left[\left(1+\left(-\frac{2}{\lambda}+\frac{2}{x}(1+\frac{4}{\lambda^2})\right)^2\right)^{-\frac{1}{2}}\right.$$
$$\left.-\left(1+\left(\frac{2}{\lambda}+\frac{2}{x}(1+\frac{4}{\lambda^2})\right)^2\right)^{-\frac{1}{2}}\right] \tag{6.189}$$

Using

$$\left(1+\frac{4}{\lambda^2}\right)^{\frac{1}{2}}\left[1+\left(-\frac{2}{\lambda}+\frac{2}{x}\left(1+\frac{4}{\lambda^2}\right)\right)^2\right]^{-\frac{1}{2}} = \frac{\lambda}{\sqrt{\left(1+\frac{4}{x^2}\right)\lambda^2-\frac{8}{x}\lambda+\frac{16}{x^2}}} \tag{6.190}$$

and the inverse Laplace transform

$$\mathscr{L}^{-1}\left(\frac{1}{\sqrt{a\lambda^2+2b\lambda+c}}\right)(y)=\frac{1}{\sqrt{a}}e^{-\frac{b}{a}y}J_0\left(y\sqrt{\frac{c}{a}-\frac{b^2}{a^2}}\right) \tag{6.191}$$

one gets

$$h(\lambda)=\frac{-2}{\Xi x t^{\alpha^2}}\frac{1}{\sqrt{(1+\frac{4}{x^2})\lambda^2-\frac{8}{x}\lambda+\frac{16}{x^2}}}$$

hence

$$\begin{aligned} C(t,r,\zeta) &= \frac{-2}{\Xi t^{\alpha^2}\sqrt{\frac{16\xi^2}{\Xi^2 t^2}+4}}\exp\left(-\frac{16\xi/\Xi t}{16\xi^2/\Xi^2t^2+4}\frac{\alpha^2r^2}{\Xi t^2}\right) \\ &\quad J_0\left(\frac{8\alpha^2r^2/\Xi t^2}{16\xi^2/\Xi^2t^2+4}\right) \\ &\sim_{\Xi\to 0} -\frac{1}{2\xi}t^{1-\alpha^2}\exp(-\frac{\alpha^2r^2}{\xi t}) \end{aligned} \tag{6.192}$$

which is (up to a constant) exactly the same expression we got for the two-point function $\langle {}_\alpha\phi^\Xi\ {}_{-\alpha}\phi^\Xi\rangle$. □

We did not manage to compute explicitly the three-point functions $\langle\psi^\Xi_{d_1}\psi^\Xi_{d_2}\psi^\Xi_{d_3}\rangle$ except in the simplest case $d_1=d_2=d_3=-1$ (see the remark after Theorem 6.8 for the definition of ψ^Ξ_{-1}). One obtains:

Theorem 6.10. *Let*

$$\mathscr{C}^\Xi := \langle\psi^\Xi_{-1}(t_1,r_1,\zeta_1)\psi^\Xi_{-1}(t_2,r_2,\zeta_2)\psi^\Xi_{-1}(t_3,r_3,\zeta_3)\rangle \tag{6.193}$$

be the three point-function of the massive field

$$\psi^\Xi_{-1} := \sum_{j=0}^{\infty} \mathrm{i}^j\frac{\Xi^{-j-\frac{1}{2}}}{j!}\Phi^{(0)}_{2j,0}$$

defined in Theorem 6.8. Then

$$\mathscr{C}^\Xi \to_{\Xi\to 0} C\left[\prod_{1\le i<j\le 3}(t_i-t_j)\right]\left[\prod_{1\le i<j\le 3}\left(2(\zeta_{ij}t_{ij}-r_{ij}^2)\right)\right]^{\frac{1}{2}}. \tag{6.194}$$

Remark. Up to the t-dependent pre-factor, this result coincides with the three-point functions for conformally covariant fields in three dimensions (see Sect. 2.4 for comments).

Proof. Let $\xi_{ij} := \zeta_{ij} - \frac{r_{ij}^2}{2t_{ij}}$ and $x_1 = \mathrm{i}\frac{\xi_{12}\xi_{13}t_{23}}{t_{12}t_{13}\xi_{23}\Xi}$, $x_2 = \mathrm{i}\frac{\xi_{23}\xi_{21}t_{31}}{t_{23}t_{21}\xi_{31}\Xi}$, $x_3 = \mathrm{i}\frac{\xi_{31}\xi_{32}t_{12}}{t_{31}t_{32}\xi_{12}\Xi}$. One must prove that $C^{\Xi} \to_{\Xi\to 0} C\left(\frac{t_{12}t_{23}t_{31}}{\xi_{12}\xi_{23}\xi_{31}}\right)^{\frac{1}{2}}$. Proposition 6.17 yields

$$\mathscr{C}^{\Xi} := \Xi^{-\frac{3}{2}} \sum_{j_1,j_2\ge 0} \frac{x_1^{j_1}x_2^{j_2}}{j_1!j_2!}(2j_1)!(2j_2)! \sum_{|j_1-j_2|\le j_3\le j_1+j_2} \frac{x_3^{j_3}}{j_3!}$$
$$\frac{(2j_3)!}{(j_1+j_3-j_2)!(j_1+j_2-j_3)!(j_2+j_3-j_1)!}. \tag{6.195}$$

Write

$$\frac{1}{(j_1+j_2-j_3)!} = (-1)^{j_3-j_1-j_2} \lim_{\varepsilon\to 0} \varepsilon\Gamma(j_3-j_1-j_2+\varepsilon). \tag{6.196}$$

This form of the complement formula for the Gamma function is valid whatever the argument. Then

$$\mathscr{C}^{\Xi} = \Xi^{-\frac{3}{2}} \sum_{j_1,j_2\ge 0} \frac{x_1^{j_1}x_2^{j_2}}{j_1!j_2!}(2j_1)!(2j_2)!\, I_3(j_1,j_2;x_3) \tag{6.197}$$

where (by using also the duplication formula for the Gamma function)

$$\begin{aligned}
&I_3(j_1,j_2;x_3) = \\
&(-1)^{j_1+j_2} \lim_{\varepsilon\to 0} \varepsilon \sum_{j_3=|j_1-j_2|}^{\infty} (-4x_3)^{j_3} \frac{\Gamma(j_3+\frac{1}{2})}{\sqrt{\pi}} \frac{\Gamma(j_3-j_1-j_2+\varepsilon)}{\Gamma(j_3+(j_1-j_2)+1)\Gamma(j_3+(j_2-j_1)+1)} \\
&= \frac{(-1)^{j_1+j_2}}{\sqrt{\pi}}(-4x_3)^{|j_1-j_2|} \sum_{j=0}^{\infty} \frac{(-4x_3)^j}{j!} \frac{\Gamma(j+|j_1-j_2|+\frac{1}{2})\Gamma(j-2\min(j_1,j_2)+\varepsilon)}{\Gamma(j+2|j_1-j_2|+1)} \\
&= \frac{(-1)^{j_1+j_2}}{\sqrt{\pi}}(-4x_3)^{|j_1-j_2|}\, {}_2\bar{F}_1(|j_1-j_2|+\frac{1}{2}, \varepsilon-2\min(j_1,j_2); 2|j_1-j_2|+1; -4x_3)
\end{aligned} \tag{6.198}$$

The symbol ${}_2\bar{F}_1$ stands for Gauss' hypergeometric function, except for a different normalization, namely,

$${}_2\bar{F}_1(a,b,c;z) = \frac{\Gamma(a)\Gamma(b)}{\Gamma(c)}\, {}_2F_1(a,b,c;z) = \sum_{n\ge 0} \frac{\Gamma(a+n)\Gamma(b+n)}{\Gamma(c+n)} \frac{z^n}{n!}. \tag{6.199}$$

Now the well-known formula connecting the behaviour around 0 with the behaviour around infinity of the hypergeometric function, see [1] for instance, yields

$$
\begin{aligned}
&{}_2\bar{F}_1(|j_1-j_2|+\frac{1}{2},\varepsilon-2\min(j_1,j_2);2|j_1-j_2|+1;-4x_3)\\
&=\Gamma\left(\varepsilon-j_1-j_2-\frac{1}{2}\right)\left(\frac{1}{4x_3}\right)^{|j_1-j_2|+\frac{1}{2}}\\
&{}_2F_1\left(|j_1-j_2|+\frac{1}{2},\frac{1}{2}-|j_1-j_2|;j_1+j_2+\frac{3}{2};-\frac{1}{4x_3}\right)\\
&+\frac{\Gamma(\varepsilon-2\min(j_1,j_2))\Gamma(j_1+j_2+\frac{1}{2})}{\Gamma(2\max(j_1,j_2)+1-\varepsilon)}\left(\frac{1}{4x_3}\right)^{-2\min(j_1,j_2)}\\
&{}_2F_1\left(-2\min(j_1,j_2),-2\max(j_1,j_2);\frac{1}{2}-j_1-j_2;-\frac{1}{4x_3}\right). \qquad (6.200)
\end{aligned}
$$

In the limit $\varepsilon\to 0$, only the second term in the right-hand side has a pole, $\Gamma(\varepsilon-2\min(j_1,j_2))\sim_{\varepsilon\to 0}\frac{1}{\Gamma(1+2\min(j_1,j_2))}\frac{1}{\varepsilon}$, hence

$$
I_3=\frac{(4x_3)^{j_1+j_2}}{\sqrt{\pi}}\frac{\Gamma(j_1+j_2+\frac{1}{2})}{\Gamma(1+2j_1)\Gamma(1+2j_2)}\,{}_2F_1\left(-2j_1,-2j_2;\frac{1}{2}-j_1-j_2;-\frac{1}{4x_3}\right) \qquad (6.201)
$$

and

$$
\begin{aligned}
\mathscr{C}^{\Xi}&=\frac{\Xi^{-\frac{3}{2}}}{\sqrt{\pi}}\sum_{j_1,j_2\geq 0}\frac{(4x_1x_3)^{j_1}(4x_2x_3)^{j_2}}{j_1!j_2!}\Gamma\left(j_1+j_2+\frac{1}{2}\right)\\
&\quad {}_2F_1\left(-2j_1,-2j_2;\frac{1}{2}-j_1-j_2;-\frac{1}{4x_3}\right). \qquad (6.202)
\end{aligned}
$$

Kummer's quadratic transformation formulas for the hypergeometric functions give (see [1], 15.3.22)

$$
\begin{aligned}
&{}_2F_1\left(-2j_1,-2j_2;\frac{1}{2}-j_1-j_2;-\frac{1}{4x_3}\right)\\
&\quad={}_2F_1\left(-j_1,-j_2;\frac{1}{2}-j_1-j_2;1-\left(1+\frac{1}{2x_3}\right)^2\right). \qquad (6.203)
\end{aligned}
$$

Now for any β

$$
\begin{aligned}
\sum_{j_1,j_2\geq 0}\frac{y_1^{j_1}}{j_1!}\frac{y_2^{j_2}}{j_2!}\Gamma(j_1+j_2+\beta)&=\sum_{j_1\geq 0}\frac{y_1^{j_1}}{j_1!}\Gamma(j_1+\beta)(1-y_2)^{-j_1-\beta}\\
&=\Gamma(\beta)(1-y_1-y_2)^{-\beta}. \qquad (6.204)
\end{aligned}
$$

Applying this formula to each term in the series expansion of the above hypergeometric function yields

$$\begin{aligned}
\mathscr{C}^{\Xi} &= \frac{\Xi^{-\frac{3}{2}}}{\sqrt{\pi}} \sum_{k\geq 0} \frac{\left(16x_3^2x_1x_2\left(\left(1+\frac{1}{2x_3}\right)^2-1\right)\right)^k}{k!} \sum_{l_1,l_2\geq 0} \frac{(4x_1x_3)^{l_1}(4x_2x_3)^{l_2}}{l_1!l_2!} \\
&\quad \times \Gamma\left(l_1+l_2+\left(k+\frac{1}{2}\right)\right) \\
&= \frac{\Xi^{-\frac{3}{2}}}{\sqrt{\pi}} \sum_{k\geq 0} \frac{\left(16x_3^2x_1x_2\left(\left(1+\frac{1}{2x_3}\right)^2-1\right)\right)^k}{k!} \\
&\quad \times \Gamma\left(k+\frac{1}{2}\right)(1-4x_1x_3-4x_2x_3)^{-k-\frac{1}{2}} \\
&= \Xi^{-\frac{3}{2}}\left(1-4x_3(x_1+x_2)+16x_3^2x_1x_2\left(1-\left(1+\frac{1}{2x_3}\right)^2\right)\right)^{-\frac{1}{2}} \\
&= \Xi^{-\frac{3}{2}}(1-4x_1x_2-4x_1x_3-4x_2x_3-16x_1x_2x_3)^{-\frac{1}{2}} \qquad (6.205)
\end{aligned}$$

hence the limit when $\Xi \to 0$.

Chapter 7
Cohomology, Extensions and Deformations

Recall that we introduced in Chap. 2 a variant of $\mathfrak{sv}$ with integer-value component indices, $\mathfrak{sv}(0)$, called *twisted Schrödinger–Virasoro algebra*; two families of deformations, $\mathfrak{sv}_\varepsilon$ and $\mathfrak{sv}_\varepsilon(0)$ such that the Y, resp. M field has conformal weight $\frac{3+\varepsilon}{2}$, resp. $1+\varepsilon$; two independent graduations, δ_1 and δ_2, on each of these algebras, which define inner or outer derivations, $\bar{\delta}_1$ and $\bar{\delta}_2$. The Virasoro cocycle extends naturally to these algebras, yielding centrally extended algebras $\widetilde{\mathfrak{sv}}_\varepsilon$, $\widetilde{\mathfrak{sv}}_\varepsilon(0)$.

The purpose of this chapter is to search for *deformations*, *central extensions* and *outer derivations* in a systematic way, using cohomological methods. Section 7.1 has been written for non-experts: we show the meaning of the lowest-dimensional cohomology groups in terms of the Lie algebra structure, and give some hints on the use of *Hochschild-Serre spectral sequences* to compute the cohomology of semi-direct products.

Let us briefly summarize the upcoming results. We concentrate on $\mathfrak{sv}(0)$ in Sects. 7.2, 7.3 and 7.4, because the generators of $\mathfrak{sv}(0)$ bear integer indices, which is more natural for computations. The main theorem is Theorem 7.1 in Sect. 7.2, which classifies all *deformations* of $\mathfrak{sv}(0)$; Theorem 7.2 shows that all the infinitesimal deformations obtained in Sect. 7.2 give rise to genuine deformations. One particularly interesting family of deformations is realized by the Lie algebras $\mathfrak{sv}_\varepsilon(0)$ ($\varepsilon \in \mathbb{R}$), which were introduced in Definition 2.2. We compute their central extensions (see Sect. 7.3), and then (see Sect. 7.4) their deformations in the particulary interesting case $\varepsilon = 1$, for which $\mathfrak{sv}_1(0)$ is the tensor product of $\mathrm{Vect}(S^1)$ with a nilpotent associative and commutative algebra. Finally, in Sect. 7.5, we come back to the original Schrödinger–Virasoro algebra and compute its deformations, as well as the central extensions of the family of deformed Lie algebras $\mathfrak{sv}_\varepsilon$.

We shall use standard techniques for Lie algebra cohomology computations in the infinite-dimensional setting, as developed in the book by D.B. Fuks [31]. Some preliminary results on the cohomology of $\mathrm{Vect}(S^1)$ will be recalled when needed.

The contents of this chapter are taken from [106].

J. Unterberger and C. Roger, *The Schrödinger-Virasoro Algebra*, Theoretical and Mathematical Physics, DOI 10.1007/978-3-642-22717-2_7,

7.1 Some Prerequisites About Lie Algebra Cohomology

To a Lie algebra $\mathfrak{g}$ and a $\mathfrak{g}$-module M, we shall associate a cochain complex known as the *Chevalley-Eilenberg complex*. The n-th space of this complex will be denoted by $C^n(\mathfrak{g}, M)$. It is trivial if $n < 0$, and if $n > 0$, it is the space of n-linear antisymmetric mappings of $\mathfrak{g}$ into M: they will be called n-cochains of $\mathfrak{g}$ with coefficients in M. The space of 0-cochains $C^0(\mathfrak{g}, M)$ reduces to M. The differential d^n will be defined by the following formula: for $c \in C^n(\mathfrak{g}, M)$, the $(n+1)$-cochain $d^n(c)$ evaluated on $x_0, x_1, \ldots, x_n \in \mathfrak{g}$ gives:

$$d^n(c)(x_0, x_1, \ldots, x_n) = \sum_{0 \le i < j \le n} (-1)^{i+j-1} c([x_i, x_j], \ldots, \check{x}_i, \ldots, \check{x}_j, \ldots) + \sum_{i=0}^{n} (-1)^{i+1} x_i . c(x_0, \ldots, \check{x}_i, \ldots, x_n) \tag{7.1}$$

where $\check{x}_i$ indicates that the term corresponding to x_i should be omitted.

One easily checks that $d^{n+1} \circ d^n = 0$ (the only difficulty lies in checking properly the various signs).

The space of *n-cocycles* of $\mathfrak{g}$ with coefficients in M is then naturally defined as the kernel of the differential,

$$Z^n(\mathfrak{g}; M) = \{c \in C^n(\mathfrak{g}, M) \mid d^n(c) = 0\}. \tag{7.2}$$

Since $d^{n+1} \circ d^n = 0$, $Z^n(\mathfrak{g}; M)$ contains the subspace of *n-coboundaries* of $\mathfrak{g}$ with coefficients in M,

$$B^n(\mathfrak{g}; M) = \{c \in C^n(\mathfrak{g}, M) \mid c = d^{n-1}(c') \text{ for some } c' \text{ in } C^{n-1}(\mathfrak{g}, M)\}. \tag{7.3}$$

The n-th cohomology space of Lie algebra $\mathfrak{g}$ with coefficients in M is then by definition the quotient
$Z^n(\mathfrak{g}; M) \,/\, B^n(\mathfrak{g}; M)$, denoted by $H^n(\mathfrak{g}; M)$.

We shall now recall classical interpretations of cohomology spaces of low degrees.

1. The space $H^0(\mathfrak{g}; M) \simeq Inv_{\mathfrak{g}}(M) := \{m \in M; \forall X \in \mathfrak{g}, X \cdot m = 0\}$ is the space of invariants.
2. The space $H^1(\mathfrak{g}; M)$ classifies derivations of $\mathfrak{g}$ with values in M modulo inner ones. This result is particularly useful when $M = \mathfrak{g}$ with the adjoint representation. In that case, a derivation is a map $\delta : \mathfrak{g} \to \mathfrak{g}$ such that

$$\delta([X, Y]) = [\delta(X), Y] + [X, \delta(Y)]$$

while an inner derivation is given by the adjoint action of some element $Z \in \mathfrak{g}$ (see also Sect. 2.1).

3. The space $H^2(\mathfrak{g}; M)$ classifies extensions of Lie algebra $\mathfrak{g}$ by M, i.e. short exact sequences of Lie algebras

$$0 \longrightarrow M \longrightarrow \hat{\mathfrak{g}} \longrightarrow \mathfrak{g} \longrightarrow 0, \tag{7.4}$$

in which M is considered as an abelian Lie algebra. We shall mainly consider two particular cases of this situation which will be extensively studied in the sequel:

(a) If M is a trivial $\mathfrak{g}$-module (typically $M = \mathbb{R}$ or $\mathbb{C}$), $H^2(\mathfrak{g}; M)$ classifies central extensions (M lying in the center of the extended Lie algebra) modulo trivial ones. Recall that a central extension of $\mathfrak{g}$ by $\mathbb{R}$ produces a new Lie bracket on $\hat{\mathfrak{g}} = \mathfrak{g} \oplus \mathbb{R}$ by setting

$$[(X,\lambda),(Y,\mu)] = ([X,Y], c(X,Y)) \tag{7.5}$$

It is trivial if the cocycle $c = d_1 l$ is a coboundary of a 1-cochain l, in which case the map $(X,\lambda) \to (X, \lambda - l(X))$ yields a Lie algebra isomorphism between $\hat{\mathfrak{g}}$ and $\mathfrak{g} \oplus \mathbb{R}$ considered as a direct sum of Lie algebras.

(b) If $M = \mathfrak{g}$ with the adjoint representation, then $H^2(\mathfrak{g}; \mathfrak{g})$ classifies infinitesimal deformations modulo trivial ones. By definition, a (formal) series

$$(X,Y) \to \Phi_\lambda(X,Y) := [X,Y] + \lambda f_1(X,Y) + \lambda^2 f_2(X,Y) + \ldots \tag{7.6}$$

is a *deformation* of the Lie bracket [,] if Φ_λ is a Lie bracket for every λ, i.e. is an antisymmetric bilinear form in X, Y and satisfies Jacobi's identity. If one sets simply

$$[X,Y]_\lambda = [X,Y] + \lambda c(X,Y), \tag{7.7}$$

c being a 2-cochain with values in $\mathfrak{g}$ and λ being a scalar, then this bracket satisfies Jacobi identity modulo terms of order $O(\lambda^2)$ if and only if c is a 2-cocycle. One thus gets what is called an *infinitesimal deformation* of the bracket of $\mathfrak{g}$, which is trivial if c is a coboundary, by which we mean (as in the case of central extensions) that an adequate linear isomorphism from $\mathfrak{g}$ to $\mathfrak{g}$ transforms the initial bracket [,] into the deformed bracket $[\ ,\ ]_\lambda$. The infinitesimal deformation associated to a cocycle c do not always give rise to an actual deformation coinciding with the infinitesimal deformation to order 1, i.e. such that $f_1 = c$, as one may check by looking inductively for functions $f_2, f_3, \ldots$ which satisfy Jacobi's identity to order 2, 3, ... Cohomological obstructions to prolongations of deformations are contained in $H^3(\mathfrak{g}; \mathfrak{g})$ (see below for details).

Let us finally say a few words on *Hochschild-Serre spectral sequences*. Assume the following exact sequence holds,

$$0 \to \mathfrak{h} \to \mathfrak{g} \to \mathfrak{g}_0 \to 0 \tag{7.8}$$

(in particular if $\mathfrak{g} = \mathfrak{g}_0 \ltimes \mathfrak{h}$, which is the case we are interested in). One has a sequence of cochain complexes, the first terms of which may be written without too many technicalities, namely,

$$E_0^{p,q} = C^q(\mathfrak{g}_0, C^p(\mathfrak{h}, M)), E_1^{p,q} = H^q(\mathfrak{g}_0, C^p(\mathfrak{h}, M)),$$
$$E_2^{p,q} = H^p(\mathfrak{h}, H^q(\mathfrak{g}_0, M)), \ldots, E_n^{p,q}, \ldots \tag{7.9}$$

with differentials $d_n^{p,q} : E_n^{p,q} \to E_n^{p+n,q-n+1}$, each one being the cohomology of the preceding one, which in the good cases stabilizes for n large enough to a complex $E_\infty^{p,q}$ with trivial differential such that (non-canonically though)

$$H^n(\mathfrak{g}; M) \simeq \oplus_{p+q=n} E_\infty^{p,q}. \tag{7.10}$$

This gives in many instances, in particular in our case, an efficient way of computing $H^n(\mathfrak{g}; M)$. The strategy is usually to compute some $E_n^{p,q}$ with n as small as possible, and prove that its cohomology is already trivial, so that $E_n^{p,q} \simeq E_\infty^{p,q}$. In most computations below one may take $n = 2$.

7.2 Classifying Deformations of $\mathfrak{sv}(0)$

We shall be interested in the classification of all formal deformations of $\mathfrak{sv}(0)$, following the now classical scheme of Nijenhuis and Richardson: deformation of a Lie algebra $\mathfrak{g}$ means that one has a formal family of Lie brackets on $\mathfrak{g}$, denoted $[\ ,\ ]_t$, inducing a Lie algebra structure on the extended Lie algebra $\mathfrak{g} \bigotimes_k k[[t]] = \mathfrak{g}[[t]]$. As explained in Sect. 7.1, one has to study the cohomology of $\mathfrak{g}$ with coefficients in the adjoint representation; the degree-two cohomology $H^2(\mathfrak{g}, \mathfrak{g})$ classifies the infinitesimal deformations (the terms of order one in the expected formal deformations) and $H^3(\mathfrak{g}, \mathfrak{g})$ contains the potential obstructions to a further prolongation of the deformations. So we shall naturally begin with the computation of $H^2(\mathfrak{sv}(0), \mathfrak{sv}(0))$ (as usual, we shall consider only local cochains, equivalently given by differential operators, or polynomial in the indices of the modes):

Theorem 7.1. *One has dim $H^2(\mathfrak{sv}(0), \mathfrak{sv}(0)) = 3$. A set of generators is provided by the cohomology classes of the cocycles c_1, c_2 and c_3, defined as follows in terms of modes (the missing components of the cocycles are meant to vanish):*
$c_1(L_n, Y_m) = -\frac{n}{2} Y_{n+m}$, $\quad c_1(L_n, M_n) = -n M_{n+m}$
$c_2(L_n, Y_m) = Y_{n+m}$ $\quad c_2(L_n, M_m) = 2M_{n+m}$
$c_3(L_n, L_m) = (m - n) M_{n+m}$

Remarks. 1. The cocycle c_1 gives rise to the family of Lie algebras $\mathfrak{sv}_\varepsilon(0)$ described in Definition 2.2.

2. The cocycle c_3 can be described globally as $c_3 : \mathrm{Vect}(S^1) \times \mathrm{Vect}(S^1) \longrightarrow \mathcal{F}_0$ given by

$$c_3(f\partial, g\partial) = \begin{vmatrix} f & g \\ f' & g' \end{vmatrix}$$

This cocycle appeared in [31] and has been used in a different context in [43].

Before entering the technicalities of the proof, we shall indicate precisely, for the comfort of the reader, some cohomological results on $\mathfrak{g} = \mathrm{Vect}(S^1)$ which will be extensively used in the sequel.

Proposition 7.1. *(see [31], or [43], Chap. IV for a more elementary approach)*

1. *$Inv_{\mathfrak{g}}(\mathcal{F}_\lambda \otimes \mathcal{F}_\mu) = 0$ unless $\mu = -1 - \lambda$ and $\mathcal{F}_\mu = \mathcal{F}_\lambda^*$; then $Inv_{\mathfrak{g}}(\mathcal{F}_\lambda \otimes \mathcal{F}_\lambda^*)$ is one-dimensional, generated by the identity mapping.*
2. *$H^i(\mathfrak{g}, \mathcal{F}_\lambda \otimes \mathcal{F}_\mu) \equiv 0$ if $\lambda \neq 1 - \mu$ and λ or μ are not integers.*
   (1) and (2) can be immediately deduced from [31], theorem 2.3.5 p. 136-137.
3. *Let W_1 be the Lie algebra of formal vector fields on the line, its cohomology represents the algebraic part of the cohomology of $\mathfrak{g} = \mathrm{Vect}(S^1)$ (see again [31], theorem 2.4.12). Then $H^1(W_1, Hom(\mathcal{F}_\lambda, \mathcal{F}_\lambda))$ is one-dimensional, generated by the cocycle $(f\partial, adx^{-\lambda}) \longrightarrow f' adx^{-\lambda}$ (cocycle I_λ in [31], p. 138).*
4. *Invariant antisymmetric bilinear operators $\mathcal{F}_\lambda \times \mathcal{F}_\mu \longrightarrow \mathcal{F}_\nu$ between densities have been determined by P. Grozman (see [39], p. 280). They are of the following type:*
   (a) The Poisson bracket for $\nu = \lambda + \mu - 1$, defined by

$$\{fdx^{-\lambda}, gdx^{-\mu}\} = (\lambda f g' - \mu f' g)dx^{-(\lambda+\mu-1)}.$$

   (b) The following three exceptional brackets :
   $\mathcal{F}_{1/2} \times \mathcal{F}_{1/2} \to \mathcal{F}_{-1}$ given by $(f\partial^{1/2}, g\partial^{1/2}) \to \frac{1}{2}(fg'' - gf'')dx$;
   $\mathcal{F}_0 \times \mathcal{F}_0 \to \mathcal{F}_{-3}$ given by $(f, g) \to (f''g' - g''f')dx^3$;
   and an operator $\mathcal{F}_{2/3} \times \mathcal{F}_{2/3} \to \mathcal{F}_{-\frac{5}{3}}$ called the Grozman bracket (see [39], p 274).

Proof of Theorem 7.1. We shall use standard techniques in Lie algebra cohomology; the proof will be rather technical, but we will spare the reader certain difficulties. Let us fix the notations: set $\mathfrak{sv}(0) = \mathfrak{g} \ltimes \mathfrak{h}$ where $\mathfrak{g} = \mathrm{Vect}(S^1)$ and $\mathfrak{h}$ is the nilpotent part of $\mathfrak{sv}(0)$.
One can consider the exact sequence

$$0 \longrightarrow \mathfrak{h} \longrightarrow \mathfrak{g} \ltimes \mathfrak{h} \longrightarrow \mathfrak{g} \longrightarrow 0 \qquad (7.11)$$

as a short exact sequence of $\mathfrak{g} \ltimes \mathfrak{h}$ modules, thus inducing a long exact sequence in cohomology:

$$\cdots \longrightarrow H^1(\mathfrak{sv}(0), \mathfrak{g}) \longrightarrow H^2(\mathfrak{sv}(0), \mathfrak{h}) \longrightarrow H^2(\mathfrak{sv}(0), \mathfrak{sv}(0)) \longrightarrow H^2(\mathfrak{sv}(0), \mathfrak{g}) \longrightarrow H^3(\mathfrak{sv}(0), \mathfrak{h}) \longrightarrow \cdots \tag{7.12}$$

So we shall consider $H^*(\mathfrak{sv}(0), \mathfrak{g})$ and $H^*(\mathfrak{sv}(0), \mathfrak{h})$ separately.

Lemma 7.2. $H^*(\mathfrak{sv}(0), \mathfrak{g}) = 0$ *for* $* = 0, 1, 2$.

Proof of Lemma 7.2. One uses the Hochschild-Serre spectral sequence associated with the exact sequence (7.11), as explained in the introductory Sect. 7.1. Let us remark first that $H^*(\mathfrak{g}, H^*(\mathfrak{h}, \mathfrak{g})) = H^*(\mathfrak{g}, H^*(\mathfrak{h}) \otimes \mathfrak{g})$ since $\mathfrak{h}$ acts trivially on $\mathfrak{g}$. So one has to understand $H^*(\mathfrak{h})$ in low dimensions; let us consider the exact sequence $0 \longrightarrow \mathfrak{n} \longrightarrow \mathfrak{h} \longrightarrow \mathfrak{q} \longrightarrow 0$, where $\mathfrak{n} = [\mathfrak{h}, \mathfrak{h}]$. As $\mathfrak{g}$-modules, these algebras are density modules, more precisely $\mathfrak{n} = \mathcal{F}_0$ and $\mathfrak{q} = \mathcal{F}_{1/2}$. So $H^1(\mathfrak{h}) = \mathfrak{q}^* = \mathcal{F}_{-3/2}$ as a $\mathcal{G}$-module. Let us recall that, as a module on itself, $\mathfrak{g} = \mathcal{F}_1$. One gets: $E_2^{p,0} = H^p(\mathfrak{g}, \mathfrak{g}) = 0$ as well-known (see [31]),

$$E_2^{1,1} = H^1(\mathfrak{g}, H^1(\mathfrak{h}) \otimes \mathfrak{g}) = H^1(\mathfrak{g}, \mathcal{F}_{-3/2} \otimes \mathcal{F}_1)$$

The determination of cohomologies of $\mathrm{Vect}(S^1)$ with coefficients in tensor products of modules of densities has been done by Fuks (see [31], Chap. 2, Theorem 2.3.5, or Proposition 7.1 (2) above), in this case everything vanishes and $E_2^{1,1} = 0$.

One has now to compute $H^2(\mathfrak{h})$ in order to get $E_2^{0,2} = Inv_{\mathfrak{g}}(H^2(\mathfrak{h}) \otimes \mathfrak{g})$. We shall use the decomposition of cochains on $\mathfrak{h}$ induced by its splitting into vector subspaces: $\mathfrak{h} = \mathfrak{n} \oplus \mathfrak{q}$. So $C^1(\mathfrak{h}) = \mathfrak{n}^* \oplus \mathfrak{q}^*$ and $C^2(\mathfrak{h}) = \Lambda^2\mathfrak{n}^* \oplus \Lambda^2\mathfrak{q}^* \oplus \mathfrak{q}^* \wedge \mathfrak{n}^*$. The coboundary ∂ is induced by the only non-vanishing part $\partial : \mathfrak{n}^* \longrightarrow \Lambda^2\mathfrak{q}^*$ which is dual to the bracket $\Lambda^2\mathfrak{q} \longrightarrow \mathfrak{n}$. So the cohomological complex splits into three subcomplexes and one deduces the following exact sequences:

$$0 \longrightarrow \mathfrak{n}^* \xrightarrow{\partial} \Lambda^2\mathfrak{q}^* \longrightarrow M_1 \longrightarrow 0$$

$$0 \longrightarrow M_2 \longrightarrow \Lambda^2\mathfrak{n}^* \xrightarrow{\partial} \Lambda^2\mathfrak{q}^* \otimes \mathfrak{n}^*$$

$$0 \longrightarrow M_3 \longrightarrow \mathfrak{q}^* \wedge \mathfrak{n}^* \xrightarrow{\partial} \Lambda^3\mathfrak{q}^*$$

and $H^2(\mathfrak{h}) = M_1 \oplus M_2 \oplus M_3$. One can then easily deduce the invariants $Inv_{\mathfrak{g}}(H^2(\mathfrak{h}) \otimes \mathfrak{g}) = \bigoplus_{i=1}^{3} Inv_{\mathfrak{g}}(M_i \otimes \mathfrak{g})$ from the cohomological exact sequences associated with the above short exact sequences. One has:

$$0 \longrightarrow Inv_{\mathfrak{g}}(M_2 \otimes \mathfrak{g}) \longrightarrow Inv_{\mathfrak{g}}(\Lambda^2\mathfrak{n}^* \otimes \mathfrak{g}) = 0$$

and

$$0 \longrightarrow Inv_{\mathfrak{g}}(M_3 \otimes \mathfrak{g}) \longrightarrow Inv_{\mathfrak{g}}(\mathfrak{q}^* \wedge \mathfrak{n}^* \otimes \mathfrak{g}) = 0$$

from Proposition 5.2;

$$\cdots \longrightarrow Inv_{\mathfrak{g}}(\Lambda^2\mathfrak{q}^* \otimes \mathfrak{g}) \longrightarrow Inv_{\mathfrak{g}}(M_1 \otimes \mathfrak{g}) \longrightarrow H^1(\mathfrak{g}, \mathfrak{n}^* \otimes \mathfrak{g})$$

$$\xrightarrow{\partial_*} H^1(\mathfrak{g}, \Lambda^2\mathfrak{q}^* \otimes \mathfrak{g}) \longrightarrow \cdots$$

From the same proposition, one gets $Inv_{\mathfrak{g}}(\Lambda^2\mathfrak{q}^* \otimes \mathfrak{g}) = 0$ and we shall see later (see the last part of the proof) that ∂_* is an isomorphism. So $Inv_{\mathfrak{g}}(H^2(\mathfrak{h}) \otimes \mathfrak{g}) = 0$ and $E_2^{0,2} = 0$. The same argument shows that $E_2^{0,1} = 0$, which ends the proof of the lemma. □

From the long exact sequence (7.12) one has now: $H^*(\mathfrak{sv}(0), \mathfrak{sv}(0)) = H^*(\mathfrak{sv}(0), \mathfrak{h})$ for $* = 0, 1, 2$. We shall compute $H^*(\mathfrak{sv}(0), \mathfrak{h})$ by using the Hochschild-Serre spectral sequence once more; there are three terms to compute.

1. First $E_2^{2,0} = H^2(\mathfrak{g}, H^0(\mathfrak{h}, \mathfrak{h}))$, but $H^0(\mathfrak{h}, \mathfrak{h}) = Z(\mathfrak{h}) = \mathfrak{n} = \mathscr{F}_0$ as $\mathfrak{g}$-module. So $E_2^{2,0} = H^2(\mathfrak{g}, \mathscr{F}_0)$ which is one-dimensional, given by $c_3(f\partial, g\partial) = \begin{vmatrix} f & g \\ f' & g' \end{vmatrix}$, or in terms of modes $c_3(L_n, L_m) = (m-n)M_{n+m}$. Hence we have found one of the classes announced in the theorem.
2. One must now compute $E_2^{1,1} = H^1(\mathfrak{g}, H^1(\mathfrak{h}, \mathfrak{h}))$. The following lemma will be useful for this purpose, and also for the last part of the proof.

Lemma 7.3 (identification of $H^1(\mathfrak{h}, \mathfrak{h})$ as a $\mathfrak{g}$-module)). *The space $H^1(\mathfrak{h}, \mathfrak{h})$ splits into the direct sum of two $\mathscr{G}$-modules $H^1(\mathfrak{h}, \mathfrak{h}) = \mathscr{H}_1 \oplus \mathscr{H}_2$ such that*

1. *$Inv_{\mathfrak{g}}\mathscr{H}_2 = 0$, $H^1(\mathfrak{g}, \mathscr{H}_2) = 0$;*
2. *$Inv_{\mathfrak{g}}\mathscr{H}_1$ is one-dimensional, generated by the 'constant multiplication' cocycle l defined by*
$$l(Y_n) = Y_n, \quad l(M_n) = 2M_n \tag{7.13}$$
3. *$H^1(\mathfrak{g}, \mathscr{H}_1)$ is two-dimensional, generated by two cocycles c_1, c_2 defined by*
$$c_1(f\partial, g\partial^{1/2}) = f'g\partial^{1/2}, \quad c_1(f\partial, g) = 2f'g$$
and
$$c_2(f\partial, g\partial^{1/2}) = fg\partial^{1/2}, \quad c_2(f\partial, g) = 2fg.$$
4. *$H^2(\mathfrak{g}, \mathscr{H}_1)$ is one-dimensional, generated by the cocycle c_{12} defined by*
$$c_{12}(f\partial, g\partial, h\partial^{1/2}) = \begin{vmatrix} f & g \\ f' & g' \end{vmatrix} h\partial^{1/2}, \quad c_{12}(f\partial, g\partial, h) = 2\begin{vmatrix} f & g \\ f' & g' \end{vmatrix} h$$

Proof of Lemma 7.3. We shall split the cochains according to the decomposition $\mathfrak{h} = \mathfrak{q} \oplus \mathfrak{n}$. Set $C^1(\mathfrak{h}, \mathfrak{h}) = C_1 \oplus C_2$, where:

$$C_1 = (\mathfrak{n}^* \otimes \mathfrak{n}) \oplus (\mathfrak{q}^* \otimes \mathfrak{q}) \quad C_2 = (\mathfrak{n}^* \otimes \mathfrak{q}) \oplus (\mathfrak{q}^* \otimes \mathfrak{n}).$$

So one readily obtains the splitting $H^1(\mathfrak{h}, \mathfrak{h}) = \mathscr{H}_1 \oplus \mathscr{H}_2$ where

$$0 \longrightarrow \mathscr{H}_1 \longrightarrow (\mathfrak{n}^* \otimes \mathfrak{n}) \oplus (\mathfrak{q}^* \otimes \mathfrak{q}) \xrightarrow{\partial} \Lambda^2\mathfrak{q}^* \otimes \mathfrak{n}$$

$$0 \longrightarrow \mathfrak{q} \xrightarrow{\partial} \mathfrak{q}^* \otimes \mathfrak{n} \longrightarrow \mathscr{H}_2 \longrightarrow 0$$

∂ being the coboundary on the space of cochains on $\mathfrak{h}$ with coefficients into itself. Its non vanishing pieces in degrees 0, 1 and 2 are the following: $\mathfrak{q} \xrightarrow{\partial} \mathfrak{q}^* \otimes \mathfrak{n}$, $\mathfrak{n}^* \otimes \mathfrak{n} \xrightarrow{\partial} \Lambda^2\mathfrak{q}^* \otimes \mathfrak{n}$, $\mathfrak{q}^* \otimes \mathfrak{q} \xrightarrow{\partial} \Lambda^2\mathfrak{q}^* \otimes \mathfrak{n}$. We can now describe the second exact sequence in terms of densities as follows:

$$0 \longrightarrow \mathscr{F}_{1/2} \longrightarrow \mathscr{F}_{-3/2} \otimes \mathscr{F}_0 \longrightarrow \mathscr{H}_2 \longrightarrow 0 \tag{7.14}$$

From Proposition 7.1, one has $Inv_{\mathfrak{g}}(\mathscr{F}_{-3/2} \otimes \mathscr{F}_0) = 0$ as well as $H^i(\mathfrak{g}, \mathscr{F}_{1/2}) = 0$, for $i = 0, 1, 2$, and $H^1(\mathfrak{g}, \mathscr{F}_{-3/2} \otimes \mathscr{F}_0) = 0$. So the long exact sequence in cohomology associated with (5.3) gives $Inv_{\mathfrak{g}}(\mathscr{H}_2) = 0$ and $H^1(\mathfrak{g}, \mathscr{H}_2) = 0$.

For $\mathscr{H}_1$, one has to analyse the cocycles by direct computation. So let $l \in C_1$ given by $l(Y_n) = a_n(k)Y_{n+k}$, $l(M_n) = b_n(k)M_{n+k}$. The cocycle conditions are given by:

$$\partial l(Y_n, Y_m) = l((m-n)M_{n+m}) - Y_n \cdot l(Y_m) + Y_m \cdot l(Y_n) = 0$$

for all $(n, m) \in \mathbb{Z}^2$. So identifying the term in M_{n+m+k}, one obtains:

$$(m-n)b_{n+m}(k) = (m-n+k)a_m(k) - (n-m+k)a_n(k)$$

so $b_{n+m}(k) = a_m(k) + a_n(k) + \frac{k}{m-n}(a_m(k) - a_n(k)) = f(n, m, k)$.

One can now determine the $a_n(k)$, remarking that the function $f(n, m, k)$ depends only on k and $(n + m)$. One then obtains that $a_n(k)$ must be affine in n so:

$$a_n(k) = n\lambda(k) + \mu(k)$$
$$b_n(k) = n\lambda(k) + k\lambda(k) + 2\mu(k)$$

So, as a vector space $\mathscr{H}_1$ is isomorphic to $\bigoplus_k \mathbb{C}(\lambda(k)) \bigoplus_k \mathbb{C}(\mu(k))$, two copies of an infinite direct sum of a numerable family of one-dimensional vector spaces.

Now we have to compute the action of $\mathfrak{g}$ on $\mathscr{H}_1$; letting $L_p \in \mathfrak{g}$, one has

$$
\begin{aligned}
(L_p \cdot l)(Y_n) &= ((n - \frac{p}{2})a_{n+p}(k) - (n + k - \frac{p}{2})a_n(k))Y_{n+p+k} \\
&= \left(n(p-k)\lambda(k) - (\frac{p^2}{2}\lambda(k) + k\mu(k)) \right) Y_{n+p+k}
\end{aligned}
$$

So if one sets $(L_p \cdot l)(Y_n) = (n(L_p \cdot \lambda)(k + p) + (L_p \cdot \mu)(k + p))Y_{n+p+k}$ one obtains:

$$(L_p \cdot \lambda)(k + p) = (p - k)\lambda(k)$$

$$(L_p \cdot \mu)(k + p) = -\frac{p^2}{2}\lambda(k) + k\mu(k)$$

Finally, $\mathscr{H}_1$ appears as an extension of modules of densities of the following type: $0 \longrightarrow \mathscr{F}_0 \longrightarrow \mathscr{H}_1 \longrightarrow \mathscr{F}_1 \longrightarrow 0$, in which $\mathscr{F}_0$ corresponds to $\bigoplus_k \mathbb{C}(\mu(k))$ and $\mathscr{F}_1$ to $\bigoplus_k \mathbb{C}(\lambda(k))$.

There is a non-trivial extension cocycle γ in $Ext^1_{\mathfrak{g}}(\mathscr{F}_1, \mathscr{F}_0) = H^1(\mathfrak{g}, Hom(\mathscr{F}_1, \mathscr{F}_0))$, given by $\gamma(f\partial)(g\partial) = f''g$; this cocycle corresponds to the term in p^2 in the above formula. In any case one has a long exact sequence in cohomology

$$\cdots \longrightarrow H^i(\mathfrak{g}, \mathscr{F}_0) \longrightarrow H^i(\mathfrak{g}, \mathscr{H}_1) \longrightarrow H^i(\mathfrak{g}, \mathscr{F}_1) \longrightarrow H^{i+1}(\mathfrak{g}, \mathscr{F}_0) \longrightarrow \cdots$$

As well-known, $H^*(\mathfrak{g}, \mathscr{F}_1) = H^*(\mathfrak{g}, \mathscr{G})$ is trivial, and finally $H^i(\mathfrak{g}, \mathscr{F}_0)$ is isomorphic to $H^i(\mathfrak{g}, \mathscr{H}_1)$. In particular $H^0(\mathfrak{g}, \mathscr{H}_1) = H^0(\mathfrak{g}, \mathscr{F}_0)$ is one-dimensional, given by the constants; a scalar μ induces an invariant cocycle as $l(Y_n) = \mu Y_n,\ l(M_n) = 2\mu M_n$.

Moreover, $H^1(\mathfrak{g}, \mathscr{F}_0)$ has dimension 2: it is generated by the cocycles $\overline{c}_1$ and $\overline{c}_2$, defined by $\overline{c}_1(f\partial) = f'$ and $\overline{c}_2(f\partial) = f$ respectively. So one obtains two generators of $H^1(\mathfrak{g}, \mathscr{H}_1)$ given by

$$c_1(f\partial, g\partial^{1/2}) = f'g\partial^{1/2}, \quad c_1(f\partial, g) = 2f'g$$

and

$$c_2(f\partial, g\partial^{1/2}) = fg\partial^{1/2}, \quad c_2(f\partial, g) = 2fg$$

respectively.

Finally $H^2(\mathfrak{g}, \mathscr{F}_0)$ is one-dimensional, with the cup-product $\overline{c}_{12}$ of $\overline{c}_1$ and $\overline{c}_2$ as generator (see [31], p. 177), so $\overline{c}_{12}(f\partial, g\partial) = \begin{vmatrix} f & g \\ f' & g' \end{vmatrix}$, and one deduces the formula for the corresponding cocycle c_{12} in $H^2(\mathfrak{g}, \mathscr{H}_1)$:

$$c_{12}(f\partial, g\partial, h\partial^{1/2}) = \begin{vmatrix} f & g \\ f' & g' \end{vmatrix} h\partial^{1/2}, \quad c_{12}(f\partial, g\partial, h) = 2\begin{vmatrix} f & g \\ f' & g' \end{vmatrix} h$$

This finishes the proof of Lemma 7.3. □

So, from Lemma 7.3, we have computed $E_2^{1,1} = H^1(\mathfrak{g}, H^1(\mathfrak{h}, \mathfrak{h}))$; it is two-dimensional, generated by c_1 and c_2, while earlier we had $H^2(\mathfrak{g}, H^0(\mathfrak{h}, \mathfrak{h})) = E_2^{2,0}$, a one-dimensional vector space generated by c_3. We have to check now that these cohomology classes shall not disappear in the spectral sequence; the only potentially non-vanishing differentials are $E_2^{0,1} \longrightarrow E_2^{2,0}$ and $E_2^{1,1} \longrightarrow E_2^{3,0}$. One has $E_2^{3,0} = H^3(\mathfrak{g}, \mathfrak{h}) = H^3(\mathfrak{g}, \mathfrak{n}) = H^3(\mathscr{G}, \mathscr{F}_0) = 0$ (see [31] p. 177); here we consider only local cohomology), then $E_2^{0,1}$ is one-dimensional determined by the constant multiplication (see above) and direct verification shows that $E_2^{0,1} \longrightarrow E_2^{2,0}$ vanishes. So we have just proved that the cocycles c_1, c_2 and c_3 defined in Theorem 7.1 represent genuinely non-trivial cohomology classes in $H^2(\mathfrak{sv}(0), \mathfrak{sv}(0))$.

3. In order to finish the proof, we still have to prove that there does not exist any other non-trivial class in the last piece of the Hochschild-Serre spectral sequence. We shall thus prove that $E_2^{0,2} = Inv_{\mathfrak{g}} H^2(\mathfrak{h}, \mathfrak{h}) = 0$. As in the proofs of the previous lemmas, we shall use decompositions of the cohomological complex of $\mathfrak{h}$ with coefficients into itself as sums of $\mathfrak{g}$-modules.

The space of adjoint cochains $C^2(\mathfrak{h}, \mathfrak{h})$ will split into six subspaces according to the vector space decomposition $\mathfrak{h} = \mathfrak{q} \oplus \mathfrak{n}$. So we can as well split the cohomological complex

$$C^1(\mathfrak{h}, \mathfrak{h}) \xrightarrow{\partial} C^2(\mathfrak{h}, \mathfrak{h}) \xrightarrow{\partial} C^3(\mathfrak{h}, \mathfrak{h})$$

into its components, and the coboundary operators will as well split into different components, as we already explained. So one obtains the following families of exact sequences of $\mathfrak{g}$-modules:

$$(\mathfrak{n}^* \otimes \mathfrak{n}) \oplus (\mathfrak{q}^* \otimes \mathfrak{q}) \xrightarrow{\partial} \Lambda^2\mathfrak{q}^* \otimes \mathfrak{n} \longrightarrow A_1 \longrightarrow 0 \tag{7.15}$$

$$0 \longrightarrow K \longrightarrow (\Lambda^2\mathfrak{q}^* \otimes \mathfrak{q}) \oplus (\mathfrak{n}^* \wedge \mathfrak{q}^* \otimes \mathfrak{q}) \xrightarrow{\partial} \Lambda^3\mathfrak{q}^* \otimes \mathfrak{n}$$

$$0 \longrightarrow \mathfrak{n}^* \otimes \mathfrak{q} \xrightarrow{\partial} K \longrightarrow A_2 \longrightarrow 0 \tag{7.16}$$

$$0 \longrightarrow A_3 \longrightarrow \mathfrak{n}^* \wedge \mathfrak{q}^* \otimes \mathfrak{q} \xrightarrow{\partial} (\Lambda^3\mathfrak{q}^* \otimes \mathfrak{q}) \oplus (\mathfrak{n}^* \wedge \Lambda^2\mathfrak{q}^*) \otimes \mathfrak{n}$$

$$0 \longrightarrow A_4 \longrightarrow \Lambda^2\mathfrak{n}^* \otimes \mathfrak{n} \xrightarrow{\partial} (\mathfrak{n}^* \wedge \Lambda^2\mathfrak{q}^*) \otimes \mathfrak{n}$$

$$0 \longrightarrow A_5 \longrightarrow \Lambda^2\mathfrak{n}^* \otimes \mathfrak{q} \xrightarrow{\partial} (\mathfrak{n}^* \wedge \Lambda^2\mathfrak{q}^*) \otimes \mathfrak{q} \oplus (\Lambda^2\mathfrak{n}^* \wedge \mathfrak{q}^*) \otimes \mathfrak{n}$$

The restrictions of coboundary operators are still denoted by ∂, and the other arrows are either inclusions of subspaces or projections onto quotients. Hence $H^2(\mathfrak{h}, \mathfrak{h}) =$

$\bigoplus_{i=1}^{5} A_i$, and our result will follow from $Inv_{\mathfrak{g}} A_i = 0$, $i = 1, \ldots, 5$. For the last three sequences, the result follows immediately from the cohomology long exact sequence by using $Inv_{\mathfrak{g}}(\mathfrak{n}^* \wedge \mathfrak{q}^* \otimes \mathfrak{q}) = 0$, $Inv_{\mathfrak{g}}(\Lambda^2\mathfrak{n}^* \otimes \mathfrak{n}) = 0$, $Inv_{\mathfrak{g}}(\Lambda^2\mathfrak{n}^* \otimes \mathfrak{q}) = 0$: there results are deduced from those of Grozman, recalled in Proposition 7.1. (Note that the obviously $\mathfrak{g}$-invariant maps $\mathfrak{n} \otimes \mathfrak{n} \longrightarrow \mathfrak{n}$ and $\mathfrak{n} \otimes \mathfrak{q} \longrightarrow \mathfrak{q}$ are not antisymmetric!) So one has $Inv_{\mathfrak{g}} A_i = 0$ for $i = 3, 4, 5$.

An analogous argument will work for K, since $Inv_{\mathfrak{g}}(\Lambda^2\mathfrak{q}^* \otimes \mathfrak{q}) = 0$ and $Inv_{\mathfrak{g}}(\mathfrak{n}^* \wedge \mathfrak{q}^*) \otimes \mathfrak{q} = 0$ from the same results. So the long exact sequence associated with the short sequence (5.5) above will give:

$$0 \longrightarrow Inv_{\mathfrak{g}}(K) \longrightarrow Inv_{\mathfrak{g}}(A_2) \longrightarrow H^1(\mathfrak{g}, \mathfrak{n}^* \otimes \mathfrak{q})$$

One has $H^1(\mathfrak{g}, \mathfrak{n}^* \otimes \mathfrak{q}) = H^1(\mathfrak{g}, \mathscr{F}_{-1} \otimes \mathscr{F}_{1/2}) = 0$ (see Proposition 7.1). So $Inv_{\mathfrak{g}}(A_2) = 0$.

For A_1, we shall need a much more subtle argument. First of all, the sequence (5.4) can be split into two short exact sequences:

$$0 \longrightarrow \mathscr{H}_1 \longrightarrow (\mathfrak{n}^* \otimes \mathfrak{n}) \oplus (\mathfrak{q}^* \otimes \mathfrak{q}) \longrightarrow B \longrightarrow 0$$

$$0 \longrightarrow B \longrightarrow \Lambda^2\mathfrak{q}^* \otimes \mathfrak{n} \longrightarrow A_1 \longrightarrow 0.$$

Let us consider the long exact sequence associated with the first one:

$$0 \longrightarrow Inv_{\mathfrak{g}}\mathscr{H}_1 \longrightarrow Inv_{\mathfrak{g}}(\mathfrak{n}^* \otimes \mathfrak{n}) \oplus Inv_{\mathfrak{g}}(\mathfrak{q}^* \otimes \mathfrak{q}) \longrightarrow Inv_{\mathfrak{g}} B \longrightarrow \cdots$$

$$\cdots \hookrightarrow H^1(\mathfrak{g}, \mathscr{H}_1) \longrightarrow H^1(\mathfrak{g}, \mathfrak{n}^* \otimes \mathfrak{n}) \oplus H^1(\mathfrak{g}, \mathfrak{q}^* \otimes \mathfrak{q}) \longrightarrow H^1(\mathfrak{g}, B) \longrightarrow \cdots$$

$$\cdots \longrightarrow H^2(\mathfrak{g}, \mathscr{H}_1) \longrightarrow H^2(\mathfrak{g}, \mathfrak{n}^* \otimes \mathfrak{n}) \oplus H^2(\mathfrak{g}, \mathfrak{q}^* \otimes \mathfrak{q}) \longrightarrow H^2(\mathfrak{g}, B) \longrightarrow \cdots$$

The case of $H^i(\mathfrak{g}, \mathscr{H}_1)$, $i = 0, 1, 2$ has been treated in Lemma 7.3, and analogous techniques can be used to study $H^i(\mathfrak{g}, \mathfrak{n}^* \otimes \mathfrak{n})$ and $H^i(\mathfrak{g}, \mathfrak{q}^* \otimes \mathfrak{q})$ for $i = 0, 1, 2$. The cohomology classes come from the inclusion $\mathscr{F}_0 \subset \mathfrak{n}^* \otimes \mathfrak{n}$, $\mathfrak{q}^* \otimes \mathfrak{q}$ or $\mathscr{H}_1$, and from the well-known computation of $H^*(\mathfrak{g}, \mathscr{F}_0)$ (Remark: as regards the results of Fuks [31], Chap. 2, one should keep in mind the fact that he computes cohomologies for W_1, the formal part of $\mathfrak{g} = \mathrm{Vect}(S^1)$. To get the cohomologies for $\mathrm{Vect}(S^1)$ one has to add the classes of differentiable order 0 (or "topological" classes), this is the reason for the occurrence of c_2 in Lemma 7.3).

So $H^i(\mathfrak{g}, \mathscr{H}_1) = H^i(\mathfrak{g}, \mathfrak{n}^* \otimes \mathfrak{n}) = H^i(\mathfrak{g}, \mathfrak{q}^* \otimes \mathfrak{q})$, $i = 0, 1, 2$, and the maps on the modules are naturally defined through the injection $\mathscr{H}_1 \longrightarrow (\mathfrak{n}^* \otimes \mathfrak{n}) \oplus (\mathfrak{q}^* \otimes \mathfrak{q})$: each generator of $H^i(\mathfrak{g}, \mathscr{H}_1)$, $i = 0, 1, 2,$, say e, will give $(e, -e)$ in the corresponding component of $H^i(\mathfrak{g}, (\mathfrak{n}^* \otimes \mathfrak{n}) \oplus (\mathfrak{q}^* \otimes \mathfrak{q}))$. So $Inv_{\mathfrak{g}} B$ and $H^2(\mathfrak{g}, B)$ are one-dimensional and $H^1(\mathfrak{g}, B)$ is two-dimensional.

Now we can examine the long exact sequence associated with:

$$0 \longrightarrow B \stackrel{\partial}{\longrightarrow} \Lambda^2\mathfrak{q}^* \otimes \mathfrak{n} \longrightarrow A_1 \longrightarrow 0,$$

which is:

$$0 \longrightarrow Inv_{\mathfrak{g}} B \stackrel{\partial^*}{\longrightarrow} Inv_{\mathfrak{g}} \Lambda^2\mathfrak{q}^* \otimes \mathfrak{n} \longrightarrow Inv_{\mathfrak{g}} A_1 \longrightarrow H^1(\mathfrak{g}, B)$$
$$\longrightarrow H^1(\mathfrak{g}, \Lambda^2\mathfrak{q}^* \otimes \mathfrak{n}) \longrightarrow \cdots$$

The generator of $Inv_{\mathfrak{g}} B$ comes from the identity map $\mathfrak{n} \longrightarrow \mathfrak{n}$, and $Inv_{\mathfrak{g}}\, \Lambda^2\mathfrak{q}^* \otimes \mathfrak{n}$ is generated by the bracket $\mathfrak{q} \wedge \mathfrak{q} \longrightarrow \mathfrak{n}$, so ∂^* is an isomorphism in this case. So one has

$$0 \longrightarrow Inv_{\mathfrak{g}} A_1 \longrightarrow H^1(\mathfrak{g}, B) \stackrel{\partial^*}{\longrightarrow} H^1(\mathfrak{g}, \Lambda^2\mathfrak{q}^* \otimes \mathfrak{n})$$

The result will follow from the fact that this ∂^* is also an isomorphism. The two generators in $H^1(\mathfrak{g}, B)$ come from the corresponding ones in $H^1(\mathfrak{g}, \mathfrak{n}^* \otimes \mathfrak{n}) \oplus H^1(\mathfrak{g}, \mathfrak{q}^* \otimes \mathfrak{y})$, modulo the classes coming from $H^1(\mathscr{G}, \mathscr{H}_1)$; so these generators can be described in terms of Yoneda extensions, since $H^1(\mathfrak{g}, \mathscr{F}_0^* \otimes \mathscr{F}_0) = Ext^1_{\mathfrak{g}}(\mathscr{F}_0, \mathscr{F}_0)$, as well as $H^1(\mathfrak{g}, \mathscr{F}_{1/2}^* \otimes \mathscr{F}_{1/2}) = Ext^1_{\mathfrak{g}}(\mathscr{F}_{1/2}, \mathscr{F}_{1/2})$.

Let us write this extension as $0 \longrightarrow \mathscr{F}_0 \longrightarrow E \longrightarrow \mathscr{F}_0 \longrightarrow 0$; the action on E can be given in terms of modes as follows:

$$e_n^1(f_a, g_b) = (a f_{n+a} + n g_{b+n}, b g_{b+n})$$

or

$$e_n^2(f_a, g_b) = (a f_{n+a} + g_{b+n}, b g_{b+n}).$$

The images of these classes in $H^1(\mathfrak{g}, \Lambda^2\mathfrak{q}^* \otimes \mathfrak{n})$ are represented by the extensions obtained through a pull-back

$$\begin{array}{ccccccc} 0 \longrightarrow \mathscr{F}_0 & \longrightarrow & E & \longrightarrow & \mathscr{F}_0 & \longrightarrow 0 \\ \| & & \uparrow & & \uparrow [\, ,\,] & \\ 0 \longrightarrow \mathscr{F}_0 & ---> & E' & ---> & \Lambda^2\mathscr{F}_{1/2} & \longrightarrow 0 \end{array}$$

where $[\ ,\ ]$ denotes the mapping given by the Lie bracket $\Lambda^2\mathscr{F}_{1/2} \longrightarrow \mathscr{F}_0$. One can easily check that these extensions are non-trivial, so finally ∂^* is injective and $Inv_{\mathfrak{g}}(A_1) = 0$, which finishes the proof of $E_2^{0,2} = Inv_{\mathfrak{g}} H^2(\mathfrak{h}, \mathfrak{h}) = 0$ and the proof of Theorem 7.1. □

Theorem 7.1 implies that we have three independent infinitesimal deformations of $\mathfrak{sv}(0)$, defined by the cocycles c_1, c_2 and c_3, so the most general infinitesimal deformation of $\mathfrak{sv}(0)$ is of the following form:

$$[\ ,\ ]_{\lambda,\mu,\nu} = [\ ,\ ] + \lambda c_1 + \mu c_2 + \nu c_3.$$

In order to study further deformations of this bracket, one has to compute the Richardson-Nijenhuis brackets of c_1, c_2 and c_3 in $H^3(\mathfrak{sv}(0), \mathfrak{sv}(0))$. By definition, the Richardson-Nijenhuis bracket $[f, g]$ of two antisymmetric bilinear mappings f and g of a vector space E into itself is a trilinear antisymmetric mapping of E into itself given by the formula $[f, g](a, b, c) = \sum_{(cycl)} f(g(a, b), c) + g(f(a, b), c)$, where the sum ranges over all cyclic permutations. Here, the cohomology classes of the brackets $[c_i, c_j]$ represent the obstruction to prolongation of corresponding deformations.

One can compute directly using our explicit formulas and finds $[c_i, c_j] = 0$ in $H^3(\mathfrak{sv}(0), \mathfrak{sv}(0))$ for $i, j = 1, 2, 3$; and even better, the bracket of the cocycles themselves vanish, not only their cohomology classes. So one has the

Theorem 7.2. *The bracket* $[\ ,\ ]_{\lambda,\mu,\nu} = [\ ,\ ] + \lambda c_1 + \mu c_2 + \nu c_3$ *where* $[\ ,\ ]$ *is the Lie bracket on* $\mathfrak{sv}(0)$ *and* c_i, $i = 1, 2, 3$ *the cocycles given in Theorem 7.1, defines a three-parameter family of Lie algebra brackets on* $\mathfrak{sv}(0)$.

For the sake of completeness, we give below the full formulas in terms of modes:

$$[L_n, L_m]_{\lambda,\mu,\nu} = (m-n)L_{n+m} + \nu(m-n)M_{n+m}$$

$$[L_n, Y_m]_{\lambda,\mu,\nu} = (m - \frac{n}{2} - \frac{\lambda n}{2} + \mu)Y_{n+m}$$

$$[L_n, M_m]_{\lambda,\mu,\nu} = (m - \lambda n + 2\mu)M_{n+m}$$

$$[Y_n, Y_m] = (n-m)M_{n+m}$$

All other terms are vanishing.

Let us comment on each of these three deformations.

1. The term with cocycle c_3 has already been considered in a slightly different context in [43].
2. The term with c_2 induces only a small change in the action on $\mathfrak{h}$: the modules $\mathcal{F}_{1/2}$ and $\mathcal{F}_0$ are changed into $\mathcal{F}_{1/2,\mu}$ and $\mathcal{F}_{0,2\mu}$ (see [31], p.127), the bracket on $\mathfrak{h}$ being fixed. This is nothing but a reparametrization of the generators in the module, and for integer values of μ, the Lie algebra given by $[\ ,\ ]_{0,\mu,0}$ is isomorphic to the original one. For *half-integer* values of μ, one obtains the original Schrödinger–Virasoro algebra, $\mathfrak{sv}$.
3. The term with c_1 changes the conformal weight of the Y and M generators, and generates the family of algebras $\mathfrak{sv}_\varepsilon(0)$ introduced in Chap. 2.

We shall focus in the sequel on the term proportional to c_1, and denote by $\mathfrak{sv}_\varepsilon(0)$ the one-parameter family of Lie algebra structures on $\mathfrak{sv}(0)$ given by $[\ ,\ ]_\varepsilon = [\ ,\ ]_{\varepsilon,0,0}$, in coherence with Definition 2.2 in Chap. 2. Inspection of the above formulas shows that $\mathfrak{sv}_\varepsilon(0)$ is a semi-direct product $\mathrm{Vect}(S^1) \ltimes \mathfrak{h}_\varepsilon$ where $\mathfrak{h}_\varepsilon$ is a deformation of $\mathfrak{h}$ as a $\mathrm{Vect}(S^1)$-module; one has $\mathfrak{h}_\varepsilon = \mathcal{F}_{\frac{1+\varepsilon}{2}} \oplus \mathcal{F}_\varepsilon$, and the bracket $\mathcal{F}_{\frac{1+\varepsilon}{2}} \times \mathcal{F}_{\frac{1+\varepsilon}{2}} \longrightarrow \mathcal{F}_\varepsilon$ is the usual one, induced by the Poisson bracket on the torus.

Now, as a by-product of the above computations, we shall determine explicitly $H^1(\mathfrak{sv}(0), \mathfrak{sv}(0))$.

Theorem 7.3. *$H^1(\mathfrak{sv}(0), \mathfrak{sv}(0))$ is three-dimensional, generated by the following cocycles, given in terms of modes by:*

$$c_1(L_n) = M_n \quad c_2(L_n) = nM_n$$
$$l(Y_n) = Y_n \quad l(M_n) = 2M_n.$$

The cocycle l already appeared in Chap. 2, when we discussed the derivations of $\mathfrak{sv}(0)$; with the notations of Definition 2.3, one has $l = 2(\delta_2 - \delta_1)$.

Proof. From Lemma 7.2 above, one has $H^1(\mathfrak{sv}(0), \mathfrak{g}) = 0$, and so $H^1(\mathfrak{sv}(0), \mathfrak{sv}(0)) = H^1(\mathfrak{sv}(0), \mathfrak{h})$. One is led to compute the H^1 of a semi-direct product, as shall be done in Sect. 8.2. The space $H^1(\mathfrak{sv}(0), \mathfrak{h})$ is made from two parts $H^1(\mathfrak{g}, \mathfrak{h})$ and $H^1(\mathfrak{h}, \mathfrak{h})$ satisfying the compatibility condition as in Theorem 8.2:

$$c([X, \alpha]) - [X, c(\alpha)] = -[\alpha, c(X)]$$

for $X \in \mathfrak{g}$ and $\alpha \in \mathfrak{h}$.

The result is then easily deduced from the previous computations : $H^1(\mathfrak{g}, \mathfrak{h}) = H^1(\mathfrak{g}, \mathfrak{n})$ is generated by $f\partial \longrightarrow f$ and $f\partial \longrightarrow f'$, which correspond in the mode decomposition to the cocycles c_1 and c_2. As a corollary, one has $[\alpha, c(X)] = 0$ for $X \in \mathfrak{g}$ and $\alpha \in \mathfrak{h}$. Hence the compatibility condition reduces to $c([X, \alpha]) = [X, c(\alpha)]$ and thus $c \in Inv_{\mathfrak{g}} H^1(\mathfrak{h}, \mathfrak{h})$. It can now be deduced from Lemma 7.3 above, that the latter space is one-dimensional, generated by l.

We shall now determine the central charges of $\mathfrak{sv}_\varepsilon(0)$; the computation will shed light on some exceptional values of ε, corresponding to interesting particular cases.

7.3 Computation of $H^2(\mathfrak{sv}_\varepsilon(0), \mathbb{R})$

We shall again make use of the exact sequence decomposition $0 \longrightarrow \mathfrak{h}_\varepsilon \longrightarrow \mathfrak{sv}_\varepsilon(0) \stackrel{\pi}{\longrightarrow} \mathfrak{g} \longrightarrow 0$, and classify the cocycles with respect to their "type" along this decomposition; trivial coefficients will make computations much easier than in the above case. First of all, $0 \longrightarrow H^2(\mathfrak{g}, \mathbb{R}) \stackrel{\pi^*}{\longrightarrow} H^2(\mathfrak{sv}_\varepsilon(0), \mathbb{R})$ is an injection. So the Virasoro class $c \in H^2(\mathrm{Vect}(S^1), \mathbb{R})$ always survives in $H^2(\mathfrak{sv}_\varepsilon(0), \mathbb{R})$.

For $\mathfrak{h}_\varepsilon$, let us use once again the decomposition $0 \longrightarrow \mathfrak{n}_\varepsilon \longrightarrow \mathfrak{h}_\varepsilon \longrightarrow \mathfrak{q}_\varepsilon \longrightarrow 0$ where $\mathfrak{n}_\varepsilon = [\mathfrak{h}_\varepsilon, \mathfrak{h}_\varepsilon]$. One has: $H^1(\mathfrak{g}, H^1(\mathfrak{h}_\varepsilon)) = H^1(\mathfrak{g}, \mathfrak{q}_\varepsilon^*) = H^1(\mathfrak{g}, \mathcal{F}^*_{\frac{1+\varepsilon}{2}}) = H^1(\mathfrak{g}, \mathcal{F}_{-(\frac{3+\varepsilon}{2})})$. The cohomologies of degree one of $\mathrm{Vect}(S^1)$ with coefficients in densities are known (see [31],Theorem 2.4.12): the space $H^1(\mathfrak{g}, \mathcal{F}_{-(\frac{3+\varepsilon}{2})})$ is trivial, except for the three exceptional cases $\varepsilon = -3, -1, 1$:

$H^1(\mathfrak{g}, \mathcal{F}_0)$ is generated by the cocycles $f\partial \longrightarrow f$ and $f\partial \longrightarrow f'$;
$H^1(\mathfrak{g}, \mathcal{F}_{-1})$ is generated by the cocycle $f\partial \longrightarrow f'' \, dx$;
$H^1(\mathfrak{g}, \mathcal{F}_{-2})$ is generated by the cocycle $f\partial \longrightarrow f''' \, (dx)^2$, corresponding to the "Souriau cocycle" associated to the central charge of the Virasoro algebra (see [43], Chap. IV).

In terms of modes, the corresponding cocycles are given by:

$$c_1(L_n, Y_m) = \delta^0_{n+m}, \quad c_2(L_n, Y_m) = n\delta^0_{n+m} \text{ for } \varepsilon = -3;$$
$$c(L_n, Y_m) = n^2\delta^0_{n+m} \text{ for } \varepsilon = -1;$$
$$c(L_n, Y_m) = n^3\delta^0_{n+m} \text{ for } \varepsilon = 1.$$

The most delicate part is the investigation of the term $E_2^{0,2} = Inv_{\mathfrak{g}} H^2(\mathfrak{h}_\varepsilon)$ (this refers of course to the Hochschild-Serre spectral sequence associated to the above decomposition). For $H^2(\mathfrak{h}_\varepsilon)$ we shall use the same short exact sequences as for $\mathfrak{h}$ in the proof of Lemma 7.2:

$$0 \longrightarrow Ker\partial \longrightarrow \Lambda^2\mathfrak{n}_\varepsilon^* \stackrel{\partial}{\longrightarrow} \Lambda^2\mathfrak{q}_\varepsilon^* \wedge \mathfrak{n}_\varepsilon^*$$

$$0 \longrightarrow Ker\partial \longrightarrow \mathfrak{q}_\varepsilon^* \wedge \mathfrak{n}_\varepsilon^* \stackrel{\partial}{\longrightarrow} \Lambda^3\mathfrak{q}_\varepsilon^*$$

$$0 \longrightarrow \underline{\mathfrak{n}}_\varepsilon^* \longrightarrow \Lambda^2\mathfrak{q}_\varepsilon^* \longrightarrow Coker\partial \longrightarrow 0$$

(where $\underline{\mathfrak{n}}_\varepsilon$ stands for $\mathfrak{n}_\varepsilon$ divided out by the space of constant functions). One readily shows that for the first two sequences one has $Inv_{\mathfrak{g}} Ker\partial = 0$. The third one is more complicated; the cohomology exact sequence yields:

$$0 \longrightarrow Inv_{\mathfrak{g}} Coker\partial \longrightarrow H^1(\mathcal{G}, \underline{\mathfrak{n}}_\varepsilon^*) \longrightarrow H^1(\mathfrak{g}, \varepsilon^2\mathfrak{q}_\varepsilon^*) \longrightarrow \cdots$$

The same result as above (see [31], Theorem 2.4.12) shows that:

$$H^1(\mathfrak{g}, \mathfrak{n}_\varepsilon^*) = H^1(\mathfrak{g}, \mathcal{F}_{(-1-\varepsilon)}) = 0 \text{ unless } \varepsilon = 1, -1, 0,$$

and one then has to investigate case by case; set $c(Y_p, Y_q) = a_p\delta^0_{p+q}$ for the potential cochains on $\mathcal{Y}_\varepsilon$. For each n one has the relation:

$$(\mathrm{ad}_{L_n} c)(Y_p, Y_q) - (q-p)\gamma(L_n)(M_{p+q}) = 0 \quad (5.6)$$

for some 1-cocycle $\gamma : \mathfrak{g} \longrightarrow \mathfrak{n}_\varepsilon^*$, and for all (p, q) such that $n + p + q = 0$; if $\gamma(L_n)(M_k) = b_n\delta^0_{n+m+k}$, one obtains in terms of modes, using $a_p = -a_{-p}$:

$$(p+q)\left(\frac{1+\varepsilon}{2}\right)(a_p - a_q) + qa_q - pa_q - (q-p)b_{-(p+q)} = 0$$

Let us now check the different cases of non-vanishing terms in $H^1(\mathcal{G}, \mathfrak{n}_\varepsilon^*)$.

For $\varepsilon = 1$ one has $b_n = n^3$, and one deduces $a_p = p^3$.

For $\varepsilon = -1$ there are two possible cases $b_n = n$ or $b_n = 1$, the above equation gives

$$qa_p - pa_q = (q-p)(\alpha(p+q)+\beta);$$

the only possible solution would be to set a_p constant, but this is not consistent with $a_p = -a_{-p}$. For $\varepsilon = 0$, one gets $b_n = n^2$ and the equation gives

$$\left(\frac{p+q}{2}\right)(a_p + a_q) + qa_p - pa_q - (q-p)(p+q)^2 = 0$$

One easily checks that there are no solutions.

Finally, one gets a new cocycle generating an independent class in $H^2(\mathfrak{sv}_1(0), \mathbb{R})$, given by the formulas:

$$c(Y_n, Y_m) = n^3 \delta_{n+m},$$
$$c(L_n, M_m) = n^3 \delta_{n+m}$$

Let us summarize our results:

Theorem 7.4. *For $\varepsilon \neq -3, -1, 1$, $H^2(\mathfrak{sv}_\varepsilon(0), \mathbb{R}) \simeq \mathbb{R}$ is generated by the Virasoro cocycle.*

For $\varepsilon = -3, -1$, $H^2(\mathfrak{sv}_\varepsilon(0), \mathbb{R}) \simeq \mathbb{R}^2$ is generated by the Virasoro cocycle and an independent cocycle of the form $c(L_n, Y_m) = \delta^0_{n+m}$ for $\varepsilon = -3$ or $c(L_n, Y_m) = n^2 \delta^0_{n+m}$ for $\varepsilon = -1$.

For $\varepsilon = 1$, $H^2(\mathfrak{sv}_1(0), \mathbb{R}) \simeq \mathbb{R}^3$ is generated by the Virasoro cocycle and the two independent cocycles c_1 and c_2 defined by (all other components vanishing)

$$c_1(L_n, Y_m) = n^3 \delta^0_{n+m};$$

$$c_2(L_n, M_m) = c_2(Y_n, Y_m) = n^3 \delta^0_{n+m}$$

Remarks. The isomorphism $H^2(\mathfrak{sv}, \mathbb{R}) \simeq \mathbb{R}$ has been proved in [50]. As we shall see in Sect. 7.4, generally speaking, local cocycles may be carried over from $\mathfrak{sv}(0)$ to $\mathfrak{sv}$ or from $\mathfrak{sv}$ to $\mathfrak{sv}(0)$ without any difficulty.

Let us look more carefully at the $\varepsilon = 1$ case. One has that $\mathfrak{h}_1 = \mathcal{F}_1 \oplus \mathcal{F}_1$ with the obvious bracket $\mathcal{F}_1 \times \mathcal{F}_1 \longrightarrow \mathcal{F}_1$; so, algebraically, $\mathfrak{h}_1 = \mathrm{Vect}(S^1) \otimes \eta\mathbb{R}[\eta]/(\eta^3 = 0)$. One deduces immediately that $\mathfrak{sv}_1(0) = \mathrm{Vect}(S^1) \otimes \mathbb{R}[\eta]/(\eta^3 = 0)$; so the cohomological result for $\mathfrak{sv}_1(0)$ can be easily reinterpreted. Let $f_\eta \partial$ and $g_\eta \partial$ be two elements in $\mathrm{Vect}(S^1) \otimes \mathbb{R}[\eta]/(\eta^3 = 0)$, and compute the Virasoro cocycle $c(f_\eta\partial, g_\eta\partial) = \int_{S^1} f_\eta''' g_\eta dt$ as a truncated polynomial in η; one has $f_\eta = f_0 + \eta f_1 + \eta^2 f_2$ and $g_\eta = g_0 + \eta g_1 + \eta^2 g_2$ so finally:

$$c(f_\eta\partial, g_\eta\partial) = \int_{S^1} f_0''' g_0 dt + \eta \int_{S^1} (f_0''' g_1 + f_1''' g_0) dt + \eta^2 \int_{S^1} (f_0''' g_2 + f_1''' g_1 + f_2''' g_0) dt.$$

In other terms: $c(f_\eta\partial, g_\eta\partial) = c_0(f_\eta\partial, g_\eta\partial) + \eta c_1(f_\eta\partial, g_\eta\partial) + \eta^2 c_2(f_\eta\partial, g_\eta\partial)$. One can easily identify the c_i, $i = 0, 1, 2$ with the cocycles defined in the above theorem, using a decomposition into modes. This situation can be described by a universal central extension

$$0 \longrightarrow \mathbb{R}^3 \longrightarrow \widehat{\mathfrak{sv}_1(0)} \longrightarrow \mathrm{Vect}(S^1) \otimes \mathbb{R}[\eta]/(\eta^3 = 0) \longrightarrow 0$$

and the formulas of the cocycles show that $\widehat{\mathfrak{sv}_1(0)}$ is isomorphic to $\mathfrak{vir} \otimes \mathbb{R}[\eta]/(\eta^3 = 0)$.

Remarks. 1. Cohomologies of Lie algebra of type $\mathrm{Vect}(S^1) \bigotimes_{\mathbb{R}} A$, where A is an associative and commutative algebra (the Lie bracket being as usual given by $[f\partial \otimes a, g\partial \otimes b] = (fg' - gf')\partial \otimes ab$), have been studied by C. Sah and collaborators (see [101]). Their result is: $H^2\left(\mathrm{Vect}(S^1) \bigotimes_{\mathbb{R}} A\right) = A'$ where $A' = Hom_{\mathbb{R}}(A, \mathbb{R})$; all cocycles are given by the Virasoro cocycle composed with the linear form on A. The isomorphism $H^2(\mathfrak{sv}_1(0), \mathbb{R}) \simeq \mathbb{R}^3$ (see Theorem 7.4) could have been deduced from this general theorem.

2. One can obtain various generalisations of our algebra $\mathfrak{h}$ as nilpotent Lie algebras with $\mathrm{Vect}(S^1)$-like brackets, such as

$$[Y_n, Y_m] = (n - m)M_{n+m} \tag{7.17}$$

by use of the same scheme. Let A be an artinian ring quotient of some polynomial ring $\mathbb{R}[t_1, \ldots, t_n]$ and $A_0 \subset A$ its maximal ideal; then $\mathrm{Vect}(S^1) \bigotimes_{\mathbb{R}} A_0$ is a nilpotent Lie algebra whose successive brackets are of the same type (7.17). One could speak of a "virasorization" of nilpotent Lie algebras. Explicit examples are provided in the Sect. 3.5 about multi-diagonal operators of the present article.

3. It is interesting in itself to look at how the dimension of $H^2(\mathfrak{sv}_\varepsilon(0), \mathbb{R})$ varies under deformations. For generic values of ε, this dimension is equal to one, and it increases for some exceptional values of ε; one can consider this as an example of so called "Fuks principle" in infinite dimension: deformations can decrease the rank of cohomologies but never increase it.

4. Analogous Lie algebra structures, of the "Virasoro-tensorized" kind, have been considered in a quite different context in algebraic topology by Tamanoi, see [111].

7.4 About Deformations of $\mathfrak{sv}_1(0)$

Let us recall that local cochains are "support-preserving" in the following sense: the support of the image of a cochain is contained in the intersection of the supports of its entries. Equivalently, according to a well-known theorem by J. Peetre, these cochains are locally given by differential operators acting on the coordinates of the vector fields.

We shall only consider the local cochains $C^*_{loc}(\mathfrak{sv}_1(0), \mathfrak{sv}_1(0))$. The Lie algebra $\mathfrak{sv}_1(0)$ admits a graduation mod 3 by the degree of polynomial in η, the Lie bracket obviously respects this graduation; this graduation induces on the space of local cochains a graduation by weight, and $C^*_{loc}(\mathfrak{sv}_1(0), \mathfrak{sv}_1(0))$ splits into direct sum of subcomplexes of homogeneous weight denoted by $C^*_{loc}(\mathfrak{sv}_1(0), \mathfrak{sv}_1(0))_{(p)}$. Moreover, as classical in computations for Virasoro algebra, one can use the adjoint action of the zero mode L_0 (corresponding geometrically to the Euler field $z\frac{\partial}{\partial z}$) to reduce cohomology computations to the subcomplexes $C^*_{loc}(\mathfrak{sv}_1(0), \mathfrak{sv}_1(0))_{(p)(0)}$ of cochains which are homogeneous of weight 0 with respect to ad L_0 (see e.g. [43], Chap. IV).

We can use the graduation in η and consider homogeneous cochains with respect to that graduation. Here is what one gets, according to the weight:

- weight 1: one has cocycles of the form

$$c(L_n, Y_m) = (m-n)M_{n+m}, \; c(L_n, L_m) = (m-n)Y_{n+m}$$

  but if $b(L_n) = Y_n$, then $c = \partial b$.
- weight 0: all cocycles are coboundaries, using the well-known result $H^*(\mathrm{Vect}(S^1), \mathrm{Vect}(S^1)) = 0$.
- weight -1: one has to consider cochains of the following from

$$\begin{aligned} c(Y_n, M_m) &= a(m-n)M_{n+m} \\ c(Y_n, Y_m) &= b(m-n)Y_{n+m} \\ c(L_n, M_m) &= e(m-n)Y_{n+m} \\ c(L_n, Y_m) &= d(m-n)L_{n+m} \end{aligned}$$

  and check that $\partial c = 0$. It readily gives $e = d = 0$.
  If one sets $\widetilde{c}(Y_n) = \alpha L_n$ and $\widetilde{c}(M_n) = \beta Y_n$, then

$$\begin{aligned} \partial\widetilde{c}(Y_n, M_m) &= (\alpha+\beta)(m-n)M_{n+m} \\ \partial\widetilde{c}(Y_n, M_m) &= (2\alpha-\beta)(m-n)Y_{n+m} \end{aligned}$$

  So all these cocycles are cohomologically trivial,
- weight -2: set

$$\begin{aligned} c(Y_n, M_m) &= \alpha(m-n)Y_{n+m} \\ c(M_n, M_m) &= \beta(m-n)M_{n+m} \\ c(L_n, M_m) &= \gamma(m-n)L_{n+m} \end{aligned}$$

Coboundary conditions give $\gamma = \alpha$ and $\beta = \gamma + \alpha$, but if $\overline{c}(M_n) = L_n$, then

$$\partial\overline{c}(M_n, M_m) = (m-n)Y_{n+m}$$
$$\partial\overline{c}(Y_n, M_m) = 2(m-n)M_{n+m}$$
$$\partial\overline{c}(L_n, M_m) = (m-n)L_{n+m}$$

- weight -3: we find for this case the only surviving cocycle.
  One readily checks that $C \in C^*_{loc}(\mathfrak{sv}_1(0), \mathfrak{sv}_1(0))_{(-3)(0)}$ defined by
  $$C(Y_n, M_m) = (m-n)L_{n+m}$$
  $$C(M_n, M_m) = (m-n)Y_{n+m}$$
  is a cocycle and cannot be a coboundary.

We can describe the cocycle C above more pleasantly by a global formula: let $f_\eta = f_0 + \eta f_1 + \eta^2 f_2$ and $g_\eta = g_0 + \eta g_1 + \eta^2 g_2$, with f_i, g_i elements of Vect(S^1). The bracket [,] of $\mathfrak{sv}_1(0)$ is then the following:

$$[f_\eta, g_\eta] = \sum_{k=0}^{2} \sum_{i+j=k} [f_i, g_i] \text{ or } (f_\eta g_\eta' - g_\eta f_\eta')|_{\eta^3=0},$$

and the deformed bracket [,] + μC will be $[f_\eta, g_\eta]_\mu = (f_\eta g_\eta' - g_\eta f_\eta')|_{\eta^3=\mu}$. So we have found that dim $H^2(\mathfrak{sv}_1(0), \mathfrak{sv}_1(0)) = 1$.

In order to construct deformations, we still have to check for the Nijenhuis-Richardson bracket $[C, C]$ in $C^3(\mathfrak{sv}_1(0), \mathfrak{sv}_1(0))$. The only possibly non-vanishing term is:

$$[C,C](M_n, M_n, M_p) = \sum_{(cycl)} C(C(M_n, M_m,), M_p) = \sum_{(cycl)} (m-n)C(Y_{n+m}, M_p)$$
$$= \sum_{(cycl)} (pm - pn + n^2 - m^2)L_{n+m+p} = 0$$

So there does not exist any obstruction and we have obtained a genuine deformation. We summarize all these results in the following

Proposition 7.4. *There exists a one-parameter deformation of the Lie algebra* $\mathfrak{sv}_1(0)$, *as* $\mathfrak{sv}_{1,\mu}(0) =$ Vect$(S^1) \otimes \mathbb{R}[\eta]/_{(\eta^3=\mu)}$. *This deformation in the only one possible, up to isomorphism.*

If one is interested in central charges, the above-mentioned theorem of C. Sah and al., see [101], shows that dim $H^2(\mathfrak{sv}_{1,\mu}(0), \mathbb{R}) = \mathbb{R}^3$ and the universal central extension $\widehat{\mathfrak{sv}_{1,\mu}(0)}$ is isomorphic to $\mathfrak{vir} \otimes \mathbb{R}[\eta]/_{(\eta^3=\mu)}$. We did not do the computations, but we conjecture that $\mathfrak{sv}_{1,\mu}(0)$ is rigid, the ring $\mathbb{R}[\eta]/_{(\eta^3=\mu)}$ being more generic than $\mathbb{R}[\eta]/_{(\eta^3=0)}$. More generally, it could be interesting to study systematically Lie algebras of type $\mathrm{Vect}(S^1) \otimes A$ where A is a commutative ring, their geometric interpretation being "Virasoro current algebras".

7.5 Coming Back to the Original Schrödinger–Virasoro Algebra

The previous results concern the 'twisted' Schrödinger–Virasoro algebra $\mathfrak{sv}(0)$ generated by the modes (L_n, Y_m, M_p) for $(n, m, p) \in \mathbb{Z}^3$, which make computations easier and allows direct application of Fuks' techniques. The original Schrödinger–Virasoro algebra $\mathfrak{sv}$ is generated by the modes (L_n, Y_m, M_p) for $(n, p) \in \mathbb{Z}^2$ but $m \in \mathbb{Z} + \frac{1}{2}$. Yet Theorem 7.1 and Theorem 7.2 on deformations of $\mathfrak{sv}(0)$ are also valid for $\mathfrak{sv}$: one has $\dim H^2(\mathfrak{sv}, \mathfrak{sv}) = 3$ with the same cocycles c_1, c_2, c_3, since these do not allow 'parity-changing' terms such as $L \times Y \to M$ or $Y \times Y \to Y$ for instance (the (L, M)-generators being considered as 'even' and the Y-generators as 'odd').

But the computation of $H^2(\mathfrak{sv}_\varepsilon, \mathbb{R})$ will yield very different results compared to Theorem 7.4, since 'parity' is not conserved for all the cocycles we found, so we must start from the beginning. Let us use the adjoint action of L_0 to simplify computations: all cohomologies are generated by cocycles c such that $adL_0 \,.\, c = 0$, i.e. such that $c(A_k, B_l) = 0$ for $k + l \neq 0$, A and B being L, Y or M. So, for non-trivial cocycles, one must have $c(Y_n, L_p) = 0, c(Y_n, M_p) = 0$ for all Y_n, L_p, M_p; in $H^2(\mathfrak{sv}_\varepsilon, \mathbb{R})$, terms of the type $H^1(\mathfrak{g}, H^1(\mathfrak{h}_\varepsilon))$ will automatically vanish. The Virasoro class in $H^2(\mathfrak{g}, \mathbb{R})$ will always survive, and one has to check what happens with the terms of type $Inv_\mathfrak{g} H^2(\mathfrak{h}_\varepsilon)$. As in the proof of Lemma 5.3, the only possibilities come from the short exact sequence

$$0 \longrightarrow \underline{\mathfrak{n}}_\varepsilon^* \longrightarrow \Lambda^2 \mathfrak{q}_\varepsilon^* \longrightarrow Coker\partial \longrightarrow 0$$

which induces: $0 \longrightarrow Inv_\mathfrak{g} Coker\partial \longrightarrow H^1(\mathfrak{g}, \underline{\mathfrak{n}}_\varepsilon^*) \longrightarrow H^1(\mathfrak{g}, \varepsilon^2 \mathfrak{q}_\varepsilon^*)$ and one obtains the same (5.6) as above:

$$(ad_{L_n} c)(Y_p, Y_q) + (q - p)\gamma(L_n)(M_{p+q}) = 0$$

If $c(Y_p, Y_q) = a_p \delta^0_{p+q}$, the equation gives:

$$-a_p(p + \frac{\varepsilon + 3}{2} n) + a_{p+n}(p - \frac{\varepsilon + 1}{2} n) - (p - q)\gamma(L_n)(M_{p+q}) = 0$$

One finds two exceptional cases with non-trivial solutions:

- for $\varepsilon = 1$, $a_p = p^3$ and $c(L_n, M_m) = n^3\delta^0_{n+m}$ gives a two-cocycle, very much analogous to the $\mathrm{Vect}(S^1) \otimes \mathbb{R}[\eta]/_{(\eta^3=0)}$ case, except that one has no term in $c(L_n, Y_p)$.
- for $\varepsilon = -3$, if $\gamma \equiv 0$, the above equation gives $pa_p = (p+n)a_{p+n}$ for every p and n. So $a_p = \frac{1}{p}$ is a solution, and one sees why this solution was not available in the twisted case.

Let us summarize:

Proposition 7.5. *The space $H^2(\mathfrak{sv}_\varepsilon, \mathbb{R})$ is one-dimensional, generated by the Virasoro cocycle, save for two exceptional values of ε, for which one has one more independent cocycle, denoted by c_1, with the following non-vanishing components:*

- *for $\varepsilon = 1$: $c_1(Y_p, Y_q) = p^3\delta^0_{p+q}$ and $c_1(L_p, M_q) = p^3\delta^0_{p+q}$,*
- *for $\varepsilon = -3$: $c_1(Y_p, Y_q) = \frac{\delta^0_{p+q}}{p}$.*

Remark. The latter case is the most surprising one, since it contradicts the well-established dogma asserting that only local classes are interesting. This principle of locality has its roots in quantum field theory (see e.g [67] for basic principles of axiomatic field theory); its mathematical status has its foundations in the famous theorem of J. Peetre, asserting that local mappings are given by differential operators, so – in terms of modes – the coefficients are polynomial in n. Moreover, there is a general theorem in the theory of cohomology of Lie algebras of vector fields (see [31]) which states that continues cohomology is in general multiplicatively generated by local cochains, called diagonal in [31]. Here our cocycle contains an anti-derivative, so there could be applications in integrable systems, considered as Hamiltonian systems, the symplectic manifold given by the dual of (usually centrally extended) infinite dimensional Lie algebras (see for example [43], Chaps. VI and X).

Chapter 8
Action of $\mathfrak{sv}$ on Schrödinger and Dirac Operators

The emphasis in the next three chapters is put on a natural action of the Schrödinger–Virasoro group on the *affine*, resp. *linear* space of *Schrödinger operators* of the form $-2\mathrm{i}\mathcal{M}\partial_t - \partial_r^2 + V(t,r)$, resp. $a(t)(-2\mathrm{i}\mathcal{M}\partial_t - \partial_r^2) + V(t,r)$[1], where V is a *periodic time-dependent potential*, and $a(t)$ a time-dependent periodic scaling coefficient. This action is essentially a restriction of the action by conjugation of the differential operators of order one $d\pi(\mathfrak{sv})$, see Chap. 1, on the space of differential operators in two variables. Let us make some general comments on these three chapters.

Chapter 8 is devoted to the basic underlying algebraic structures (definition of the action, affine cocycles...), mostly for the restricted action on a subspace of the affine space of Schrödinger operators, $\mathcal{S}^{\mathrm{aff}}_{\leq 2}$, called the space of *generalized harmonic oscillators* – for which $V(t,r) = V_0(t) = V_1(t)r + V_2(t)r^2$ is quadratic in the space coordinate. The results are taken from [106].

Chapter 9 gives the detailed *orbit structure* for the action of $\mathfrak{sv}$ on generalized harmonic oscillators, in particular, an explicit classification by means of *normal forms*. It relies heavily on the study of the *isotropy groups* that allowed A.A. Kirillov [70] to classify the orbits of the action of the group of diffeomorphisms on the space of *Hill operators*; our results may actually be seen as a quantization of the latter. These operators have been solved independently by physicists (using the so-called *Ermakov-Lewis invariants*[2]) and by mathematicians (using a semi-classical approach[3]). Algebraic, geometric, analytic and physical tools merge nicely in our study and provide a thorough picture of these operators, including a computation of the *monodromy operator*. The restricted action is shown to be Hamiltonian for a certain Poisson structure at the end of the chapter.

---

[1]Note the replacement of $\mathcal{M}$ by $-\mathrm{i}\mathcal{M}$ in these three chapters with respect to all others, made necessary by the use of spectral theory for Schrödinger operators in Chap. 9, which does not allow complex masses.

[2]LewRie.

[3]Hag.

J. Unterberger and C. Roger, *The Schrödinger-Virasoro Algebra*, Theoretical and Mathematical Physics, DOI 10.1007/978-3-642-22717-2_8,

In Chap. 10 (which is independent from Chap. 9), we come back to the unrestricted action on the whole linear space of Schrödinger operators, and show it is Hamiltonian for another unrelated Poisson structure inherited from the much larger loop space over the pseudo-differential algebra. Our approach there is more in the spirit of infinite-dimensional Poisson geometry, and may hopefully give rise to new integrable systems.

The short Sect. 8.3 (introducing an action of $\mathfrak{sv}$ on *Dirac* operators) looks isolated in the picture. Yet it fits in well within our general idea of considering actions of general reparametrization groups on operators coming from the mathematical physics. One can imagine many kinds of possible generalizations and hope to use (as in Chap. 9) algebraic structures to classify other classes of time-dependent operators. This section is also interesting in that it gives another example of a 'concrete' coinduced representation (see Chap. 5).

8.1 Definitions and Notations

Let $\partial = \frac{\partial}{\partial x}$ be the derivation operator on the torus $\mathbb{T} = [0, 2\pi]$. A *Hill operator* is by definition a second order operator on $\mathbb{T}$ of the form $\partial^2 + u$, $u \in C^\infty(\mathbb{T})$. Let π_λ be the representation of Diff(S^1) on the space of $(-\lambda)$-densities $\mathcal{F}_\lambda$ (see Definition 1.8). One identifies the vector spaces $C^\infty(\mathbb{T})$ and $\mathcal{F}_\lambda$ in the natural way, by associating to $f \in C^\infty(\mathbb{T})$ the density $f dx^{-\lambda}$. Then, for any couple $(\lambda, \mu) \in \mathbb{R}^2$, one has an action $\Pi_{\lambda,\mu}$ of Diff(S^1) on the space of differential operators on $\mathbb{T}$ through the left-and-right action

$$\Pi_{\lambda,\mu}(\phi) : D \to \pi_\lambda(\phi) \circ D \circ \pi_\mu(\phi)^{-1},$$

with corresponding infinitesimal action

$$d\Pi_{\lambda,\mu}(\phi) : D \to d\pi_\lambda(\phi) \circ D - D \circ d\pi_\mu(\phi).$$

For a particular choice of λ, μ, namely, $\lambda = -\frac{3}{2}, \mu = \frac{1}{2}$, this representation preserves the affine space of Hill operators; more precisely,

$$\pi_{-3/2}(\phi) \circ (\partial^2 + u) \circ \pi_{1/2}(\phi)^{-1} = \partial^2 + (\phi')^2 (u \circ \phi') + \frac{1}{2} S(\phi) \tag{8.1}$$

where S stands for the Schwarzian derivative. In other words, u transforms as an element of $\mathfrak{vir}^*_{\frac{1}{2}}$ (see Sect. 3.1). One may also – taking an opposite point of view – say that Hill operators define a Diff(S^1)-equivariant morphism from $\mathcal{F}_{\frac{1}{2}}$ into $\mathcal{F}_{-\frac{3}{2}}$.

The purpose of this chapter is to find out similar left-and-right actions of SV on several affine spaces of differential operators. This will lead us to introduce several representations of SV that may all be obtained by the general method of coinduction (see Chap. 5).

Definition 8.1 (*affine subspace of Schrödinger operators*). Let $\mathscr{S}^{lin}$ be the vector space of second order operators on $\mathbb{R}^2$ defined by

$$D \in \mathscr{S}^{lin} \Leftrightarrow D = h(t)\Delta_0 + V(t,r), \quad h, V \in C^\infty(\mathbb{R}^2),$$

where

$$\Delta_0 := -2\mathrm{i}\mathscr{M}\partial_t - \partial_r^2 \tag{8.2}$$

is the *free Schrödinger operator*, and $\mathscr{S}^{aff} \subset \mathscr{S}^{lin}$ be the affine subspace of 'Schrödinger operators' given by the hyperplane $h = 1$.

In other words, an element of $\mathscr{S}^{aff}$ is the sum of the free Schrödinger operator $\Delta_0 = -2\mathrm{i}\mathscr{M}\partial_t - \partial_r^2$ and of a potential V.

The following theorem proves that there is a natural family of actions of the group SV on the space $\mathscr{S}^{lin}$: more precisely, for every $\lambda \in \mathbb{R}$, and $g \in SV$, there is a 'scaling function' $F_{g,\lambda} \in C^\infty(S^1)$ such that

$$\pi_\lambda(g)(\Delta_0 + V)\pi_\lambda(g)^{-1} = F_{g,\lambda}(t)(\Delta_0 + V_{g,\lambda}) \tag{8.3}$$

where $V_{g,\lambda} \in C^\infty(\mathbb{R}^2)$ is a 'transformed potential' depending on g and on λ (see Sect. 1.2 for the definition of π_λ). Taking the infinitesimal representation of $\mathfrak{sv}$ instead, this is equivalent to demanding that the 'adjoint' action of the vector field representation $d\pi_\lambda(\mathfrak{sv})$ (see Introduction) preserve $\mathscr{S}^{lin}$, namely

$$[d\pi_\lambda(X), \Delta_0 + V](t,r) = f_{X,\lambda}(t)(\Delta_0 + V_{X,\lambda}), \quad X \in \mathfrak{sv} \tag{8.4}$$

for a certain infinitesimal 'scaling' function $f_{X,\lambda}$ and with a transformed potential $V_{X,\lambda}$.

We shall actually prove that this last property even characterizes in some sense the differential operators of order one that belong to $d\pi_\lambda(\mathfrak{sv})$.

Theorem 8.1. *1. The Lie algebra of differential operators of order one $\mathscr{X}$ on $\mathbb{R}^2$ preserving the space $\mathscr{S}^{lin}$, i.e., such that*

$$[\mathscr{X}, \mathscr{S}^{lin}] \subset \mathscr{S}^{lin}$$

is equal to the image of $\mathfrak{sv}$ by the representation $d\pi_\lambda$ (modulo the addition to $\mathscr{X}$ of operators of multiplication by an arbitrary function of t).

2. The action of $d\pi_{\lambda+1/4}(\mathfrak{sv})$ on the free Schrödinger operator Δ_0 is given by

$$[d\pi_{\lambda+1/4}(L_f), \Delta_0] = f'\Delta_0 - \frac{\mathscr{M}^2}{2} f'''r^2 - 2\mathrm{i}\mathscr{M}\lambda f'' \tag{8.5}$$

$$[d\pi_{\lambda+1/4}(Y_g), \Delta_0] = -2\mathscr{M}^2 r g'' \tag{8.6}$$

$$[d\pi_{\lambda+1/4}(M_h), \Delta_0] = -2\mathcal{M}^2 h' \tag{8.7}$$

Proof. Let $\mathcal{X} = f\partial_t + g\partial_r + h$ preserving the space $\mathcal{S}^{lin}$: this is equivalent to the existence of two functions $\phi(t)$, $V(t,r)$ such that $[\mathcal{X}, \Delta_0] = \phi(\Delta_0 + V)$. It is clear that $[h, \mathcal{S}_{lin}] \subset \mathcal{S}_{lin}$ if h is a function of t only.

By considerations of degree, one must then have $[\mathcal{X}, \partial_r] = a(t,r)\partial_r + b(t,r)$, hence f is a function of t only. Then

$$[f\partial_t, -2\mathrm{i}\mathcal{M}\partial_t - \partial_r^2] = 2\mathrm{i}\mathcal{M} f'\partial_t \tag{8.8}$$

$$[g\partial_r, -2\mathrm{i}\mathcal{M}\partial_t - \partial_r^2] = 2\mathrm{i}\mathcal{M}\partial_t g\partial_r + 2\partial_r g\partial_r^2 + \partial_r^2 g\partial_r \tag{8.9}$$

$$[h, -\partial_r^2] = 2\partial_r h\partial_r + \partial_r^2 h \tag{8.10}$$

so, necessarily,

$$f' = 2\partial_r g = -\phi$$

and

$$(-2\mathrm{i}\mathcal{M}\partial_t - \partial_r^2)g = -2\partial_r h.$$

Putting together these relations, one gets points 1 and 2 simultaneously. □

Using a left-and-right action of $\mathfrak{sv}$ that combines $d\pi_\lambda$ and $d\pi_{1+\lambda}$, one gets a new family of representations $d\sigma_\lambda$ of $\mathfrak{sv}$ which map the affine space $\mathcal{S}^{aff}$ into differential operators of order zero (that is to say, into functions):

Proposition 8.2. *1. Let $d\sigma_\lambda : \mathfrak{sv} \to Hom(\mathcal{S}^{lin}, \mathcal{S}^{lin})$ be defined by the left-and-right infinitesimal action*

$$d\sigma_\lambda(X) : D \to d\pi_{1+\lambda}(X) \circ D - D \circ d\pi_\lambda(X).$$

Then $d\sigma_\lambda$ is a representation of $\mathfrak{sv}$ and $d\sigma_\lambda(\mathfrak{sv})(\mathcal{S}^{aff}) \subset C^\infty(\mathbb{R}^2)$.

2. The infinitesimal representation $d\sigma_\lambda$ exponentiates to a left-and-right action of the Schrödinger–Virasoro group on the affine space $\mathcal{S}^{aff}$ defined by

$$\sigma_\lambda(g)(D) = \pi_{\lambda+1}(g)D\pi_\lambda(g^{-1}). \tag{8.11}$$

Proof. Let $X_1, X_2 \in \mathfrak{sv}$, and put $d\bar{\pi}_{S^1}(X_i) = f_i(t)$, $i = 1, 2$; then, with a slight abuse of notations, $d\sigma_\lambda(X_i) = \mathrm{ad}\, d\pi_\lambda(X_i) + f_i'$, so

$$[d\sigma_\lambda(X_1), d\sigma_\lambda(X_2)] = [\mathrm{ad}(d\pi_\lambda(X_1)) + f_1', \mathrm{ad}(d\pi_\lambda(X_2)) + f_2'] \tag{8.12}$$

$$= \mathrm{ad}\, d\pi_\lambda([X_1, X_2]) + \big([d\pi_\lambda(X_1), f_2'(t)] - [d\pi_\lambda(X_2), f_1'(t)]\big). \tag{8.13}$$

Now ad $d\pi_\lambda$ commutes with operators of multiplication by any function of time $g(t)$ if $X \in \mathfrak{h}$, and

$$[d\pi_\lambda(L_f), g(t)] = [f(t)\partial_t, g(t)] = f(t)g'(t)$$

so as a general rule

$$[d\pi_\lambda(X_i), g(t)] = f_i(t)g'(t).$$

Hence

$$\begin{aligned}
[d\sigma_\lambda(X_1), d\sigma_\lambda(X_2)] &= \mathrm{ad}\, d\pi_\lambda([X_1, X_2]) + (f_1(t)f_2''(t) - f_2(t)f_1''(t)) &(8.14)\\
&= \mathrm{ad}\, d\pi_\lambda([X_1, X_2]) + (f_1 f_2' - f_2 f_1')'(t) &(8.15)\\
&= d\sigma_\lambda([X_1, X_2]). &(8.16)
\end{aligned}$$

By the preceding Theorem, it is now clear that $d\sigma_\lambda(\mathfrak{sv})$ sends $\mathcal{S}^{\mathrm{aff}}$ into differential operators of order zero. □

Remark. Choosing $\lambda = \frac{1}{4}$ leads to a representation of Sch preserving the kernel of Δ_0 (as already known from Proposition 1.5). So, in some sense, $\lambda = \frac{1}{4}$ is the 'best' choice.

The following formulas hold:

Proposition 8.3 (affine action on Schrödinger operators).

1. *The action σ_λ is written as follows*

$$\sigma_\mu(\phi; 0).(\Delta_0 + V(t,r)) =$$
$$\Delta_0 + \dot\phi(t)V(\phi(t), r\sqrt{\dot\phi(t)}) + 2\mathrm{i}\left(\mu - \frac{1}{4}\right)\mathcal{M}\frac{\ddot\phi}{\dot\phi} + \frac{1}{2}\mathcal{M}^2 r^2 S(\phi)(t)$$

$$\sigma_\mu(1; (\alpha, \beta)).(\Delta_0 + V(t,r)) =$$
$$\Delta_0 + V(t, r - \alpha(t)) - 2\mathcal{M}^2 r\ddot\alpha(t) - \mathcal{M}^2(2\dot\beta(t) - \alpha(t)\ddot\alpha(t)) \quad (8.17)$$

where $S : \phi \to \frac{\dddot\phi}{\dot\phi} - \frac{3}{2}\left(\frac{\ddot\phi}{\dot\phi}\right)^2$ is the Schwarzian derivative.

2. *The infinitesimal action $d\sigma_\mu : X \to \frac{d}{dt}\big|_{t=0}\left(\tilde\sigma_\mu(\exp tX)\right)$ of $\mathfrak{sv}$ gives*

$$d\tilde\sigma_\mu(\mathcal{L}_f)(\Delta_0 + V(t,r)) =$$
$$-f\dot V - \frac{1}{2}\dot f r\frac{\partial V}{\partial r} - 2\mathrm{i}\left(\mu - \frac{1}{4}\right)\mathcal{M}\ddot f - \frac{1}{2}\mathcal{M}^2\dddot f r^2 - \dot f V$$

$$d\sigma_\mu(\mathcal{Y}_g)(\Delta_0 + V(t,r)) = -g\frac{\partial V}{\partial r} - 2\mathcal{M}^2\ddot g r$$

$$d\sigma_\mu(\mathcal{M}_h)(\Delta_0 + V(t,r)) = -2\mathcal{M}^2\dot h \quad (8.18)$$

Let us now introduce the affine space of *generalized harmonic oscillators*, $\mathcal{S}^{aff}_{\leq 2}$. The previous formulas imply immediately that $\mathcal{S}^{aff}_{\leq 2}$ is preserved by the Schrödinger–Virasoro actions $d\sigma_\lambda$.

Definition 8.4 (space of generalized harmonic oscillators). Let $\mathcal{S}^{aff}_{\leq 2} \subset \mathcal{S}^{aff}$ be the affine subspace of Schrödinger operators with potentials that are at most quadratic in r, that is,

$$D \in \mathcal{S}^{aff}_{\leq 2} \Leftrightarrow D = -2i\mathcal{M}\partial_t - \partial_r^2 + V_2(t)r^2 + V_1(t)r + V_0(t). \qquad (8.19)$$

is mapped into potentials of the same form under $d\sigma_\lambda(SV)$.

Let us use the same vector notation for elements of $\mathcal{S}^{aff}_{\leq 2}$ and for potentials that are at most quadratic in r (what is precisely meant will be clearly seen from the context). Set $D = \begin{pmatrix} V_2 \\ V_1 \\ V_0 \end{pmatrix}$ and $V = \begin{pmatrix} V_2 \\ V_1 \\ V_0 \end{pmatrix}$ for $D = \Delta_0 + V_2(t)r^2 + V_1(t)r + V_0(t) \in \mathcal{S}^{aff}_{\leq 2}$ and $V(t,r) = V_2(t)r^2 + V_1(t)r + V_0(t) \in C^\infty(\mathbb{R}^2)$, respectively. Then one can give an explicit formula for the action of $d\sigma_\lambda$ on $\mathcal{S}^{aff}_{\leq 2}$:

Proposition 8.5. *1. Let* $D = \begin{pmatrix} V_2 \\ V_1 \\ V_0 \end{pmatrix} \in \mathcal{S}^{aff}_{\leq 2}$ *and* $f_0, f_1, f_2 \in C^\infty(\mathbb{R})$. *Then the following formulas hold:*

$$d\sigma_{\lambda+1/4}(\mathcal{L}_{f_0})(D) = -\begin{pmatrix} \frac{\mathcal{M}^2}{2} f_0''' + 2f_0'V_2 + f_0V_2' \\ f_0V_1' + \frac{3}{2}f_0'V_1 \\ f_0V_0' + f_0'V_0 + 2i\mathcal{M}\lambda f_0'' \end{pmatrix} \qquad (8.20)$$

$$d\sigma_{\lambda+1/4}(\mathcal{Y}_{f_1})(D) = -\begin{pmatrix} 0 \\ 2f_1V_2 + 2\mathcal{M}^2 f_1'' \\ f_1V_1 \end{pmatrix} \qquad (8.21)$$

$$d\sigma_{\lambda+1/4}(\mathcal{M}_{f_2})(D) = \begin{pmatrix} 0 \\ 0 \\ -2\mathcal{M}^2 f_2' \end{pmatrix} \qquad (8.22)$$

2. Consider the restriction of $d\sigma_{1/4}$ *to* $\mathrm{Vect}(S^1) \subset \mathfrak{sv}$. *Then* $d\sigma_{1/4}|_{\mathrm{Vect}(S^1)}$ *acts diagonally on the 3-vectors* $\begin{pmatrix} V_2 \\ V_1 \\ V_0 \end{pmatrix}$ *and its restriction to the subspaces* $\mathcal{S}^{aff}_i := \{\Delta_0 + g(t)r^i \mid g \in C^\infty(\mathbb{R})\}$, $i = 0, 1, 2$, *is equal to the coadjoint action of* $\mathrm{Vect}(S^1)$ *on the affine hyperplane* $\mathfrak{vir}^*_{1/4}$ $(i = 2)$, *and to the usual action* *of* Vect(S^1) *on* $\mathscr{F}_{-3/2} \simeq \mathscr{F}^*_{1/2}$ *(when* $i = 1$*), respectively on* $\mathscr{F}_{-1} \simeq \mathscr{F}^*_0$ *(when* $i = 0$*). Taking* $\lambda \neq 0$ *leads to an affine term proportional to* f_0'' *on the third coordinate, corresponding to the non-trivial affine cocycle in* $H^1(\mathrm{Vect}(S^1), \mathscr{F}_{-1})$.

In other words, if one identifies $\mathscr{S}^{aff}_{\leq 2}$ with $\mathfrak{sv}^*_{\frac{1}{4}}$ by

$$\left\langle \begin{pmatrix} V_2 \\ V_1 \\ V_0 \end{pmatrix}, \begin{pmatrix} f_0 & f_1 & f_2 \end{pmatrix} \right\rangle_{\mathscr{S}^{aff}_{\leq 2} \times \mathfrak{sv}} = \sum_{i=0}^{2} \int_{S^1} (V_{2-i} f_i)(z)\, dz,$$

then the restriction of $d\sigma_{1/4}$ to Vect(S^1) is equal to the restriction of the coadjoint action of $\mathfrak{sv}$ on $\mathfrak{sv}^*_{\frac{1}{4}}$.

Note, however, that $d\sigma_{1/4}$ is *not* equal to the coadjoint action of $\mathfrak{sv}$.

Proof. Point 2 is more or less obvious, and we shall only give some of the computations for the first one. One has $[d\pi_{\lambda+1/4}(\mathscr{L}_{f_0}), \Delta_0] = f_0' \Delta_0 + \mathrm{i}\frac{\mathscr{M}}{2} f_0''' r^2 - 2\mathrm{i}\mathscr{M}\lambda f_0''$, $[d\pi_{\lambda+1/4}(\mathscr{L}_{f_0}), V_0(t)] = -f_0(t)V_0'(t)$, $[d\pi_{\lambda+1/4}(\mathscr{L}_{f_0}), V_1(t)r] = -(f_0(t)V_1'(t) + \frac{1}{2} f_0'(t) V_1(t))r$, $[d\pi_{\lambda+1/4}(\mathscr{L}_{f_0}), V_2(t)r^2] = -(f_0(t)V_2'(t) + f_0'(t)V_2(t))r^2$, so

$$\begin{aligned} d\sigma_{\lambda+1/4}(\mathscr{L}_{f_0})(D) &= -f_0'.D + [d\pi_{\lambda+1/4}(\mathscr{L}_{f_0}), D] \\ &= -\left(f_0 V_0' + f_0' V_0 + 2\mathrm{i}\mathscr{M}\lambda f_0''\right) - \left(f_0 V_1' + \frac{3}{2} f_0' V_1\right) r \\ &\quad - \left(-\frac{\mathscr{M}^2}{2} f_0''' + 2 f_0' V_2 + f_0 V_2'\right) r^2. \end{aligned}$$

Hence the result for $d\sigma_{\lambda+1/4}(\mathscr{L}_{f_0})$. The other computations are similar though somewhat simpler. □

This representation is easily integrated to a representation σ of the group SV. We let $S(\phi) = \frac{\phi'''}{\phi'} - \frac{3}{2}\left(\frac{\phi''}{\phi'}\right)^2$ ($\phi \in \mathrm{Diff}(S^1)$) be the Schwarzian derivative of the function ϕ.

Proposition 8.6. *Let* $D = \begin{pmatrix} V_2 \\ V_1 \\ V_0 \end{pmatrix} \in \mathscr{S}^{aff}_{\leq 2}$, *then*

$$\sigma_{\lambda+1/4}(\phi; (a,b))D = \sigma_{\lambda+1/4}(1; (a,b))\sigma_{\lambda+1/4}(\phi; 0)D \tag{8.23}$$

$$\sigma_{\lambda+1/4}(\phi;1).D = \begin{pmatrix} (\phi')^2.(V_2\circ\phi) + \frac{\mathcal{M}^2}{2}S(\phi) \\ (\phi')^{\frac{3}{2}}(V_1\circ\phi) \\ \phi'(V_0\circ\phi) + 2\mathrm{i}\mathcal{M}\lambda\left(\frac{\phi'}{\phi}\right)' \end{pmatrix} \tag{8.24}$$

$$\sigma_{\lambda+1/4}(1;(a,b)).D = \begin{pmatrix} V_2 \\ V_1 - 2aV_2 - 2\mathcal{M}^2a'' \\ V_0 - aV_1 + a^2V_2 - \mathcal{M}^2(2b' - aa'') \end{pmatrix} \tag{8.25}$$

defines a representation of SV that integrates $d\sigma$, and maps the affine space $\mathcal{S}^{aff}_{\leq 2}$ into itself.

In other words, elements of $\mathcal{S}^{aff}_{\leq 2}$ define an SV-equivariant morphism from $\mathcal{H}_\lambda$ into $\mathcal{H}_{\lambda+1}$, where $\mathcal{H}_\lambda$, respectively $\mathcal{H}_{\lambda+1}$, is the space $C^\infty(\mathbb{R}^2)$ of functions of t, r that are at most quadratic in r, equipped with the action π_λ, respectively $\pi_{\lambda+1}$ (see Sect. 1.2).

Proof. Put $SV = G \ltimes H$. Then the restrictions $\sigma|_G$ and $\sigma|_H$ define representations (this is a classical result for the first action, and may be checked by direct computation for the second one). The associated infinitesimal representation of $\mathfrak{sv}$ is easily seen to be equal to $d\sigma$. □

In particular, the orbit of the free Schrödinger operator Δ_0 is given by the remarkable formula

$$\sigma_{\lambda+1/4}(\phi;(a,b))\Delta_0 = \begin{pmatrix} \frac{\mathcal{M}^2}{2}S(\phi) \\ -2\mathcal{M}^2a'' \\ -\mathcal{M}^2(2b' - aa'') \end{pmatrix} + \lambda\begin{pmatrix} 0 \\ 0 \\ 2\mathrm{i}\mathcal{M}\left(\frac{\phi'}{\phi}\right)' \end{pmatrix}, \tag{8.26}$$

mixing a third-order cocycle with coefficient $-\mathcal{M}^2$ which extends the Schwarzian cocycle $\phi \to S(\phi)$ with a second-order cocyle with coefficient $2\mathrm{i}\mathcal{M}\lambda$ which extends the well-known cocycle $\phi \to \frac{\phi''}{\phi}$ in $H^1(\mathrm{Vect}(S^1), \mathcal{F}_{-1})$. The following section shows that all affine cocycles of $\mathfrak{sv}$ with coefficients in the (linear) representation space $\mathcal{S}^{aff}_{\leq 2}$ are of this form.

8.2 Affine Schrödinger Cocycles

The representations of groups and Lie algebras described above are of affine type; if one has linear representations of the group G and Lie algebra $\mathfrak{g}$ on a module M, one can deform these representations into affine ones, using the following construction.

Let $C : G \to M$ (resp $c : \mathfrak{g} \longrightarrow M$) be a 1-cocycle in $Z^1_{diff}(G, M)$ (resp $Z^1(\mathfrak{g}, M)$); it defines an affine action of G (resp $\mathfrak{g}$) by deforming the linear action as follows:

$$g * m = g.m + C(g)$$
$$\xi * m = \xi.m + c(\xi)$$

respectively. Here the dot indicates the original linear action and $*$ the affine action. One deduces from the formulas given in Propositions 8.5 and 8.6 that the above representations are of this type; the first cohomology of SV (resp $\mathfrak{sv}$) with coefficients in the module $\mathcal{S}^{aff}_{\leq 2}$ (equipped with the linear action) classifies all the affine deformations of the action, up to isomorphism. They are given by the following theorem:

Theorem 8.2. *The degree-one cohomology of the group SV (resp Lie algebra $\mathfrak{sv}$) with coefficients in the module $\mathcal{S}^{aff}_{\leq 2}$ (equipped with the linear action) is two-dimensional and can be represented by the following cocycles:*

- for SV:

$$C_1(\phi,(a,b)) = \begin{pmatrix} -\frac{1}{2}S(\phi) \\ 2a'' \\ 2b' - aa'' \end{pmatrix}$$

$$C_2(\phi,(a,b)) = \begin{pmatrix} 0 \\ 0 \\ \left(\frac{\phi'}{\phi}\right)' \end{pmatrix}$$

- for $\mathfrak{sv}$:

$$c_1(\mathcal{L}_{f_0} + \mathcal{Y}_{f_1} + \mathcal{M}_{f_2}) = \begin{pmatrix} \frac{1}{2}f_0''' \\ -2f_1'' \\ -2f_2' \end{pmatrix}$$

$$c_2(\mathcal{L}_{f_0} + \mathcal{Y}_{f_1} + \mathcal{M}_{f_2}) = \begin{pmatrix} 0 \\ 0 \\ f_0'' \end{pmatrix}$$

One recognizes the cocycles evidenced in Propositions 8.5 and 8.6.

Proof. One shall first make the computations for the Lie algebra and then try to integrate explicitly; here, the "heuristical" version of Van-Est theorem, generalized to the infinite-dimensional case, guarantees the isomorphism between the H^1 groups for SV and $\mathfrak{sv}$ (see [43], Chap. IV).

So let us compute $H^1(\mathfrak{sv}, M) \simeq H^1(\mathfrak{g} \ltimes \mathfrak{h}, M)$ for $M = \mathcal{S}^{aff}_{\leq 2}$ (equipped with the linear action). Let $c : \mathfrak{g} \times \mathfrak{h} \longrightarrow M$ be a cocycle and set $c = c' + c''$ where $c' = c|_{\mathfrak{g}}$ and $c'' = c|_{\mathfrak{h}}$. One has $c' \in Z^1(\mathfrak{g}, M)$ and $c'' \in Z^1(\mathfrak{h}, M)$, and these two cocycles are linked together by the compatibility relation

$$c''([X,\alpha]) - X.(c''(\alpha)) + \alpha.(c'(X)) = 0 \tag{8.27}$$

As a $\mathfrak{g}$-module, $M = \mathcal{F}_{-2} \oplus \mathcal{F}_{-3/2} \oplus \mathcal{F}_{-1}$, so one determines easily that $H^1(\mathfrak{g}, \mathcal{F}_{-2})$ and $H^1(\mathfrak{g}, \mathcal{F}_{-1})$ are one-dimensional, generated by $\mathcal{L}_{f_0} \longrightarrow f_0''' dx^2$

and $\mathcal{L}_{f_0} \longrightarrow f_0'' dx$ respectively, and $H^1(\mathfrak{g}, \mathcal{F}_{-3/2}) = 0$ (see [43], Chap. IV). One can now readily compute the 1-cohomology of the nilpotent part $\mathfrak{h}$; one easily remarks that the linear action on $\mathcal{S}^{aff}_{\leq 2}$ is defined as follows:

$$(\mathcal{Y}_{f_1} + \mathcal{M}_{f_2}) \, . \begin{pmatrix} V_2 \\ V_1 \\ V_0 \end{pmatrix} = \begin{pmatrix} 0 \\ 2f_1 V_2 \\ f_1 V_1 \end{pmatrix}$$

Direct computation shows that there are several cocycles, but compatibility condition (8.27) destroys all but one of them:

$$c''(\mathcal{Y}_{f_1} + \mathcal{M}_{f_2}) = \begin{pmatrix} 0 \\ f_1'' \\ f_2' \end{pmatrix}$$

The compatibility condition gives

$$c''([\mathcal{L}_{f_0}, \mathcal{Y}_{f_1} + \mathcal{M}_{f_2}]) - \mathcal{L}_{f_0}\, (c''(\mathcal{Y}_{f_1} + \mathcal{M}_{f_2})) = -\frac{1}{2} f_1 f_0'''$$

On the other hand

$$(\mathcal{Y}_{f_1} + \mathcal{M}_{f_2}).(c'(\mathcal{L}_{f_0})) = (\mathcal{Y}_{f_1} + \mathcal{M}_{f_2}) \, . \begin{pmatrix} f_0''' \\ 0 \\ f_0'' \end{pmatrix} = \begin{pmatrix} 0 \\ 2f_1 f_0''' \\ 0 \end{pmatrix}$$

Hence the result, with the right proportionality coefficients, and one obtains the formula for c_1. One also remarks that the term with $f_0'' dx$ disappears through the action of $\mathfrak{h}$, so it will induce an independent generator in $H^1(\mathfrak{sv}, \mathcal{S}^{aff})$, precisely c_2.

Finally the cocycles C_1 and C_2 in $H^1(SV, \mathcal{S}^{aff}_{\leq 2})$ are not very difficult to compute, once we have determined the unipotent action of H on $\mathcal{S}^{aff}_{\leq 2}$:

$$(a,b). \begin{pmatrix} V_2 \\ V_1 \\ V_0 \end{pmatrix} = \begin{pmatrix} V_2 \\ V_1 + 2aV_2 \\ V_0 + aV_1 - a^2 V_2 \end{pmatrix}$$

□

8.3 Action on Dirac-Lévy-Leblond Operators

J.-J. Lévy-Leblond introduced in [76] an order one matrix differential operator $\mathscr{D}_0$ on $\mathbb{R}^{d+1}$ (with coordinates $t, r_1, \ldots, r_d$), called *Dirac-Lévy-Leblond operator*. It is similar to the *Dirac operator*, whose square is equal to $-\Delta_0 \otimes \mathrm{Id}$ $= -\begin{pmatrix} \Delta_0 & & \\ & \ddots & \\ & & \Delta_0 \end{pmatrix}$ for $\Delta_0 = -2\mathrm{i}\mathscr{M}\partial_t - \sum_{i=1}^d \partial_{r_i}^2$. So, in some sense, $\mathscr{D}_0$ is a square-root of the free Schrödinger operator, just as the Dirac operator is a square-root of the D'Alembertian. The group of Lie invariance of $\mathscr{D}_0$ has been studied in [32], where it is seen to be isomorphic to $\mathrm{Sch}(d)$ with a realization that is close in spirit to that of Proposition 1.12; it exponentiates $d\pi^{\sigma}_{\frac{1}{2}}$ (see Definition 8.8 below).

Let us restrict to the case $d = 1$ (see Sect. 11.1 for details). Then $\mathscr{D}_0$ acts on spinors, or pairs of functions $\begin{pmatrix} \phi_1 \\ \phi_2 \end{pmatrix}$ of two variables t, r, and may be written as

$$\mathscr{D}_0 = \begin{pmatrix} \partial_r & 2\mathrm{i}\mathscr{M} \\ \partial_t & -\partial_r \end{pmatrix}. \tag{8.28}$$

One checks immediately that $\mathscr{D}_0^2 = -\Delta_0 \otimes \mathrm{Id}$.

From the explicit realization of $\mathfrak{sch}$ on spinors (see again Sect. 11.1), one may easily guess a realization of $\mathfrak{sv}$ that extends the action of $\mathfrak{sch}$, and, perhaps more interestingly, acts on an affine space $\mathscr{D}^{aff}$ of Dirac-Lévy-Leblond operators with potential, in the same spirit as in the previous section. More precisely, one has the following theorem (we need to introduce some notations first).

Definition 8.7. Let $\mathscr{D}^{lin}$ be the vector space of first order matrix operators on $\mathbb{R}^2$ defined by

$$D \in \mathscr{D}^{lin} \Leftrightarrow D = h(t)\mathscr{D}_0 + \begin{pmatrix} 0 & 0 \\ V(t,r) & 0 \end{pmatrix}, \quad h, V \in C^{\infty}(\mathbb{R}^2)$$

and $\mathscr{D}^{aff}$, $\mathscr{D}^{aff}_{\leq 2}$ be the affine subspaces of $\mathscr{D}^{lin}$ such that

$$D \in \mathscr{D}^{aff} \Leftrightarrow D = \mathscr{D}_0 + \begin{pmatrix} 0 & 0 \\ V(t,r) & 0 \end{pmatrix},$$

$$D \in \mathscr{D}^{aff}_{\leq 2} \Leftrightarrow D = \mathscr{D}_0 + \begin{pmatrix} 0 & 0 \\ V_2(t)r^2 + V_1(t)r + V_0(t) & 0 \end{pmatrix}.$$

We shall say that a matrix of the form $\begin{pmatrix} 0 & 0 \\ V(t,r) & 0 \end{pmatrix}$, where $V \in C^\infty(\mathbb{R}^2)$, is a *Dirac potential.*

Definition 8.8. Let $d\pi_\lambda^\sigma$ ($\lambda \in \mathbb{C}$) be the infinitesimal representation of $\mathfrak{sv}$ on the space $\widetilde{\mathscr{H}}_\lambda^\sigma \simeq (C^\infty(\mathbb{R}^2))^2$ with coordinates t, r, defined by

$$d\pi_\lambda^\sigma(L_f) = \left(-f(t)\partial_t - \frac{1}{2}f'(t)r\partial_r + \frac{\mathrm{i}}{4}\mathscr{M} f''(t)r^2\right) \otimes \mathrm{Id}$$

$$-f'(t) \otimes \begin{pmatrix} \lambda - \frac{1}{4} & 0 \\ 0 & \lambda + \frac{1}{4} \end{pmatrix} - \frac{1}{2}f''(t)r \otimes \begin{pmatrix} 0 & 0 \\ 1 & 0 \end{pmatrix}; \quad (8.29)$$

$$d\pi_\lambda^\sigma(Y_g) = (-g(t)\partial_r + \mathrm{i}\mathscr{M} g'(t)r) \otimes \mathrm{Id} - g'(t) \otimes \begin{pmatrix} 0 & 0 \\ 1 & 0 \end{pmatrix}; \quad (8.30)$$

$$d\pi_\lambda^\sigma(M_h) = \mathrm{i}\mathscr{M} h(t) \otimes \mathrm{Id}. \quad (8.31)$$

Theorem 8.3. *1. Let* $d\sigma : \mathfrak{sv} \to \mathrm{Hom}(\mathscr{D}^{lin}, \mathscr{D}^{lin})$ *be defined by the left-and-right infinitesimal action*

$$d\sigma(X) : D \to d\pi_1^\sigma(X) \circ D - D \circ d\pi_{\frac{1}{2}}^\sigma(X).$$

Then $d\sigma$ *maps* $\mathscr{D}_{\le 2}^{aff}$ *into the vector space of Dirac potentials.*

2. Representing the Dirac potential $V = \begin{pmatrix} 0 & 0 \\ V_2(t)r^2 + V_1(t)r + V_0(t) & 0 \end{pmatrix}$ *or, equivalently, the Dirac operator* $\mathscr{D}_0 + V$, *by the vector* $\begin{pmatrix} V_2 \\ V_1 \\ V_0 \end{pmatrix}$, *we see that the action of* $d\sigma$ *on* $\mathscr{D}_{\le 2}^{aff}$ *is given by the same formula as in Proposition 8.5, except for the affine terms (with coefficient proportional to* $\mathscr{M}$ *or* $\mathscr{M}^2$*) that should all be divided by* $-2\mathrm{i}\mathscr{M}$.

The proof is straightforward, similar to Theorem 8.1, Proposition 8.2 and Proposition 8.6).

Note that, as in the previous section, one may define a 'shifted' action

$$d\sigma_\lambda(X) : D \to d\pi_{\lambda+1}^\sigma(X) \circ D - D \circ d\pi_{\lambda+\frac{1}{2}}^\sigma(X) \quad (8.32)$$

which will modify only the coefficients of the affine cocycles.

To conclude, let us emphasize two points (see introduction to Sect. 8.1, and also Chap. 10 for a comparison):

- contrary to the case of the Hill operators, there is a free parameter λ in the left-and-right actions on the affine space of Schrödinger or Dirac operators;
- looking at the *shift of indices* between the left action and the right action, one realizes that Schrödinger operators are of order one, while Dirac operators are of order one-half ! (recall the difference of indices was $2 = \frac{3}{2} - (-\frac{1}{2})$ in the case of the Hill operators, which was the signature of operators of order 2 – see [19]).

So Schrödinger operators are somehow reminiscent of the operators $\partial + u$ of order one on the line, which intertwine $\mathcal{F}_\lambda$ with $\mathcal{F}_{1+\lambda}$ for any value of λ. The case of the Dirac operators, on the other hand, has no counterpart whatsoever for differential operators on the line.

On the other hand, for the *unrestricted linear action* of $\mathfrak{sv}$ on the space of Schrödinger operators which is seen in Chap. 10 to be Hamiltonian, the shift of indices is *two*, so that, in this respect, Schrödinger operators behave more naturally as *second-order operators*.

Chapter 9
Monodromy of Schrödinger Operators

This chapter is adapted from [106].

9.1 Introduction

This chapter – which makes use of the objects and concepts introduced in Chap. 8 – is mainly concerned with the determination of the *monodromy* of Schrödinger operators in $\mathcal{S}_{\le 2}^{aff} := \{-2i\partial_t - \partial_r^2 + V_2(t)r^2 + V_1(t)r + V_0(t)\}$, where V_0, V_1 and V_2 are 2π-periodic. The *monodromy*, for an ordinary differential equation with periodic coefficients, expresses the "rotation" of the general solution over a period. Typically, for Hill equations, the monodromy is a two-by-two matrix in $SL(2,\mathbb{R})$. Here we are concerned with partial differential equations, so the monodromy is an *operator* instead of a *matrix*. The problem is solved by using two main ingredients: the first one is the classification of the orbits of the action of the Schrödinger–Virasoro group on $\mathcal{S}_{\le 2}^{aff}$, which results in particular in a finite family of explicit normal forms; the second one is the exact solution of the Cauchy problem associated with any Schrödinger operator $D \in \mathcal{S}_{\le 2}^{aff}$. This was established independently by several authors: Lewis and Riesenfeld [80] in one dimension, and Hagedorn [46] in several dimensions.

Let us explain the general motivation underlying this work.

The Cauchy problem $D\psi = (-2i\partial_t - \partial_r^2)\psi = 0$[1] with initial condition $\psi(0,r) = \psi_0(r)$ has a unique solution. The usual method used in mathematical physics to solve such a time-dependent problem is to consider the adiabatic approximation: if one puts formally a small coefficient ε in front of ∂_t, the problem is equivalent by dilating the time coordinate to the equation $(-2i\partial_t - \partial_r^2 + V(\varepsilon t, r))\psi = 0$,

[1]The value of the mass parameter has been chosen equal to 1 in the whole chapter, in order to simplify the formulas.

J. Unterberger and C. Roger, *The Schrödinger-Virasoro Algebra*, Theoretical and Mathematical Physics, DOI 10.1007/978-3-642-22717-2_9,

so that V is a potential that is slowly varying in time. Suppose that $\Delta_\varepsilon(t) := -\partial_r^2+V(\varepsilon t,r)$ has a pure point spectrum $\{\lambda_n(t), n\in\mathbb{N}\}$ for every t, where λ_n is C^∞ in t, say, and let $\psi_n(t)$ be a normalized eigenfunction of $\Delta_\varepsilon(t)$ satisfying the *gauge-fixing condition* $\langle\psi_n(t),\dot\psi_n(t)\rangle = 0$. Then there exists a *parallel transport operator* $W(s,t)$ carrying the eigenspace with eigenvalue $\lambda_n(s)$ to the eigenspace with eigenvalue $\lambda_n(t)$, and a *phase operator* $\Phi(s,t)$, given simply by the multiplication by a phase $e^{\frac{i}{2}\int^t\lambda_n(s)\,ds}$ on each eigenspace, such that the solution of the Schrödinger equation is given at first order in ε by the composition of W and Φ. One may see the solutions formally as flat sections for a connection (called *Berry connection*) related in simple terms to the phase operator (see [13]). This scheme may be iterated, giving approximate solutions to the Schrödinger equation that are correct to any order in ε (see for instance Joye [62]), but it seldom happens that one can give *exact* solutions. By considering the related classical problem, G. Hagedorn (see [46]) constructs a set of raising and lowering operators (generalizing those associated to the usual harmonic oscillator) for general Schrödinger operators in $\mathcal{S}^{\mathit{aff}}_{\le 2}$, and uses them to solve the equation explicitly. The same set of operators had been considered previously by two quantum physicists, H.R. Lewis and W.B. Riesenfeld (see [80]), and found by looking for an exact invariant, i.e. for a time-dependent operator $I(t)$ (not including the time-derivative) such that $\frac{dI}{dt} = \frac{\partial I}{\partial t} + \mathrm{i}[I(t),\frac{1}{2}(\partial_r^2 - V(t,r))] = 0$. They find for each operator D in $\mathcal{S}^{\mathit{aff}}_{\le 2}$ a family of invariants (called sometimes the *Ermakov-Lewis invariants*, see [96]) depending on an arbitrary real solution ξ of a certain differential equation of order 3 (see Proposition 9.18), constructed out of generalized raising and lowering operators and spectrally equivalent to the standard harmonic oscillator $-\frac{1}{2}(\partial_r^2-r^2)$. These invariants have been used to solve quite a few physical problems, ranging from quantum mechanics for charged particles to cosmology (see [40], [41], [96], [102] for instance). It turns out that very few Schrödinger operators have an exact invariant of the type $I(t) = f_2(t,r)\partial_r^2 + f_1(t,r)\partial_r + f_0(t,r)$. These may be expressed, as shown by H. Lewis and P. Leach (see [77]), in terms of three arbitrary functions of time (the exact expression is complicated). Exact invariants allow in principle to solve explicitly the original problem, at least if one knows how to diagonalize them (which is the case here). Hence (provided one requires that an exact invariant exists) the space $\mathcal{S}^{\mathit{aff}}_{\le 2}$ is maximal.

This scheme has been successfully used to understand the asymptotic scattering associated with such time-dependent Schrödinger operators. Namely, supposing that $D(t)$ converges in the limits $t\to\pm\infty$ to harmonic oscillators $-2\mathrm{i}\partial_t - \omega_\pm^2\partial_r^2$, the problem is to understand the relation between the asymptotic bases of excited states at $t=-\infty$ and $t=+\infty$.

The problem we consider here is of a different nature. We suppose that the potential of the operator D is 2π*-periodic* in time. Then (as follows from Floquet's theory) there exists a bounded operator $\mathcal{M}$ in $\mathcal{B}(L^2(\mathbb{R}))$, called the *monodromy operator*, connecting solutions at time t with solutions at time $t+2\pi$ for any t, namely:

$$\mathcal{M}\psi(t) = \psi(t+2\pi) \tag{9.1}$$

for any L^2 solution ψ of the Schrödinger equation $D\psi = 0$.

Suppose $D = -2i\partial_t - V_2(t)\partial_r^2$ has a time-periodic quadratic potential. The computation of the monodromy of the associated classical equation

$$(\partial_t^2 + u(t))\,\psi(t) = 0 \tag{9.2}$$

– which is a Hill equation – is a classical problem. It turns out that Hill operators are classified by their *lifted monodromy* (see Khesin-Wendt [68]). Let us explain briefly this object. If (ψ_1, ψ_2) is a basis of solutions of the ordinary differential equation $(\partial_t^2 + u(t))\psi(t) = 0$, then (by Floquet's theory)

$$\begin{pmatrix} \psi_1(t_0 + 2\pi) \\ \psi_2(t_0 + 2\pi) \end{pmatrix} = M. \begin{pmatrix} \psi_1(t_0) \\ \psi_2(t_0) \end{pmatrix}, \tag{9.3}$$

where M is some matrix (called *monodromy matrix*) with determinant 1 which does not depend on the base point t_0. If M is *elliptic*, i.e. conjugate to a rotation, then the eigenvectors for M are multiplied by a phase $e^{i\theta}$. If M is *hyperbolic*, i.e. conjugate to a Lorentz shift $\begin{pmatrix} e^{\lambda} & \\ & e^{-\lambda} \end{pmatrix}$, then the eigenvectors are multiplied by a real factor $e^{\pm\lambda}$, so that the solutions of the Hill equation are unstable, aymptotic either to zero or to infinity when $t \to \pm\infty$. A nice way to see it (and made rigorous in Sect. 9.2) is to imagine the vector $\begin{pmatrix} \psi_1(t) \\ \psi_2(t) \end{pmatrix}$ as 'rotating' in the plane (it may also change norm but never vanishes). The curve described by this vector may be lifted to the Riemann surface of the logarithm for instance (obtained from the cut plane $\mathbb{C} \setminus \mathbb{R}_-$), so that it rotates by an angle unambiguously defined in $\mathbb{R}$. This gives the lifted monodromy.

In Sect. 9.2 below (see Sect. 9.2.4), we classify the orbits of SV in $\mathcal{S}_{\le 2}^{aff}$. The classification is mainly an extension of Kirillov's results on the classification of the orbits of the space of *Hill operators* under the Virasoro group. These are operators of the type $\partial_t^2 + u(t)$. It is well-known (see for instance Guieu [42] or [43]) that the group of orientation-preserving diffeomorphisms $\mathrm{Diff}(\mathbb{R}/2\pi\mathbb{Z})$ of the circle exponentiates the centerless Virasoro algebra, and acts on the affine space of Hill operators. Now the remarkable fact is that, despite the apparent differences between the two problems, the action of the Virasoro group $\mathrm{Diff}(\mathbb{R}/2\pi\mathbb{Z}) \subset \mathrm{SV}$ on the quadratic part of the potential, $V_2(t)r^2$, is equivalent to that of $\mathrm{Diff}(\mathbb{R}/2\pi\mathbb{Z})$ on the Hill operator $\partial_t^2 + V_2(t)$. This follows from the fact that the Hill operator is the corresponding classical problem in the semi-classical limit (see Sect. 9.3.2). Hence part of the classification may be borrowed directly from the work of Kirillov (see [70]). A.A. Kirillov obtains his classification by studying the isotropy algebra $Lie(\mathrm{Stab}_u) := \{X \in Lie(\mathrm{Diff}(\mathbb{R}/2\pi\mathbb{Z})) \mid X.(\partial^2 + u) = 0\}$.

The space $\mathcal{S}_{\le 2}^{aff}$ of Schrödinger operators with time-periodic potentials which are *at most quadratic* in the space coordinate has been considered independently by mathematicians and physicists, with similar motivations but different methods. However, they are seen to be equivalent, ultimately. The general idea was to solve the evolution problem associated with $D \in \mathscr{S}^{aff}_{\leq 2}$, i.e. to show that there are three new features here:

- *the action of the Schrödinger–Virasoro group* on $\mathscr{S}^{aff}_{\leq 2}$ (which is essentially a conjugate action, leaving all invariant quantities unchanged, for instance the spectrum and the monodromy) makes it possible to reduce the study to five families of operators, with qualitatively different properties (see Sect. 9.2.4). They are characterized mainly by the monodromy of the associated Hill operator $\partial_t^2 + V_2(t)$, but there also appear some non-generic orbits in cases when the quantum and linear parts of the potential are 'resonant'. The non-periodic case is much simpler, since (*locally* in time) all Schrödinger operators in $\mathscr{S}^{aff}_{\leq 2}$ are formally equivalent (see Sect. 9.3.1 below). The coefficients of the Ermakov-Lewis invariants are related in a very simple way to the invariants of the orbits;
- to investigate Schrödinger operators with *time-periodic potential*, one may consider (as in the case of ordinary differential operators, see above) the monodromy, which is a bounded operator acting on $L^2(\mathbb{R})$. The monodromy operator is given explicitly and shown to be closely related to the classical monodromy of the related Hill operator;
- the computation of the monodromy in the case when the associated Hill operator is *hyperbolic* (see above) requires the use of an Ermakov- Lewis invariant associated to a *purely imaginary* function ξ, which is equivalent to the standard harmonic 'repulsor' $-\frac{1}{2}(\partial_r^2 + r^2)$. The reason (explained more precisely in Sect. 9.3.1 below) is that the usual Ermakov-Lewis invariants are defined only if $I_{V_2}(\xi) > 0$, where the invariant quantity $I_{V_2}(\xi)$ (*quadratic* in ξ) is associated to the Hill operator $\partial_t^2 + V_2(t)$ and its stabilizer $\xi(t)$ in *Lie*(Diff($\mathbb{R}/2\pi\mathbb{Z}$)). The stabilizer satisfies a linear differential equation of order 3 and has generically only one periodic solution (up to a constant). If one does not require ξ to be *periodic*, then $I_u(\xi)$ may be chosen to be positive; this is perfectly suitable for a local study (in time) but is of little practical use for the computation of the monodromy. However, if one requires that ξ be periodic, then $I_u(\xi)$ is *negative* in the hyperbolic case, unless one chooses ξ to be purely imaginary. Hence one is naturally led to use the spectral decomposition of the harmonic 'repulsor' (which has an absolutely continuous spectrum equal to the whole real line). Usually there is no adiabatic scheme, hence no phase operator, in the case when eigenvalues are not separated by a gap. But in this very particular case, such a phase operator may be computed and is very analogous to that obtained in the elliptic case, for which the spectrum is discrete. There exists also some non-generic cases (corresponding to a unipotent monodromy matrix for the underlying Hill operator) for which $I_{V_2}(\xi) = 0$. The natural invariant is then spectrally equivalent either to the bare Laplacian $-\frac{1}{2}\partial_r^2$ or to the Airy operator $-\frac{1}{2}(\partial_r^2 - r)$.

One of the main results may be stated as follows (see Sects. 9.3.4, 9.3.5, 9.3.6): the monodromy operator is unitarily equivalent to the unitary multiplication operator $f(k) \to e^{\mathrm{i}kT - \mathrm{i}\pi\gamma} f(k)$, where γ is some constant and k is the spectral parameter of the *model operator* $-\frac{1}{2}(\partial_r^2 + \kappa r^2)$ ($\kappa = \pm 1, 0$) or $-\frac{1}{2}(\partial_r^2 - r)$, and

$T = \int_0^{2\pi} \frac{du}{\xi(u)}$ (ξ real) or $\mathrm{i}\int_0^{2\pi} \frac{du}{\xi(u)}$ (ξ imaginary). The above integrals must be understood in a generalized sense if ξ has some zeros, in which case a complex contour deformation is needed. Comparing this phase to the usual Berry phase $e^{\frac{i}{2}\int_0^{2\pi} \lambda_k(s)\, ds}$, one sees that the eigenvalue $\lambda_k(t) = -2k$ is *constant*, but that the natural (possibly singular) time-scale is $\tau := \int^t \frac{du}{\xi(u)}$.

The chapter is organized as follows.

Section 9.2 is dedicated to the classification of the orbits and of the isotropy subgroups $G_D := \{g \in SV \mid g.D = D\}$, $D \in \mathcal{S}^{aff}_{\leq 2}$ (see Sect. 9.2.4). It contains long but unavoidable preliminaries on the action of the Virasoro group on Hill operators. The connection to the results of U. Niederer is made in the last paragraph.

We solve the monodromy problem for the Schrödinger operators of the form $-2\mathrm{i}\partial_t - \partial_r^2 + V_2(t)r^2 + \gamma$ (γ constant) in Sect. 9.3. We study first the corresponding classical problem given by the associated Hill operator, $\ddot{x} + V_2(t)x = 0$ (an ordinary differential equation). The solution of the quantum problem is then easily deduced from that of the classical problem. In both cases, the monodromy is obtained by relating the Ermakov-Lewis invariants to the orbit data.

Finally, we show in Sect. 9.4 how to parametrize a general Schrödinger operator $-2\mathrm{i}\partial_t - \partial_r^2 + V_2(t)r^2 + V_1(t)r + V_0(t) \in \mathcal{S}^{aff}_{\leq 2}$ by means of a *three-dimensional invariant* $(\xi(t), \delta_1(t), \delta_2(t))$ (see Definition 9.25). The parametrization is one-to-one or 'almost' one-to-one depending on the orbit class of the potential V_2 (SV-orbits in $\mathcal{S}^{aff}_{\leq 2}$ have generically codimension 2, whereas adjoint orbits corresponding to the invariant have generically codimension 2 or 3). The action of the Schrödinger–Virasoro group on $\mathcal{S}^{aff}_{\leq 2}$, once written in terms of the invariant, becomes much simpler, and is easily shown to be Hamiltonian for a natural symplectic structure. A generalized Ermakov-Lewis invariant may also be written in terms of this three-dimensional invariant. We then solve the monodromy cases for the 'resonant' cases left from Sect. 9.3.

For the convenience of the reader, let us gather a few notations that are scattered throughout the text. Time and space coordinates are usually (at least starting from Sect. 9.2) denoted by θ and x (see explanations before Lemma 9.1 for the passage to Laurent coordinates (t, r)). The angle coordinate θ varies in $\mathbb{R}/2\pi\mathbb{Z}$ instead of S^1, hence the notation $\mathrm{Diff}(\mathbb{R}/2\pi\mathbb{Z})$ (instead of the usual one, $\mathrm{Diff}(S^1)$) for the group of orientation-preserving diffeomorphisms. Stabilizers in $Lie(\mathrm{Diff}(\mathbb{R}/2\pi\mathbb{Z}))$ of the Hill operator $\partial_\theta^2 + V_2(\theta)$ are usually denoted by ξ (which is either real or purely imaginary). If ξ is purely imaginary, then one sets $\xi := \mathrm{i}\eta$. We write the (operator) invariants of the Schrödinger operators (see Sect. 9.4) as $\frac{1}{2}\left[a(\theta)x^2 - b(\theta)\partial_x^2 - \mathrm{i}c(\theta)(x\partial_x + \partial_x x) + d(\theta)(-\mathrm{i}\partial_x) + e(\theta)x + f(\theta)\right]$. The correspondence between the vector invariant $(\xi, \delta_1, \delta_2)$ and the operator (generalized Ermakov-Lewis) invariant is given in Theorem 9.7.

The family of actions $\sigma_\lambda : D \mapsto \pi_{\lambda+1}(g) D \pi_\lambda(g^{-1})$ of the Schrödinger–Virasoro group on Schrödinger operators has been defined in Chap. 8. Looking at the formulas in Definition 8.3, one sees that the subspace $\mathcal{S}^{aff}_2 := \{D = -2\mathrm{i}\partial_t - \partial_r^2 + V_2(t)r^2\}$ of operators with quadratic potential is preserved by $\sigma_\lambda|_{\mathrm{Diff}(\mathbb{R}/2\pi\mathbb{Z})}$

if and only if $\lambda = 1/4$, which corresponds to the 'scaling dimension' of the Schrödingerian field in one dimension (see Chap. 1). This is absolutely crucial for the sequel, *so we fix* $\lambda = 1/4$.

We shall occasionally use the time-reparametrization

$$\phi : \mathbb{R}/2\pi\mathbb{Z} \to S^1 \simeq U(1), \quad \theta \to t = e^{i\theta} \tag{9.4}$$

from the one-dimensional torus to the unit circle. It allows us to switch from the Fourier coordinate θ to the Laurent coordinate t. In particular,

$$\mathscr{L}_{t^{n+1}} = \pi_{1/4}(\phi;0)\, \mathscr{L}_{e^{in\theta}}\, \pi_{1/4}(\phi;0)^{-1}, \tag{9.5}$$

$$\mathscr{Y}_{t^{n+\frac{1}{2}}} = \pi_{1/4}(\phi;0)\, \mathscr{Y}_{e^{in\theta}}\, \pi_{1/4}(\phi;0)^{-1}, \quad \mathscr{M}_{t^n} = \pi_{1/4}(\phi;0)\, \mathscr{M}_{e^{in\theta}}\, \pi_{1/4}(\phi;0)^{-1}. \tag{9.6}$$

If n is an integer, $\mathscr{Y}_{t^{n+\frac{1}{2}}}$ should be understood to be acting on the two-fold covering of the complex plane where the square-root is defined; conversely, if n is a half-integer, then $\mathscr{Y}_{e^{in\theta}}$ acts on 4π-periodic functions. In other words, the 'natural' choice for $\mathfrak{sv}$ should be the genuine Schrödinger–Virasoro algebra $\mathfrak{sv}$, resp. the 'twisted' Schrödinger–Virasoro algebra $\mathfrak{sv}(0)$ in the Laurent, resp. Fourier coordinates (see Sect. 2.1).

Applying formally the formulas of Chap. 8, one gets

$$\left(\pi_{1/4}(\phi;0)^{-1} f\right)(\theta, x) = \left(ie^{-i\theta}\right)^{1/4} e^{-\frac{1}{4}ix^2} f\left(e^{i\theta}, \pm x e^{i(\frac{\theta}{2}+\frac{\pi}{4})}\right) \tag{9.7}$$

(with some ambiguity in the sign) which is an 8π-periodic function. Applying now (still formally) Proposition 8.3 yields the following result, which can be checked by direct computation.

Lemma 9.1. *Let* $f(t,r)$ *be a solution of the Schrödinger equation*

$$(-2i\partial_t - \partial_r^2 + V(t,r)) f(t,r) = 0.$$

Then

$$\tilde{f} : (\theta, x) \to e^{-i\theta/4} e^{-\frac{1}{4}ix^2} f\left(e^{i\theta}, x e^{i(\theta/2+\pi/4)}\right) \tag{9.8}$$

is a solution of the transformed Schrödinger equation

$$\left[-2i\partial_\theta - \partial_x^2 + \frac{1}{4}x^2 + ie^{i\theta} V\left(e^{i\theta}, x e^{i(\theta/2+\pi/4)}\right)\right] \tilde{f}(\theta, x) = 0. \tag{9.9}$$

In the following sections, unless otherwise stated, we shall work with the Lie algebra $\mathfrak{sv}(0)$ *in the Fourier coordinates* θ, x *(i.e. the Lie algebra generated by the* $\mathscr{L}_f$, $\mathscr{Y}_g$ *and* $\mathscr{M}_h$ *with* 2π*-periodic functions* f, g, h*), and write* $\mathfrak{sv}$ *instead of* $\mathfrak{sv}(0)$ *for simplicity.*

9.2 Classification of the Schrödinger Operators in $\mathcal{S}^{aff}_{\le 2}$

Henceforth, we shall concentrate on the affine subspace $\mathcal{S}^{aff}_{\le 2}$ of Schrödinger operator with potentials which are at most quadratic in the space coordinate. As mentioned in the introduction to this chapter, this subspace is invariant under the action of SV. The purpose of this section is to classify the corresponding orbits.

9.2.1 *Statement of the Problem and Connection with the Classification of Hill Operators*

We introduced at the beginning of this chapter the following affine subspaces of Schrödinger operators,

$$\mathcal{S}^{aff}_{\le 2} = \{-2i\partial_\theta - \partial_x^2 + V_2(\theta)x^2 + V_1(\theta)x + V_0(\theta)\}$$

and

$$\mathcal{S}^{aff}_{2} = \{-2i\partial_\theta - \partial_x^2 + V_2(\theta)x^2\} \subset \mathcal{S}^{aff}_{\le 2}.$$

We do not assume that V_2 is positive. Hence what we really consider are harmonic 'oscillators-repulsors', corresponding to the quantization of a classical oscillator-repulsor with time-dependent Hamiltonian $\frac{1}{2}(p^2 + V_2(\theta)x^2 + V_1(\theta)x + V_0(\theta))$. If $V_1 \equiv 0$, then the classical equation of motion $\frac{d^2x}{d\theta^2} = -V_2(\theta)x - \frac{1}{2}V_1(\theta)$ has 0 as an attractive, resp. repulsive fixed point depending on the sign of V_2. If V_2 is not of constant sign, things can be complicated; it is not clear a priori whether solutions are stable or unstable. We shall come back to this problem (which turns out to be more or less equivalent to the a priori harder quantum problem, at least as far as monodromy in concerned) in Sect. 9.3.2.

The first subspace $\mathcal{S}^{aff}_{\le 2}$ is preserved by the action of SV and is in some sense minimal (the SV-orbit of the free Schrödinger equation, or of the standard harmonic oscillator $-2i\partial_\theta - \partial_x^2 + a^2x^2$, contains 'almost' all potentials which are at most quadratic in x). As we shall prove below, the orbits in $\mathcal{S}^{aff}_{\le 2}$ have finite codimension.

For the convenience of the reader, let us write down the restriction of the action of $\sigma_{1/4}$ to $\mathcal{S}^{aff}_{\le 2}$. Letting $D = -2i\partial_\theta - \partial_x^2 + V_2(\theta)x^2 + V_1(\theta)x + V_0(\theta)$, with $S(\phi) = \frac{\phi'''}{\phi'} - \frac{3}{2}\left(\frac{\phi''}{\phi'}\right)^2$ the Schwarzian derivative,

$$\sigma_{1/4}(\phi;0)(D) = -2i\partial_\theta - \partial_x^2 + \left(\phi'^2 \,.\, V_2 \circ \phi + \frac{1}{2}S(\phi)\right)x^2 + \left(\phi'^{3/2} \,.\, V_1 \circ \phi\right)x + \phi' \,.\, V_0 \circ \phi, \qquad (9.10)$$

and

$$\sigma_{1/4}(1;(a,b))(D) = -2i\partial_\theta - \partial_x^2 + V_2 x^2 + \left(V_1 - 2aV_2 - 2a''\right)x \\ + \left(V_0 - aV_1 + a^2 V_2 - 2b' + aa''\right), \quad (9.11)$$

while the infinitesimal action is given by

$$d\sigma_{1/4}(\mathscr{L}_f)(D) = -\left(\frac{1}{2}f''' + 2f'V_2 + fV_2'\right)x^2 - \left(fV_1' + \frac{3}{2}f'V_1\right)x \\ -\left(fV_0' + f'V_0\right), \quad (9.12)$$

$$d\sigma_{1/4}(\mathscr{Y}_g + \mathscr{M}_h)(D) = -2\left(g'' + gV_2\right)x - \left(2h' + gV_1\right). \quad (9.13)$$

These four formulas are fundamental for most computations below, and we shall refer to them throughout.

Similarly, $\mathscr{S}_2^{aff}$ is preserved by the $\sigma_{1/4}$-action of $\mathrm{Diff}(\mathbb{R}/2\pi\mathbb{Z})$ (see Proposition 8.3). It turns out that the orbit theory for this space is equivalent to that of the Hill operators under the Virasoro group. Let us first give some notations and recall basic facts concerning Hill operators (see also Sect. 8.1).

Definition 9.2. A *Hill operator* is a Sturm-Liouville operator on the one-dimensional torus, i.e. a second-order operator of the form $\partial_\theta^2 + u(\theta)$ where $u(\theta) \in C^\infty(\mathbb{R}/2\pi\mathbb{Z})$ is a 2π-periodic function.

The action of the group of time-reparametrizations on a Hill operator may be constructed as follows. Starting from the simple action of diffeomorphisms on functions,

$$\psi \to \psi \circ \phi, \quad \phi \in \mathrm{Diff}(\mathbb{R}/2\pi\mathbb{Z}),$$

one sees that $(\partial^2 + u)(\psi) = 0$ is equivalent to the transformed equation $(\partial^2 + p(\theta)\partial + q(\theta))(\psi \circ \phi) = 0$ if one sets $p = -\frac{\phi''}{\phi'}$ and $q = \phi'^2 \,.\, u \circ \phi$. Then one uses the following:

Definition 9.3 (Wilczinsky's semi-canonical form). (see Magnus-Winkler, [84], 3.1, or Guieu, [42], Proposition 2.1.1)

If ψ is a solution of the second-order equation $(\partial^2 + p(\theta)\partial + q(\theta))\psi = 0$, then $\tilde{\psi} := \lambda(\theta)\psi$ is a solution of the Hill equation $(\partial^2 + u(\theta))\tilde{\psi} = 0$ provided

$$\lambda(\theta) = \exp\left(\frac{1}{2}\int_{\theta_0}^{\theta} p(s)\, ds\right) \quad (9.14)$$

for some θ_0 and

$$u = -\frac{1}{2}p' - \frac{1}{4}p^2 + q. \quad (9.15)$$

One obtains in this case $\lambda = (\phi')^{-1/2}$, and the transformed operator reads: $\partial^2 + (\phi')^2 \,.\, u \circ \phi + \frac{1}{2}S(\phi)$, where S is the Schwarzian derivative. The presence of this last term shows that this transformation defines a *projective* action of $\mathrm{Diff}(\mathbb{R}/2\pi\mathbb{Z})$. Summarizing, one obtains:

Proposition 9.4 (see Guieu, [42] or Guieu-Roger, [43]). *The transformation*

$$\partial^2 + u \to \phi_* \left(\partial^2 + u\right) := \partial^2 + \left(\phi'\right)^2 \,.\, u \circ \phi + \frac{1}{2}S(\phi)$$

defines an action of $\mathrm{Diff}(\mathbb{R}/2\pi\mathbb{Z})$ *on the space of Hill operators, which is equivalent to the affine coadjoint action on* $\mathfrak{vir}^*_{\frac{1}{2}}$ *(i.e. with central charge* $c = \frac{1}{2}$*). A solution of the transformed equation may be obtained from a solution* ψ *of the initial equation* $(\partial^2 + u)\psi = 0$ *by setting* $\phi_*\psi = (\phi')^{-\frac{1}{2}}\psi \circ \phi$*. In other words, the solutions of the Hill equations behave as* $(-\frac{1}{2})$*-densities.*

Let us now state the following important

Lemma 9.5. *The above action of* $\mathrm{Diff}(\mathbb{R}/2\pi\mathbb{Z})$ *on the space of Hill operators is equivalent to the* $\sigma_{1/4}$*-action of* $\mathrm{Diff}(\mathbb{R}/2\pi\mathbb{Z})$ *on the space* $\mathcal{S}^{aff}_2$*.*

Namely, Proposition 8.3 above (see also (9.10)) shows that

$$\sigma_{1/4}(\phi)\left(-2\mathrm{i}\partial_\theta - \partial_x^2 + V_2(\theta)x^2\right) = -2\mathrm{i}\partial_\theta - \partial_x^2 + \tilde{V}_2(\theta)x^2, \qquad (9.16)$$

where the potential $\tilde{V}_2$ is the image of V_2 (viewed as the potential of a Hill operator in the coordinate θ) under the diffeomorphism ϕ, i.e. $\phi_*(\partial_\theta^2 + V_2(\theta)) = \partial_\theta^2 + \tilde{V}_2(\theta)$. Once again, this should not come as a surprise since the Hill equation is the semi-classical limit of the Schrödinger operator (see Sect. 9.3.2). □

So we shall need to recall briefly the classification of the orbits of Hill operators under the Virasoro group. There are mainly three a priori different classifications, which of course turn out in the end to be equivalent: the first one is by the *lifted monodromy* of the solutions (see for instance B. Khesin and R. Wendt, [68]); the second one consists in looking for normal forms for the solutions, of either an exponential form for non-vanishing solutions, or of a standard form for a dynamical system associated with the repartition of the zeros (see the article by V.F. Lazutkin and T.F. Pankratova, [75]); the third classification, due to A.A. Kirillov (see [70]), proceeds in a more indirect way by looking at the isotropy groups. For our purposes, we shall need the first and the third classifications. They are the subject of the two upcoming subsections.

9.2.2 Classification of Hill Operators by the Lifted Monodromy

Let us now turn to the classification of the orbits under the Virasoro group of the space of Hill operators.

Consider a pair (ψ_1, ψ_2) of linearly independent solutions of the Hill equation $(\partial^2 + u)\psi = 0$. It is a classical result (a particular case of Floquet's theory for Schrödinger equations with (space)-periodic potential) that

$$\begin{pmatrix} \psi_1(\theta + 2\pi) \\ \psi_2(\theta + 2\pi) \end{pmatrix} = M(u) \cdot \begin{pmatrix} \psi_1(\theta) \\ \psi_2(\theta) \end{pmatrix} \tag{9.17}$$

for a certain matrix $M(u) \in SL(2,\mathbb{R})$ (independent of θ), called the *monodromy matrix*. Starting from a different basis $\begin{pmatrix} \tilde{\psi}_1 \\ \tilde{\psi}_2 \end{pmatrix}$, one obtains a conjugate matrix $\tilde{M}(u)$. The above action of the Virasoro group on the Hill equation leaves the monodromy matrix unchanged, as can be seen from the transformed solutions $\phi_*\psi_1, \phi_*\psi_2$. Hence the conjugacy class of the monodromy matrix is an invariant of the Hill operator under the action of the diffeomorphism group.

Floquet's theory, together with the orbit theory for $SL(2,\mathbb{R})$, imply that $\partial^2 + u$ is *stable* (meaning that all solutions are bounded) if $|\mathrm{Tr} M| < 2$ or equivalently, if M is *elliptic*, i.e. conjugate to a rotation matrix. It is *unstable* (meaning that all solutions are unbounded) if $|\mathrm{Tr} M| > 2$ or equivalently, if M is *hyperbolic*, i.e. conjugate to a Lorentz shift $\begin{pmatrix} e^{\lambda} & \\ & e^{-\lambda} \end{pmatrix}$, $\lambda > 0$. If $|TrM| = 2$, then M can be shown to be conjugate either to $\pm\mathrm{Id}$ or to the *unipotent matrix* $\pm \begin{pmatrix} 1 & 2\pi \\ 0 & 1 \end{pmatrix}$; in the latter case, $\partial^2 + u$ is *semi-stable*, with stable *and* unstable solutions. Two linearly independent 2π- or 4π-periodic solutions exist when $M = \pm\mathrm{Id}$; only one in the unipotent case; and none in the remaining cases.

An important result due to V.F. Lazutkin and T.F. Pankratova (see [75]) is that all stable Hill operators are conjugate by a suitable time-reparametrization to a Hill operator with constant potential $\partial^2 + \alpha$, $\alpha > 0$. They also distinguish between *oscillating* and *non-oscillating* equations (oscillating equations have solutions with infinitely many zeros, while non-oscillating equations have solutions with at most one zero), but we shall not need to go further into this. Let us just remark that (as they also show) non-oscillating operators are also conjugate to a Hill operator with constant potential $\partial^2 + \alpha$, with $\alpha \le 0$ this time. Hence operators of type II, resp. III of Kirillov's classification (see Definition 9.10 below) are exactly the unstable, resp. semi-stable *oscillating* operators.

A *complete classification* of the orbits under the action of $\mathrm{Diff}(\mathbb{R}/2\pi\mathbb{Z})$ may be obtained by considering the *lifted monodromy*. Set $\begin{pmatrix} \psi_1(\theta) \\ \psi_2(\theta) \end{pmatrix} = M(u)(\theta) \begin{pmatrix} \psi_1(0) \\ \psi_2(0) \end{pmatrix}$. The path $\theta \to M(u)(\theta) \in SL(2,\mathbb{R})$ may be lifted uniquely to a path $\theta \to \tilde{M}(u)(\theta) \in \widetilde{SL}(2,\mathbb{R})$ such that $M(u)(0) = \mathrm{Id}$, where $\widetilde{SL}(2,\mathbb{R})$ is the universal covering of $SL(2,\mathbb{R})$. This procedure defines a unique lifted monodromy matrix $\tilde{M}(u) := \tilde{M}(u)(2\pi)$ modulo conjugacy.

The following arguments (see [68]) show briefly why this invariant suffices to characterize the orbit of u under diffeomorphisms. Set $\begin{pmatrix} \psi_1(\theta) \\ \psi_2(\theta) \end{pmatrix} = \sqrt{\xi(\theta)} \begin{pmatrix} \cos\omega(\theta) \\ \sin\omega(\theta) \end{pmatrix}$. The Wronskian

$$W := \psi_1 \psi_2' - \psi_1' \psi_2$$

(a constant of motion) is equal to $\omega'(\theta)\xi(\theta)$, hence $\omega' = \frac{W}{\xi}$ is of constant sign, say > 0 (by choosing $W > 0$). By the action of $\mathrm{Diff}(\mathbb{R}/2\pi\mathbb{Z})$, one can arrange that ω' is constant, while $\omega(0)$ and $\omega(2\pi)$ remain related by the homographic action of $M(u)$, viz. $\mathrm{cotan}\omega(2\pi) = \frac{a\,\mathrm{cotan}\omega(0)+b}{c\,\mathrm{cotan}\omega(0)+d}$ if $M(u) = \begin{pmatrix} a & b \\ c & d \end{pmatrix} \in SL(2,\mathbb{R})$. The lifting of the monodromy produces a supplementary invariant: the *winding number* $n := \lfloor(\omega(2\pi) - \omega(0))/2\pi\rfloor = \lfloor \frac{W}{2\pi} \int_0^{2\pi} \frac{d\theta}{\xi(\theta)} \rfloor$ ($\lfloor\ .\ \rfloor$=integer part), namely, the integer number of complete rotations made by the angle ω.

This change of function is particularly relevant in the *elliptic* case. Choose a basis $\begin{pmatrix} \psi_1 \\ \psi_2 \end{pmatrix}$ such that $M = \begin{pmatrix} \cos\lambda & -\sin\lambda \\ \sin\lambda & \cos\lambda \end{pmatrix}$. Then $\pm\lambda = \omega(2\pi) - \omega(0) = W \int_0^{2\pi} \frac{d\theta}{\xi(\theta)}\ [2\pi]$.

On the other hand, if $M = \begin{pmatrix} e^{\lambda} & \\ & e^{-\lambda} \end{pmatrix}$ is *hyperbolic*, set

$$\psi_1^2(\theta) = \frac{1}{2}|\xi(\theta)|e^{2\omega(\theta)}, \quad \psi_2^2(\theta) = \frac{1}{2}|\xi(\theta)|e^{-2\omega(\theta)} \tag{9.18}$$

with $\xi(\theta) = 2(\psi_1\psi_2)(\theta)$, so that $\pm\lambda = \omega(2\pi) - \omega(0)\ [2i\pi]$. Then one finds $\omega' = -\frac{W}{\xi}$, hence $\omega = -W \int \frac{d\theta}{\xi(\theta)}$. The functions $\frac{1}{\xi}$ and ω are not well-defined if ψ_1 or ψ_2 has some zeros. Supposing u is analytic, the functions ψ_1, ψ_2 may be extended analytically to some strip $\Omega = \{|\mathrm{Im}\,\theta| < \varepsilon\}$. Choose some contour $\Gamma \subset \Omega$ that avoids the zeros of ψ_1 and ψ_2, such that (assuming $\xi(0) \neq 0$, otherwise use a translation) $\Gamma(0) = 0$ and $\Gamma(2\pi) = 2\pi$. The idea is that Γ should be *real* outside of some symmetric neighbourhood U_ε of the zeros. One completes the path with half-circles centered on the real axis of radius ε around each zero, taken equivalently in the upper- or lower-half plane (compare with Sect. 9.3.2 below where more care is needed). Suppose $\psi_1(\theta_0) = 0$ for instance, so $\psi_1'(\theta_0) = a \neq 0$ and $\psi_2(\theta_0) = -\frac{W}{a}$. Then

$$\begin{aligned} -W \int_{\Gamma\cap[\theta_0-\varepsilon,\theta_0+\varepsilon]} \frac{d\theta}{\xi(\theta)} &= -W \int_{\theta_0-\varepsilon}^{\theta_0+\varepsilon} \frac{d\theta}{\theta - \theta_0 \pm i0} \frac{\theta - \theta_0}{\xi(\theta)} \\ &= -W\ p.v. \int_{\theta_0-\varepsilon}^{\theta_0+\varepsilon} \frac{d\theta}{\xi(\theta)} \pm i\frac{\pi}{2} \end{aligned} \tag{9.19}$$

(depending on the position of the half-circle with respect to the real axis) since $\frac{1}{\theta-\theta_0\pm i0} = p.v.\frac{1}{\theta-\theta_0} \mp i\pi\delta_{\theta_0}$ (see for instance [34]) and the residues of $\frac{1}{\xi(\theta)}$ at the zeros of ξ are $\pm\frac{1}{2W}$. It is clear from the above definitions that ξ has only simple zeros, and that there is an *even* number of them. Hence $-W\int_\Gamma \frac{d\theta}{\xi(\theta)} \equiv -W\ p.v.\int_0^{2\pi} \frac{d\theta}{\xi(\theta)} \equiv \lambda\ [i\pi]$. By exponentiating, one obtains a monodromy matrix in $PSL(2,\mathbb{R}) = SL(2,\mathbb{R})/\{\pm 1\}$.

Finally, if M is *unipotent*, $M = \pm\begin{pmatrix} 1 & a \\ 0 & 1\end{pmatrix}$ in some basis $\begin{pmatrix}\psi_1 \\ \psi_2\end{pmatrix}$, set $\psi_1(\theta) = \omega\psi_2(\theta)$ and $\xi = \psi_2^2$, so that $\omega(2\pi) = \omega(0) + a$. Then $\omega' = -\frac{W}{\xi}$, so ω is once again defined as $-W\int\frac{d\theta}{\xi(\theta)}$ if ξ does not have any zero. In the contrary case, one uses a deformation of contour as in the hyperbolic case, to obtain

$$-W\int_{\Gamma\cap[\theta_0-\varepsilon,\theta_0+\varepsilon]}\frac{d\theta}{\xi(\theta)} = -W\int_{\theta_0-\varepsilon}^{\theta_0+\varepsilon}\frac{d\theta}{(\theta-\theta_0\pm i0)^2}\frac{(\theta-\theta_0)^2}{\psi_2^2(\theta)}. \tag{9.20}$$

Since $\frac{1}{(\theta-\theta_0\pm i0)^2} = p.v.\frac{1}{(\theta-\theta_0)^2} \pm i\pi\delta'_{\theta_0}$ and $\frac{(\theta-\theta_0)^2}{\psi_2^2(\theta)} = 1 + O((\theta-\theta_0)^2)$ – since $\psi_2''(\theta_0) = -V_2(\theta_0)\psi_2(\theta_0) = 0$ –, the Dirac term does not make any contribution at all this time, hence

$$a = \omega(2\pi) - \omega(0) = -W\int_\Gamma\frac{d\theta}{\xi(\theta)}, \tag{9.21}$$

where $\Gamma : [0,2\pi] \to \mathbb{C}$ is an arbitrary contour as defined above.

Summarizing:

Proposition 9.6 (see [68] for (ii)).

(i) The lifted monodromy of the operator $\partial^2 + u$ is characterized by the (correctly normalized) quantity $\int_0^{2\pi}\frac{d\theta}{\xi(\theta)}$ or $\int_\Gamma\frac{d\theta}{\xi(\theta)}$, where $\xi\in \mathrm{Stab}_u$.

(ii) The orbits under the diffeomorphism group of the space of Hill operators are characterized by the conjugacy class of their lifted monodromy. More precisely, the lifted monodromy defines a bijection from the set of orbits onto the space of conjugacy classes of $\left(\widetilde{SL}(2,\mathbb{R})\setminus\{\pm1\}\right)/\{\pm1\}$ (an element $M\in\widetilde{SL}(2,\mathbb{R})$ has to be identified with its opposite $-M$).

9.2.3 Kirillov's Classification of Hill Operators by Isotropy Subgroups

Another classification, also useful for our purposes (and more explicit in some sense), is due to Kirillov. Introduce first

Definition 9.7. Let $Stab_u$, $u \in C^\infty(\mathbb{R}/2\pi\mathbb{Z})$ be the *isotropy subgroup* (or *stabilizer*) of $\partial^2 + u$ in $\mathrm{Diff}(\mathbb{R}/2\pi\mathbb{Z})$, namely,

$$Stab_u := \{\phi \in \mathrm{Diff}(\mathbb{R}/2\pi\mathbb{Z}) \mid \phi_*(\partial^2 + u) = \partial^2 + u\}. \tag{9.22}$$

Proposition 9.8 (definition of the first integral I). *(see [42])*

1. *Let $\xi \in C^\infty(\mathbb{R}/2\pi\mathbb{Z})$: then $\xi \in Lie(Stab_u)$ if and only if ξ satisfies*

$$\frac{1}{2}\xi''' + 2u\xi' + u'\xi = 0. \tag{9.23}$$

2. *Let $I_u(\xi) := \xi\xi'' - \frac{1}{2}\xi'^2 + 2u\xi^2$. Then $I_u(\xi)$ is a constant of motion if $\xi \in Lie(Stab_u)$.*
3. *Consider $\phi \in \mathrm{Diff}(\mathbb{R}/2\pi\mathbb{Z})$ and the transformed potential $\tilde{u}$ such that $\phi_*(\partial^2 + u) = \partial^2 + \tilde{u}$. Then*

$$I_{\tilde{u}}\left(\phi'^{-1} . \xi \circ \phi\right) = I_u(\xi). \tag{9.24}$$

4. *Consider the Hill equation $(\partial^2 + u)\psi(\theta) = 0$. If (ψ_1, ψ_2) is a basis of solutions of this equation, then $\xi := a_{11}\psi_1^2 + 2a_{12}\psi_1\psi_2 + a_{22}\psi_2^2$ $(a_{11}, a_{12}, a_{22} \in \mathbb{R})$ satisfies the equation*

$$\frac{1}{2}\xi''' + 2u\xi' + u'\xi = 0 \tag{9.25}$$

In other terms, $\xi \in Stab_u$ is in the isotropy subgroup of the Hill operator $\partial^2 + u$. Conversely, any solution of (9.25) can be obtained in this way.

5. *(same notations) consider in particular $\xi = \psi_1^2 + \psi_2^2$. Then $I_u(\xi) = W^2$ if W is the Wronskian of (ψ_1, ψ_2), namely, $W = \psi_1\psi_2' - \psi_1'\psi_2$ (constant of the motion).*

Note (see 3.) that $\left(\phi'^{-1} . \xi \circ \phi\right)\partial$ is the conjugate of the vector $\xi\partial \in Vect(S^1)$ by the diffeomorphism ϕ. Hence one may say that the first integral I is invariant under the (adjoint-and-coadjoint) action of $\mathrm{Diff}(\mathbb{R}/2\pi\mathbb{Z})$.

Consider now the (adjoint) orbit of ξ under $\mathrm{Diff}(\mathbb{R}/2\pi\mathbb{Z})$. Clearly, $\int_0^{2\pi} \frac{d\theta}{\xi(\theta)}$ (if well-defined, i.e. if ξ has no zero) does not depend on the choice of the point on the orbit since $\int_0^{2\pi} \frac{d\theta}{\phi'^{-1}(\theta)\xi\circ\phi(\theta)} = \int_0^{2\pi} \frac{du}{\xi(u)}$. It is easy to see from Proposition 9.8 (2) that ξ either never vanishes (*case I*), or has an even number of simple zeros (*case II*), or has a finite number of double zeros (*case III*). Cases II, III correspond to a hyperbolic, resp. unipotent monodromy matrix (see discussion in Sect. 9.2.2). In case II, $I_u(\xi) = -\frac{1}{2}\xi'(t_0)^2 < 0$ if t_0 is any zero. The principal value integral $p.v. \int_0^{2\pi} \frac{dt}{\xi(t)}$ is well-defined. In case III, $I_u(\xi) = 0$ and the regularized integral $\int_\Gamma \frac{d\theta}{\xi(\theta)}$ (see above) is well-defined and independent of the choice of the contour Γ. Note that A. Kirillov uses instead the following regularization, $\lim_{\varepsilon\to 0} \int_{[0,2\pi]\setminus U_\varepsilon} \frac{dt}{\xi(t)} - \frac{C}{\varepsilon}$ (where U_ε is a symmetric ε-neighbourhood of the zeros) with C chosen so that the limit is finite. The two regularizations are different. Both are perfectly satisfactory to define an invariant of the orbits, but computations show that the Berry phase is proportional to $\int_\Gamma \frac{d\theta}{\xi(\theta)}$.

Now the integral $\int_0^{2\pi} \frac{d\theta}{\xi(\theta)}$ (case I) and its variants for case II, III are invariants under the diffeomorphism group. The discussion in Sect. 9.2.2 shows that they characterize the lifted monodromy of $\partial^2 + u$. The value of the invariant $I_u(\xi)$ is also needed to fix u uniquely in case I (see Proposition 9.8(2)) since ξ stabilizes all operators of the type $\partial^2 + u + \frac{C}{\xi^2}$ ($C \in \mathbb{R}$). It turns out that $\int_0^{2\pi} \frac{d\theta}{\xi(\theta)}$ – or its variants – and $I_u(\xi)$, together with a discrete invariant $n \in \mathbb{N}$, suffice to distinguish between the different adjoint orbits of stabilizers (note that general adjoint orbits may be much more complicated, see [43]). One has the following *Kirillov classification*:

Proposition 9.9 (Kirillov's classification of the orbits). *(see Kirillov [70])*

1. *Case I: ξ is conjugate by a diffeomorphism ϕ to a (non-zero) constant $a\partial_\theta$, $a \neq 0$. Hence $\phi'^{-1} . \xi \circ \phi \in Lie(Stab_{\partial^2+\alpha})$ for a certain constant α. The stabilizer $Stab_{\partial^2+\alpha}$ is:*
   - *(i) (non-generic case) either isomorphic to $\widetilde{SL}^{(n)}(2,\mathbb{R})$ (the n-fold covering of $SL(2,\mathbb{R})$), with $Lie(Stab_{\partial^2+\alpha}) = \mathbb{R}\partial_\theta \oplus \mathbb{R}\cos n\theta\partial_\theta \oplus \mathbb{R}\sin n\theta\partial_\theta$ if $\alpha = \frac{n^2}{4}$ for some $n \in \mathbb{N}^*$; then the monodromy in $PSL(2,\mathbb{R}) = SL(2,\mathbb{R})/\{\pm 1\}$ is trivial, while the lifted monodromy matrix is the central element in $\widetilde{SL}(2,\mathbb{R})/\{\pm 1\}$ corresponding to a rotation of an angle πn;*
   - *(ii) or (generic case) one-dimensional, equal to the rotation group $Rot \subset$ Diff$(\mathbb{R}/2\pi\mathbb{Z})$ generated by the constant field ∂ in the remaining cases.*

     The invariants are given by $I_u(\xi) = 2\alpha a^2$, $\int_0^{2\pi} \frac{d\theta}{\xi(\theta)} = \frac{2\pi}{a}$. The monodromy can be in any conjugacy class of PSL$(2,\mathbb{R})$ except $\pm$Id.

2. *Case II: ξ is conjugate to the field $a \sin n\theta(1 + \alpha \sin n\theta)\partial_\theta$, $n = 1, 2 \ldots, 0 \leq \alpha < 1$, which stabilizes $\partial^2 + u_{n,\alpha}$, where*

$$u_{n,\alpha}(\theta) := \frac{n^2}{4}\left[\frac{1 + 6\alpha \sin n\theta + 4\alpha^2 \sin^2 n\theta}{(1 + \alpha \sin n\theta)^2}\right]. \tag{9.26}$$

   The monodromy matrix is hyperbolic. The invariants take the values $I_u(\xi) = -2a^2n^2 < 0$, p.v.$\int_0^{2\pi} \frac{d\theta}{\xi(\theta)} = \frac{2\pi\alpha}{a\sqrt{1-\alpha^2}}$.

3. *Case III: ξ is conjugate to $\xi_{\pm,n,\alpha} := \pm(1 + \sin n\theta)(1 + \alpha \sin n\theta)\partial$, $0 \leq \alpha < 1$, corresponding to a potential $v_{n,\alpha}$,*

$$v_{n,\alpha}(\theta) = \frac{n^2}{4}\left[\frac{(\alpha - 1)^2 + 2\alpha(3 - \alpha)\sin n\theta + 4\alpha^2 \sin^2 n\theta}{(1 + \alpha \sin n\theta)^2}\right] \tag{9.27}$$

   The monodromy matrix is unipotent. The invariant $I_u(\xi)$ vanishes, while $\int_\Gamma \frac{d\theta}{\xi_+(\theta)} = \frac{-2\pi}{(1-\alpha)\sqrt{1-\alpha^2}}$. The discrete invariant n suffices to characterize the orbit of $\partial^2 + u$.

In cases II and III (provided $\alpha > 0$), the stabilizer is one-dimensional, generated by $\xi\partial_\theta$.

In the generic cases (case I, $\alpha \neq n^2/4$, $n = 0, 1, \ldots$ or case II) the monodromy matrix is elliptic, resp. hyperbolic, if and only if $I_u(\xi) > 0$, resp. $I_u(\xi) < 0$. In cases I ($\alpha = 0$) and III (with unipotent monodromy), $I_u(\xi) = 0$.

There is a mistake in Lemma 3 of [70] (the potential $u_{n,\alpha}$ given there is not correct). The potential $v_{n,\alpha}$ was missing, together with the value of $\int_\Gamma \frac{d\theta}{\xi_\pm(\theta)}$. Both are obtained by straightforward computations.

This classification is also natural when one thinks of the behaviour of the solutions (see Lazutkin-Pankratova [75] and Sect. 9.2.2). In particular, case II (resp. III) correspond to operators with *unstable* (resp. *semi-stable*), *oscillating* solutions, while case I corresponds to operators with *stable, oscillating* solutions ($\alpha > 0$), resp. *unstable, non-oscillating* solutions ($\alpha < 0$), resp. *semi-stable, non-oscillating* solutions ($\alpha = 0$).

Note that in the case I generic, the three-dimensional isotropy subalgebra contains fields ξ of type I, II ($\alpha = 0$) and III ($\alpha = 0$), hence the following nomenclature:

Definition 9.10 (nomenclature for Hill operators). If $\partial^2 + u$ has a stabilizer ξ of type I, or of type II, III with $\alpha = 0$, then $\partial^2 + u$ may be turned into a Hill operator with constant potential, and we shall say that the operator $\partial^2 + u$ (or the potential u) is of type I. If $\partial^2 + u$ has a stabilizer of type II, resp. III with $\alpha \neq 0$, then we shall say that $\partial^2 + u$ and u are of type II, resp. type III.

Similarly, we shall say that the Schrödinger operator $-2i\partial_\theta - \partial_x^2 + V_2(\theta)x^2 + V_1(\theta)x + V_0(\theta)$ is of type I (resp. II, III) if the Hill operator $\partial_\theta^2 + V_2(\theta)$ is of the corresponding type.

Note that the cases I generic ($\alpha \neq \frac{n^2}{4}, n = 0, 1, \ldots$) and II are generic (i.e. dense in $\mathscr{S}^{aff}_{\leq 2}$).

Now the eigenvalues of the monodromy matrix (and also the lifted monodromy) can easily be obtained once one knows the values of the invariants $\int_0^{2\pi} \frac{d\theta}{\xi(\theta)}$ and $I_u(\xi)$. The following Lemma gives the link between the two classifications:

Lemma 9.11. *Suppose $D = \partial^2 + u$ is of type I (with $\alpha \neq 0$) or II (i.e. its monodromy is either elliptic or hyperbolic). If D is of type I non generic, conjugate to $\partial^2 + n^2/4$ for some $n \geq 1$, choose ξ to be conjugate to some non-zero multiple of ∂_θ. Now (in all cases) normalize ξ by requiring that $I_u(\xi) = 2$, so that ξ is real in the elliptic case and purely imaginary in the hyperbolic case. Then the eigenvalues of the monodromy matrix are given by* $\exp \pm i \int_0^{2\pi} \frac{d\theta}{\xi(\theta)}$ *or* $\exp \pm i$ *p.v.* $\int_0^{2\pi} \frac{d\theta}{\xi(\theta)}$.

Proof. Coming back to the discussion in Sect. 9.2.2, one checks easily (with the normalization chosen there) that $I_u(\xi) = 2W^2$ in the elliptic case, and $I_u(\xi) = -2W^2$ in the hyperbolic case. Choose a basis of solutions (ψ_1, ψ_2) such that $W = 1$ and multiply ξ by i in the hyperbolic case. Then (in both cases) the eigenvalues of the monodromy matrix ($\pm i\lambda$ in the elliptic case, and $\pm\lambda$ in the hyperbolic case) are given by $\exp \pm i \int_0^{2\pi} \frac{d\theta}{\xi(\theta)}$ or the exponential of the corresponding principal value integral. □

9.2.4 Classification of the SV-Orbits in $\mathcal{S}^{aff}_{\leq 2}$

This problem can be solved by extending the above results, which may be interpreted as the decomposition of $\mathcal{S}_2^{aff}$ into $\mathrm{Diff}(\mathbb{R}/2\pi\mathbb{Z})$-orbits. Let us first compute the stabilizers of some operators that will be shown later to be representatives of all the orbits. We choose to present the results in the Fourier coordinates (θ, x). The orbits of type I, resp. III split into orbits of type (i), (i)bis, resp. (iii), (iii)bis due to the presence of the linear term $V_1(\theta)x$ in the potential.

The computations depend on the formulas of Proposition 8.3, see formulas (9.10), (9.11), (9.12), (9.13) for more convenience.

Definition 9.12. If $D \in \mathcal{S}^{aff}_{\leq 2}$, we denote by G_D the stabilizer of D in the Schrödinger–Virasoro group SV, i.e. $G_D = \{g \in \mathrm{SV} \mid \sigma_{1/4}(g).D = D\}$.

Using the semi-direct product structure $G := SV = G_0 \ltimes H = \mathrm{Diff}(\mathbb{R}/2\pi\mathbb{Z}) \ltimes H$ (see Chap. 1), one defines in the same way the stabilizing subgroups $(G_0)_D$, H_D.

Recall the notation Stab_u, $u \in C^\infty(\mathbb{R}/2\pi\mathbb{Z})$ is used for the stabilizer in $\mathrm{Diff}(\mathbb{R}/2\pi\mathbb{Z})$ of the corresponding Hill operator.

Note that $\mathcal{M}_1 = M_0$ (whose exponential amounts to the multiplication of the wave functions ψ by a constant phase) acts trivially on any operator D, hence $\mathcal{M}_1 \in G_D$ always. The rotation group $\theta \to \theta + \theta_0$ generated by $d\sigma_{1/4}(\mathcal{L}_1) = d\sigma_{1/4}(L_0) = -\partial_\theta$ will be denoted by Rot.

In the following classification, we shall call *harmonic oscillators* (resp. *harmonic repulsors*) operators with *elliptic*, resp. *hyperbolic* monodromy.

(i) *Time-independent harmonic oscillators or repulsors*
Set $D_{\alpha,\gamma} := -2i\partial_\theta - \partial_x^2 + \alpha x^2 + \gamma$ ($\alpha, \gamma \in \mathbb{R}$). It is clear that $L_{-1} = \partial_\theta$ leaves $D_{\alpha,\gamma}$ invariant in all cases. Suppose first for simplicity that $\gamma = 0$. Then $G_D = (G_0)_D \ltimes H_D$ is a semi-direct product, so one retrieves Kirillov's results (see Proposition 9.9, case I) for $(G_0)_D$; to be specific, $Lie((G_0)_{D_{n^2/4,0}}) = \mathbb{R}\partial_\theta \oplus \mathbb{R}\mathcal{L}_{\sin n\theta} \oplus \mathbb{R}\mathcal{L}_{\cos n\theta}$ if $n \in \mathbb{N}^*$, and $Lie((G_0)_{D_{\alpha,0}}) = \mathbb{R}\partial_\theta$ otherwise.

Now $(1; (a, b)) \in H_{D_{\alpha,0}}$ if and only if $b' = 0$ and $a'' = -\alpha a$. The latter equation has a non-trivial solution if and only if $\alpha = 0$ (in which case $Lie(H_D) = \mathbb{R}\mathcal{Y}_1 \oplus \mathbb{R}\mathcal{M}_1$) or $\alpha = n^2/4$, $n \geq 1$ with n *even*, in which case $Lie(H_D) = \mathbb{R}\mathcal{Y}_{\cos n\theta/2} \oplus \mathbb{R}\mathcal{Y}_{\sin n\theta/2} \oplus \mathcal{M}_1$. Then $\exp \frac{1}{n}\mathcal{L}_1 \subset \widetilde{SL}^{(n)}(2, \mathbb{R})$ is the rotation of angle 2π, while $\exp \frac{1}{n}\mathrm{ad}\mathcal{L}_1|_{[Lie(\mathcal{H}_D), Lie(\mathcal{H}_D)]}$ is a rotation of angle π.

The isotropy groups G_D are the same in the case $\gamma \neq 0$, except for a different embedding involving sometimes complicated components in the nilpotent part of SV which do not change the commutation relations (so that G_D is no more a semi-direct product $(G_0)_D \ltimes H_D$).

All together, one has proved:

Theorem 9.1. *1. If $\alpha = n^2/4$, where $n \geq 2$ is an even integer, then $G_{D_{n^2/4,\gamma}} \simeq \widetilde{SL}^{(n)}(2,\mathbb{R}) \ltimes \mathcal{H}_1$ is isomorphic to an n-covering of the Schrödinger group; the semi-direct action of $\widetilde{SL}^{(n)}(2,\mathbb{R})$ quotients out into an action of the two-fold covering $\widetilde{SL}^{(2)}(2,\mathbb{R})$. The Lie algebra of the group $\widetilde{SL}^{(n)}(2,\mathbb{R})$ acts as $\mathbb{R}\partial_\theta \oplus \mathbb{R}(\mathcal{L}_{\sin n\theta} + \mathcal{M}_{-\frac{1}{2}\gamma \sin n\theta}) \oplus \mathbb{R}(\mathcal{L}_{\cos n\theta} + \mathcal{M}_{-\frac{1}{2}\gamma \cos n\theta})$. After transformation to the Laurent coordinates (t,r) (and supposing $\gamma = 0$), $G_{D_{n^2/4,0}}$ is the connected Lie group with Lie algebra $\langle L_0, L_{\pm n}\rangle \ltimes \langle Y_{\pm n/2}, M_0\rangle \subset \mathfrak{sv}(0)$.*
2. If $\alpha = n^2/4$, where $n \geq 1$ is odd, then $G_{D_{n^2/4,0}} \simeq \widetilde{SL}^{(n)}(2,\mathbb{R}) \times \exp \mathbb{R}\mathcal{M}_1$.
3. If $\alpha = 0$, then $G_{D_{0,\gamma}} = Rot \times \exp(\mathbb{R}\mathcal{Y}_1 \oplus \mathbb{R}\mathcal{M}_1) \simeq (\mathbb{R}/2\pi\mathbb{Z}) \times \mathbb{R} \times (\mathbb{R}/2\pi\mathbb{Z})$ is the commutative group of constant translations-phases. After transformation to the Laurent coordinates (t,r), it is the connected Lie group with Lie algebra $\langle L_0, Y_0, M_0\rangle \subset \mathfrak{sv}(0)$.
4. In the generic case $\alpha \neq n^2/4$, $n = 0, 1, \ldots$ one has simply $G_D =$ Rot $\times \exp \mathbb{R}\mathcal{M}_1 \simeq (\mathbb{R}/2\pi\mathbb{Z})^2$.

It is natural in view of these results to consider the two-fold covering $\tilde{H}^{(2)}$ of H obtained by considering 4π-periodic fields. Then the stabilizer in $\widetilde{SV}^{(2)} := G_0 \ltimes \tilde{H}^{(2)}$ of $D_{n^2/4,0}$ ($n \geq 1$ odd) is isomorphic to $\widetilde{SL}^{(n)}(2,\mathbb{R}) \ltimes \mathcal{H}_1$ as in the case of an even index n. This time $Lie(\mathcal{H}_1) = \langle Y_{\pm n/2}, M_0\rangle \subset \mathfrak{sv}$.

The best-known case is $\alpha = 1/4$ ($n = 1$), $\gamma = 0$. In the Laurent coordinates (t,r), $D_{1/4,0}$ is equal to $-2\mathrm{i}\partial_t - \partial_r^2$, namely, it is the free Schrödinger equation. Then Sch $= SL(2,\mathbb{R}) \ltimes \mathcal{H}_1$ acts on $D_{1/4,0}$ by the original Schrödinger representation $\pi_{1/4}(\text{Sch})$ (see Introduction) in the Laurent coordinates.

(i)bis *Special time-independent harmonic oscillators with added resonant oscillating drift*
Consider

$$D = -2\mathrm{i}\partial_\theta - \partial_x^2 + n^2x^2 + C\cos(n\theta - \sigma/2)\,.\,x + \gamma$$

$(C,\sigma,\gamma \in \mathbb{R},\ C \neq 0,\ n \geq 1$ integer). Then computations show that $G_D \simeq \mathbb{R} \times \mathbb{R} \times \mathbb{R}/2\pi\mathbb{Z}$ is three-dimensional, generated by

$$\mathcal{L}_{1-\cos(2n\theta-\sigma)} + \mathcal{Y}_{\frac{C}{8n}\sin 3(n\theta-\sigma/2)} - \mathcal{M}_{\frac{C^2}{32n^2}\left(\frac{\cos 4n(\theta-\sigma/2)}{4} + \frac{\cos(2n\theta-\sigma)}{2} + \frac{\gamma}{2}\cos(2n\theta-\sigma)\right)}, \tag{9.28}$$

$$\mathcal{Y}_{C\sin(n\theta-\sigma/2)} + \mathcal{M}_{\frac{C^2}{8n}\cos(2n\theta-\sigma)} \tag{9.29}$$

and $\mathscr{M}_1$. One checks (by direct computation) that the value of the associated invariant $I_{n^2}(1 - \cos(2n\theta - \sigma))$ is 0.

(ii) *Time-dependent harmonic repulsors of type II*
Consider

$$D_{n,\alpha,\gamma} = -2i\partial_\theta - \partial_x^2 + u_{n,\alpha}(\theta)x^2 + \gamma, \quad n = 1, 2, \ldots, \ \alpha \in (0, 1) \quad (9.30)$$

where

$$u_{n,\alpha}(\theta) = \frac{n^2}{4}\left[\frac{1 + 6\alpha \sin n\theta + 4\alpha^2 \sin^2 n\theta}{(1 + \alpha \sin n\theta)^2}\right]. \quad (9.31)$$

Then (see Proposition 9.9 (2)) $\mathscr{L}_\xi - \frac{\gamma}{2}\mathscr{M}_\xi \in Lie(G_D)$ ($\xi \neq 0$) if and only if ξ is proportional to $\xi_{n,\alpha}$, with $\xi_{n,\alpha} = \sin n\theta(1 + \alpha \sin n\theta)\partial_\theta$. Now

$$d\sigma_{1/4}(\mathscr{Y}_{f_1} + \mathscr{M}_{f_2}).D = 0$$

if and only if $f_2' = 0$ and $f_1'' + u_{n,\alpha} f_1 = 0$. The latter equation is known under the name of *Ince's equation* (see Magnus and Winkler, [84]). The change of variable and function $\theta \to \delta(\theta) = \frac{\pi}{4} - n\frac{\theta}{2}$, $f_1(\theta) \to y(\delta) = (1 + \alpha \cos 2\delta)^{b/4\alpha} f_1(\delta(\theta))$ with $b = -2\alpha[1 + \frac{i\alpha}{\sqrt{1-\alpha^2}}]$ turns the above equation into the standard form

$$(1 + a \cos 2\delta)y'' + b \sin 2\delta y' + (c + d \cos 2\delta)y = 0$$

with $a = \alpha$, $c = 1 - \frac{\alpha^2}{1-\alpha^2}$, $d = \alpha[3 + \frac{\alpha^2}{1-\alpha^2} \mp \frac{2i\alpha}{\sqrt{1-\alpha^2}}]$. Conditions for the coexistence of two independent periodic solutions of Ince's equation have been studied in detail. In our case, there is no periodic solution since $\partial^2 + u_{n,\alpha}$ is unstable (see discussion in Sect. 9.2.2). Hence

$$G_D = \exp\left(\mathbb{R}\left(\mathscr{L}_{\xi_{n,\alpha}} - \frac{\gamma}{2}\mathscr{M}_{\xi_{n,\alpha}}\right) \oplus \mathbb{R}\mathscr{M}_1\right) \simeq \mathbb{R} \times (\mathbb{R}/2\pi\mathbb{Z}). \quad (9.32)$$

(iii) *Non-resonant time-dependent Schrödinger operators of type III*
Consider

$$D_{n,\alpha,\gamma} = 2i\partial_\theta - \partial_x^2 + v_{n,\alpha}(\theta)x^2 + \gamma$$

($n = 1, 2, \ldots, \alpha \in (0, 1)$). See (9.27). Similarly to case (ii), $\mathscr{L}_\xi - \frac{\gamma}{2}\mathscr{M}_\xi \in Lie(G_D)$ ($\xi \neq 0$) if and only if ξ is proportional to $\xi_{\pm,n,\alpha}$, where $\xi_{\pm,n,\alpha} = \pm(1 + \sin n\theta)(1 + \alpha \sin n\theta)\partial_\theta$. Then $d\sigma_{1/4}(\mathscr{Y}_{f_1} + \mathscr{M}_{f_2}) \cdot D = 0$ if and only if $f_2' = 0$ and $f_1'' + v_{n,\alpha} f_1 = 0$. This is once again Ince's equation, with parameters $a = \alpha$, $b = -2\alpha$, $c = 1 - 2\alpha$, $d = 3\alpha$. One verifies immediately that $y(\delta) = \cos\delta$ is the unique (up to a constant) periodic solution of this semi-stable Hill equation, corresponding to $f_1(\theta) := (1+\alpha\sin n\theta)^{\frac{1}{2}}\cos(\frac{\pi}{4} - n\frac{\theta}{2})$. Note that $\xi_{+,n,\alpha} = f_1^2$ (so that f_1 is - up to a sign - the unique C^∞ square-root of $\xi_{+,n,\alpha}$). Hence

$$G_D = \exp\left(\mathbb{R}\left(\mathcal{L}_{\xi_{\pm,n,\alpha}} - \frac{\gamma}{2}\mathcal{M}_{\xi_{\pm,n,\alpha}}\right) \oplus \mathbb{R}\mathcal{Y}_{f_1} \oplus \mathbb{R}\mathcal{M}_1\right) \simeq \mathbb{R}\times\mathbb{R}\times(\mathbb{R}/2\pi\mathbb{Z}) \tag{9.33}$$

(iii)bis *Schrödinger operators of type III with added resonant drift*

Consider

$$D = -2i\partial_\theta - \partial_x^2 + v_{n,\alpha}(\theta)x^2 + C(1+\alpha\sin n\theta)^{1/2}\cos\left(\frac{\pi}{4} - n\frac{\theta}{2}\right)x + \gamma$$

($C \neq 0$) with $v_{n,\alpha}$ as in case (iii). Set $\xi(\theta) = (1+\sin n\theta)(1+\alpha\sin n\theta)$ and $f(\theta) = (1+\alpha\sin n\theta)^{\frac{1}{2}}\cos(\pi/4 - n\theta/2)$. Recall $\xi = f^2$.

Suppose $\mathcal{L}_\xi + \mathcal{Y}_{f_1} + \mathcal{M}_{f_2}$ stabilizes D. Then (see (9.12 and (9.13))

$$2\left(f_1'' + v_{n,\alpha}f_1\right) = C\left(\xi f' + \frac{3}{2}\xi' f\right) = 4Cf^2f' \tag{9.34}$$

and

$$f_2' = -\frac{1}{2}(\gamma\xi' + Cf_1f). \tag{9.35}$$

The kernel of the operator $\partial^2 + v_{n,\alpha}$ is one-dimensional, generated by f. Hence the above (9.34) has a solution if and only if $\int_0^{2\pi}(\xi f' + \frac{3}{2}\xi' f)f\,d\theta = 0$, which is true since $(\xi f' + \frac{3}{2}\xi' f)f = (f^4)'$. Now (9.35) has a solution if and only if f_1 is chosen to be the unique solution orthogonal to the kernel of $\partial^2 + v_{n,\alpha}$, namely, if $\int_0^{2\pi} f_1 f\,d\theta = 0$.

Now

$$d\sigma_{1/4}\left(\mathcal{Y}_{g_1} + \mathcal{M}_{g_2}\right).D = -2\left(g_1'' + v_{n,\alpha}g_1\right)x - fg_1 - 2g_2' \tag{9.36}$$

vanishes if and only if $g_1 = f$ (up to a multiplicative constant) and $\int_0^{2\pi} g_1 f\,d\theta = 0$. The two conditions are clearly incompatible.

All together one has proved that

$$G_D = \exp\left(\mathbb{R}\left(\mathcal{L}_{(1+\sin n\theta)(1+\alpha\sin n\theta)} + \mathcal{Y}_{f_1} + \mathcal{M}_{f_2}\right) \oplus \mathbb{R}\mathcal{M}_1\right) \simeq \mathbb{R}\times\mathbb{R}/2\pi\mathbb{Z} \tag{9.37}$$

(with f_1, f_2 solving (9.34), (9.35)) is commutative two-dimensional.

Explicit but cumbersome formulas for f_1, f_2 may be derived from the integration of (9.115) (actually, by definition, $f_1 = \delta_1$ (up to a coefficient), see Definition 9.25, and $\delta_1 = -d/2$, see notations at the end of Sect. 9.1 and Theorem 9.7). We shall not need them.

It remains to prove that we have classified *all* the orbits in $\mathcal{S}^{aff}_{\le 2}$.

Theorem 9.2. *Any Schrödinger operator D in $\mathscr{S}^{aff}_{\le 2}$ belongs to the orbit of one of the above operators. In other words, the above operators are normal forms for the action of the Schrödinger–Virasoro group.*

Proof. Let $D \in \mathscr{S}^{aff}_{\le 2}$. Suppose first that V_2 is of type I. Then one may assume (by a time-reparametrization) that $V_2 = \alpha$ is a constant. The operator D belongs to the orbit of $D_{\alpha,\gamma}$ (case (i)) for some γ if and only if $V_1 = 2(a'' + \alpha a)$. If $\alpha \ne n^2$ (or $n^2/4$ if one considers the $\widetilde{SV}^{(2)}$-orbits) then this equation has a unique solution for every V_1. If $\alpha = n^2$, then a Fourier series $V_1 = \sum_k c_k e^{ik\theta}$ is in the image of $\partial_\theta^2 + \alpha$ if and only if $c_{\pm n} = 0$. This analysis accounts for the two cases (i), (i)bis.

Suppose now V_2 is of type II. By a time-reparametrization one may choose $V_2 = u_{n,\alpha}$. The operator D belongs to the orbit of $D_{n,\alpha,\gamma}$ (see case (ii)) for some γ, provided $V_1 = 2(a'' + u_{n,\alpha} a)$. Since $\partial^2 + u_{n,\alpha}$ (acting on $C^\infty(\mathbb{R}/2\pi\mathbb{Z})$) has a trivial kernel, it has a bounded inverse and the unique solution of the above equation is C^∞. Hence D belongs to the orbit of $D_{n,\alpha,\gamma}$.

Finally, suppose V_2 is of type III. One is led to solve the equation $V_1 = 2(a'' + v_{n,\alpha} a)$. Recall $V_1(\theta) = (1 + \alpha \sin n\theta)^{1/2} \cos(\frac{\pi}{4} - n\frac{\theta}{2})$ solves the equation $f_1'' + v_{n,\alpha} f_1 = 0$. Hence $V_1 = 2(a'' + v_{n,\alpha} a)$ has a solution if and only if $\int_0^{2\pi} V_1(\theta)$ $(1 + \alpha \sin n\theta)^{1/2} \cos(\frac{\pi}{4} - n\frac{\theta}{2})\, d\theta = 0$, which accounts for cases (iii), (iii)bis. □

Thus we have obtained a complete family of *normal forms* for Schrödinger operators in $\mathscr{S}^{aff}_{\le 2}$. We shall compute their monodromy later on.

Note that Schrödinger operators of type III are generically of type (iii)bis, and Schrödinger operators of type I with $\alpha = n^2, n = 1, 2, \ldots$ are generically of type (i)bis.

Corollary 9.13. *For generic orbits (type (i) with $\alpha \ne \frac{n^2}{4}$, $n \ge 0$, or type (ii)), the isotropy group is two-dimensional, given by* $\exp \mathbb{R}(\mathscr{L}_\xi + \mathscr{Y}_{\delta_1} + \mathscr{M}_{\delta_2}) \oplus \mathbb{R}\mathscr{M}_1 \simeq \mathbb{R} \times \mathbb{R}/2\pi\mathbb{Z}$ *or* $\mathbb{R}/2\pi\mathbb{Z} \times \mathbb{R}/2\pi\mathbb{Z}$ *for some triple* $(\xi, \delta_1, \delta_2)$ *with* $\xi \ne 0$.

Let us conclude with a remark. Consider a potential $V_2(\theta)x^2 + V_1(\theta)x + V_0(\theta)$ of type (i), (ii) or (iii). As we shall see in the next section, the monodromy of the corresponding Schrödinger operator depends only on the conjugacy class of the invariant ξ and the value of the constant γ (which acts as a simple energy shift). Computing the invariant ξ is a difficult task in general, but, supposing we have done so, how can we determine the constant γ? We give an answer for generic elliptic or hyperbolic potentials of type (i).

Lemma 9.14. *Let $D = -2i\partial_\theta - \partial_x^2 + V_2(\theta)x^2 + V_1(\theta)x + V_0(\theta)$ be of type (i), elliptic or hyperbolic, generic, so that D is conjugate to a unique operator $D_{\alpha,\gamma} = -2i\partial_\theta - \partial_x^2 + \alpha x^2 + \gamma$ ($\alpha \in \mathbb{R}$, $\alpha \ne \frac{n^2}{4}$, $n = 0, 1, \ldots$). Then γ may be recovered from*

$$\gamma = \frac{1}{2\pi} \int_0^{2\pi} \left(V_0 - \frac{1}{4} V_1 W_1 \right)(\theta)\, d\theta \tag{9.38}$$

where W_1 is the unique solution of the equation $(\partial^2 + V_2) W_1 = V_1$.

Proof. Start from the model operator $D_{\alpha,\gamma}$, with stabilizer $\xi = 1$, and apply successively $\sigma_{1/4}(\phi; (0,0))$ and $\sigma_{1/4}(1; (g,h))$. Then one obtains the operator

$$D := -2\mathrm{i}\partial_\theta - \partial_x^2 + V_2(\theta)x^2 - 2\left(\left(\partial^2 + V_2\right) g\right) x + \left(\gamma\dot{\phi} + g\left(\partial^2 + V_2\right) g - 2\dot{h}\right)$$

(see formulas (9.10), (9.11)). Now $\int_0^{2\pi} \dot{\phi}(\theta)\, d\theta = 2\pi$ since $\phi \in \mathrm{Diff}(\mathbb{R}/2\pi\mathbb{Z})$. Hence the result. □

9.2.5 *Connection to U. Niederer's Results*

We are referring to a classical paper by U. Niederer (see [90, 91]) concerning the maximal groups of Lie symmetries of Schrödinger equations with arbitrary potentials. One may rephrase his main result as follows (though the Schrödinger–Virasoro group had not been introduced at that time). U. Niederer shows that any transformation

$$\psi(t,r) \to \tilde{\psi}(t,r) = \exp \mathrm{i} f_g\left(g^{-1}(t,r)\right) \psi\left(g^{-1}(t,r)\right),$$

where $g : (t,r) \to (t',r')$ is an arbitrary coordinate transformation and f_g an arbitrary 'companion function' corresponding to a projective action), that carries the space of solutions of the Schrödinger equation

$$\left(-2\mathrm{i}\partial_t - \partial_r^2 + V(t,r)\right)\psi(t,r) = 0 \tag{9.39}$$

into itself is necessarily of the form $\sigma_{1/4}(g)$ for some $g \in \mathrm{SV}$. This provides, incidentally, an elegant way of introducing the SV group in the first place. Then Niederer gives a necessary and sufficient condition for $g \in \sigma_{1/4}(SV)$ to leave (9.39) invariant, and produces some physically interesting examples. Let us analyze some of these examples from our point of view. It should be understood that Niederer's examples are given in the Laurent coordinates (t,r) and should hence be transformed by using Lemma 9.1 to compare with our results.

(i) $V = 0$ (free Schrödinger equation): after the transformation of Lemma 9.1 to the potential $V(\theta,x) = \frac{1}{4}x^2$, this case corresponds to invariance under the full Schrödinger group (see case (i) in Sect. 9.2.4, with $\alpha = 1/4$ and $\gamma = 0$).
(ii) $V = -gr$ (free fall) corresponds to $V(\theta,x) = \frac{1}{4}x^2 - g e^{\mathrm{i}(\theta/2+3\pi/4)} x$ (a 4π-periodic potential), which belongs to the same orbit as case (i) (free Schrödinger equation in the Laurent coordinates).
(iii) $V = \frac{1}{2}\omega^2 r^2$ (harmonic oscillator) may be obtained from the free Schrödinger equation by the time reparametrization $t(u) = \tan \omega u$ for which the Schwarzian derivative is a constant, $\Theta(t) = 2\omega^2$ (see formulas in Proposition 8.3).

(iv) $V = k/r^2$ (inverse-square potential), corresponding to the operator $-2i\partial_\theta - \partial_x^2 + \frac{x^2}{4} + kx^{-2}$ (harmonic oscillator with added inverse-square potential) in the Fourier coordinate. The operator is not in $\mathcal{S}^{aff}_{\leq 2}$, but the (time-independent) inverse-square potential is interesting in that it is the only potential left invariant by all transformations $V(t,r) \to \phi'(t)V(\phi(t), r\sqrt{\phi'(t)})$ (see formulas in Proposition 8.3). So this equation is invariant under the kernel of the Schwarzian derivative, i.e. homographic transformations.

9.3 Monodromy of Time-Dependent Schrödinger Operators of Non-resonant Types and Ermakov-Lewis Invariants

In this section, we use the Ermakov-Lewis invariants (to be introduced below) to solve all Schrödinger operators in $\mathcal{S}^{aff}_{\leq 2}$ of class (i), (ii) or (iii). Since any such operator is conjugate to an operator of the form $-2i\partial_\theta - \partial_x^2 + V_2(\theta)x^2 + \gamma$ (γ constant), and γ corresponds to a simple energy shift, we shall implicitly assume that the potential is simply quadratic ($V_1 = V_0 = 0$).

Lemma 9.16 and Proposition 9.18 yield explicitly an evolution operator $U(\theta, \theta_0)$, i.e. a unitary operator on $L^2(\mathbb{R})$ which gives the evolution of the solutions of the Schrödinger equation from time θ_0 to time θ. This operator gives the unique solution to the Cauchy problem and allows to compute the (exact) Berry phase. The arguments in Lemmas 9.15, 9.16 and Proposition 9.18 are reproduced from the article of Lewis and Riesenfeld ([80]). Unfortunately this method gives the monodromy only in the *elliptic* case (i.e. for operators of class (i) with $\alpha > 0$). So we generalize their invariants to the hyperbolic and unipotent case; the invariant we must choose in order to be able to compute the monodromy is not a harmonic oscillator any more, but an operator with absolutely continuous spectrum. Nevertheless, it turns out that there does exist a phase operator, given in terms of the (possibly regularized) integral $\int_0^{2\pi} \frac{d\theta}{\xi(\theta)}$ for a certain stabilizer ξ of the quadratic part of the potential. The key point is that, in order to get the whole picture, one must build the bridge between Kirillov's results and the Ermakov-Lewis invariants.

9.3.1 *Ermakov-Lewis Invariants and Schrödinger–Virasoro Invariance*

Let $H = \frac{1}{2}(-\partial_x^2 + V_2(\theta)x^2)$ be the (quantum) Hamiltonian corresponding to a time-dependent harmonic oscillator. The evolution of the wave function $\psi(\theta, x)$ is given by: $i\partial_\theta\psi(\theta,x) = H\psi(\theta,x)$, or $D\psi = 0$ where $D = -2i\partial_\theta + 2H = -2i\partial_\theta - \partial_x^2 + V_2(\theta)x^2$.

The *Ermakov-Lewis dynamical invariants* were invented in order to find the solutions of the above equation. The idea is simple. Suppose $I(\theta, x)$ is a *time-dependent* hermitian operator of the form $\sum_{j=0}^{N} I_j(\theta, x)\partial_x^j$, which is an invariant of the motion, i.e. $\frac{d}{dt}I = \partial_t I + \frac{1}{\mathrm{i}}[I, H] = 0$. Suppose also that, for every fixed value of θ, $I(\theta, x)$ (defined on an appropriately defined dense subspace of $L^2(\mathbb{R}, dx)$, for instance on the space of test functions) is essentially self-adjoint and has a purely point spectrum. For simplicity, we shall assume that all multiplicities are one, and that one may choose normalized eigenvectors which depend regularly on θ, namely,

$$I(\theta, x)h_n(\theta, x) = \lambda_n(\theta)h_n(\theta, x) \tag{9.40}$$

and $\int_{\mathbb{R}} |h_n(\theta, x)|^2\, dx = 1$. The fact that I is an invariant of the motion implies by definition that $I\psi$ is a solution of the Schrödinger equation if ψ is. The following lemma shows how to solve the Schrödinger equation by means of the invariant I:

Lemma 9.15 (see Lewis-Riesenfeld[80]).

1. The eigenvalues $\lambda_n(\theta)$ are constants, i.e. they do not depend on time.
2. If $n \neq m$, then $\langle h_m(\theta), (\mathrm{i}\partial_\theta - H)h_n(\theta)\rangle = 0$.

Proof. (i) Applying the invariance property $\frac{\partial I}{\partial \theta} + \frac{1}{\mathrm{i}}[I(\theta), H(\theta)] = 0$ to the eigenvector $h_n(\theta)$ yields

$$\frac{\partial I}{\partial \theta}h_n(\theta) + \frac{1}{\mathrm{i}}(I(\theta) - \lambda_n(\theta))H(\theta)h_n(\theta) = 0.$$

Taking the scalar product with $h_m(\theta)$ gives a first equation,

$$\left\langle h_m(\theta), \frac{\partial I}{\partial \theta}h_n(\theta)\right\rangle + \frac{1}{\mathrm{i}}(\lambda_m(\theta) - \lambda_n(\theta))\langle h_m(\theta), H(\theta)h_n(\theta)\rangle = 0. \tag{9.41}$$

The eigenvalue equation $I(\theta)h_n(\theta) = \lambda_n(\theta)h_n(\theta)$ gives after time differentiation a second equation, namely

$$\frac{\partial I}{\partial \theta}h_n(\theta) + (I(\theta) - \lambda_n(\theta))\dot{h}_n(\theta) = \dot{\lambda}_n(\theta)h_n(\theta). \tag{9.42}$$

Combining (9.41) and (9.42) for $n = m$ yields $\dot{\lambda}_n(\theta) = 0$.
(ii) Combining this time (9.41) and (9.42) for $n \neq m$ yields the desired equality.

The above Lemma shows that one may *choose* eigenvectors $h_n(\theta)$ that satisfy the Schrödinger equation by multiplying them by an appropriate time-dependent phase, which is the content of the following Lemma.

Lemma 9.16. *Let, for each n, $\alpha_n(\theta)$ be a solution of the equation*

$$\frac{d\alpha_n}{d\theta} = \langle h_n(\theta), \quad (\mathrm{i}\partial_\theta - H)h_n(\theta)\rangle\,. \tag{9.43}$$

Then the gauge-transformed eigenvectors for the invariant I

$$\tilde{h}_n(\theta) = e^{i\alpha_n(\theta)} h_n(\theta) \tag{9.44}$$

are solutions of the Schrödinger equation.

In other words, the general solution of the Schrödinger equation is:

$$\psi(\theta) := \sum_n c_n e^{i\alpha_n(\theta)} h_n(\theta) \tag{9.45}$$

where c_n are constant (time-independent) coefficients.

Let us specialize to the case when H is a time-dependent harmonic oscillator as above, i.e. $H = \frac{1}{2}(-\partial_x^2 + V_2(\theta)x^2)$. A natural idea is to assume the following Ansatz

$$I(\theta) = \frac{1}{2}\left[-b(\theta)\partial_x^2 + a(\theta)x^2 - ic(\theta)(x\partial_x + \partial_x x)\right].$$

This problem has a unique family of non-trivial solutions:

Definition 9.17 (Pinney-Milne equation). The non-linear equation

$$\ddot{\zeta} + f(\theta)\zeta - \frac{K}{\zeta^3} = 0 \tag{9.46}$$

($K > 0$) is called a Pinney-Milne equation. If $K = 1$, then we shall say that (9.46) is a *normalized* Pinney-Milne equation.

Of course, every Pinney-Milne equation can easily be normalized by multiplying the function by the constant factor $K^{1/4}$.

The following Proposition summarizes results due to H.R. Lewis and W.B. Riesenfeld (see [80]).

Proposition 9.18 (Ermakov-Lewis invariants for time-dependent harmonic oscillators).

1. *The second-order operator $\mathscr{E}\mathscr{L}(\zeta^2)$*

$$\mathscr{E}\mathscr{L}(\zeta^2)(\theta) = \frac{1}{2}\left[\frac{x^2}{\zeta^2} + \left(i\zeta(\theta)\partial_x + \dot{\zeta}(\theta)x\right)^2\right] \tag{9.47}$$

is an invariant of the time-dependent harmonic oscillator $-2i\partial_\theta - \partial_x^2 + V_2(\theta)x^2$ provided ζ is a solution of the following normalized Pinney-Milne equation:

$$\ddot{\zeta} + V_2(\theta)\zeta - \frac{1}{\zeta^3} = 0. \tag{9.48}$$

Setting $\xi = \zeta^2$, one may also write equivalently

$$\begin{aligned} \mathscr{E}\mathscr{L}(\xi)(\theta) &= \frac{1}{2\xi}\left[x^2 + \left(\mathrm{i}\xi\partial_x + \frac{1}{2}\dot{\xi}x\right)^2\right] \\ &= \frac{1}{2\xi}\left[-\xi^2\partial_x^2 + \left(1 + \frac{\dot{\xi}^2}{4}\right)x^2 + \frac{\mathrm{i}}{2}\xi\dot{\xi}\left(x\partial_x + \partial_x x\right)\right]. \end{aligned} \tag{9.49}$$

2. *Set*

$$a(\theta) := \frac{1}{\sqrt{2}}\left[\frac{x}{\zeta(\theta)} - \left(\zeta(\theta)\partial_x + \mathrm{i}\dot{\zeta}(\theta)x\right)\right]$$

and

$$a^*(\theta) = \frac{1}{\sqrt{2}}\left[\frac{x}{\zeta(\theta)} + \left(\zeta(\theta)\partial_x + \mathrm{i}\dot{\zeta}(\theta)x\right)\right]$$

(formal adjoint of the operator $a(\theta)$). Then

$$\mathscr{E}\mathscr{L}(\xi)(\theta) = a^*(\theta)a(\theta) + \frac{1}{2}. \tag{9.50}$$

In other words, for every fixed value of θ, the operators $a(\theta), a^*(\theta)$ play the roles of an annihilation, resp. creation operator for the (time-dependent) harmonic oscillator $\mathscr{E}\mathscr{L}(\xi)$.

3. *The normalized ground state of the operator $a(\theta)$ is*

$$h_0(\theta) = \frac{1}{2\sqrt{\pi}}\frac{1}{\sqrt{\xi(\theta)}}\exp\left(\left(-\frac{1}{2}\frac{1}{\xi(\theta)} + \frac{\mathrm{i}}{2}(\dot{\xi}/\xi)(\theta)\right)x^2\right). \tag{9.51}$$

4. *The solutions of (9.43) giving the phase evolution of the solutions of the Schrödinger equation are given by*

$$\alpha_n(\theta) = -\left(n + \frac{1}{2}\right)\int^{\theta}\frac{d\theta'}{\xi(\theta')} \tag{9.52}$$

provided one chooses the time-evolution of the eigenstates h_n by setting

$$\langle h_n, \partial_\theta h_n\rangle = \frac{\mathrm{i}}{2}\left(n + \frac{1}{2}\right)\left(\zeta\ddot{\zeta} - \dot{\zeta}^2\right). \tag{9.53}$$

The above choice for the time-evolution of the eigenstates appears natural if one requires the standard lowering and raising relations $a(\theta)h_n(\theta) = n^{\frac{1}{2}}h_{n-1}(\theta)$, $a^*(\theta)h_n(\theta) = (n+1)^{\frac{1}{2}}h_{n+1}(\theta)$. Then computations show that

$$\langle h_n, \partial_\theta h_n\rangle = \langle h_0, \partial_\theta h_0\rangle + \mathrm{i}\frac{n}{2}\left(\zeta\ddot{\zeta} - \dot{\zeta}^2\right). \tag{9.54}$$

Hence it remains only to choose the time-evolution of the ground-state h_0. This particular choice leads to the $(n+\frac{1}{2})$-factor typical of the spectrum of the harmonic oscillator. Note that the $h_n(\theta)$ do *not* satisfy the gauge-fixing condition typical of the adiabatic approximation (see A. Joye [62] for instance). But this phase choice leads to a nice interpretation of the phases α_n (up to a constant) as a *canonical coordinate* conjugate to the classical invariant $\mathscr{E}\mathscr{L}_{cl}$ (see Lemma 9.21 below) for the corresponding classical problem, in the generalized symplectic formalism for which time is a coordinate, so that the problem becomes autonomous (see Lewis-Riesenfeld [80]; see also Sect. 9.4 for the symplectic formalism). Also, as mentioned in the introduction, the natural time-scale (both for the classical and the quantum problem) is $\tau(\theta) := \int^{\theta} \frac{du}{\xi(u)}$.

The connection with the preceding sections is given by the following classical lemma (see [84], Chap. 3), which is an easy corollary of Proposition 9.8:

Lemma 9.19. *1. Let ξ be a (non-necessarily periodic) solution of the equation*

$$\frac{1}{2}\xi''' + 2u\xi' + u'\xi = 0, \tag{9.55}$$

so that ξ stabilizes $\partial^2 + u$. Then $\zeta := \sqrt{\xi}$ is a solution of the Pinney-Milne equation

$$\zeta'' + u(\theta)\zeta - \frac{I_u(\xi)/2}{\zeta^3} = 0 \tag{9.56}$$

where $I_u(\xi) := \xi\xi'' - \frac{1}{2}\xi'^2 + 2u\xi^2$ is the constant defined in Proposition 9.8 (2).

In particular, if $\xi = \psi_1^2 + \psi_2^2$, where (ψ_1, ψ_2) is a basis of solutions of the Hill equation $(\partial^2 + u)\psi = 0$, and $\zeta = \sqrt{\xi}$, then

$$\zeta'' + u(\theta)\zeta - \frac{W^2}{\zeta^3} = 0 \tag{9.57}$$

where $W := \psi_1\psi_2' - \psi_1'\psi_2$ is the Wronskian of the two solutions.

2. *Consider $\xi \in Stab_u$ such that $\zeta = \sqrt{\xi}$ satisfies the Pinney-Milne equation (9.57), and a time-reparametrization ϕ. Then $\tilde{\xi} := \phi'^{-1} \,.\, \xi \circ \phi$ is a stabilizer of $\partial^2 + \tilde{u} := \phi_*(\partial^2 + u)$ and $\tilde{\zeta} := \sqrt{\tilde{\xi}}$ satisfies the transformed Pinney-Milne equation*

$$\tilde{\zeta}'' + \tilde{u}\tilde{\zeta} - \frac{W^2}{\tilde{\zeta}^3} = 0$$

for the same constant W.

The interesting point now is that one can choose the Ermakov-Lewis invariant in such a way that the invariant associated to the image of the time-dependent harmonic oscillator D by a time reparametrization (through the representation $\sigma_{1/4}$) is its image by a very natural transformation (essentially, by the corresponding change of coordinates). This provides an elegant, natural explanation for the complicated-looking phase appearing in the formulas for $\sigma_{1/4}$.

Theorem 9.3. *Let* $D := -2\mathrm{i}\partial_\theta - \partial_x^2 + V_2(\theta)x^2$ *be a time-dependent harmonic oscillator,* ζ *satisfy the Pinney equation* $\zeta'' + V_2\zeta - \frac{1}{\zeta^3} = 0$, *and* $\mathscr{E}\mathscr{L}(\zeta^2) = \frac{1}{2}\left[\left(\frac{x}{\zeta}\right)^2 + (\mathrm{i}\zeta\partial_x + \zeta' x)^2\right]$ *be the associated Ermakov-Lewis invariant.*

Let $\phi \in \mathrm{Diff}(\mathbb{R}/2\pi\mathbb{Z})$ *be a time-reparametrization and* $\tilde{V}_2$ *be the image of* V_2 *through* ϕ, *defined by* $\sigma_{1/4}(\phi).D = -2\mathrm{i}\partial_\theta - \partial_r^2 + \tilde{V}_2(\theta)x^2$.

Then:

1. $\tilde{\zeta} := (\phi' \circ \phi^{-1})^{\frac{1}{2}} \,.\, \zeta \circ \phi^{-1}$ *satisfies the transformed Pinney equation* $\tilde{\zeta}'' + \tilde{V}_2\tilde{\zeta} - \frac{1}{\tilde{\zeta}^3} = 0$.
2. *Consider the transformed Ermakov-Lewis invariant*

$$\widetilde{\mathscr{E}\mathscr{L}}(\tilde{\zeta}^2)(x) := \frac{1}{2}\left[\left(\frac{\tilde{x}}{\tilde{\zeta}}\right)^2 + (\mathrm{i}\tilde{\zeta}\partial_{\tilde{x}} + \frac{d\tilde{\zeta}}{d\tilde{\theta}}\tilde{x})^2\right] \tag{9.58}$$

where $(\tilde{\theta}, \tilde{x}) = (\phi(\theta), x\sqrt{\phi'(\theta)})$ *are the transformed coordinates.*
Then

$$\widetilde{\mathscr{E}\mathscr{L}}(\tilde{\zeta}^2) = \pi_{1/4}(\phi)\mathscr{E}\mathscr{L}(\zeta^2)\pi_{1/4}(\phi)^{-1}. \tag{9.59}$$

In particular, $\widetilde{\mathscr{E}\mathscr{L}}(\tilde{\zeta}^2)$ *is an Ermakov-Lewis invariant for* $\sigma_{1/4}(\phi)D$.

Proof. 1. Follows from Lemma 9.19 (2). This implies that $\widetilde{\mathscr{E}\mathscr{L}}(\tilde{\zeta}^2)$ is an Ermakov-Lewis invariant for $\sigma_{1/4}(\phi)\,.\,D$. Supposing one has proved that $\widetilde{\mathscr{E}\mathscr{L}}(\tilde{\zeta}^2)$ is the conjugate of $\mathscr{E}\mathscr{L}(\zeta^2)$ by $\pi_{1/4}(\phi)$, then it follows once again that $\widetilde{\mathscr{E}\mathscr{L}}(\tilde{\zeta}^2)$ is an invariant for $\sigma_{1/4}(\phi)\,.\,D$ since

$$\begin{aligned}
&\left(\sigma_{1/4}(\phi)\,.\,D\right)\widetilde{\mathscr{E}\mathscr{L}}(\tilde{\zeta}^2) - \widetilde{\mathscr{E}\mathscr{L}}(\tilde{\zeta}^2)\left(\sigma_{1/4}(\phi)\,.\,D\right) \\
&= \phi'\,\pi_{1/4}(\phi)D\,.\,\mathscr{E}\mathscr{L}(\zeta^2)\,.\,\pi_{1/4}(\phi)^{-1} - \pi_{1/4}(\phi)\,.\,\mathscr{E}\mathscr{L}(\zeta^2)\,.\,\phi' D\,\pi_{1/4}(\phi)^{-1} \\
&= 0
\end{aligned} \tag{9.60}$$

(the function of time ϕ' commutes with the operator $\mathscr{E}\mathscr{L}(\zeta^2)$).

So all there remains to show is that $\widetilde{\mathscr{E}\mathscr{L}}(\tilde{\zeta}^2)$ is indeed conjugate to $\mathscr{E}\mathscr{L}(\zeta^2)$. This is actually true for both terms appearing inside parentheses in the expression for the Ermakov-Lewis invariant (and trivial for the first one). Set $\mathscr{E} = \mathrm{i}\zeta\partial_x + \zeta' x$ and $\tilde{\mathscr{E}} = \mathrm{i}\tilde{\zeta}\partial_{\tilde{x}} + \frac{d\tilde{\zeta}}{d\tilde{\theta}}\tilde{x}$. Then a simple computation shows that

$$\tilde{\mathscr{E}} = \mathrm{i}\zeta\partial_x + x\zeta' + \frac{1}{2}x\frac{\phi''}{\phi'}\zeta.$$

On the other hand,

$$
\begin{aligned}
\left(\pi_{1/4}(\phi)\mathscr{E}\pi_{1/4}(\phi)^{-1}\right)\psi(\tilde{\theta},\tilde{x}) &= (\phi'(\theta))^{-1/4} e^{\frac{1}{4}\mathrm{i}\frac{\phi''(\theta)}{\phi'(\theta)}x^2}\mathscr{E}\pi_{1/4}(\phi)^{-1}\psi(\theta,x) \\
&= \frac{1}{2}(\phi'(\theta))^{-1/4} e^{\frac{1}{4}\mathrm{i}\frac{\phi''(\theta)}{\phi'(\theta)}x^2}\left(\mathrm{i}\zeta(\theta)\partial_x + \zeta'(\theta)x\right) \\
&\quad \times \left(\phi'(\theta)^{1/4} e^{-\frac{1}{4}\mathrm{i}\frac{\phi''(\theta)}{\phi'(\theta)}x^2}\psi(\phi(\theta), x\sqrt{\phi'(\theta)})\right).
\end{aligned}
\tag{9.61}
$$

Hence $\pi_{1/4}(\phi)\mathscr{E}\pi_{1/4}(\phi)^{-1} = \tilde{\mathscr{E}}$. □

We now want to be able to write the general solution of the Schrödinger equation as

$$\psi(\theta) = \int_{\Sigma} e^{\mathrm{i}\alpha_k(\theta)} c_k h_k(\theta) d\sigma(k)$$

(for some spectral measure σ on a set Σ, a discrete measure in the case studied by Lewis and Riesenfeld) with *periodic* eigenstates h_k and a phase α_k with *periodic derivative, i.e. given by integrating a periodic function*, so that

$$\psi(\theta + 2\pi) = \int e^{\mathrm{i}\lambda_k} e^{\mathrm{i}\alpha_k(\theta)} c_k h_k(\theta) d\sigma(k)$$

where the $\lambda_k := \alpha_k(\theta + 2\pi) - \alpha_k(\theta)$ are *constants* and measure the rotation of the eigenstates h_k after a time 2π. Then the monodromy operator is unitarily equivalent to the multiplication operator $f(k) \to f(k)e^{\mathrm{i}\lambda_k}$ on $L^2(\Sigma, d\sigma)$.

Consider any Schrödinger operator with quadratic potential $V_2(\theta)x^2$ and an associated non-zero vector field $\xi \in Stab(V_2)$ as before. (We postpone the discussion of 'resonant' operators (classes (i)bis and (iii)bis) to the next section.) It turns out that the eigenstates h_k and the measure σ can be taken as the (possibly generalized) eigenfunctions and spectral measure of one of the three following 'model' operators H, depending on the *sign of the invariant* $I_u(\xi)$:

(i) $(I_u(\xi) > 0)$: take for H the standard harmonic oscillator

$$H = -\frac{1}{2}\left(\partial_x^2 - a^2x^2\right) \quad (a \in \mathbb{R});$$

this case corresponds to harmonic oscillators of type (i), i.e. Schrödinger operators of type (i) conjugate to $-2\mathrm{i}\partial_\theta - \partial_x^2 + a^2x^2$ with $a^2 > 0$;

(ii) $(I_u(\xi) < 0)$: take for H the 'standard harmonic repulsor'

$$H = -\frac{1}{2}\left(\partial_x^2 + a^2x^2\right) \quad (a \in \mathbb{R});$$

this case corresponds to harmonic repulsors of type (i), i.e. operators of type (i) conjugate to $-2i\partial_\theta - \partial_x^2 - a^2x^2$ ($-a^2 < 0$), and operators of type (ii);

(iii) $(I_u(\xi) = 0)$: take for H the usual one-dimensional Laplacian,

$$H = -\frac{1}{2}\partial_x^2;$$

this case corresponds to operators of type (i) conjugate to the free Schrödinger operator $-2i\partial_\theta - \partial_x^2$, and operators of type (iii).

Note that this classification is equivalent to the classification of the (conjugacy classes of) monodromy matrices for the associated Hill operators $\partial_\theta^2 + V_2(\theta)$ into elliptic, hyperbolic and unipotent elements.

The next section circumvents the spectral analysis technicalities by solving the associated classical problem. The essential prerequisites for understanding the (operator-valued) monodromy for the quantum problem are already contained in the study of the ($SL(2,\mathbb{R})$-valued) monodromy of the ordinary differential equation $\ddot{x} = -V_2(\theta)x$, so we found this short digression convenient for the reader. Then we study the spectral decomposition of the above model operators; finally, we solve the quantum problem for a quadratic potential $V_2(\theta)x^2$ and compute the monodromy operator. The general case $D \in \mathscr{S}^{aff}_{\leq 2}$ may be reduced to the quadratic case $D \in \mathscr{S}^{aff}_2$ after applying some transformation in SV, except for the operators of type (i)bis and (iii)bis; these will be treated in the last section.

9.3.2 *Solution of the Associated Classical Problem*

The associated classical problem (obtained for instance as the lowest-order term in $\hbar$ in the usual semi-classical expansion) is a Hill equation.

Definition 9.20 (associated classical problem). Let H be the classical hamiltonian $H = \frac{1}{2}(p^2 + V_2(\theta)x^2)$.

The associated motion in phase space reads $\dot{x} = \partial_p H = p$, $\dot{p} = -\partial_x H = -V_2$, and is equivalent to the Hill equation $(\partial_\theta^2 + V_2)x(\theta) = 0$.

Lemma 9.21. *1. Suppose V_2 is of type I with $\alpha \neq 0$ or of type II, and choose $\xi \in \mathrm{Stab}\, V_2$ so that $I_u(\xi) = 2$ (ξ is real in the elliptic case and purely imaginary in the hyperbolic case). Then*

$$\mathscr{E}\mathscr{L}_{cl}(\xi)(x) := \frac{1}{2}\left[\frac{x^2}{\xi} + \xi\left(\dot{x} - \frac{1}{2}\frac{\dot{\xi}}{\xi}x\right)^2\right] \tag{9.62}$$

is an invariant of the motion.

2. *Suppose* V_2 *is of type I with* $\alpha = 0$ *or of type III (so that the associated monodromy is unipotent), and take any* $\xi \in \mathrm{Stab}\, V_2$, $\xi \neq 0$. *Then*

$$\mathscr{E}\mathscr{L}_{cl}(\xi)(x) := \frac{1}{2}\left[\xi\left(\dot{x} - \frac{1}{2}\frac{\dot{\xi}}{\xi}x\right)^2\right] \tag{9.63}$$

is an invariant of the motion.

Proof. Simple computation ($\mathscr{E}\mathscr{L}_{cl}$ may be obtained from the quantum Ermakov-Lewis invariant by letting $\hbar$ go to zero). □

Assuming V_2 is elliptic, i.e. of type I with $\alpha > 0$, one may choose $\xi > 0$. Then the equation $\mathscr{E}\mathscr{L}_{cl}(\xi)(x) = C$, C constant is equivalent to $\left(\frac{dz}{d\tau}\right)^2 + z^2 = C$ after changing the function and time to $\tau(\theta) = \int^\theta \frac{d\theta'}{\xi(\theta')}$ and $x(\theta) = \xi^{\frac{1}{2}}(\theta)z(\tau(\theta))$, respectively, with obvious solutions $\cos\tau$, $\sin\tau$. Hence a basis of solutions of the equation of motion is given by

$$x_1(\theta) = \xi^{\frac{1}{2}}(\theta)\cos\int^\theta \frac{d\theta'}{\xi(\theta')},\quad x_2(\theta) = \xi^{\frac{1}{2}}(\theta)\sin\int^\theta \frac{d\theta'}{\xi(\theta')} \tag{9.64}$$

Assume for instance that $\dot{\xi}(0) = 0$, and choose $\int^\theta \frac{d\theta'}{\xi(\theta')} = \int_0^\theta \frac{d\theta'}{\xi(\theta')}$. Then $\begin{pmatrix} x_1 \\ x_2 \end{pmatrix}(2\pi) = \begin{pmatrix} \cos T & -\sin T \\ \sin T & \cos T \end{pmatrix} \cdot \begin{pmatrix} x_1 \\ x_2 \end{pmatrix}(0)$ with $T = \int_0^{2\pi} \frac{d\theta'}{\xi(\theta')}$. Hence the eigenvalues of the monodromy matrix are given by $\pm \mathrm{i}T$.

In the hyperbolic case (type I with $\alpha < 0$, or type II), $\xi := \mathrm{i}\eta$ is purely imaginary. The above formulas (9.64) give solutions of the Hill equation on either side of any zero of ξ (note that the normalization $I_u(\xi) = 2$ implies $\xi(\theta) \sim_{\theta\to\theta_0} \pm 2\mathrm{i}(\theta - \theta_0)$ near any zero, so that (9.64) defines a continuous function, as should be, of course), but the easiest way to define the solutions x_1, x_2 globally is to use a deformation of contour. One may always assume that ξ is analytic on some complex neighbourhood of $\mathbb{R}$ (it is conjugate by a time-reparametrization to some $u_{n,\alpha}$ which is entire, see Prop. 9.9). Define a contour Γ from 0 to 2π which avoids the zeros of ξ by going around them along half-circles of small radii centered on the real axis. This time (see discussion in Sect. 9.2.2), the half-circles must be chosen alternatively in the upper- and lower-half planes so that $\mathrm{Re}\,\xi(z) \geq 0$ on Γ. Then $\begin{pmatrix} x_1 \\ x_2 \end{pmatrix}(2\pi) = \begin{pmatrix} \cos T & -\sin T \\ \sin T & \cos T \end{pmatrix} \cdot \begin{pmatrix} x_1 \\ x_2 \end{pmatrix}(0)$ as before, with $T = \int_\Gamma \frac{d\theta'}{\xi(\theta')}$. Note that, in this case, T is purely imaginary.

Finally, in the unipotent case (type I with $\alpha = 0$, or type III), normalize ξ by setting for instance $\xi(0) = \mathrm{i}$, $\dot{\xi}(0) = 0$, so that ξ is purely imaginary. The same function- and time-change yields $\left(\frac{dz}{d\tau}\right)^2 = C$, hence a natural basis of solutions is given by $x_1(\theta) = \xi^{\frac{1}{2}}(\theta)$, $x_2(\theta) = \xi^{\frac{1}{2}}(\theta)\int_0^\theta \frac{d\theta'}{\xi(\theta')}$. To obtain globally defined solutions, one avoids the double zeros of ξ by drawing half-circles in the upper half-plane. Then the monodromy matrix is $\begin{pmatrix} 1 & T \\ 0 & 1 \end{pmatrix}$, with $T = \int_\Gamma \frac{d\theta'}{\xi(\theta')} = \int_\Gamma \frac{d\theta'}{x_1^2(\theta')}$.

9.3.3 Spectral Decomposition of the Model Operators

We shall need below the spectral decomposition of the three model operators $-\frac{1}{2}(\partial_x^2 - a^2x^2)$, $-\frac{1}{2}(\partial_x^2 + a^2x^2)$, $-\frac{1}{2}\partial_x^2$ introduced above. They are essentially self-adjoint on $C_0^\infty(\mathbb{R})$ by the classical Sears theorem (see [11], Theorem 1.1 Chap. 2 for instance), so the spectral theorem applies. The first operator has a pure point spectrum, while the second and the third have an absolutely continuous spectrum. Note that $-\frac{1}{2}\partial_x^2$ is non-negative, while the spectrum of $-\frac{1}{2}(\partial_x^2 + a^2x^2)$ is the whole real line, as the following Lemma proves.

Lemma 9.22. *1. (elliptic case)*
The spectral decomposition of $L^2(\mathbb{R})$ for the operator $-\frac{1}{2}(\partial_x^2 - a^2x^2)$ is given by

$$L^2(\mathbb{R}) = \oplus_{n\geq 0} L^2_{a(n+\frac{1}{2})} \tag{9.65}$$

where $L^2_{a(n+\frac{1}{2})}$ is one-dimensional, generated by the normalized Hermite functions $Ca^{1/4}e^{-ax^2/2}He_n(x\sqrt{a})$ for some constant C (see [1] for the notations and normalization).

2. (hyperbolic case)
Set, for $\lambda \in \mathbb{R}$,

$$\psi_\lambda^\pm(x) := \left(\frac{2}{a}\right)^{1/4} e^{\lambda/8a} e^{-iax^2/2} \cdot \left[\frac{1}{\Gamma(\frac{3}{4} + \frac{i\lambda}{4a})} {}_1F_1\left(\frac{1}{4}\left(1 + \frac{i\lambda}{a}\right), \frac{1}{2}; iax^2\right) \right.$$
$$\left. \pm \frac{2\sqrt{a}}{\Gamma(\frac{1}{4} + \frac{i\lambda}{4a})} e^{i\pi/4} x \cdot {}_1F_1\left(\frac{1}{4}\left(3 + \frac{i\lambda}{a}\right), \frac{3}{2}; iax^2\right)\right] \tag{9.66}$$

where ${}_1F_1$ is the usual confluent hypergeometric function. Then $H\psi_\lambda^\pm = \lambda\psi_\lambda^\pm$, and the $(\psi_\lambda^\pm, \lambda \in \mathbb{R})$ form a complete orthonormal system of generalized eigenfunctions of the operator $H = -\frac{1}{2}(\partial_x^2 + a^2x^2)$, so that any function $f \in L^2(\mathbb{R})$ decomposes uniquely as

$$f(x) = \int_{\mathbb{R}} \psi_\lambda^+(x)\overline{g^+(\lambda)}\, d\lambda + \int_{\mathbb{R}} \psi_\lambda^-(x)\overline{g^-(\lambda)}\, d\lambda \tag{9.67}$$

with $g^\pm(\lambda) = \int_{\mathbb{R}} f(x)\overline{\psi_\lambda^\pm(x)}\, dx$. In particular, the following Parseval identity holds,

$$\int_{\mathbb{R}} |f(x)|^2\, dx = \int_{\mathbb{R}} |g^+(\lambda)|^2\, d\lambda + \int_{\mathbb{R}} |g^-(\lambda)|^2\, d\lambda. \tag{9.68}$$

3. *(unipotent case)*
Set, for $\lambda > 0$, $\psi_\lambda^\pm(x) = e^{\pm \mathrm{i} x\sqrt{2\lambda}}$. *Then* $H\psi_\lambda^\pm = \lambda\psi_\lambda^\pm$ *and the* $\psi_\lambda^\pm$, $\lambda > 0$, *form a complete orthonormal system of generalized eigenfunctions of the operator* $H = -\frac{1}{2}\partial_x^2$, *with the usual Parseval-Bessel identity.*

Proof. (1) is classical and (3) is straightforward by Fourier inversion and the usual Parseval-Bessel identity. Case (2) is less common, though it can certainly be found in the literature. Let us explain briefly how to obtain its spectral decomposition for $a = 1$. The easiest way is to remark that $H = A\Lambda A^{-1}$ where $\Lambda = \frac{\mathrm{i}}{2}(x\partial_x + \partial_x x) = \mathrm{i}(x\partial_x + \frac{1}{2})$ and A is the image of the rotation matrix $\begin{pmatrix} \cos\pi/4 & -\sin\pi/4 \\ \sin\pi/4 & \cos\pi/4 \end{pmatrix}$ by the metaplectic representation. The operator A is unitary. Explicit formulas found for instance in [44] show that

$$(Af)(x) = \mathrm{i}\sqrt{2}e^{\mathrm{i}\pi/4}e^{\mathrm{i}\pi x^2}\int_0^\infty e^{-\mathrm{i}\pi(x\sqrt{2}-y)^2} f(y)\, dy. \tag{9.69}$$

As for the operator Λ, it is conjugate to $\mathrm{i}(\partial_y + \frac{1}{2})$ after the obvious change of variable $x = \pm e^y$, hence its spectral decomposition is given by Fourier inversion on either half-lines, $\Lambda\phi_\lambda^\pm = \lambda\phi_\lambda^\pm$ ($\lambda \in \mathbb{R}$) with $\phi_\lambda^\pm(x) = x_\pm^{-\frac{1}{2}-\mathrm{i}\lambda}$ constituting an orthonormal basis of generalized eigenfunctions. Finally, $\psi_\lambda^\pm := A\phi_\lambda^\pm$ may be obtained by applying the following formula (see [38])

$$\int_0^\infty x^{\nu-1}e^{-\beta x^2-\gamma x}\, dx = (2\beta)^{-\nu/2}\Gamma(\nu)e^{\gamma^2/8\beta}D_{-\nu}\left(\frac{\gamma}{\sqrt{2\beta}}\right)$$

($\mathrm{Re}\,\beta, \mathrm{Re}\,\nu > 0$) where D_ν is a parabolic cylinder function, also given by

$$D_\nu(z) = 2^{\nu/2}e^{-z^2/4}\left\{\frac{\sqrt{\pi}}{\Gamma(\frac{1}{2}(1-\nu))}\,{}_1F_1\left(-\frac{\nu}{2}, \frac{1}{2}; z^2/2\right)\right.$$
$$\left. -z\,\frac{\sqrt{2\pi}}{\Gamma(-\nu/2)}\,{}_1F_1\left(\frac{1-\nu}{2}, \frac{3}{2}; z^2/2\right)\right\} \tag{9.70}$$

(see [27], 8.2. (4) p. 117). □

9.3.4 Monodromy of Non-resonant Harmonic Oscillators (Elliptic Case)

We assume here that $D \in \mathscr{S}_{\leq 2}^{aff}$ is of class (i) with $\alpha > 0$. Then D is conjugate by a transformation in SV to an operator of the type $-2\mathrm{i}\partial_\theta - \partial_x^2 + a^2x^2 + \gamma$ where $a > 0$ and γ is a constant. Choose $\xi = \frac{1}{a}$ so that $\sqrt{\xi}$ satisfies a normalized Pinney-Milne equation. Then Proposition 9.18 shows the following:

Theorem 9.4. *The solution of the Schrödinger equation with arbitrary initial state*

$$\psi(0) := \sum_{n\geq 0} c_n h_n(0) \tag{9.71}$$

is given by

$$\psi(\theta) := \sum_{n\geq 0} c_n e^{-\mathrm{i}(n+\frac{1}{2})a\theta - \mathrm{i}\gamma\theta/2} h_n(\theta). \tag{9.72}$$

The monodromy operator is given by the 'infinite-dimensional' monodromy matrix $M_D := \mathrm{diag}(e^{\mathrm{i}\lambda_n}, n \in \mathbb{N})$, *with* $\lambda_n := -2\pi(n+\frac{1}{2})a - \pi\gamma$.

9.3.5 Monodromy of Harmonic Repulsors (Hyperbolic Type)

One assumes now that $D \in \mathcal{S}^{\mathrm{aff}}_{\leq 2}$ is either of class (i) with $\alpha < 0$ or of class (ii). Consider again the Ermakov-Lewis invariant

$$\mathcal{EL}(\xi) = \frac{1}{2\xi}\left[x^2 + \left(\mathrm{i}\xi\partial_x + \frac{1}{2}\dot{\xi}x\right)^2\right] \tag{9.73}$$

where one has assumed that $\xi = \mathrm{i}\eta$ is *purely imaginary* this time, and $I_{V_2}(\xi) = 2$. Note that $\mathcal{EL}(\mathrm{i}\eta)$ is *anti-hermitian*. Then

$$\frac{\mathcal{EL}(\mathrm{i}\eta) - \mathrm{i}k}{\mathrm{i}\eta} = -\frac{1}{2}\left[\partial_x^2 - \mathrm{i}\frac{\dot{\eta}}{\eta}x\partial_x + \frac{1-\frac{1}{4}\dot{\eta}^2}{\eta^2}x^2 - \frac{1}{2}\mathrm{i}\frac{\dot{\eta}}{\eta} + \frac{2k}{\eta}\right]. \tag{9.74}$$

Suppose $\psi_k \neq 0$ is an eigenvector of the Ermakov-Lewis operator, $\mathcal{EL}(\mathrm{i}\eta)\psi_k = \mathrm{i}k\psi_k$. Then Proposition 9.3 implies that $\tilde{\psi}_k := \exp -\frac{\mathrm{i}}{4}\frac{\dot{\eta}}{\eta}x^2 \,.\, \psi_k$ is a generalized eigenfunction of the model harmonic repulsor, namely

$$-\frac{1}{2}\left(\partial_x^2 + \frac{x^2}{\eta^2}\right)\tilde{\psi}_k = \frac{k}{\eta}\tilde{\psi}_k. \tag{9.75}$$

Hence:

Lemma 9.23. *1. The equation* $(\mathcal{EL}(\mathrm{i}\eta) - \mathrm{i}k)\psi_k = 0$ $(k \in \mathbb{R})$ *has two linearly independent solutions,*

$$\psi^k_{even}(\theta, x)$$
$$= \sqrt{2}(2\mathrm{i}\eta)^{1/4}e^{k/4}e^{\frac{\mathrm{i}}{4}\frac{\dot{\eta}}{\eta}x^2}e^{-\frac{\mathrm{i}}{2\eta}x^2} \cdot \frac{1}{\Gamma(\frac{1}{4}+\mathrm{i}\frac{k}{2})}\,{}_1F_1\left(\frac{1}{4}(1+2\mathrm{i}k), \frac{1}{2}; \frac{\mathrm{i}x^2}{\eta}\right)$$

and

$$\psi^k_{odd}(\theta, x)$$
$$= 2\sqrt{2}(2\mathrm{i}\eta)^{1/4} e^{k/4} e^{\frac{\mathrm{i}}{4}\frac{\dot\eta}{\eta}x^2} e^{-\frac{\mathrm{i}}{2\eta}x^2} \frac{1}{\Gamma(\frac{3}{4}+\mathrm{i}\frac{k}{2})} x \,.\, {}_1F_1\left(\frac{1}{4}(3+2\mathrm{i}k), \frac{3}{2}; \frac{\mathrm{i}x^2}{\eta}\right).$$

The functions $((\psi^k_{even}, \psi^k_{odd}), k \in \mathbb{R})$ *constitute a complete orthonormal system for the operator* $\mathscr{E}\mathscr{L}(\mathrm{i}\eta)$.

2. *One has*

$$D\psi^k_{even}(x) = \left(\frac{2k}{\eta} - \mathrm{i}\frac{\dot\eta}{\eta}\right)\psi^k_{even}(x)$$

and

$$D\psi^k_{odd}(x) = \left(\frac{2k}{\eta} - 2\mathrm{i}\frac{\dot\eta}{\eta}\right)\psi^k_{odd}(x).$$

Hence $x \to \frac{1}{\sqrt{\xi}}\exp\left(k\int^\theta \frac{d\theta'}{\xi(\theta')}\right)\psi^k_{even}(x)$ *and* $x \to \frac{1}{\xi}\exp\left(k\int^\theta \frac{d\theta'}{\xi(\theta')}\right)\psi^k_{odd}(x)$ *are solutions of the Schrödinger equation.*

Proof. 1. is a direct application of Lemma 9.22, while 2. follows from an easy computation using the confluent hypergeometric differential equation $z\frac{d^2}{dz^2}\,{}_1F_1(a,c;z) + (c-z)\frac{d}{dz}\,{}_1F_1(a,c,;z) - a\,{}_1F_1(a,c;z) = 0$. □

The eigenfunctions $\psi^k_{even}, \psi^k_{odd}$ depend analytically on ξ for $\xi \in \mathbb{C}\setminus\mathbb{R}_-$. If the operator D is of type I (so that ξ has no zero), say with $\gamma = 0$, then the phase $\exp\left(k\int^\theta \frac{d\theta'}{\xi(\theta')}\right)$ gives the monodromy. If D is of type II, then one must resort to a deformation of contour in order to avoid the singularities, as in the classical case, see Sect. 9.3.2. Mind that the deformation of contour may change drastically the behaviour of the functions $\psi^k_{even}, \psi^k_{odd}$ for large x or large k (for instance, ψ^k_{even} and ψ^k_{odd} become exponentially increasing for large x). Hence, in order to be able to follow the phase shift of the eigenfunctions $\psi^k_{even}, \psi^k_{odd}$ along the contour Γ without getting divergent integrals, it is better to assume to begin with that the 'Fourier transform' (with respect to the spectral decomposition of $\mathscr{E}\mathscr{L}(\xi)$) of the solution has compact support. In other words, the solution of the Schrödinger equation with initial state

$$\psi(0,x) := \int_{\mathbb{R}} \bar c_+(k)\psi^k_{even}(0,x)\,dk + \int_{\mathbb{R}} \bar c_-(k)\psi^k_{odd}(0,x)\,dk$$

for $z \in \Gamma$ (complex time), where c_+, c_- are assumed to be compactly supported, is given by

$$\psi(z,x) = \sqrt{\frac{\xi(0)}{\xi(z)}} \int_{\mathbb{R}} \bar{c}_+(k) e^{k\int_0^z \frac{dz'}{\xi(z')} - i\gamma\theta/2} \psi^k_{even}(z,x)\, dk$$
$$+ \frac{\xi(0)}{\xi(z)} \int_{\mathbb{R}} \bar{c}_-(k) e^{k\int_0^z \frac{dz'}{\xi(z')} - i\gamma\theta/2} \psi^k_{odd}(z,x)\, dk$$

An immediate corollary is:

Theorem 9.5. *Let $\psi(0) \in L^2(\mathbb{R})$, with decomposition*

$$\psi(0,x) := \int_{\mathbb{R}} \bar{c}_+(k) \psi^k_{even}(0,x)\, dk + \int_{\mathbb{R}} \bar{c}_-(k) \psi^k_{odd}(0,x)\, dk. \tag{9.76}$$

Then the solution of any type (ii) Schrödinger equation with initial state $\psi(0)$ is given at time $\theta = 2\pi$ by

$$\psi(2\pi,x) = \int_{\mathbb{R}} \bar{c}_+(k) e^{kT - i\pi\gamma} \psi^k_{even}(0,x)\, dk + \int_{\mathbb{R}} \bar{c}_-(k) e^{kT - i\pi\gamma} \psi^k_{odd}(0,x)\, dk \tag{9.77}$$

where $T = \int_0^{2\pi} \frac{du}{\xi(u)}$ or $\int_\Gamma \frac{du}{\xi(u)}$ (depending on the class of V_2), with Γ chosen as in Sect. 9.3.2, is purely imaginary. The associated monodromy operator in $\mathscr{B}(L^2(\mathbb{R}), L^2(\mathbb{R}))$ is unitarily equivalent to the multiplication by the function $k \to e^{kT - i\pi\gamma}$ with modulus one.

9.3.6 Monodromy of Non-resonant Operators of Unipotent Type

Suppose now $D \in \mathscr{S}^{aff}_{\le 2}$ is of class (i), $\alpha = 0$ or (iii). Then

$$\mathscr{E}\mathscr{L}(\xi)(\theta) := \frac{1}{2\xi}\left[\left(i\xi\partial_x + \frac{1}{2}\dot{\xi}x\right)^2\right] \tag{9.78}$$

($\xi \in \mathrm{Stab}_{V_2}$) is an invariant of D (note the difference with respect to Proposition 9.18). Case (i), $\alpha = 0$ is trivial, for it is conjugate to the free Schrödinger equation. So assume $D = -2i\partial_\theta - \partial_x^2 + V_2 x^2$ is of class (iii). Take $\xi = i\eta$ with $\eta \ge 0$ as in Sect. 9.3.2. Then (if $k > 0$)

$$\frac{\mathscr{E}\mathscr{L}(\xi) - ik}{\xi} = -\frac{1}{2}\left(\partial_x - \frac{i}{2}\frac{\dot{\eta}}{\eta}x\right)^2 - \frac{k}{\eta}. \tag{9.79}$$

So

$$\psi_{k,\pm}(x) := \exp\frac{i}{4}\frac{\dot{\eta}}{\eta}x^2 \,.\, \exp \pm i\sqrt{\frac{2k}{\eta}}x \tag{9.80}$$

constitute a complete orthonormal system for $\mathscr{E}\mathscr{L}(\mathrm{i}\eta)$ (the same statement holds true for potentials of class (i), in which case $\eta = 1$ and the exponential prefactor is trivial). A short computation shows that

$$D\psi_{k,\pm} = \left(\frac{2k}{\eta} - \frac{\mathrm{i}}{2}\frac{\dot\eta}{\eta}\right)\psi_{k,\pm}.$$

Hence one has the following:

Theorem 9.6. *Let $\psi(0) \in L^2(\mathbb{R})$, with decomposition*

$$\psi(0,x) := \int_{\mathbb{R}_+} \bar c_+(k)\psi_{k,+}(x)\, dk + \int_{\mathbb{R}_+} \bar c_-(k)\psi_{k,-}(x)\, dk. \tag{9.81}$$

Then the solution of any type (i), $\alpha = 0$ or type (iii) Schrödinger equation with initial state $\psi(0)$ is given at time $\theta = 2\pi$ by

$$\psi(2\pi,x) = \int_{\mathbb{R}_+} \bar c_+(k)e^{kT-\mathrm{i}\pi\gamma}\psi_{k,+}(x)\, dk + \int_{\mathbb{R}_+} \bar c_-(k)e^{kT-\mathrm{i}\pi\gamma}\psi_{k,-}(x)\, dk, \tag{9.82}$$

where $T = \int_0^{2\pi} \frac{du}{\xi(u)}$ or $\int_\Gamma \frac{du}{\xi(u)}$, Γ chosen as in Sect. 9.3.2 (depending on the class of V_2) is purely imaginary. The associated monodromy operator in $\mathscr{B}(L^2(\mathbb{R}), L^2(\mathbb{R}))$ is unitarily equivalent to the unitary operator on $L^2(\mathbb{R}_+)$ given by the multiplication by the function $k \to e^{kT-\mathrm{i}\pi\gamma}$.

9.4 Symplectic Structures and General Solution of the Schrödinger Equation

The general emphasis in this section is, so to speak, on the non-quadratic part of the potential, namely, on V_0 and V_1 if $D = -2\mathrm{i}\partial_\theta - \partial_x^2 + V_2(\theta)x^2 + V_1(\theta)x + V_0(\theta)$. It contains somewhat loosely related results: a definition of a three-dimensional invariant $(\xi, \delta_1, \delta_2)$; a generalization of the Ermakov-Lewis invariants to general potentials; a symplectic structure on a space 'containing' $\mathscr{S}^{aff}_{\le 2}$ such that the SV-action becomes naturally Hamiltonian; finally, the computation of the monodromy for the 'resonant' operators of type (i)bis, (iii)bis.

Definition 9.24. We shall say that $D \in \mathscr{S}^{aff}_{\le 2}$ is of *generic type* if: D is of class (i), D conjugate to $D_{\alpha,\gamma} = -2\mathrm{i}\partial_\theta - \partial_x^2 + \alpha x^2 + \gamma$ with $\alpha \ne n^2/4$, $n = 0, 1, \ldots$; or D is of class (ii), D conjugate to $D_{n,\alpha,\gamma} = -2\mathrm{i}\partial_\theta - \partial_x^2 + u_{n,\alpha}(\theta)x^2 + \gamma$.

Denote by $\mathscr{S}^{aff}_{\le 2,gen}$ the set of operators of generic type; it is a disjoint union of SV-orbits.

Note (see Corollary 9.13) that the isotropy group of an operator D of generic type is generated by $\mathcal{M}_1$ and some $\mathcal{L}_\xi + \mathcal{Y}_{f_1} + \mathcal{M}_{f_2}$ with $\xi \neq 0$.

Definition 9.25 (vector invariant). Let $D = -2i\partial_\theta - \partial_x^2 + V_2(\theta)x^2 + V_1(\theta)x + V_0(\theta) \in \mathcal{S}^{aff}_{\le 2,gen}$ be of generic type.

Define:

(i) $\xi(D)$ to be the unique (up to a sign) periodic vector field such that $\xi(D) \in \mathrm{Stab}_{V_2}$ and $I_{V_2}(\xi(D)) = 2$ (ξ real in the elliptic case, purely imaginary in the hyperbolic case);

(ii) $\delta_1(D)$ to be the unique periodic function such that

$$\ddot{\delta}_1(D) + V_2\delta_1(D) = -\frac{1}{2}\left(\dot{V}_1\xi(D) + \frac{3}{2}V_1\dot{\xi}(D)\right); \tag{9.83}$$

(iii) $\delta_2(D)$ to be the unique periodic function (up to a constant) such that

$$\delta_2(D) = -\frac{1}{2}\int^\theta V_1(\theta')\delta_1(D)(\theta')d\theta' - \frac{1}{2}V_0\xi(D) \tag{9.84}$$

Observe that $\mathcal{L}_\xi + \mathcal{Y}_{\delta_1} + \mathcal{M}_{\delta_2} \in Lie(G_D)$ is indeed unique (up to the addition of a constant times $\mathcal{M}_1$) as follows from Corollary 9.13. The ambiguity in the definition of δ_2 may be solved by choosing for each SV-orbit an arbitrary base-point, an invariant $(\xi, \delta_1, \delta_2)$ for this base-point, and transforming $(\xi, \delta_1, \delta_2)$ covariantly by the adjoint action along the orbit. Some non-local formulas fixing δ_2 more explicitly can probably be found, at least for potentials of type (i) (see Lemma 9.14), but we shall not need them.

Another problem comes from the fact that the map $(V_2, V_1, V_0) \to (\xi, \delta_1, \delta_2)$ is not one-to-one (nor onto). Suppose one has some triple of functions $(\xi, \delta_1, \delta_2)$. Under some conditions that we shall not write explicitly (depending on the class of the potential), $(\xi, \delta_1, \delta_2)$ is an invariant for some potential (V_2, V_1, V_0); the quadratic part V_2 is given (by definition) by $V_2 = \frac{1}{2\xi^2}(2 - \xi\ddot{\xi} + \frac{1}{2}\dot{\xi}^2)$. (Supposing ξ has only a finite number of zeros, all of which are simple or double, one has some rather straightforward conditions on the values of $\dot{\xi}$ and $\dddot{\xi}$ at the zeros of ξ that ensure that $\xi \in \mathrm{Stab}_{V_2}$ for some potential V_2). But V_1 is not determined uniquely if ξ does not vanish on the torus, since $\xi^{-3/2}$ is in the kernel of the operator $\xi\partial + \frac{3}{2}\dot{\xi}$ (see formula (9.83)). This can easily be explained by supposing (by conjugating by some element $g \in SV$) that D is the model operator $D = -2i\partial_\theta - \partial_x^2 + \alpha x^2 + \gamma$ (α generic). Then ξ is proportional to the constant vector field $\mathcal{L}_1$ which commutes with $\mathcal{Y}_1$, hence the invariant $(\xi, \delta_1, \delta_2)$ is left unchanged by space-translations, whereas the operator D (and also the generalized Ermakov-Lewis invariant defined in Theorem 9.7 below) is not. Hence the vector invariant $(\xi, \delta_1, \delta_2)$ parametrizes Schrödinger operators of type (i) 'up to space-translations'. On the other hand, the map $(V_2, V_1, V_0) \to (\xi, \delta_1, \delta_2)$ is one-to-one for operators of type (ii) (up to a sign for ξ).

It is not *a priori* self-evident that δ_2 defined by equation (9.84) is a periodic function. Considering the 'inverse problem', i.e. supposing that the invariant $(\xi, \delta_1, \delta_2)$ is given, and supposing ξ does not vanish on the torus, one must also check that every choice for V_1 gives a function δ_2 which is periodic. This is the content of the following lemma:

Lemma 9.26. *One has:*

$$\frac{d}{d\theta}\left(\xi\frac{d}{d\theta}(\xi^{-\frac{1}{2}}\delta_1)\right) = -\frac{1}{2}\frac{d}{d\theta}(\xi^{3/2}V_1) - \xi^{-3/2}\delta_1. \tag{9.85}$$

This formula implies: $\int_0^{2\pi}\xi^{-3/2}\delta_1 = 0$; $\int_0^{2\pi} V_1\delta_1 = 0$.

Proof. Using the invariant equations $\xi\ddot{\xi} - \frac{1}{2}\dot{\xi}^2 + 2V_2\xi^2 = 2$ and $\ddot{\delta}_1 + V_2\delta_1 = -\frac{1}{2}(\dot{V}_1\xi + \frac{3}{2}V_1\dot{\xi})$, one obtains

$$\frac{d}{d\theta}\left(\xi^{\frac{1}{2}}\dot{\delta}_1\right) = \frac{1}{2}\frac{d}{d\theta}\left(\xi^{-\frac{1}{2}}\dot{\xi}\delta_1\right) - \frac{1}{2}\frac{d}{d\theta}\left(\xi^{3/2}V_1\right) - \xi^{-3/2}\delta_1,$$

hence the first equation, which implies immediately: $\int_0^{2\pi}\xi^{-3/2}(\theta)\delta_1(\theta)\,d\theta = 0$. Hence (considering the inverse problem), *if* some potential V_1 verifies $\int_0^{2\pi} V_1(\theta)\delta_1(\theta)\,d\theta = 0$ (so that δ_2 is well-defined), then this is also true for all possible potentials V_1. Now, integrating the first equation, one gets

$$\xi\frac{d}{d\theta}\left(\xi^{-\frac{1}{2}}\delta_1\right) + \frac{1}{2}\xi^{3/2}V_1 = -\int^{\theta}\xi^{-3/2}(\theta')\delta_1(\theta')d\theta',$$

hence

$$\xi^{\frac{1}{2}}V_1 = -2\left[\frac{d}{d\theta}\left(\xi^{-\frac{1}{2}}\delta_1\right) + \frac{1}{\xi}\int^{\theta}\xi^{-3/2}(\theta')\delta_1(\theta')\,d\theta'\right].$$

Hence

$$\begin{aligned}\int^{\theta}V_1(\theta')\delta_1(\theta')\,d\theta' &= \int^{\theta}(\xi^{\frac{1}{2}}V_1)(\theta')(\xi^{-\frac{1}{2}}\delta_1)(\theta')\,d\theta'\\ &= -\left[\left(\xi^{-\frac{1}{2}}(\theta)\delta_1(\theta)\right)^2 + \left(\int^{\theta}\xi^{-3/2}(\theta')\delta_1(\theta')\,d\theta'\right)^2\right]\end{aligned} \tag{9.86}$$

and the integral over a period is zero. □

The following covariance result is an extension of Theorem 9.3.

Theorem 9.7. *Let $D \in \mathcal{S}^{aff}_{\le 2,gen}$ be of generic type, with associated invariant ($\xi = \xi(D), \delta_1 = \delta_1(D), \delta_2 = \delta_2(D)$). Then:*

1.

$$\mathcal{EL}(D) := \frac{1}{2}\left[\frac{1}{\xi}\left(1+\frac{1}{4}\dot{\xi}^2\right)x^2 - \xi\partial_x^2 + \frac{i}{2}\dot{\xi}\,(x\partial_x + \partial_x x)\right.$$
$$\left. + \left(-2\delta_1(-i\partial_x) + (V_1\xi + 2\dot{\delta}_1)x\right) + 2\left(\delta_2 + \frac{1}{2}V_0\xi\right)\right] \quad (9.87)$$

is an invariant for the Schrödinger operator D.

2. *Let $(\phi; (a,b)) \in \mathrm{SV}$ and $g : (\theta, x) \to (\theta', x') = (\phi(\theta), x\sqrt{\dot{\phi}(\theta)} - a(\theta))$ be the associated coordinate change. Then*

$$\pi_{1/4}(\phi; (a,b))\mathcal{EL}(D)\pi_{1/4}(\phi; (a,b))^{-1} = \widetilde{\mathcal{EL}}(D) \quad (9.88)$$

where $\widetilde{\mathcal{EL}}(D)$ is obtained by applying the transformation g to the coordinates, changing the potentials V_0 and V_1 by the $\sigma_{1/4}$-action of SV, and transforming the invariant as follows:

$$\tilde{\xi} = \phi' \,.\, \xi \circ \phi^{-1}; \quad (9.89)$$

$$\tilde{\delta_1} = \phi'^{\frac{1}{2}} \,.\, \delta_1 \circ \phi^{-1} + \left(\tilde{\xi}\dot{a} - \frac{1}{2}a\dot{\tilde{\xi}}\right); \quad (9.90)$$

$$\tilde{\delta_2} = \delta_2 \circ \phi^{-1} + \left(\delta_1\dot{a} - a\dot{\delta_1}\right) + \tilde{\xi}\dot{b} + \left(\tilde{\xi}(\dot{a}^2 - a\ddot{a}) - \dot{\tilde{\xi}}a\dot{a} - \ddot{\tilde{\xi}}a^2\right). \quad (9.91)$$

Furthermore, $(\tilde{\xi}, \tilde{\delta_1}, \tilde{\delta_2})$ is (up to the addition of an arbitrary constant to δ_2) the invariant associated to $\sigma_{1/4}(D)$.

Proof. 1. Look for an invariant of the form

$$\frac{1}{2}\left[a(\theta)x^2 - b(\theta)\partial_x^2 - ic(\theta)(x\partial_x + \partial_x x) + d(\theta)(-i\partial_x) + e(\theta)x + f(\theta)\right] \quad (9.92)$$

and solve in a, b, c, d, e, f. One obtains the following constraints:

$$\dot{a} = 2V_2 c, \quad \dot{b} = -2c, \quad \dot{c} = -a + V_2 b \quad (9.93)$$

– whose general solution is in Proposition 9.18 above, namely, $a = \frac{1}{\xi}(1 + \frac{1}{4}\dot{\xi}^2)$, $b = \xi$, $c = -\frac{1}{2}\dot{\xi}$ – and the set of following equations:

$$\dot{d} = V_1 b - e, \quad \dot{e} = V_1 c + V_2 d, \quad \dot{f} = \frac{1}{2}dV_1 \quad (9.94)$$

which implies the compatibility condition

$$\ddot{d} + V_2 d = \dot{V}_1 \xi + \frac{3}{2} V_1 \dot{\xi}.$$

2. Since (assuming $I_u(\xi) = 2$ is fixed) there is a unique (up to the addition of a constant times $\mathcal{M}_1$) invariant for operators of generic type, one necessarily has

$$\mathcal{L}_{\tilde{\xi}} + \mathcal{Y}_{\tilde{\delta}_1} + \mathcal{M}_{\tilde{\delta}_2} = Ad(\phi; (a,b)).(\mathcal{L}_\xi + \mathcal{Y}_{\delta_1} + \mathcal{M}_{\delta_2}) \bmod \mathcal{M}_1 \tag{9.95}$$

which gives the above formulas for $(\tilde{\xi}, \tilde{\delta}_1, \tilde{\delta}_2)$.

It remains to check (9.88). Consider first the covariance under a time-reparametrization ϕ. It has already been proved for the quadratic part of the Ermakov-Lewis operator, see Theorem 3.1.6. The linear part $-2(-\mathrm{i}\delta_1\partial_x + (V_1\xi - \dot{\delta}_1)x)$ transforms covariantly under ϕ since (see proof of Theorem 9.3)

$$\tilde{V}_1 \tilde{\xi} - \frac{d\tilde{\delta}_1}{d\theta'} = \phi'^{-\frac{1}{2}} \left(V_1.\xi - \dot{\delta}_1 - \frac{1}{2}\frac{\ddot{\phi}}{\dot{\phi}} \delta_1 \right), \quad -\mathrm{i}\tilde{\delta}_1 \partial_{x'} = -\mathrm{i}\delta_1 \partial_x \tag{9.96}$$

and

$$\begin{aligned} &\left(\pi_{1/4}(\phi)(-\mathrm{i}\delta_1(\theta)\partial_x + (V_1\xi - \dot{\delta}_1)x)\pi_{1/4}(\phi^{-1})\right)\psi \\ &= \frac{1}{2}\dot{\phi}^{-1/4} e^{\frac{\mathrm{i}}{4}\frac{\ddot{\phi}}{\dot{\phi}}x^2} \left(-\mathrm{i}\delta_1\partial_x + (V_1\xi - \dot{\delta}_1)x\right) \dot{\phi}^{1/4} e^{-\frac{\mathrm{i}}{4}\frac{\ddot{\phi}}{\dot{\phi}}x^2} \psi\left(\phi(\theta), x\sqrt{\dot{\phi}(\theta)}\right) \\ &= \left(-\mathrm{i}\tilde{\delta}_1\partial_{x'} + (\tilde{V}_1\tilde{\xi} - \dot{\tilde{\delta}}_1)x'\right)\psi\left(\theta', x'\right). \end{aligned} \tag{9.97}$$

As for the zero-order term $-\frac{1}{2}(\delta_2 + \frac{1}{2}V_0\xi)$, it is obviously invariant under the conjugate action of $\pi_{1/4}(\phi)$. Since $\tilde{V}_0\tilde{\xi} = (V_0\xi) \circ \phi^{-1}$, this implies also $\tilde{\delta}_2 = \delta_2 \circ \phi^{-1}$.

Consider now the covariance under an infinitesimal nilpotent transformation $\mathcal{Y}_{f_1} + \mathcal{M}_{f_2}$. One has

$$\begin{aligned} [a\partial_x + \mathrm{i}\dot{a}x + b, \mathcal{E}\mathcal{L}(\xi, \delta_1, \delta_2)] &= \frac{1}{2}\left[a\partial_x + \mathrm{i}\dot{a}x, \frac{1}{\xi}\left(1 + \frac{1}{4}\dot{\xi}^2\right)x^2 - \xi\partial_x^2\right. \\ &\quad \left. + \mathrm{i}\dot{\xi}x\partial_x - 2\left(\delta_1(-\mathrm{i}\partial_x) + \left(V_1\xi - \dot{\delta}_1\right)x\right) d^X\right] \\ &= \frac{1}{2}\left\{\left(\frac{2a}{\xi}\left(1 + \frac{1}{4}\dot{\xi}^2\right) + \dot{a}\dot{\xi}\right)x - \left(a\dot{\xi} + 2\dot{a}\xi\right)\right. \\ &\quad \left.(-\mathrm{i}\partial_x) - 2a\left(V_1\xi - \dot{\delta}_1\right) + 2\dot{a}\delta_1\right\}, \end{aligned} \tag{9.98}$$

to be compared with the infinitesimal change of $\mathcal{EL}$ under the transformation $x \to x + \varepsilon a$, $\delta_1 \to \delta_1 + \varepsilon(\xi\dot{a} - \frac{1}{2}a\dot{\xi})$, $\delta_2 \to \delta_2 + \varepsilon((\delta_1\dot{a} - a\dot{\delta}_1) + \xi\dot{b})$, $V_1 \to V_1 - 2\varepsilon(\ddot{a} + V_2 a)$. This is a straightforward computation, which requires the use of the equation defining ξ, namely, $\ddot{\xi} = \frac{1}{\xi}(1 + \frac{1}{4}\dot{\xi}^2) - 2V_2\xi$. □

Using the parametrization of $\mathcal{S}^{aff}_{\leq 2,gen}$ by the vector invariant $(\xi, \delta_1, \delta_2)$, one can easily define a natural symplectic structure on a linear space Ω and a hamiltonian action of SV on Ω reproducing the SV-action on $\mathcal{S}^{aff}_{\leq 2,gen}$.

Definition 9.27. Let $\Omega \simeq C^\infty(\mathbb{R}/2\pi\mathbb{Z}, \mathbb{R}^4)$ be the linear manifold consisting of all 2π-periodic vector-valued C^∞ functions $\boldsymbol{X}(\tau) := (p, q, E, t)(\tau)$, $\tau \in \mathbb{R}/2\pi\mathbb{Z}$ with singular Poisson structure defined by

$$\{p(\tau), q(\tau')\} = \delta(\tau - \tau'), \quad \{E(\tau), t(\tau')\} = \delta(\tau - \tau') \tag{9.99}$$

See for instance [43], Chap. X for some remarks on distribution-valued singular Poisson structures on infinite-dimensional spaces. The energy E is canonically conjugate to t, which allows us to consider generalized canonical transformations for which t is a coordinate. This usual trick for Hamiltonian systems with time-dependent Hamiltonians can for instance be found in [37]. Hamiltonian vector fields $\mathcal{X}_H$, for $H = H(p, q, E, t)$, act separately on each fiber τ =constant, namely,

$$(\mathcal{X}_H f)(\tau) := \{(\partial_p H \partial_q - \partial_q H \partial_p + \partial_E H \partial_p - \partial_t H \partial_E) f\}(\tau). \tag{9.100}$$

Definition 9.28 (associated functional). Let $(\xi, \delta_1, \delta_2)$ be a triple of 2π-periodic functions. Define $\Phi := \Phi(\xi, \delta_1, \delta_2)$ to be the following functional on Ω,

$$\langle \Phi, \boldsymbol{X} \rangle = \oint \Big\{ \xi(t(\tau))E(\tau) + \frac{1}{2}\dot{\xi}(t(\tau))p(\tau)q(\tau) + \delta_1(t(\tau))p(\tau) - \dot{\delta}_1(t(\tau))q(\tau) + \delta_2(t(\tau)) \Big\}\, d\tau. \tag{9.101}$$

Theorem 9.8. *Represent $\mathcal{L}_f + \mathcal{Y}_g + \mathcal{M}_h \in \mathfrak{sv}$ by the hamiltonian vector field $\mathcal{X}_{H(f,g,h)}$ associated to*

$$H(f, g, h) := -\left(f(t)E + \frac{1}{2}\dot{f}(t)pq + \frac{1}{4}\ddot{f}q^2 \right) - (g(t)p + \dot{g}(t)q) - h(t). \tag{9.102}$$

Then the action of $\mathcal{X}_H$ on the functional $\Phi(\xi, \delta_1, \delta_2)$ coincides with that given in Theorem 9.7.

Proof. Observe that the map from $\mathfrak{sv}$ to the Lie algebra of vector fields on Ω given by $\mathcal{L}_f + \mathcal{Y}_g + \mathcal{M}_h \to \mathcal{X}_{H(f,g,h)}$ is a Lie algebra homomorphism. The vector field $\mathcal{X}_H$ is given explicitly by

$$\mathscr{X}_{H(f,g,h)} = -\left[\frac{1}{2}\dot{f}(t)(q\partial_q - p\partial_p) + f(t)\partial_t - \frac{1}{2}\ddot{f}(t)q\partial_p\right] - [g(t)\partial_q - \dot{g}(t)\partial_p]$$
$$+\left[\left(\frac{1}{2}\ddot{f}(t)pq + \dot{f}(t)E + \frac{1}{4}\dddot{f}(t)q^2\right) + (\dot{g}p + \ddot{g}q) + \dot{h}(t)\right]\partial_E. \tag{9.103}$$

The rest is a straightforward computation. □

Let us conclude this section by computing the monodromy for 'resonant' operators of type (i)bis and (iii)bis.

Consider any resonant operator D. The associated classical monodromy is unipotent. We choose $\xi \in \mathrm{Stab}_{V_2}$ to be purely imaginary, $\xi := \mathrm{i}\eta$ as before (see Sect. 9.3.2). A generalized Ermarkov-Lewis invariant may then be defined as

$$\mathscr{E}\mathscr{L}(D) = \frac{1}{2\xi}\left[\left(\mathrm{i}\xi\partial_x + \frac{1}{2}\dot{\xi}x\right)^2\right] + \frac{\mathrm{i}}{2}\left[d(-\mathrm{i}\partial_x) + ex + f\right], \tag{9.104}$$

where d, e, f are defined as in Theorem 9.7 but with ξ replaced by η (see (9.92) for notations). Hence

$$\frac{\mathscr{E}\mathscr{L}(D) - \mathrm{i}k}{\xi} = -\frac{1}{2}\left[\left(\partial_x - \frac{\mathrm{i}}{2}\frac{\dot{\eta}}{\eta}x\right)^2 - \frac{d}{\eta}(-\mathrm{i}\partial_x) - \frac{e}{\eta}x - \frac{f}{\eta}\right] - \frac{k}{\eta}. \tag{9.105}$$

Suppose $\mathscr{E}\mathscr{L}(D)\psi_k = \mathrm{i}k\psi_k$ and set

$$\tilde{\psi}_k = \exp\left(-\frac{\mathrm{i}}{4}\frac{\dot{\eta}}{\eta}x^2 + \frac{\mathrm{i}}{2}\frac{d}{\eta}x\right)\psi_k. \tag{9.106}$$

Then a simple calculation gives

$$\left[\partial_x^2 - \left(\frac{1}{2}d\frac{\dot{\eta}}{\eta^2} + \frac{e}{\eta}\right)x + \frac{-f + 2k}{\eta} + \frac{1}{4}\left(\frac{d}{\eta}\right)^2\right]\tilde{\psi}_k = 0 \tag{9.107}$$

If D is of type (i)bis, then d, e, f (easy to obtain from Theorem 9.7 and the isotropy algebra given in Sect. 9.2) satisfy $\frac{1}{2}d\frac{\dot{\eta}}{\eta^2} + \frac{e}{\eta} = 0$ identically, so the model operator is (up to a constant) the Laplacian as for case (iii). Then the monodromy can be computed along the same lines as in Sect. 9.3.6, with a time-independent shift in k due to the function $-f + \frac{1}{4}\frac{d^2}{\eta} = \frac{3C^2}{128n^2}$.

Lemma 9.29. *Let* $D = -2\mathrm{i}\partial_\theta - \partial_x^2 + n^2x^2 + C\cos n(\theta - \sigma/2)x + \gamma$ *be a Schrödinger operator of type (i)bis. Set*

$$\psi_{k,\pm}(\theta,x) = e^{\frac{\mathrm{i}}{4}\frac{\dot\eta}{\eta}x^2 - \frac{\mathrm{i}}{2}\frac{d}{\eta}x} \, . \, e^{\pm \mathrm{i}\sqrt{\frac{2k'}{\eta}}x}, \tag{9.108}$$

with $d = -\frac{C}{4n}\sin 3n(\theta - \sigma/2)$, $\eta = 1 - \cos(2n\theta - \sigma)$, $k' = k + 3\left(\frac{C}{16n}\right)^2$. *Then*

$$D\psi_{k,\pm} = \left(\frac{2k'}{\eta} + \frac{1}{4}\left(\frac{d}{\eta}\right)^2 \mp d\eta^{-3/2}\sqrt{2k'} - \frac{\mathrm{i}}{2}\frac{\dot\eta}{\eta} + \gamma\right)\psi_{k,\pm}. \tag{9.109}$$

Proof. Tedious computations. □

Apart from the time-periodic shift $\frac{1}{4}\left(\frac{d}{\eta}\right)^2 = \left(\frac{C}{16n}\right)^2 \frac{\sin^2 3n(\theta-\sigma/2)}{\sin^4 n(\theta-\sigma/2)}$ (which is integrable on the contour Γ) and the time-independent shift in k, one is left once again with a phase proportional to k/η (note that the term in $d\eta^{-3/2}\sqrt{2k'}$ is irrelevant since $\int_0^{2\pi}(d\eta^{-3/2})(\theta)\,d\theta = 0$ by Lemma 4.3; recall $d = -2\delta_1$ by Theorem 9.7).

Hence one obtains:

Theorem 9.9. *Let* $\psi(0) \in L^2(\mathbb{R})$, *with decomposition*

$$\psi(0,x) := \int_{\mathbb{R}_+} \bar c_+(k)\psi_{k,+}(0,x)\,dk + \int_{\mathbb{R}_+} \bar c_-(k)\psi_{k,-}(0,x)\,dk. \tag{9.110}$$

Then the solution of the type (i)bis Schrödinger equation

$$\left(-2\mathrm{i}\partial_\theta + \partial_x^2 + n^2x^2 + C\cos(n\theta - \sigma/2).x + \gamma\right)\psi = 0 \tag{9.111}$$

with initial state $\psi(0)$ *is given at time* $\theta = 2\pi$ *by*

$$\psi(2\pi,x) = \int_{\mathbb{R}_+} \bar c_+(k)e^{k'T - \mathrm{i}\pi\tilde\gamma}\psi_{k,+}(0,x)\,dk + \int_{\mathbb{R}_+} \bar c_-(k)e^{k'T-\mathrm{i}\pi\tilde\gamma}\psi_{k,-}(0,x)\,dk, \tag{9.112}$$

where $k' = k + 3\left(\frac{C}{16n}\right)^2$, $T = \int_0^{2\pi}\frac{du}{\xi(u)}$ *(T is purely imaginary) and*

$$\tilde\gamma = \gamma + \frac{1}{4}\int_\Gamma \left(\frac{d}{\eta}\right)^2(\theta)d\theta.$$

The associated monodromy operator in $\mathcal{B}(L^2(\mathbb{R}), L^2(\mathbb{R}))$ *is unitarily equivalent to the unitary operator on* $L^2(\mathbb{R}_+)$ *given by the multiplication by the function* $k \to e^{kT - \mathrm{i}\pi\tilde\gamma}$.

Suppose now D is of type (iii)bis. Then the x-coefficient in the transformed Ermakov-Lewis operator (9.107) does not vanish, so one must take for 'model operator' $-\partial_x^2 + x$, whose eigenfunctions are related to the Airy function. The solution of the monodromy problem will be given by a series of lemmas. In the sequel, $\eta = (1 + \sin n\theta)(1 + \alpha\sin n\theta)$ is the (real-valued and non-negative) invariant, and $\eta^{1/2} = (1 + \alpha \sin n\theta)^{1/2} \cos(\frac{\pi}{4} - n\frac{\theta}{2})x$ is the smooth square-root of η chosen in Sect. 9.2.

Lemma 9.30. *Let Ai be the entire function, solution of the Airy differential equation $(-\partial_x^2 + x)Ai(x) = 0$, defined on the real line as*

$$Ai(x) = \frac{1}{\pi} \int_0^\infty \cos\left(\frac{t^3}{3} + xt\right) dt. \tag{9.113}$$

It is (up to a constant) the only solution of the Airy differential equation which do not increase exponentially on $\mathbb{R}_+$. The functions $f_k(x) := Ai(x - k)$, $k \in \mathbb{R}$ define (up to a coefficient) a complete orthonormal system of generalized eigenfunctions of the self-adjoint closure of the Airy operator $-\partial_x^2 + x$ with core $C_0^\infty(\mathbb{R}) \subset L^2(\mathbb{R})$.

Proof. Easy by using a Fourier transform. □

Lemma 9.31. *The x-coefficient in the transformed Ermakov-Lewis invariant (9.107) reads*

$$\frac{1}{2} d \frac{\dot{\eta}}{\eta^2} + \frac{e}{\eta} = -C_\alpha \eta^{-3/2} \tag{9.114}$$

where $C_\alpha = (1-\alpha)(1+\alpha/2)\sqrt{1-\alpha^2}$.

Proof. Computations similar to that of Lemma 9.26 (with the simple difference that $\xi\ddot{\xi} - \frac{1}{2}\dot{\xi}^2 + 2V_2\xi^2 = 0$ here) yield

$$\frac{d}{d\theta}(\eta^{-\frac{1}{2}} d) = \eta - \frac{C_\alpha}{\eta} \tag{9.115}$$

where C_α is some constant which must be chosen in order that the right-hand side be 2π-periodic. Note that the singularities in the above equation are only apparent; one may avoid them altogether by using a contour Γ in the upper-half plane as in Sect. 9.3.2. Since $\int_0^{2\pi} \eta = 2\pi(1 + \alpha/2)$ and $\int_\Gamma \frac{d\theta'}{\eta(\theta')} = -\frac{2\pi}{(1-\alpha)\sqrt{1-\alpha^2}}$ (see Proposition 9.9), this gives $C_\alpha = (1-\alpha)(1+\alpha/2)\sqrt{1-\alpha^2}$. Then a straightforward computation yields formula (9.114). □

Lemma 9.32. *Set*

$$\psi_k(\theta, x) = \exp\left(\frac{\mathrm{i}}{4}\frac{\dot{\eta}}{\eta}x^2 - \frac{\mathrm{i}}{2}\frac{d}{\eta}x\right) . \eta^{-\frac{1}{4}} Ai\left(x C_\alpha^{1/3} \eta^{-\frac{1}{2}} - C_\alpha^{-2/3}\left(-f + 2k + \frac{1}{4}\frac{d^2}{\eta}\right)\right) \tag{9.116}$$

Then

$$D\psi_k(\theta,x) = \left(\frac{2k}{\eta} + \left(\frac{\mathrm{i}}{2}\frac{\dot\eta}{\eta} + \frac{1}{2}\left(\frac{d}{\eta}\right)^2 - \frac{f}{\eta}\right)\right)\psi_k(\theta,x). \tag{9.117}$$

Proof. The ψ_k are obtained as in Sect. 9.3.5 (about the monodromy of hyperbolic operators) by taking a complete orthonormal system of generalized eigenfunctions $\tilde\psi_k$ for the transformed Ermakov-Lewis invariant (9.107) and going back to the functions ψ_k. Then (9.117) is proved by a direct tedious computation. □

One may now conclude:

Lemma 9.33. *Let $\psi(0) \in L^2(\mathbb{R})$, with decomposition*

$$\psi(0,x) := \int_{\mathbb{R}} \bar c(k)\psi_k(0,x)\, dk. \tag{9.118}$$

Then the solution of the type (iii)bis Schrödinger equation

$$\left(-2\mathrm{i}\partial_\theta + \partial_x^2 + v_{n,\alpha}x^2 + C(1+\alpha\sin n\theta)^{\frac{1}{2}}\cos\left(\frac{\pi}{4} - n\frac{\theta}{2}\right)x + \gamma\right)\psi = 0 \tag{9.119}$$

with initial state $\psi(0)$ is given at time $\theta = 2\pi$ by

$$\psi(2\pi,x) = \int_{\mathbb{R}} \bar c(k) e^{kT - \mathrm{i}\pi\tilde\gamma}\psi_k(0,x)\, dk, \tag{9.120}$$

where $T = \int_0^{2\pi} \frac{du}{\xi(u)}$ (T is purely imaginary) and

$$\tilde\gamma = \gamma + \int_\Gamma \left(-\frac{f}{\eta} + \frac{1}{2}\left(\frac{d}{\eta}\right)^2\right)(\theta)\, d\theta.$$

The associated monodromy operator in $\mathcal{B}(L^2(\mathbb{R}), L^2(\mathbb{R}))$ is unitarily equivalent to the unitary operator on $L^2(\mathbb{R})$ given by the multiplication by the function $k \to e^{kT - \mathrm{i}\pi\tilde\gamma}$.

Chapter 10
Poisson Structures and Schrödinger Operators

This chapter – which uses once again the definitions and concepts introduced in Chap. 8, but is unrelated with Chap. 9 – is adapted from [107].

10.1 Introduction

Recall from Proposition 8.2 that there exists a family of actions of the Schrödinger–Virasoro group SV on the affine space of periodic time-dependent Schrödinger operators $\mathcal{S}^{aff} := \{-2i\mathcal{M}\partial_t - \partial_r^2 + V(t,r)\}$, denoted by σ_λ, $\lambda \in \mathbb{R}$. In Chaps. 8 and 9 we mainly considered the restriction of these actions to $\mathcal{S}^{aff}_{\leq 2}$. Here we shall work with the whole *linear* space $\mathcal{S}^{lin} := \{a(t)(-2i\mathcal{M}\partial_t - \partial_r^2) + V(t,r)\}$ and consider the following action of the Schrödinger–Virasoro group on it:

Definition 10.1. Let $\tilde{\sigma}_\lambda$, $\lambda \in \mathbb{R}$ be the family of left-and-right actions of the Schrödinger–Virasoro group on the linear space of Schrödinger operators $\mathcal{S}^{lin}$ defined by

$$\tilde{\sigma}_\lambda(g)(D) = \pi_{\lambda+2}(g) D \pi_\lambda(g)^{-1}. \tag{10.1}$$

Because of the shift $\lambda \rightsquigarrow \lambda+2$ instead of $\lambda \rightsquigarrow \lambda+1$ (compare with the σ_λ-action from Chap. 8), the *affine* space $\mathcal{S}^{aff}$ is not preserved by this action.

Let us give explicit formulas for the action $\tilde{\sigma}_\lambda$ (compare with the formulas of Proposition 8.3):

Proposition 10.2. linear action on Schrödinger operators

1. The action $\tilde{\sigma}_\lambda$ is written as follows

$$\tilde{\sigma}_\lambda(\phi;0).(a(t)(-2i\mathcal{M}\partial_t - \partial_r^2) + V(t,r))$$
$$= \dot{\phi}(t)a(\phi(t))(-2i\mathcal{M}\partial_t - \partial_r^2) + \dot{\phi}^2(t)V(\phi(t),$$

J. Unterberger and C. Roger, *The Schrödinger-Virasoro Algebra*, Theoretical and Mathematical Physics, DOI 10.1007/978-3-642-22717-2_10,

$$r\sqrt{\dot\phi(t)}) + a\left(2\mathrm{i}(\lambda - \frac{1}{4})\mathscr{M}\frac{\ddot\phi}{\dot\phi} + \frac{1}{2}\mathscr{M}^2 r^2 S(\phi)(t)\right)$$

$$\tilde\sigma_\lambda(1;(\alpha,\beta)).(-2\mathrm{i}\mathscr{M}\partial_t - \partial_r^2 + V(t,r))$$
$$= -2\mathrm{i}\mathscr{M}\partial_t - \partial_r^2 + V(t, r-\alpha(t)) + a\left(-2\mathscr{M}^2 r\ddot\alpha(t) - \mathscr{M}^2(2\dot\beta(t) - \alpha(t)\ddot\alpha(t))\right) \tag{10.2}$$

where $S : \phi \to \frac{\dddot\phi}{\dot\phi} - \frac{3}{2}\left(\frac{\ddot\phi}{\dot\phi}\right)^2$ *is the Schwarzian derivative.*

2. *Let* $\Delta_0 := -2\mathrm{i}\mathscr{M}\partial_t - \partial_r^2$ *be the free Schrödinger operator. The infinitesimal action* $d\tilde\sigma_\lambda : X \to \frac{d}{dt}\big|_{t=0}(\tilde\sigma_\lambda(\exp tX))$ *of* $\mathfrak{sv}$ *gives (recall* $V' := \partial_r V$*):*

$$d\tilde\sigma_\lambda(\mathscr{L}_f)(a(t)\Delta_0 + V(t,r)) = -(a\dot f + f\dot a)\Delta_0 - f\dot V - \frac{1}{2}\dot f r V'$$
$$+ a\left(-2\mathrm{i}(\lambda - \frac{1}{4})\mathscr{M}\ddot f - \frac{1}{2}\mathscr{M}^2\dddot f r^2\right) - 2\dot f V$$

$$d\tilde\sigma_\lambda(\mathscr{Y}_g)(a(t)\Delta_0 + V(t,r)) = -gV' - 2\mathscr{M}^2 a\ddot g r$$

$$d\tilde\sigma_\lambda(\mathscr{M}_h)(a(t)\Delta_0 + V(t,r)) = -2\mathscr{M}^2 a\dot h \tag{10.3}$$

The main result of this chapter is the following (see Theorem 10.2).

Theorem. *There exists a Poisson structure on* $\mathscr{S}^{lin} = \{a(t)(-2\mathrm{i}\mathscr{M}\partial_t - \partial_r^2) + V(t,r)\}$ *for which the infinitesimal action* $d\tilde\sigma_\lambda$ *of* $\mathfrak{sv}$ *is Hamiltonian.*

The analogue in the case of Hill operators is well-known (see for instance [43]). Namely, the action of the Virasoro group on the space $\mathscr{H}$ of Hill operators is equivalent to its affine coadjoint action with central charge $c = \frac{1}{2}$, with the identification $\partial_t^2 + u(t) \to u(t)dt^2 \in \mathfrak{vir}^*_{\frac{1}{2}}$, where $\mathfrak{vir}^*_c$ is the affine hyperplane $\{(X,c) \mid X \in \mathrm{Vect}(S^1)^*\}$. Hence this action preserves the canonical KKS (Kirillov-Kostant-Souriau) structure on $\mathfrak{vir}^*_{\frac{1}{2}} \simeq \mathscr{H}$. As well-known, one may exhibit a bi-Hamiltonian structure on $\mathfrak{vir}^*$ which provides an integrable system on $\mathscr{H}$ associated to the Korteweg-De Vries equation.

The above identification does not hold true any more in the case of the Schrödinger action of SV on the space of Schrödinger operators, which is *not* equivalent to its coadjoint action (see [106], Sect. 3.2). Hence the existence of a Poisson structure for which the action on Schrödinger operators is Hamiltonian has to be proved in the first place. It turns out that the action on Schrödinger operators is more or less the restriction of the coadjoint action of a much larger Lie algebra $\mathfrak{g}$ on its dual. The Lie algebra $\mathfrak{g}$ is introduced in Definition 10.15.

Our approach to finding this Lie algebra $\mathfrak{g}$ has been a bit tortuous.

The first idea (see Sect. 2.3, or Chap. 11 for superized versions of this statement) was to see $\mathfrak{sv}$ as a *subquotient* of an algebra $D\Psi D$ of *extended* pseudodifferential symbols on the line: one easily checks that the assignment $\mathcal{L}_f \to -f(\xi)\partial_\xi$, $\mathcal{Y}_g \to -g(\xi)\partial_\xi^{\frac{1}{2}}$, $\mathcal{M}_h \to -\frac{1}{2}h(\xi)$ yields a linear application $\mathfrak{sv} \to D\Psi D := \mathbb{R}[\xi, \xi^{-1}]] \, [\partial_\xi^{\frac{1}{2}}, \partial_\xi^{-\frac{1}{2}}]]$ which respects the Lie brackets of both Lie algebras, up to unpleasant terms which are pseudodifferential symbols of *negative* order. Define $D\Psi D_{\leq\kappa}$ as the subspace of pseudodifferential symbols with order $\leq \kappa$. Then $D\Psi D_{\leq 1}$ is a Lie subalgebra of $D\Psi D$, $D\Psi D_{\leq -\frac{1}{2}}$ is an ideal, and the above assignment defines an isomorphism $\mathfrak{sv} \simeq D\Psi D_{\leq 1}/D\Psi D_{\leq -\frac{1}{2}}$.

The second idea (sketched in [113]) was to use a non-local transformation $\Theta : D\Psi D \to \Psi D$ (ΨD being the usual algebra of pseudo-differential symbols) which maps $\partial_\xi^{\frac{1}{2}}$ to ∂_r and ξ to $\frac{1}{2}r\partial_r^{-1}$ (see Definition 10.7). The transformation Θ is formally an integral operator, simply associated to the heat kernel, which maps the first-order differential operator $-2\mathrm{i}\mathcal{M}\partial_t - \partial_\xi$ into $-2\mathrm{i}\mathcal{M}\partial_t - \partial_r^2$. The operator $-2\mathrm{i}\mathcal{M}\partial_t - \partial_\xi$ (which is simply the $\partial_{\bar{z}}$-operator in complex coordinates) is now easily seen to be invariant under an infinite-dimensional Lie algebra which generates (as an associative algebra) an algebra isomorphic to $D\Psi D$. One has thus defined a natural action of $D\Psi D$ on the space of solutions of the free Schrödinger equation $(-2\mathrm{i}\mathcal{M}\partial_t - \partial_r^2)\psi = 0$.

The crucial point now is that (after conjugation with Θ, i.e. coming back to the usual (t, r)-coordinates) the action of $D\Psi D_{\leq 1}$ coincides *up to pseudodifferential symbols of negative order* with the vector field representation $d\pi_0$ (see Introduction) of the generators L_n, Y_m, M_p ($n, p \in \mathbb{Z}, m \in \frac{1}{2} + \mathbb{Z}$). In other words, loosely speaking, the abstract isomorphism $\mathfrak{sv} \simeq D\Psi D_{\leq 1}/D\Psi D_{\leq -\frac{1}{2}}$ has received a concrete interpretation, and one has somehow reduced a problem concerning *differential operators in two variables t, r* into a problem concerning *time-dependent pseudodifferential operators in one variable*, which is a priori much simpler.

Integrable systems associated to Poisson structures on the loop algebra $\mathfrak{L}_t(\Psi D)$ over ΨD (with the usual Kac-Moody cocycle $(X, Y) \to \oint \mathrm{Tr}\dot{X}(t)Y(t)\, dt$, where Tr is Adler's trace on ΨD) have been studied by A.G. Reiman and M.A. Semenov-Tyan-Shanskii [103]. By considering Poisson structures on a looped Virasoro algebra, one may also construct a Kadomtsev-Petviashvili type equation with two space variables, see [94]. In our case, computations show that the $\mathfrak{sv}$-action on Schrödinger operators is related to the coadjoint action of $\widetilde{\mathcal{L}_t((\Psi D_r)_{\leq 1})}$, where $\widetilde{\mathcal{L}_t((\Psi D_r)_{\leq 1})}$ is a central extension of $\mathcal{L}_t((\Psi D_r)_{\leq 1})$ which is unrelated to the Kac-Moody cocycle.

Actually, the above scheme works out perfectly well only for the restriction of the $\mathfrak{sv}$-action to the nilpotent part of $\mathfrak{sv}$. For reasons explained in Sects. 10.3 and 10.4, the generators of $\mathrm{Vect}(S^1) \hookrightarrow \mathfrak{sv}$ play a particular rôle. So the action $d\sigma_\lambda$ of $\mathfrak{sv}$ is really obtained through the *projection* on the second component of the coadjoint action of an extended Lie algebra $\mathfrak{g} := \mathrm{Vect}(S^1) \ltimes \widetilde{\mathcal{L}_t((\Psi D_r)_{\leq 1})}$. The definition of $\mathfrak{g}$ requires in itself some work and is given only at the end of Sect. 10.5.

It is natural to expect that there should exist some bi-Hamiltonian structure on $\mathcal{S}^{lin}$ allowing to define some unknown integrable system. We hope to answer this question in the future.

Here is the outline of the chapter. Section 10.2 on pseudo-differential operators is mainly introductive, except for the definition of the non-local transformation Θ. The realization of $D\Psi D_{\le 1}$ as symmetries of the free Schrödinger equation is explained in Sect. 10.3. Sections 10.4 and 10.5 are devoted to the construction of the extended Lie algebra $\widetilde{\mathcal{L}_t((\Psi D_r)_{\le 1})}$ and its extension $\mathfrak{g}$. The action $d\tilde{\sigma}_\mu$ of $\mathfrak{sv}$ on Schrödinger operators is obtained as part of the coadjoint action of $\mathfrak{g}$ restricted to a stable submanifold $\mathcal{N} \subset \mathfrak{g}^*$ defined in Sect. 10.6, where the main theorem is stated and proved. Finally, an explicit rewriting in terms of the underlying Poisson formalism is given in Sect. 10.7.

Notation: In the sequel, the derivative with respect to r, resp. t will always be denoted by a prime ($'$), resp. by a dot, namely, $V'(t,r) := \partial_r V(t,r)$ and $\dot{V}(t,r) := \partial_t V(t,r)$.

10.2 Algebras of Pseudodifferential Symbols

Definition 10.3 (algebra of formal pseudodifferential symbols). Let $\Psi D := \mathbb{R}[z, z^{-1}]] [\partial_z, \partial_z^{-1}]]$ be the associative algebra of Laurent series in z, ∂_z with defining relation $[\partial_z, z] = 1$.

Using the coordinate $z = e^{i\theta}$, $\theta \in \mathbb{R}/2\pi\mathbb{Z}$, one may see elements of ΨD as formal pseudodifferential operators with periodic coefficients.

The algebra ΨD comes with a trace, called *Adler's trace*, defined in the Fourier coordinate θ by

$$\mathrm{Tr}\left(\sum_{q=-\infty}^{N} f_q(\theta)\partial_\theta^q\right) = \frac{1}{2\pi}\int_0^{2\pi} f_{-1}(\theta)\, d\theta. \tag{10.4}$$

Coming back to the coordinate z, this is equivalent to setting

$$\mathrm{Tr}(a(z)\partial_z^q) = \delta_{q,-1} \cdot \frac{1}{2i\pi}\oint a(z)dz \tag{10.5}$$

where $\frac{1}{2i\pi}\oint$ is the Cauchy integral giving the residue a_{-1} of the Laurent series $\sum_{p=-\infty}^{N} a_p z^p$.

For any $n \le 1$, the vector subspace generated by the pseudo-differential operators $D = f_n(z)\partial_z^n + f_{n-1}(z)\partial_z^{n-1} + \ldots$ of degree $\le n$ is a Lie subalgebra of ΨD that we shall denote by $\Psi D_{\le n}$. We shall sometimes write $D = O(\partial_z^n)$ for a pseudodifferential operator of degree $\le n$. Also, letting $OD = \Psi D_{\ge 0} =$

$\{\sum_{k=0}^{n} f_k(z)\partial_z^k,\ n \geq 0\}$ (differential operators) and $\mathfrak{volt} = \Psi D_{\leq -1}$ (called: *Volterra algebra*), we shall denote by (D_+, D_-) the decomposition of $D \in \Psi D$ along the direct sum $OD \oplus \mathfrak{volt}$, and call D_+ the *differential part* of D.

We shall also need to introduce the following 'extended' algebra of formal pseudodifferential symbols.

Definition 10.4 (algebra of extended pseudodifferential symbols). Let $D\Psi D$ be the extended pseudo-differential algebra generated as an associative algebra by ξ, ξ^{-1} and $\partial_\xi^{\frac{1}{2}}, \partial_\xi^{-\frac{1}{2}}$.

Let $D \in D\Psi D$. As in the case of the usual algebra of pseudodifferential symbols, we shall write $D = O(\partial_z^\kappa)$ $(\kappa \in \frac{1}{2}\mathbb{Z})$ for an extended pseudodifferential symbol with degree $\leq \kappa$, and denote by $D\Psi D_{\leq \kappa}$ the Lie subalgebra $\mathrm{span}(f_j(\xi)\partial_\xi^j\ ;\ j = \kappa, \kappa - \frac{1}{2}, \kappa - 1, \ldots)$ if $\kappa \leq 1$.

The Lie algebra $D\Psi D$ contains two interesting subalgebras for our purposes:

(i) $\mathrm{span}(f_1(\xi)\partial_\xi, f_0(\xi);\ f_1, f_0 \in C^\infty(S^1))$ which is isomorphic to $\mathrm{Vect}(S^1) \ltimes \mathcal{F}_0$;
(ii) $D\Psi D_{\leq 1} := \mathrm{span}(f_\kappa(\xi)\partial_\xi^\kappa;\ \kappa = 1, \frac{1}{2}, 0, -\frac{1}{2}, \ldots, f_\kappa \in C^\infty(S^1))$, which is also the Lie algebra generated by $\mathrm{span}(f_1(\xi)\partial_\xi, f_{\frac{1}{2}}(\xi)\partial_\xi^{1/2}, f_0(\xi);\ f_1, f_{\frac{1}{2}}, f_0 \in C^\infty(S^1))$.

As has been proved in Sect. 2.3, the Schrödinger–Virasoro Lie algebra $\mathfrak{sv}$ is isomorphic to a subquotient of $D\Psi D$. Let us give explicit formulas:

Proposition 10.5 ($\mathfrak{sv}$ as a subquotient of $D\Psi D$). *Let p be the projection of $D\Psi D_{\leq 1}$ onto $D\Psi D_{\leq 1}/D\Psi D_{\leq -\frac{1}{2}}$, and j be the linear morphism from $\mathfrak{sv}$ to $D\Psi D_{\leq 1}$ defined by*

$$\mathcal{L}_f \longrightarrow -\frac{\mathrm{i}}{2\mathcal{M}} f(-2\mathrm{i}\mathcal{M}\xi)\partial_\xi,\ \ \mathcal{Y}_g \longrightarrow -g(-2\mathrm{i}\mathcal{M}\xi)\partial_\xi^{\frac{1}{2}},\ \ \mathcal{M}_h \longrightarrow \mathrm{i}\mathcal{M}h(-2\mathrm{i}\mathcal{M}\xi). \tag{10.6}$$

Then the composed morphism $p \circ j : \mathfrak{sv} \to D\Psi D_{\leq 1}/D\Psi D_{\leq -\frac{1}{2}}$ is a Lie algebra isomorphism.

Proof. Straightforward computation. (Formulas look simpler with the normalization $-2\mathrm{i}\mathcal{M} = 1$.) □

The Lie algebra $\Psi D_{\leq 1}$ may be integrated to a group in the following way. Consider first the pronilpotent Lie group $\mathrm{Volt} := \exp \mathfrak{volt} = \{1 + f_{-1}(\xi)\partial_\xi^{-1} + \ldots\}$ obtained by the formal exponentiation of pseudo-differential symbols, $\exp V = \sum_{k\geq 0} \frac{V^k}{k!}$, $V \in \mathfrak{volt}$. It is easily extended to the semi-direct product group $\overline{\mathrm{Volt}} = \exp \mathcal{F}_0 \ltimes \mathrm{Volt}$ (where $\exp \mathcal{F}_0 = \exp C^\infty(S^1) \simeq \{f \in C^\infty(S^1) \mid \forall \theta \in [0, 2\pi], f(e^{\mathrm{i}\theta}) \neq 0\}$) which integrates $\Psi D_{\leq 0} \simeq \mathcal{F}_0 \ltimes \mathfrak{volt}$. Finally, $\mathrm{Diff}(S^1)$ acts naturally on $\overline{\mathrm{Volt}}$, which yields a Lie group $\mathrm{Diff}(S^1) \ltimes \overline{\mathrm{Volt}}$ integrating $\Psi D_{\leq 1} \simeq \mathrm{Vect}(S^1) \ltimes \Psi D_{\leq 0}$.

This explicit construction does not work for $D\Psi D_{\leq 1}$ because the formal series $\sum_{k\geq 0}\frac{V^k}{k!}$ is not in $C^\infty[\xi,\xi^{-1}]]\,[\partial_\xi,\partial_\xi^{-1}]]$ if $V = f_{1/2}(\xi)\partial_\xi^{\frac{1}{2}} + O(\partial_\xi^0)$, $f_{1/2}\not\equiv 0$. Yet the Campbell-Hausdorff formula makes it possible to integrate $D\Psi D_{\leq 1}$ by a similar procedure into an abstract group $DG_{\leq 1}$:

Lemma 10.6. *The Lie algebra $D\Psi D_{\leq 1}$ may be exponentiated into a group $DG_{\leq 1}$.*

Proof. First exponentiate $D\Psi D_{\leq 1/2} = \mathrm{span}(f_\kappa(\xi)\partial_\xi^\kappa\ ;\ \kappa = \frac{1}{2},0,-\frac{1}{2},\ldots)$ by defining $DG_{\leq\frac{1}{2}} := \exp D\Psi D_{\leq\frac{1}{2}}$ with multiplication given by the Campbell-Hausdorff formula

$$\begin{aligned}\exp\left(f(\xi)\partial_\xi^{\frac{1}{2}} + D_1\right)\exp\left(g(\xi)\partial_\xi^{\frac{1}{2}} + D_2\right)\\ = \exp\Bigg\{\left((f(\xi)+g(\xi))\partial_\xi^{\frac{1}{2}} + D_1 + D_2 + \ldots\right)\\ +\frac{1}{2}\left[f(\xi)\partial_\xi^{\frac{1}{2}} + D_1, g(\xi)\partial_\xi^{\frac{1}{2}} + D_2\right] + \ldots\Bigg\}\end{aligned} \tag{10.7}$$

($D_1, D_2 \in D\Psi D_{\leq 0}$); the first Lie bracket is $D\Psi D_{\leq 0}$-valued, and the successive iterated brackets belong to $D\Psi D_{\leq\kappa_1}$, $D\Psi D_{\leq\kappa_2}$, ... where $(\kappa_n)_{n\in\mathbb{N}^*}$ is a strictly decreasing sequence (with $\kappa_1 = -\frac{1}{2}$), hence the series converges.

Then define the semi-direct product $DG_{\leq 1} := \mathrm{Diff}(S^1)\ltimes DG_{\leq\frac{1}{2}}$ by the following natural action ρ of $\mathrm{Diff}(S^1)$ on $DG_{\leq\frac{1}{2}}$:

– let $\rho' : \mathrm{Diff}(S^1) \to Lin(D\Psi D_{\leq\frac{1}{2}})$ be the linear action defined by

$$\rho'(\phi)(f\,\partial_\xi^\kappa) = (f\circ\phi^{-1})\,.\,(\phi'\circ\phi^{-1}\,.\,\partial_\xi)^\kappa,\quad \kappa\leq\frac{1}{2}$$

where $(\phi'\circ\phi^{-1}\,.\,\partial_\xi)^{\frac{1}{2}} = \sqrt{\phi'\circ\phi^{-1}}\partial_\xi^{\frac{1}{2}} + \ldots$ is the usual square root of operators (recall $\phi' > 0$ by definition), and

$$\begin{aligned}(\phi'\circ\phi^{-1}\,.\,\partial_\xi)^\kappa &= \left[(\phi'\circ\phi^{-1}\,.\,\partial_\xi)^{\frac{1}{2}}\right]^{2\kappa}\\ &= \left(\partial_\xi^{-\frac{1}{2}}(\phi'\circ\phi^{-1})^{-\frac{1}{2}}\,.\,(1+\partial_\xi^{-\frac{1}{2}}(\phi'\circ\phi^{-1})^{-\frac{1}{2}}O(\partial_\xi^{-\frac{1}{2}}))^{-1}\right)^{-2\kappa}\\ &\in O(\partial_\xi^\kappa)\end{aligned} \tag{10.8}$$

if $\kappa\leq 0$;

– if $\phi\in\mathrm{Diff}(S^1)$, one lets $\rho(\phi)\exp D := \exp(\rho'(\phi)D)\in DG_{\leq\frac{1}{2}}$.

□

It turns out that a certain non-local transformation gives an isomorphism between $D\Psi D$ and ΨD. For the sake of the reader, we shall in the sequel add the name of the variable as an index when speaking of algebras of (extended or not) pseudodifferential symbols.

Definition 10.7 (non-local transformation Θ). Let $\Theta : D\Psi D_\xi \to \Psi D_r$ be the associative algebra isomorphism defined by

$$\partial_\xi^{\frac{1}{2}} \to \partial_r, \quad \partial_\xi^{-\frac{1}{2}} \to \partial_r^{-1}$$
$$\xi \to \frac{1}{2} r\partial_r^{-1}, \quad \xi^{-1} \to 2\partial_r r^{-1} \tag{10.9}$$

The inverse morphism $\Theta^{-1} : \partial_r \to \partial_\xi^{\frac{1}{2}}$, $r \to 2\xi\partial_\xi^{\frac{1}{2}}$ is easily seen to be an algebra isomorphism because the defining relation $[\partial_r, r] = 1$ is preserved by Θ^{-1}. It may be seen formally as the integral transformation $\psi(r) \to \tilde{\psi}(\xi) := \int_{-\infty}^{+\infty} \frac{e^{-r^2/4\xi}}{\sqrt{\xi}} \psi(r)\, dr$ (one verifies straightforwardly for instance that $r\partial_r\psi$ goes to $2\xi\partial_\xi\tilde{\psi}$ and that $\partial_r^2\psi$ goes to $\partial_\xi\tilde{\psi}$). In other words, assuming $\psi \in L^1(\mathbb{R})$, one has $\tilde{\psi}(\xi) = (P_\xi\psi)(0)$ $(\xi \geq 0)$ where $(P_\xi, \xi \geq 0)$ is the usual heat semi-group. Of course, this does not make sense at all for $\xi < 0$.

Remark. Denote by $\mathscr{E}_r = [r\partial_r, .]$ the Euler operator. Let $\Psi D_{(0)}$, resp. $\Psi D_{(1)}$ be the vector spaces generated by the operators $D \in \Psi D$ such that $\mathscr{E}_r(D) = nD$ where n is even, resp. odd. Then $\Psi D_{(0)}$ is an (associative) subalgebra of ΨD, and one has

$$[\Psi D_{(0)}, \Psi D_{(0)}] = \Psi D_{(0)}, \quad [\Psi D_{(0)}, \Psi D_{(1)}] = \Psi D_{(1)},$$
$$[\Psi D_{(1)}, \Psi D_{(1)}] = \Psi D_{(0)}.$$

Now, the inverse image of $D \in \Psi D_r$ by Θ^{-1} belongs to $\Psi D_\xi \subset D\Psi D_\xi$ if and only if $D \in (\Psi D_r)_{(0)}$.

Lemma 10.8 (pull-back of Adler's trace). *The pull-back by Θ of Adler's trace on ΨD_r yields a trace on $D\Psi D$ defined by*

$$\mathrm{Tr}_{D\Psi D_\xi}(a(\xi)\partial_\xi^q) := Tr_{\Psi D_r}\left(\Theta(a(\xi)\partial_\xi^q)\right) = 2\delta_{q,-1} \cdot \frac{1}{2i\pi} \oint a(\xi)d\xi. \tag{10.10}$$

Proof. Note first that the Lie bracket of ΨD_r, resp. $D\Psi D_\xi$ is graded with respect to the adjoint action of the Euler operator $\mathscr{E}_r := [r\partial_r, .]$, resp. $\mathscr{E}_\xi := [\xi\partial_\xi, .]$, and that $\Theta \circ \mathscr{E}_\xi = \frac{1}{2}\mathscr{E}_r \circ \Theta$. Now $\mathrm{Tr}_{\Psi D_r} D = 0$ if $D \in \Psi D_r$ is not homogeneous of degree 0 with respect to $\mathscr{E}_r$, hence the same is true for $\mathrm{Tr}_{D\Psi D_\xi}$. Consider $D := \xi^j\partial_\xi^j = \Theta^{-1}((\frac{1}{2}r\partial_r^{-1})^j\partial_r^{2j})$: then $\mathrm{Tr}_{D\Psi D_\xi}(D) = 0$ if $j \geq 0$ because (as one checks easily by an explicit computation) $\Theta(D) \in \mathcal{O}D$; and $\mathrm{Tr}_{D\Psi D_\xi}(D) = 0$ if $j \leq -2$ because $\Theta(D) = O(\partial_r^{-2})$. □

In order to obtain time-dependent equations, one needs to add an extra dependence on a formal parameter t of all the algebras we introduce. One obtains in this way loop algebras, whose formal definition is as follows:

Definition 10.9 (loop algebras). *Let* $\mathfrak{g}$ *be a Lie algebra. Then the* loop algebra *over* $\mathfrak{g}$ *is the Lie algebra*

$$\mathfrak{L}_t\mathfrak{g} := \mathfrak{g}[t, t^{-1}]]. \tag{10.11}$$

Elements of $\mathfrak{L}_t\mathfrak{g}$ may also be considered as Laurent series $\sum_{n=-\infty}^{N} t^n X_n$ ($X_n \in \mathfrak{g}$), or simply as functions $t \to X(t)$, where $X(t) \in \mathfrak{g}$.

The transformation Θ yields immediately (by lacing with respect to the time-variable t) an algebra isomorphism

$$\mathfrak{L}_t\Theta : \mathscr{L}_t(D\Psi D_\xi) \to \mathfrak{L}_t(\Psi D_r), \quad D \to (t \to \Theta(D(t))). \tag{10.12}$$

10.3 Time-Shift Transformation and Symmetries of the Free Schrödinger Equation

In order to define extended symmetries of the Schrödinger equation, one must first introduce the following time-shift transformation.

Definition 10.10 (time-shift transformation). Let $\mathscr{T}_t : D\Psi D_\xi \to \mathscr{L}_t(D\Psi D_\xi)$ be the linear transformation defined by

$$\mathscr{T}_t\left(f(\xi)\partial_\xi^\kappa\right) = (\mathscr{T}_t f(\xi))\partial_\xi^\kappa \tag{10.13}$$

where:

$$\mathscr{T}_t P(\xi) = P\left(\frac{\mathrm{i}}{2\mathscr{M}}t + \xi\right) \tag{10.14}$$

for *polynomials* P, and

$$\begin{aligned}\mathscr{T}_t\xi^{-k} &= \left(\frac{\mathrm{i}}{2\mathscr{M}}t + \xi\right)^{-k} \\ &:= \left(\frac{\mathrm{i}}{2\mathscr{M}}t\right)^{-k} \sum_{j=0}^{\infty}(-1)^j \frac{k(k+1)\ldots(k+j-1)}{j!}(-2\mathrm{i}\mathscr{M}\xi/t)^j. \end{aligned} \tag{10.15}$$

In other words, for any Laurent series $f \in \mathbb{C}[\xi, \xi^{-1}]]$,

$$\mathscr{T}_t f(\xi) = \sum_{j=0}^{\infty} \frac{f^{(j)}(\frac{\mathrm{i}}{2\mathscr{M}}t)}{j!}\xi^j.$$

Then $\mathscr{T}_t$ is an injective Lie algebra homomorphism, with left inverse $\mathscr{S}_t$ given by

$$\mathscr{S}_t(g(t,\xi)) = \frac{1}{2i\pi}\oint g(-2i\mathscr{M}\xi, t)\frac{dt}{t}. \tag{10.16}$$

Proof. Straightforward. □

Now comes an essential remark (see Sect. 10.1) which we shall first explain in an informal way. The free Schrödinger equation reads in the 'coordinates' (t,ξ)

$$(-2i\mathscr{M}\partial_t - \partial_\xi)\tilde{\psi}(t,\xi) = 0. \tag{10.17}$$

In the complex coordinates $z = t - 2i\mathscr{M}\xi$, $\bar{z} = t + 2i\mathscr{M}\xi$, one simply gets (up to a constant) the $\bar{\partial}$-operator, whose algebra of Lie symmetries is span$(f(t - 2i\mathscr{M}\xi)\partial_\xi, g(t-2i\mathscr{M}\xi)\partial_t)$ for arbitrary functions f, g. An easy but crucial consequence of these considerations is the following:

Definition 10.11 ($\mathscr{X}_f^{(i)}$-generators and Θ_t-homomorphism). Let, for $f \in \mathbb{C}[\xi, \xi^{-1}]]$ and $j \in \frac{1}{2}\mathbb{Z}$,

$$\mathscr{X}_f^{(j)} = \Theta_t\left(-f(-2i\mathscr{M}\xi)\partial_\xi^j\right) \in \mathfrak{L}_t(\Psi D_r) \tag{10.18}$$

where Θ_t is the composition of the non-local transformation Θ and the time-shift $\mathscr{T}_t$,

$$\Theta_t := \Theta \circ \mathscr{T}_t. \tag{10.19}$$

In other words,

$$\mathscr{X}_f^{(j)} = \mathfrak{L}_t(\Theta)\left(-f(t-2i\mathscr{M}\xi)\partial_\xi^j\right) = -f\left(t - i\mathscr{M}r\partial_r^{-1}\right)\partial_r^{2j} \tag{10.20}$$

(at least if f is a polynomial).The homomorphism Θ_t will play a key role in the sequel.

Lemma 10.12 (invariance of the Schrödinger equation).

(i) The free Schrödinger equation $\Delta_0\psi(t,r) = 0$ is invariant under the Lie algebra of transformations generated by $\mathscr{X}_f^{(i)}$, $i \in \frac{1}{2}\mathbb{Z}$.

(ii) Denote by $\dot{f}, \ddot{f}, \dddot{f}$ the time-derivatives of f of order 1, 2, 3, then

$$\mathscr{X}_f^{(1)} = -f(t)\partial_r^2 + i\mathscr{M}\dot{f}(t)r\partial_r + \frac{1}{2}\mathscr{M}^2\ddot{f}(t)r^2 - \left(\frac{1}{2}\mathscr{M}^2\ddot{f}(t)r + \frac{i}{6}\mathscr{M}^3\dddot{f}r^3\right)\partial_r^{-1} + O(\partial_r^{-2}); \tag{10.21}$$

$$\mathscr{X}_g^{(1/2)} = -g(t)\partial_r + \mathrm{i}\mathscr{M}\dot{g}(t)r + \frac{\mathscr{M}^2}{2}\ddot{g}(t)r^2\partial_r^{-1} + O(\partial_r^{-2}); \quad (10.22)$$

$$\mathscr{X}_h^{(0)} = -h(t) + O(\partial_r^{-1}). \quad (10.23)$$

In particular, denoting by D_+ the differential part of a pseudo-differential operator D, i.e. its projection onto $\mathcal{O}D$, the operators $(\mathscr{X}_g^{(1/2)})_+, (\mathscr{X}_h^{(0)})_+$ coincide (see Definition 1.31) with $d\pi_0(\mathscr{Y}_g)$, resp. $d\pi_0(\mathscr{M}_h)$, while

$$2\mathrm{i}\mathscr{M}\, d\pi_0(\mathscr{L}_f) = \left(\mathscr{X}_f^{(1)}\right)_+ - f(t)\left(2\mathrm{i}\mathscr{M}\partial_t - \partial_r^2\right). \quad (10.24)$$

Proof. (i) One has

$$\Theta^{-1}\left(\mathscr{X}_f^{(j)}\right) = \mathscr{T}_t\left(\xi \to f\left(-2\mathrm{i}\mathscr{M}\xi\right).\partial_\xi^j\right)$$
$$= \sum_{k=0}^{\infty} f^{(k)}\left(\frac{\mathrm{i}}{2\mathscr{M}}t\right)\frac{(-2\mathrm{i}\mathscr{M}\xi)^k}{k!}\,.\,\partial_\xi^j, \quad (10.25)$$

which is easily seen by a straightforward computation to commute with the Schrödinger operator $\Theta^{-1}(-2\mathrm{i}\mathscr{M}\partial_t - \partial_r^2) = -2\mathrm{i}\mathscr{M}\partial_t - \partial_\xi$, hence preserves the free Schrödinger equation. Note that, when f is a polynomial, $\Theta^{-1}(\mathscr{X}_f^{(j)}) = -f(t - 2\mathrm{i}\mathscr{M}\xi)\partial_\xi^j$ obviously commutes with $-2\mathrm{i}\mathscr{M}\partial_t - \partial_\xi$, see (10.17) and following lines.

(ii) Straightforward computations.

□

In other words (up to constant multiplicative factors), the projection $(\mathscr{X}_f^{(k)})_+$ of $\mathscr{X}_f^{(k)}, k = 1, \frac{1}{2}, 0$ onto $\mathcal{O}D$ forms a Lie algebra which coincides with the realization $d\pi_0$ of the Schrödinger–Virasoro algebra, *apart* from the fact that $-2\mathrm{i}\mathscr{M}\partial_t$ is substituted by ∂_r^2 in the formula for $\mathscr{X}_f^{(1)}$. This discrepancy is not unduly alarming, since $-2\mathrm{i}\mathscr{M}\partial_t \equiv \partial_r^2$ on the kernel of the free Schrödinger operator. As we shall see below, one may alter the $\mathscr{X}_f^{(1)}$ in order to make them 'begin with' $-f(t)\partial_t$ as expected, but then the $\mathscr{X}_f^{(1)}$ appear to have a specific algebraic status.

10.4 From Central Cocycles of $(\Psi D_r)_{\leq 1}$ to the Kac-Moody Algebra $\mathfrak{g}$

The above symmetry generators of the free Schrödinger equation, $\mathscr{X}_f^{(i)}, i \geq 1$ may be seen as elements of $\mathscr{L}_t(\Psi D_r)$. The original idea (following the scheme for Hill operators recalled in the Introduction) was to try to embed the space of Schrödinger operators $\mathscr{S}^{aff}$ into the dual of $\mathscr{L}_t(\Psi D_r)$ and realize the action $d\sigma_\lambda$ of Proposition 10.2 as part of the coadjoint representation of an appropriate central extension of $\mathscr{L}_t(\Psi D_r)$.

Unfortunately this scheme is a little too simple: it allows to retrieve only the action of the Y- and M-generators, as could have been expected from the remarks at the end of Sect. 3. It turns out that the $\mathscr{X}_f^{(i)}$, $i \leq \frac{1}{2}$ may be seen as elements of $\mathscr{L}_t((\Psi D_r)_{\leq 1})$, while the vector field representation $d\pi_0(\mathscr{L}_f)$ (see Introduction) of the generators in $\mathrm{Vect}(S^1) \subset \mathfrak{sv}$ involve *outer derivations* of this looped algebra. Then the above scheme works correctly, provided one chooses the right central extension of $\mathscr{L}_t((\Psi D_r)_{\leq 1})$. As explained below, there are many possible families of central extensions, and the correct one is obtained by 'looping' a cocycle $c_3 \in H^2((\Psi D_r)_{\leq 1}, \mathbb{R})$ which does *not* extend to the whole Lie algebra ΨD_r.

In this section and the following ones, we shall formally assume the coordinate $r = e^{i\theta}$ to be on the circle S^1. If $f(r) = \sum_{k\in\mathbb{Z}} f_k r^k$, the Cauchy integral $\frac{1}{2i\pi}\oint_{S^1} f(r)dr$ selects the residue f_{-1}. Alternatively, we shall sometimes use the angle coordinate θ in the next paragraph, so $f(r) = \sum_{k\in\mathbb{Z}} f_k e^{ik\theta}$ may be seen as a 2π-periodic function.

10.4.1 Central Cocycles of $(\Psi D_r)_{\leq 1}$

We shall (almost) determine $H^2(\Psi D_{\leq 1}, \mathbb{R})$, using its natural semi-direct product structure $\Psi D_{\leq 1} = \mathrm{Vect}(S^1) \ltimes \Psi D_{\leq 0}$. We choose to work with periodic functions $f = f(\theta)$ in this paragraph.

One has (by using the Hochschild-Serre spectral sequence, see Chap. 7 or [31]):

$$H^2(\Psi D_{\leq 1}, \mathbb{R}) = H^2(\mathrm{Vect}(S^1), \mathbb{R}) \oplus H^1(\mathrm{Vect}(S^1), H^1(\Psi D_{\leq 0}, \mathbb{R})) \oplus Inv_{\mathrm{Vect}(S^1)} H^2(\Psi D_{\leq 0}, \mathbb{R}). \tag{10.26}$$

The one-dimensional space $H^2(\mathrm{Vect}(S^1), \mathbb{R})$ is generated by the Virasoro cocycle, which we shall denote by c_0.

For the second piece, elementary computations give $[\Psi D_{\leq 0}, \Psi D_{\leq 0}] = \Psi D_{\leq -2}$. So $H_1(\Psi D_{\leq 0}, \mathbb{R})$ is isomorphic to $\Psi D_{\leq 0}/\Psi D_{\leq -2}$, i.e. to the space of symbols of type $f_0 + f_{-1}\partial^{-1}$. In terms of density modules, one has $H_1(\Psi D_{\leq 0}) = \mathscr{F}_0 \oplus \mathscr{F}_{-1}$. So $H^1(\Psi D_{\leq 0}, \mathbb{R}) = (\mathscr{F}_0 \oplus \mathscr{F}_{-1})^* = \mathscr{F}_{-1} \oplus \mathscr{F}_0$ by the standard duality $\mathscr{F}_\lambda^* \simeq \mathscr{F}_{-1-\lambda}$, and $H^1(\mathrm{Vect}(S^1), H^1(\Psi D_{\leq 0}, \mathbb{R})) = H^1(\mathrm{Vect}(S^1), \mathscr{F}_{-1} \oplus \mathscr{F}_0) = H^1(\mathrm{Vect}(S^1), \mathscr{F}_{-1}) \oplus H^1(\mathrm{Vect}(S^1), \mathscr{F}_0)$. From the results of Fuks [31], one knows that $H^1(\mathrm{Vect}(S^1), \mathscr{F}_1)$ is one-dimensional, generated by $f\partial_\theta \longrightarrow f''d\theta$, and $H^1(\mathrm{Vect}(S^1), \mathscr{F}_0)$ is two-dimensional, generated by $f\partial_\theta \longrightarrow f$ and $f\partial \longrightarrow f'$. So we have proved that $H^1(\mathrm{Vect}(S^1), H^1(\Psi D_{\leq 0}, \mathbb{R}))$ is three-dimensional, with generators c_1, c_2 and c_3 as follows:

$$c_1\left(g\partial_\theta, \sum_{k=-\infty}^{0} f_k\partial_\theta^k\right) = \frac{1}{2\pi}\int g'' f_0\, d\theta \tag{10.27}$$

$$c_2\left(g\partial_\theta, \sum_{k=-\infty}^{0} f_k\partial_\theta^k\right) = \frac{1}{2\pi}\int g f_{-1}\, d\theta \tag{10.28}$$

$$c_3\left(g\partial_\theta, \sum_{k=-\infty}^{0} f_k\partial_\theta^k\right) = \frac{1}{2\pi}\int g' f_{-1}\, d\theta \tag{10.29}$$

Let us finally consider the third piece $Inv_{\mathrm{Vect}(S^1)} H^2(\Psi D_{\leq 0}, \mathbb{R})$.We shall once more make use of a decomposition into a semi-direct product: setting $\mathfrak{volt} = \Psi D_{\leq -1}$, one has $\Psi D_{\leq 0} = \mathscr{F}_0 \ltimes \mathfrak{volt}$, where $\mathscr{F}_0$ is considered as an abelian Lie algebra, acting non-trivially on $\mathfrak{volt}$. We do not know how to compute the cohomology of $\mathfrak{volt}$, because of its "pronilpotent" structure, but we shall make the following:

Conjecture.

$$Inv_{\mathscr{F}_0} H^2(\mathfrak{volt}, \mathbb{R}) = 0. \tag{10.30}$$

We shall now work out the computations modulo this conjecture.
One first gets $H^2(\Psi D_{\leq 0}, \mathbb{R}) = H^2(\mathscr{F}_0, \mathbb{R}) \oplus H^1(\mathscr{F}_0, H^1(\mathfrak{volt}, \mathbb{R}))$. Then

$$\begin{aligned} Inv_{\mathrm{Vect}(S^1)} H^2(\Psi D_{\leq 0}, \mathbb{R}) &= Inv_{\mathrm{Vect}(S^1)} H^2(\mathscr{F}_0, \mathbb{R}) \\ &\quad \oplus Inv_{\mathrm{Vect}(S^1)} H^1(\mathscr{F}_0, H^1(\mathfrak{volt}, \mathbb{R})). \end{aligned} \tag{10.31}$$

Since $\mathscr{F}_0$ is abelian, one has $H^2(\mathscr{F}_0, \mathbb{R}) = \Lambda^2(\mathscr{F}_0^*)$, and $Inv_{\mathrm{Vect}(S^1)}(\Lambda^2(\mathscr{F}_0^*))$ is one-dimensional, generated by the well-known cocycle

$$c_4(f, g) = \frac{1}{2\pi}\int \left(g' f - f' g\right)\, d\theta. \tag{10.32}$$

A direct computation then shows that $[\mathfrak{volt}, \mathfrak{volt}] = \Psi D_{\leq -3}$, so $H_1(\mathfrak{volt}, \mathbb{R}) = \mathscr{F}_{-1} \oplus \mathscr{F}_{-2}$ and $H^1(\mathfrak{volt}, \mathbb{R}) = \mathscr{F}_0 \oplus \mathscr{F}_1$ as $\mathrm{Vect}(S^1)$-module. Then $H^1(\mathscr{F}_0, H^1(\mathfrak{volt}, \mathbb{R}))$ is easily determined by direct computation, as well as $Inv_{\mathrm{Vect}(S^1)} H^1(\mathscr{F}_0, H^1(\mathfrak{volt}, \mathbb{R}))$; the latter is one-dimensional, generated by the following cocycle:

$$c_5\left(g, \sum_{k=-\infty}^{0} f_k\partial_\theta^k\right) = \frac{1}{2\pi}\int g f_{-1}\, d\theta, \tag{10.33}$$

Let us summarize our results in the following:

Proposition 10.13. *Assuming conjecture (10.30) holds true, the space $H^2(\Psi D_{\leq 1}, \mathbb{R})$ is six-dimensional, generated by the cocycles c_i, $i = 0, \ldots, 5$, defined above.*

Remarks. 1. If conjecture (10.30) turned out to be false, it could only add some supplementary generators; in any case, we have proved that $H^2(\Psi D_{\leq 1}, \mathbb{R})$ is at least six-dimensional.

2. The natural inclusion $i : \Psi D_{\leq 1} \longrightarrow \Psi D$ induces $i^* : H^2(\Psi D, \mathbb{R}) \longrightarrow H^2(\Psi D_{\leq 1}, \mathbb{R})$; one may then determine the image by i^* of the two generators of $H^2(\Psi D, \mathbb{R})$ determined by B. Khesin and O. Kravchenko [69]. Set $c_{KK_1}(D_1, D_2) = \mathrm{Tr}[\log\theta, D_1]D_2$ and $c_{KK_2}(D_1, D_2) = \mathrm{Tr}[\log\partial, D_1]D_2$. Then $i^*c_{KK_1} = c_2$ and $i^*c_{KK_2} = c_0 + c_1 + c_4$.

The right cocycle for our purposes turns out to be c_3: coming back to the radial coordinate r, one gets a centrally extended Lie algebra of pseudodifferential symbols $\widetilde{\Psi D}_{\leq 1}$ as follows.

Definition 10.14. Let $\widetilde{\Psi D}_{\leq 1}$ be the central extension of $\Psi D_{\leq 1}$ associated with the cocycle cc_3 ($c \in \mathbb{R}$), where $c_3 : \Lambda^2 \Psi D_{\leq 1} \to \mathbb{C}$ verifies

$$c_3\left(f\partial_r, g\partial_r^{-1}\right) = c_3\left(f\partial_r^{-1}, g\partial_r\right) = \frac{1}{2i\pi}\oint f'g\,dr \tag{10.34}$$

(all other relations being trivial).

10.4.2 Introducing the Kac-Moody Type Lie Algebra $\mathfrak{g}$

Let us introduce now the looped algebra $\mathcal{L}_t(\widetilde{(\Psi D_r)_{\leq 1}})$ in order to allow for time-dependence. An element of $\mathcal{L}_t(\widetilde{(\Psi D_r)_{\leq 1}})$ is a pair $(D(t), \alpha(t))$ where $\alpha \in \mathbb{C}[t, t^{-1}]]$ and $D(t) \in \mathcal{L}_t((\Psi D_r)_{\leq 1})$. By a slight abuse of notation, we shall write $c_3(D_1, D_2)$ $(D_1, D_2 \in \mathcal{L}_t((\Psi D_r)_{\leq 1}))$ for the function $t \to c_3(D_1(t), D_2(t))$, so now c_3 has to be seen as a function-valued central cocycle of $\mathcal{L}_t((\Psi D_r)_{\leq 1})$. In other words, we consider the looped version of the exact sequence

$$0 \longrightarrow \mathbb{R} \longrightarrow \widetilde{(\Psi D_r)_{\leq 1}} \longrightarrow (\Psi D_r)_{\leq 1} \longrightarrow 0, \tag{10.35}$$

namely,

$$0 \longrightarrow \mathbb{R}[t, t^{-1}]] \longrightarrow \mathcal{L}_t(\widetilde{(\Psi D_r)_{\leq 1}}) \longrightarrow \mathcal{L}_t((\Psi D_r)_{\leq 1}) \longrightarrow 0. \tag{10.36}$$

As mentioned in the Introduction, $\mathcal{L}_t(\widetilde{(\Psi D_r)_{\leq 1}})$ is naturally equipped with the Kac-Moody cocycle

$$\begin{aligned}\mathcal{L}_t(\widetilde{(\Psi D_r)_{\leq 1}}) \times \mathcal{L}_t(\widetilde{(\Psi D_r)_{\leq 1}}) \to \mathbb{C}, ((D_1(t), \lambda_1(t)), (D_2(t), \lambda_2(t)))\\ \to \mathrm{Tr}\, D_1(t)\dot{D}_2(t).\end{aligned} \tag{10.37}$$

However this further central extension is irrelevant here. On the other hand, we shall need to incorporate into our scheme *time derivations* $f(t)\partial_t$ (which are outer Lie derivations of $\mathcal{L}_t(\widetilde{(\Psi D_r)_{\leq 1}})$, as is the case of any looped algebra), obtaining thus a Lie algebra $\mathfrak{g}$ which is the main object of this article.

Definition 10.15 (Kac-Moody type Lie algebra $\mathfrak{g}$). Let $\mathfrak{g} \simeq \mathrm{Vect}(S^1)_t \ltimes \mathcal{L}_t(\widetilde{(\Psi D_r)_{\leq 1}})$ be the Kac-Moody type Lie algebra obtained from $\mathcal{L}_t(\widetilde{(\Psi D_r)_{\leq 1}})$ by including the outer Lie derivations

$$f(t)\partial_t \,.\, (D(t), \alpha(t)) = \left(f(t)\dot{D}(t), f(t)\dot{\alpha}(t)\right). \tag{10.38}$$

10.5 Construction of the Embedding I of $(D\Psi D_\xi)_{\leq 1}$ into $\mathfrak{g}$

This section, as explained in the introduction to Sect. 4, is devoted to the construction of an explicit embedding, denoted by I, of the abstract algebra of extended pseudodifferential symbols $(D\Psi D_\xi)_{\leq 1}$ into $\mathfrak{g}$. Loosely speaking, the image $I((D\Psi D_\xi)_{\leq 1})$ is made up of the $\mathscr{X}_f^{(j)}$, $j \leq \frac{1}{2}$ and the $\mathscr{X}_f^{(1)}$ *with* ∂_r^2 *substituted by* $-2\mathrm{i}\mathcal{M}\partial_t$ (see end of Sect. 3). More precisely, $I\big|_{(D\Psi D_\xi)_{\leq \frac{1}{2}}}$ maps an operator D into its image by Θ_t (see Definition 10.11), namely, $\Theta_t(D)$, viewed as an element of the centrally extended Lie algebra $\mathcal{L}_t(\widetilde{(\Psi D_r)_{\leq 1}})$. On the other hand, the operator of degree one $-f(-2\mathrm{i}\mathcal{M}\xi)\partial_\xi$ will not be mapped to $\Theta_t(-f(-2\mathrm{i}\mathcal{M}\xi)\partial_\xi) = \mathscr{X}_f^{(1)}$ (which is of degree 2 in ∂_r), but to some element in the product $\mathfrak{g} = \mathrm{Vect}(S^1)_t \ltimes \mathcal{L}_t(\widetilde{(\Psi D_r)_{\leq 1}})$ with both components non-zero, as described in the following

Theorem 10.1 (homomorphism I). *Let* $I : (D\Psi D_\xi)_{\leq 1} \simeq \mathrm{Vect}(S^1)_\xi \ltimes (D\Psi D_\xi)_{\leq \frac{1}{2}} \hookrightarrow \mathfrak{g} = \mathrm{Vect}(S^1)_t \ltimes \mathcal{L}_t(\widetilde{(\Psi D_r)_{\leq 1}})$ *be the mapping defined by*

$$I\left((0, D)\right) = (0, \Theta_t(D))\,; \tag{10.39}$$

$$I\left(\left(-\frac{\mathrm{i}}{2\mathcal{M}} f(-2\mathrm{i}\mathcal{M}\xi)\partial_\xi, 0\right)\right) = \left(-f(t)\partial_t, \frac{\mathrm{i}}{2\mathcal{M}}\left(\mathscr{X}_f^{(1)}\right)_{\leq 1}\right) \tag{10.40}$$

where

$$\begin{aligned}\left(\mathscr{X}_f^{(1)}\right)_{\leq 1} &= \left(\Theta_t(-f(-2\mathrm{i}\mathcal{M}\xi)\partial_\xi)\right)_{\leq 1} \\ &= \mathrm{i}\mathcal{M}\dot{f}(t)r\partial_r + \frac{1}{2}\mathcal{M}^2\ddot{f}(t)r^2 - \left(\frac{1}{2}\mathcal{M}^2\ddot{f}(t)r + \frac{\mathrm{i}}{6}\mathcal{M}^3\dddot{f}r^3\right)\partial_r^{-1} + \ldots\end{aligned} \tag{10.41}$$

(see Lemma 10.12) is $\mathcal{X}_f^{(1)}$ shunted of its term of order ∂_r^2, i.e. the projection of $\mathcal{X}_f^{(1)}$ onto $\mathcal{L}_t((\Psi D_r)_{\leq 1})$.

Then I is a Lie algebra homomorphism.

Proof. First of all, the cocycle c_3 (see Definition 10.14) vanishes on the product of two operators of the form $\frac{\mathrm{i}}{2\mathcal{M}}(\mathcal{X}_f^{(1)})_{\leq 1} + \Theta_t(D)$ belonging to the image of $(D\Psi D_\xi)_{\leq 1}$ by I, see (10.39) and (10.40), because these involve only non-negative powers of r. Hence I may be seen as a map ($\bar{I}$, say) with values in $\mathrm{Vect}(S^1)_t \ltimes \mathcal{L}_t((\Psi D_r)_{\leq 1})$ (discarding the central extension). Now the Lie bracket $[(-f_1(t)\partial_t, W_1), (-f_2(t)\partial_t, W_2)]$ in $\mathrm{Vect}(S^1)_t \ltimes \mathcal{L}_t((\Psi D_r)_{\leq 1})$ coincides with the usual Lie bracket of the Lie algebra $\Psi D_{t,r}$ of pseudo-differential symbols in two variables, t and r, hence $I((D\Psi D_\xi)_{\leq 1})$ may be seen as sitting in $\Psi D_{t,r}$. Then

$$\bar{I}\left(-\frac{\mathrm{i}}{2\mathcal{M}} f(-2\mathrm{i}\mathcal{M}\xi)\partial_\xi\right) = -f(t)\partial_t + \frac{\mathrm{i}}{2\mathcal{M}}\left(\mathcal{X}_f^{(1)}\right)_{\leq 1} = \frac{\mathrm{i}}{2\mathcal{M}}\left(\mathcal{L}'_f + \mathcal{X}_f^{(1)}\right), \tag{10.42}$$

where $\mathcal{L}'_f := -f(t)(-2\mathrm{i}\mathcal{M}\partial_t - \partial_r^2)$ is an independent copy of $\mathrm{Vect}(S^1)$, by which we mean that $[\mathcal{L}'_f, \mathcal{L}'_g] = \mathcal{L}'_{\{f,g\}} = \mathcal{L}'_{f'g-fg'}$ and $[\mathcal{L}'_f, \mathcal{X}_f^{(i)}] = 0$ for all i. This is immediate in the 'coordinates' (t,ξ) since $\Theta^{-1}(\mathcal{L}'_f) = -f(t)(-2\mathrm{i}\mathcal{M}\partial_t - \partial_\xi)$ commutes with $\Theta^{-1}(\mathcal{X}_f^{(i)}) = -f(t-2\mathrm{i}\mathcal{M}\xi)\partial_\xi^i$ as shown in Lemma 10.12. Hence $\bar{I}$ is a Lie algebra homomorphism. □

As we shall see in the next two sections, the coadjoint representation of the semi-direct product $\mathfrak{g}$ is the key to define a Poisson structure on $\mathcal{S}^{lin}$ for which the action of SV is Hamiltonian.

10.6 The Action of $\mathfrak{sv}$ on Schrödinger Operators as a Coadjoint Action

Here and in the sequel, an element of $\mathfrak{g} = \mathrm{Vect}(S^1)_t \ltimes \widetilde{\mathcal{L}_t((\Psi D_r)_{\leq 1})}$ will be denoted by $(w(t)\partial_t, (W(t,r), \alpha(t)))$ (see Sect. 4.2) or simply by the triplet $(w(t)\partial_t; W, \alpha(t))$. Since an element of $\mathfrak{h} = \widetilde{\mathcal{L}_t((\Psi D_r)_{\leq 1})}$is of the form $(W(t), \alpha(t))$ with $W(t) \in (\Psi D_r)_{\leq 1}$ (for every fixed t), it is natural (using Adler's trace) to represent an element of the restricted dual $\mathfrak{g}^*$ as a triplet $(v(t)dt^2; Vdt, a(t)dt)$ with $v \in C^\infty(S^1)$, $V \in \mathfrak{L}_t((\Psi D_r)_{\geq -2})$ and $a \in C^\infty(S^1)$. The coupling between $\mathfrak{g}$ and its dual $\mathfrak{g}^*$ writes then

$$\langle (v(t)dt^2; Vdt, a(t)dt), (w(t)\partial_t; W, \alpha(t))\rangle_{\mathfrak{g}^*\times\mathfrak{g}} = \frac{1}{2\mathrm{i}\pi}\oint [v(t)w(t) + \mathrm{Tr}_{\Psi D_r}(V(t)W(t)) + a(t)\alpha(t)]\, dt. \tag{10.43}$$

This section is devoted to the proof of the main Theorem announced in the Introduction, which we may now state precisely:

Theorem 10.2. *Let $\widetilde{(\Psi D_r)_{\leq 1}}$ be the central extension of $(\Psi D_r)_{\leq 1}$ associated with the cocycle cc_3 (see Definition 10.14) with $c = 2$; $\mathfrak{h} = \widetilde{\mathcal{L}_t((\Psi D_r)_{\leq 1})}$ the corresponding looped algebra, and $\mathfrak{g} = \mathrm{Vect}(S^1)_t \ltimes \widetilde{\mathcal{L}_t((\Psi D_r)_{\leq 1})}$ the corresponding Kac-Moody type extension by outer derivations (see Definition 10.15). Let also $\mathcal{N}$ be the affine subspace $\mathrm{Vect}(S^1)_t^* \ltimes \{([V_{-2}(t,r)\partial_r^{-2} + V_0(t)\partial_r^0]\,dt, a(t)dt)\} \subset \mathfrak{g}^*$ (note that V_0 is assumed to be a function of t only). Then:*

(i) the coadjoint action $\mathrm{ad}^*_{\mathfrak{g}}$, restricted to the image $I((D\Psi D_\xi)_{\leq 1})$, preserves $\mathcal{N}$, and quotients out into an action of $\mathfrak{sv}$;
(ii) decompose $d\tilde{\sigma}_0(X)(a(t)\Delta_0 + V(t,r))$, $X \in \mathfrak{sv}$ into $d\tilde{\sigma}_0^{op}(X)(a)\Delta_0 + d\tilde{\sigma}_0^{pot}(X)(a,V)$ (free Schrödinger operator depending only on a, plus a potential depending on (a,V)). Then it holds

$$\begin{aligned}&\mathrm{ad}^*_{\mathfrak{g}}(\mathcal{L}_f).\left(v(t)dt^2;[V_{-2}(t,r)\partial_r^{-2} + V_0(t)\partial_r^0]\,dt, a(t)dt\right)\\ &= \left(\left[-\frac{1}{2}\dddot{f}(\oint rV_{-2}dr) - (f\dot{v} + 2\dot{f}v)\right]dt^2;\right.\\ &\left[d\tilde{\sigma}_0^{pot}(\mathcal{L}_f)(a,V_{-2})\partial_r^{-2} + (-f\dot{V}_0 - \dot{f}V_0 + a\dot{f})\partial_r^0\right]dt,\\ &\left.d\tilde{\sigma}_0^{op}(\mathcal{L}_f)(a)dt\right); \end{aligned} \tag{10.44}$$

$$\begin{aligned}&\mathrm{ad}^*_{\mathfrak{g}}(\mathcal{Y}_g).\left(v(t)dt^2;[V_{-2}(t,r)\partial_r^{-2} + V_0(t)\partial_r^0]\,dt, a(t)dt\right)\\ &= \left(-\dot{g}(\oint V_{-2}dr)dt^2; \left(d\tilde{\sigma}_0^{pot}(\mathcal{Y}_g)(a,V_{-2})\right)\partial_r^{-2}dt, 0\right); \end{aligned} \tag{10.45}$$

$$\begin{aligned}&\mathrm{ad}^*_{\mathfrak{g}}(\mathcal{M}_h).\left(v(t)dt^2;[V_{-2}(t,r)\partial_r^{-2} + V_0(t)\partial_r^0]\,dt, a(t)dt\right)\\ &= \left(0; \left(d\tilde{\sigma}_0^{pot}(\mathcal{M}_h)(a,V_{-2})\right)\partial_r^{-2}dt, 0\right). \end{aligned} \tag{10.46}$$

In other words (disregarding the ∂_r^0-component in $\mathfrak{h}^*$ and the dt^2-component in $\mathrm{Vect}(S^1)^*$) the restriction of the coadjoint action of $\mathrm{ad}^*_{\mathfrak{g}}|_{\mathfrak{sv}}$ to $\mathcal{N}$ coincides with the infinitesimal action $d\tilde{\sigma}_0$ of $\mathfrak{sv}$ on $\mathcal{S}^{lin} = \{a(t)(-2i\mathcal{M}\partial_t - \partial_r^2) + V_{-2}(t,r)\}$.

Remark. The term $a\dot{f}\partial_r^0$ in (10.44) shows that the subspace of $\mathcal{N}$ with vanishing coordinate $V_0 \equiv 0$ is not stable by the action of $\mathfrak{sv}$. The V_0-component is actually important since the terms proportional to $\mathcal{M}$ or $\mathcal{M}^2$ in the action $d\tilde{\sigma}_0(\mathfrak{sv})$, see Proposition 10.2 (which are affine terms for the affine representation $d\sigma_0$) will be obtained in the next section as the image by the Hamiltonian operator of functionals of V_0.

Proof of the Theorem. Recall $\mathfrak{sv} \simeq D\Psi D_{\leq 1}/D\Psi D_{\leq -\frac{1}{2}}$ (see Lemma 10.5). The first important remark is that the coadjoint action $\mathrm{ad}^*_{\mathfrak{g}}$ of $I\left((D\Psi D_\xi)_{\leq 1}\right)$ on elements $\left(v(t)dt^2; \left[V_{-2}(t,r)\partial_r^{-2} + V_0(t)\partial_r^0\right] dt, a(t)dt\right) \in \mathscr{N}$ quotients out into an action of the Schrödinger–Virasoro group. Namely, let $-\kappa \leq -\frac{1}{2}$ and $(w; W, \alpha(t)) = (w(t)\partial_t; \sum_{j\leq 1} W_j(t,r)\partial_r^j, \alpha(t)) \in \mathfrak{g}$, then

$$\begin{aligned}
&\Big\langle \mathrm{ad}^*_{I(f(-2\mathrm{i}\mathscr{M}\xi)\partial_\xi^{-\kappa})} \left(v(t)dt^2; \left[V_{-2}(t,r)\partial_r^{-2} + V_0(t)\partial_r^0\right] dt, a(t)dt\right), \\
&\quad (w(t)\partial_t; W, \alpha(t))\Big\rangle_{\mathfrak{g}^*\times\mathfrak{g}} \\
&= -\Bigg\langle \left(\left[V_{-2}(t,r)\partial_r^{-2} + V_0(t)\partial_r^0\right] dt, a(t)dt\right), \\
&\quad \left[f(t)\partial_r^{-2\kappa} + O(\partial_r^{-2\kappa-1}), \sum_{j\leq 1} W_j(t,r)\partial_r^j \right]_{\mathfrak{h}} \Bigg\rangle_{\mathfrak{h}^*\times\mathfrak{h}} \\
&\quad + \left\langle \left[V_{-2}(t,r)\partial_r^{-2} + V_0(t)\partial_r^0\right] dt, w(t)\dot{f}(t)\partial_r^{-2\kappa} + O(\partial_r^{-2\kappa-1}) \right\rangle \\
&= 0
\end{aligned} \tag{10.47}$$

since the Lie bracket in $\mathfrak{g}$ produces (i) no term along the central charge (namely, the coefficient of ∂_r^{-1} is constant in r, see Definition 10.14); (ii) if $-\kappa = -\frac{1}{2}$ only, a term of order -1 coming from $-[f(t)\partial_r^{-1}, W_1\partial_r] + w(t)\dot{f}(t)\partial_r^{-1} = \left(f(t)W_1' + w(t)\dot{f}(t)\right)\partial_r^{-1} + \ldots$, whose coupling with the potential yields $\int\int V_0(t)\left(f(t)W_1' + w(t)\dot{f}(t)\right) dtdr = 0$ (total derivative in r); (iii) a pseudodifferential operator of degree ≤ -2 which does not couple to the potential.

Denote by $p \circ j$ the isomorphism from $\mathfrak{sv}$ to $D\Psi D_{\leq 1}/D\Psi D_{\leq -\frac{1}{2}}$, as in Lemma 10.5. With a slight abuse of notation, we shall write ad^*_X instead of $\mathrm{ad}^*_{p\circ j(X)}$ for $X \in \mathfrak{sv}$ and consider $\mathrm{ad}^* \circ p \circ j$ as a "coadjoint action" of $\mathfrak{sv}$.

Let us now study successively the "coadjoint action" of the Y, M and L generators of $\mathfrak{sv}$ on elements $\left(v(t)dt^2; \left[V_{-2}(t,r)\partial_r^{-2} + V_0(t)\partial_r^0\right] dt, a(t)dt\right) \in \mathfrak{g}^*$.

Recall from the Introduction that the derivative with respect to r, resp. t is denoted by $'$, resp. by a dot, namely, $V'(t,r) := \partial_r V(t,r)$ and $\dot{V}(t,r) := \partial_t V(t,r)$.

Action of the Y-Generators

Let $W = \sum_{j\leq 1} W_j(t,r)\partial_r^j \in \mathfrak{L}_t((\Psi D_r)_{\leq 1})$ and $\alpha(t) \in C^\infty(S^1)$ as before. A computation gives (see (10.22))

$$\left\langle \mathrm{ad}^*_{\mathscr{Y}_g}\left(v(t)dt^2;\left[V_{-2}(t,r)\partial_r^{-2}+V_0(t)\partial_r^0\right]dt,a(t)dt\right),(w(t)\partial_t;W,\alpha(t))\right\rangle_{\mathfrak{g}^*\times\mathfrak{g}}$$

$$=-\left\langle\left(\left[V_{-2}(t,r)\partial_r^{-2}+V_0(t)\right]dt,a(t)dt\right),\right.$$

$$\left[-g(t)\partial_r+\mathrm{i}\mathscr{M}\dot{g}(t)r+\frac{\mathscr{M}^2}{2}\ddot{g}(t)r^2\partial_r^{-1}+O(\partial_r^{-2}),\right.$$

$$\left.\left.W_1(t,r)\partial_r+W_0(t,r)\partial_r^0+W_{-1}(t,r)\partial_r^{-1}+O(\partial_r^{-2})\right]_{\mathfrak{h}}-w\left(-\dot{g}\partial_r+\frac{\mathscr{M}^2}{2}\dddot{g}r^2\partial_r^{-1}\right)\right\rangle_{\mathfrak{h}^*\times\mathfrak{h}}$$

$$=-\left\langle\left(V_{-2}(t,r)\partial_r^{-2}dt,a(t)dt\right),\left(-(gW_1'-\dot{g}w)\partial_r,c\mathscr{M}^2\ddot{g}\cdot\frac{1}{2\mathrm{i}\pi}\oint rW_1dr\right)\right\rangle$$

$$+\left\langle V_0(t)\partial_r^0dt,g(t)W_{-1}'+\left(\frac{\mathscr{M}^2}{2}\ddot{g}r^2W_1\right)'\right\rangle$$

$$=\int\int V_{-2}(gW_1'-\dot{g}w)\,dt\,dr-c\mathscr{M}^2.\frac{1}{2\mathrm{i}\pi}\int\int a\ddot{g}rW_1\,dt\,dr. \tag{10.48}$$

The coupling of V_0 with W vanishes, as may be seen in greater generality as follows (this will be helpful later when looking at the action of the L-generators): the term of order -1 comes from a bracket of the type $[A(t,r)\partial_r,B(t,r)\partial_r^{-1}]=(A'B+AB')\partial_r^{-1}+\ldots$; this is a total derivative in r, hence (since $V_0'\equiv 0$ by hypothesis) $\langle V_0(t)\partial_r^0dt,[A\partial_r,B\partial_r^{-1}]\rangle=0$.

Generally speaking (by definition of the duality given by Adler's trace), the terms in the above expression that depend on W_i, $i=1,0,\ldots$ give the projection of $\mathrm{ad}^*_{\mathscr{Y}_g}(v(t)dt^2;Vdt,a(t)dt)$ on the component ∂_r^{-i-1}, while the term depending on w gives the projection on the $\mathrm{Vect}(S^1)$-component.

Hence altogether one has proved:

$$\mathrm{ad}^*_{\mathscr{Y}_g}\left((v(t)dt^2;\left[V_{-2}(t,r)\partial_r^{-2}+V_0(t)\partial_r^0\right]dt,a(t)dt)\right)$$

$$=\left(-\dot{g}(\oint V_{-2}dr)dt^2;-\left(g(t)V_{-2}'+c\mathscr{M}^2a\ddot{g}(t)r\right)\partial_r^{-2}dt,0\right) \tag{10.49}$$

which gives the expected result for $c=2$.

Action of the M-Generators

It may be deduced from that of the Y-generators since the Lie brackets of the Y-generators generate all M-generators.

Action of the Virasoro Part

One computes (see (10.21) or (10.41)):

$$\left\langle \mathrm{ad}^*_{\mathcal{L}_f}(v(t)dt^2; [V_{-2}(t,r)\partial_r^{-2} + V_0(t)\partial_r^0]dt, a(t)dt), (w(t)\partial_t; W, \alpha(t))\right\rangle_{\mathfrak{g}^*\times\mathfrak{g}}$$

$$= -\left\langle \mathrm{ad}^*_{\mathrm{Vect}(S^1)} f(t)\partial_t.v(t)dt^2, w(t)\partial_t \right\rangle_{\mathrm{Vect}(S^1)^*\times\mathrm{Vect}(S^1)}$$

$$-\Bigg\langle \left(\left[V_{-2}(t,r)\partial_r^{-2} + V_0(t)\partial_r^0\right]dt, a(t)dt\right), -f(t)\partial_t.$$

$$(W_1(t,r)\partial_r + W_0(t,r) + W_{-1}(t,r)\partial_r^{-1} + O(\partial_r^{-2}))$$

$$+\frac{\mathrm{i}}{2\mathcal{M}}\left[\mathrm{i}\mathcal{M}\dot{f}(t)r\partial_r + \frac{\mathcal{M}^2}{2}r^2\ddot{f}(t) - \left(\frac{\mathcal{M}^2}{2}\ddot{f}(t)r + \frac{\mathrm{i}}{6}\mathcal{M}^3\dddot{f}r^3\right)\partial_r^{-1}\right.$$

$$\left.+O(\partial_r^{-2}), W_1(t,r)\partial_r + W_0(t,r) + W_{-1}(t,r)\partial_r^{-1} + O(\partial_r^{-2})\right]_{\mathfrak{h}}\Bigg\rangle_{\mathfrak{h}^*\times\mathfrak{h}}$$

$$+\int dt\ \mathrm{Tr}(V_{-2}(t,r)\partial_r^{-2} + V_0(t)).w(t).$$

$$\times\left(-\frac{1}{2}\ddot{f}r\partial_r + \left(-\frac{\mathrm{i}}{4}\mathcal{M}\dddot{f}r - \frac{\mathcal{M}^2}{12}\frac{d^4 f}{dt^4}r^3\right)\partial_r^{-1}\right)$$

$$= -\int (f\dot{v} + 2\dot{f}v)w\,dt + \Bigg\langle (V_{-2}(t,r)\partial_r^{-2}dt, a(t)dt),$$

$$\left(\left(f(t)\dot{W}_1 + \frac{1}{2}\dot{f}(t)(rW_1' - W_1)\right)\partial_r, -f\dot{\alpha} + \frac{c}{2\mathrm{i}\pi}\frac{\mathrm{i}}{2\mathcal{M}}\right.$$

$$\left.\left(-\mathrm{i}\mathcal{M}\dot{f}(t)\oint W_{-1}dr + \frac{\mathcal{M}^2}{2}\ddot{f}(t)\oint W_1 dr + \frac{\mathrm{i}}{2}\mathcal{M}^3\dddot{f}(t)\oint r^2 W_1 dr\right)\right)\Bigg\rangle_{\mathfrak{h}^*\times\mathfrak{h}}$$

$$+\langle V_0(t)\partial_r^0 dt, f(t)\dot{W}_{-1}\partial_r^{-1}\rangle_{\mathfrak{h}^*\times\mathfrak{h}} - \frac{1}{2}\int\int w\ddot{f}rV_{-2}\,dt\,dr. \qquad (10.50)$$

A term of the form $\langle V_0(t)\partial_r^0 dt, [A\partial_r, B\partial_r^{-1}]\rangle$ (which vanishes after integration as above, see computations for the action of the Y-generators) has been left out. The term depending on α gives the projection on the a-coordinate.

Hence:

$$\mathrm{ad}^*_{\mathcal{L}_f}((v(t)dt^2; \left[V_{-2}(t,r)\partial_r^{-2} + V_0(t)\partial_r^0\right]dt, a(t)dt))$$

$$= \left(\left[-\frac{1}{2}\ddot{f}(\oint rV_{-2}dr) - (f\dot{v} + 2\dot{f}v)\right]dt^2;\right.$$

$$\left[\left(-f(t)\dot{V}_{-2}-\frac{1}{2}\dot{f}(t)\left(rV'_{-2}+4V_{-2}\right)+ca(t)\left(\frac{\mathrm{i}\mathcal{M}}{4}\ddot{f}(t)-\frac{\mathcal{M}^2}{4}r^2\dddot{f}(t)\right)\right)\partial_r^{-2}\right.$$

$$\left.+\left(-f\dot{V}_0-\dot{f}V_0+\frac{c}{2}a\dot{f}\right)\partial_r^0\right]dt,-(a\dot{f}+f\dot{a})dt\Bigg) \tag{10.51}$$

which gives the expected result for $c = 2$. □

Remark. By modifying as follows the relation defining the non-local transformation Θ (see Definition 10.7)

$$\partial_\xi^{\frac{1}{2}} \longrightarrow \partial_r, \quad \xi \longrightarrow \frac{1}{2}r\partial_r^{-1} + \nu\partial_r^{-2} \tag{10.52}$$

for an arbitrary real parameter ν, one may obtain all the actions in the family $d\tilde{\sigma}_\mu$, $\mu \in \mathbb{R}$ (as detailed but straightforward computations show). Note that

$$\left(\frac{1}{2}r\partial_r^{-1}\right)^* = -\frac{1}{2}\partial_r^{-1}r = -\frac{1}{2}r\partial_r^{-1} + \frac{1}{2}\partial_r^{-2}$$

so the operators $\frac{1}{2}r\partial_r^{-1} + \nu\partial_r^{-2}$, $\nu \in \mathbb{R}$ correspond to various (and mainly harmless) symmetrizations of $\frac{1}{2}r\partial_r^{-1}$.

10.7 Connection with the Poisson Formalism

The previous results suggest by the Kirillov-Kostant-Souriau formalism that $d\sigma_0(X)$, $X \in \mathfrak{sv}$ is a Hamiltonian vector field, image of some function F_X by the Hamiltonian operator. It is the purpose of this section to write down properly the Hamiltonian operator H and to spell out for every $X \in \mathfrak{sv}$ a function F_X such that $H_{F_X} = X$.

Identify $\mathfrak{h}^*$ as a subspace of $\mathcal{L}_t((\Psi D_r)_{\geq -2}) \oplus \mathcal{F}_{-1}$ through the pairing given by Adler's trace as in the first lines of Sect. 6, so that an element of $\mathfrak{h}^*$ writes generically $\left(\sum_{k\geq -2} V_k\partial_r^k \,.\, dt, a(t)dt\right)$. Consider similarly to [43] the space $\mathcal{F}_{loc}$ of local functionals on $\mathcal{L}_t((\Psi D_r)_{\geq -2})$, $\mathcal{F}_{loc} := \hat{\mathcal{F}}_{loc}/\mathrm{span}\left(\frac{d}{dt}\hat{\mathcal{F}}_{loc}, \frac{d}{dr}\hat{\mathcal{F}}_{loc}\right)$, with $\hat{\mathcal{F}}_{loc} = C^\infty(S^1 \times S^1) \otimes \mathbb{C}[(\partial_t^i\partial_r^j V_k)_{k\geq -2, i,j\geq 0}]$. An element F of $\mathcal{F}_{loc}$ defines by integration a $\mathbb{C}$-valued function $\int\int F(t,r)\,dt dr$ on $\mathcal{L}_t((\Psi D_r)_{\geq -2})$. The classical Euler-Lagrange variational formula yields the variational derivative

$$\frac{\delta F}{\delta V_k} = \sum_{i,j=0}^{\infty} (-1)^{i+j}\partial_t^i\partial_r^j\left(\frac{\partial F}{\partial(\partial_t^i\partial_r^j V_k)}\right). \tag{10.53}$$

Local vector fields are then formally derivations of $\hat{\mathscr{F}}_{loc}$ commuting with $\frac{d}{dt}$ and $\frac{d}{dr}$, so that they define linear morphisms $X : \mathscr{F}_{loc} \to \mathscr{F}_{loc}$. It is also possible to represent X more geometrically as a vector field on $\mathscr{L}_t((\Psi D_r)_{\geq -2})$; since $\mathscr{L}_t((\Psi D_r)_{\geq -2})$ is linear, X is a mapping $X : \mathscr{L}_t((\Psi D_r)_{\geq -2}) \to \mathscr{L}_t((\Psi D_r)_{\geq -2})$ with some additional requirements due to locality. Set $X(D) = \sum_{k \geq -2} A_k(D)\partial^k$, then (as a derivation of $\hat{\mathscr{F}}_{loc}$) it holds $X = \sum_{k\in\mathbb{Z}} \sum_{i,j\geq 0} \partial_t^i \partial_r^j a_k \ . \ \partial/\partial(\partial_t^i \partial_r^j V_k)$. Now the differential dF of a function $F \in \mathscr{F}_{loc}$ verifies by definition $dF(X) = X(F) = \sum_k a_k \frac{\delta F}{\delta V_k}$. Choose $D \in \mathscr{L}_t((\Psi D_r)_{\geq -2})$: then the differential of F at D should be a linear evaluation $\langle d_D F, X(D)\rangle = \int Tr d_D F(t) X(D)(t)\, dt$, hence (using once again the pairing given by Adler's trace) one has the following representation: $d_D F = \sum_k \partial^{-k-1} \frac{\delta F}{\delta V_k}(D) \in \mathfrak{h}$. Formally, one may simply write $dF = \sum_k \partial^{-k-1} \frac{\delta F}{\delta V_k}$.

Similar considerations apply to local functionals on $\text{Vect}(S^1)^*$ or $\mathscr{F}_{-1}$, with the difference that the variable r is absent. We refer once again to [43] for this very classical case. Since the generic element of $\text{Vect}(S^1)$, resp. $\text{Vect}(S^1)^*$, is denoted by $w(t)\partial_t$, resp. $v(t)dt^2$, the differential of a functional $F = F(v)$ will be denoted by $dF = \frac{\delta F}{\delta v}\partial_t$, while a vector field becomes $X(v) = A_{\mathscr{F}_{-2}}(v)dt^2$. Similarly, the differential of a functional $F = F(a)$ will be denoted by $dF = \frac{\delta F}{\delta a}$, while a vector field is $X(a) = A_{\mathscr{F}_{-1}}(a)dt$. Note that (considering e.g. thc case of $\text{Vect}(S^1)^*$) such a functional may be seen as a particular case of a 'mixed-type local functional' $\Phi(v, (V_k)_{k\geq -2})$ by setting $\Phi(v, (V_k)_{k\geq -2}) = r^{-1}F(v)$ (integrating with respect to r yields $\oint r^{-1} dr = 1$), but we shall not need such mixed-type functionals. We shall restrict to (i) local functionals on $\mathscr{L}_t((\Psi D_r)_{\geq -2})$, (ii) local functionals on $\text{Vect}(S^1)^*$ and (iii) local functionals on $\mathscr{F}_{-1}$, which are sufficient for our purposes.

It is now possible to write down explicitly the Poisson bracket of local functionals of the above three types on $\mathfrak{g}^*$; we shall restrict to the affine subspace

$$\text{Vect}(S^1)^* \ltimes \left\{\left(\left[V_{-2}(t,r)\partial_r^{-2} + V_0(t,r)\partial_r^0\right]dt, a(t)dt\right)\right\} \subset \mathfrak{g}^*$$

(note that we allow a dependence on r of the potential V_0 for the time being). Denote by $V = (V_{-2}, V_0)$ the element $V_{-2}(t,r)\partial_r^{-2} + V_0(t,r)\partial_r^0$. Consider first local functionals F, G on $\mathscr{L}_t((\Psi D_r)_{\geq -2})$. By the Kirillov-Kostant-Souriau construction,

$$\begin{aligned}
&\left\{\iint F\, dtdr, \iint G\, dtdr\right\}\left(\left(v(t)dt^2; \left[V_{-2}(t,r)\partial_r^{-2} + V_0(t,r)\partial_r^0\right]dt, a(t)dt\right)\right) \\
&\quad = \left\langle \left(\left[V_{-2}\partial_r^{-2} + V_0\right]dt, a(t)dt\right), [d_V F, d_V G]_{\mathfrak{h}} \right\rangle_{\mathfrak{h}^* \times \mathfrak{h}} \\
&\quad = \int \left\{ \text{Tr}\left((V_{-2}\partial_r^{-2} + V_0).[d_V F, d_V G]_{\mathscr{L}_t((\Psi D_r)_{\leq 1})}\right) + cc_3(d_V F, d_V G)\right\} dt.
\end{aligned} \tag{10.54}$$

Recall from the previous considerations that $dF = \partial_r \frac{\delta F}{\delta V_{-2}} + \frac{\delta F}{\delta V_{-1}} + \partial_r^{-1} \frac{\delta F}{\delta V_0} + \dots$ The operation of taking the trace leaves out only the bracket $\left[\partial_r \frac{\delta F}{\delta V_{-2}}, \partial_r \frac{\delta G}{\delta V_{-2}}\right]$ which couples to $V_{-2}\partial_r^{-2}$, and the mixed brackets $\left[\partial_r \frac{\delta F}{\delta V_{-2}}, \partial_r^{-1} \frac{\delta G}{\delta V_0}\right]$ and $\left[\partial_r \frac{\delta G}{\delta V_{-2}}, \partial_r^{-1} \frac{\delta F}{\delta V_0}\right]$ which couple to V_0, while the central extension couples only the coefficients of ∂_r and ∂_r^{-1}. All together one obtains

$$\left\{\iint F\, dt dr, \iint G\, dt dr\right\} \left((v(t)dt^2; \left[V_{-2}(t,r)\partial_r^{-2} + V_0(t,r)\partial_r^0\right] dt, a(t)dt)\right)$$

$$= \iint V_{-2}\left[\left(\frac{\delta G}{\delta V_{-2}}\right)' \frac{\delta F}{\delta V_{-2}} - \left(\frac{\delta F}{\delta V_{-2}}\right)' \frac{\delta G}{\delta V_{-2}}\right] dt\, dr$$

$$+ \iint V_0 \left[\frac{\delta G}{\delta V_0}\frac{\delta F}{\delta V_{-2}} - \frac{\delta G}{\delta V_{-2}}\frac{\delta F}{\delta V_0}\right]' dt\, dr$$

$$+c \iint \left[\left(\frac{\delta F}{\delta V_0}\right)' \cdot \left(\frac{\delta G}{\delta V_{-2}}\right) + \left(\frac{\delta F}{\delta V_{-2}}\right)' \cdot \left(\frac{\delta G}{\delta V_0}\right)\right] a(t)\, dt\, dr. \tag{10.55}$$

Assume now that F is a functional on $\mathrm{Vect}(S^1)^*$ and G a functional on $\mathscr{L}_t((\Psi D_r)_{\geq -2})$; then

$$\left\{\int F\, dt, \iint G\, dt dr\right\} \left((v(t)dt^2; \left[V_{-2}(t,r)\partial_r^{-2} + V_0(t,r)\partial_r^0\right] dt, a(t)dt)\right)$$

$$= \left\langle \left(\left[V_{-2}\partial_r^{-2} + V_0\right] dt, a(t)dt\right), \frac{\delta F}{\delta v}\partial_t . d_V G\right\rangle_{\mathfrak{h}^*\times\mathfrak{h}}$$

$$= \iint \frac{\delta F}{\delta v} \cdot \left(V_{-2}\frac{d}{dt}\left(\frac{\delta G}{\delta V_{-2}}\right) + V_0 \frac{d}{dt}\left(\frac{\delta G}{\delta V_0}\right)\right) dt\, dr. \tag{10.56}$$

Similarly, if F is a functional on $\mathrm{Vect}(S^1)^*$ and G a function on $\mathscr{F}_{-1}$, then

$$\left\{\int F\, dt, \int G\, dt\right\} \left((v(t)dt^2; \left[V_{-2}(t,r)\partial_r^{-2} + V_0(t,r)\partial_r^0\right] dt, a(t)dt)\right)$$

$$= \int a(t) \frac{\delta F}{\delta v}\frac{d}{dt}\left(\frac{\delta G}{\delta a}\right) dt. \tag{10.57}$$

Finally, if both F and G are functionals on $\mathrm{Vect}(S^1)^*$, then (as is classical)

$$\left\{\int F\, dt, \int G\, dt\right\} \left((v(t)dt^2; \left[V_{-2}(t,r)\partial_r^{-2} + V_0(t,r)\partial_r^0\right] dt, a(t)dt)\right)$$

$$= \int v(t) \left[\frac{\delta F}{\delta v}\frac{d}{dt}\left(\frac{\delta G}{\delta v}\right) - \frac{\delta G}{\delta v}\frac{d}{dt}\left(\frac{\delta F}{\delta v}\right)\right] dt. \tag{10.58}$$

Consider now the Hamiltonian operator $F \to H_F$. Set $H_F = A_{\mathscr{F}_{-2}} dt^2 + \sum_{k\geq -2} A_k \partial^k + A_{\mathscr{F}_{-1}} dt$, then

$$dG(H_F)(v(t)dt^2; Vdt, a(t)dt) = H_F(G)(v(t)dt^2; Vdt, a(t)dt)$$
$$= \{F, G\}(v(t)dt^2; Vdt, a(t)dt) \quad (10.59)$$

writes $\int\int \sum_{k\geq -2} A_k \frac{\delta G}{\delta V_k}(V)\ dtdr$ if G is a functional on $\mathcal{L}_t((\Psi D_r)_{\geq -2})$, $\int A_{\mathcal{F}_{-2}} \frac{\delta G}{\delta v}(v)\ dt$ if G is a functional on $\mathrm{Vect}(S^1)^*$, and $\int A_{\mathcal{F}_{-1}} \frac{\delta G}{\delta a}(a)\ dt$ if G if a functional on $\mathcal{F}_{-1}$, hence

$$\begin{aligned} &H_F\left(v(t)dt^2; \left[V_{-2}(t,r)\partial_r^{-2} + V_0(t,r)\partial_r^0\right] dt, a(t)dt\right) \\ &= \left(-\oint dr \left[V_{-2}\frac{d}{dt}\left(\frac{\delta F}{\delta V_{-2}}\right) + V_0 \frac{d}{dt}\left(\frac{\delta F}{\delta V_0}\right)\right] dt^2;\right. \\ &\left(-2V_{-2}\left(\frac{\delta F}{\delta V_{-2}}\right)' - V'_{-2}\frac{\delta F}{\delta V_{-2}} + ca(t)\left(\frac{\delta F}{\delta V_0}\right)' + V'_0\frac{\delta F}{\delta V_0}\right)\partial_r^{-2} \\ &\left. + \left(ca(t)\left(\frac{\delta F}{\delta V_{-2}}\right)' - V'_0\frac{\delta F}{\delta V_{-2}}\right)\partial_r^0, 0\right) \end{aligned} \quad (10.60)$$

if F is a functional on $\mathcal{L}_t((\Psi D_r)_{\geq -2})$, and

$$\begin{aligned} &H_F\left(v(t)dt^2; \left[V_{-2}(t,r)\partial_r^{-2} + V_0(t,r)\partial_r^0\right] dt, a(t)dt\right) \\ &= \left(\left[-2v\frac{d}{dt}\left(\frac{\delta F}{\delta v}\right) - \dot v \frac{\delta F}{\delta v}\right] dt^2; -\frac{d}{dt}\left(V_{-2}\frac{\delta F}{\delta v}\right)\partial_r^{-2}\right. \\ &\left. -\frac{d}{dt}\left(V_0 \frac{\delta F}{\delta v}\right)\partial_r^0, -\frac{d}{dt}(a\frac{\delta F}{\delta v})dt\right) \end{aligned} \quad (10.61)$$

if F is a functional on $\mathrm{Vect}(S^1)^*$.

Let F be a functional on $\mathcal{L}_t((\Psi D_r)_{\geq -2})$ depending only on V_0 and V_{-2}; note that H_F preserves the affine subspace $\mathcal{N}$ if and only if

$$F = F_0(V_0) + \sum_{i,j=0}^{\infty} \partial_t^i \partial_r^j V_{-2}. \sum_{k=0}^{j+1} r^k f_{ijk}(V_0), \quad (10.62)$$

where F_0 is any functional depending on V_0 and t, r, and $(f_{ijk})_{i,j,k}$ any set of functionals depending on V_0 and only on t. In particular, such a functional is affine in V_2 and its derivatives, and the coefficient of V_{-2} affine in r.

Lemma 10.16. *The coadjoint action* $\mathrm{ad}^*_{\mathfrak{g}}(X)$ *of* $X = \mathcal{L}_f$, *resp.* $\mathcal{Y}_g$, *resp.* $\mathcal{M}_h \in \mathfrak{sv}$ *on* $\mathcal{N}$ *may be identified with the Hamiltonian vector field* H_{F_X} *with*

$$\begin{aligned} F_{\mathcal{L}_f}(v, (V_k)_{k\in\mathbb{Z}}) &= \int f v\, dt + \frac{1}{2}\iint r\dot f V_{-2}\, dt\, dr \\ &+ \iint \left(\mathrm{i}\frac{\mathcal{M}}{4} r\ddot f - \frac{\mathcal{M}^2}{12} r^3 \dddot f\right) V_0\, dt\, dr; \end{aligned} \quad (10.63)$$

$$F_{\mathcal{Y}_g}((V_k)_{k\in\mathbb{Z}}) = \iint gV_{-2}\,dt\,dr - \frac{\mathcal{M}^2}{2}\iint \ddot{g}r^2 V_0\,dt\,dr; \quad (10.64)$$

$$F_{\mathcal{M}_h}((V_k)_{k\in\mathbb{Z}}) = \mathcal{M}^2 \iint r\dot{h}V_0\ dt\ dr. \quad (10.65)$$

Furthermore, $\{F_X, F_Y\} = F_{[X,Y]}$ *if* $X, Y \in \mathfrak{sv}$*, except for the Poisson brackets*

$$\{F_{\mathcal{Y}_g}, F_{\mathcal{M}_h}\} = \mathcal{M}^2 \iint \dot{h}gV_0\,dtdr; \quad (10.66)$$

$$\{F_{\mathcal{L}_f}, F_{\mathcal{Y}_g}\} = F_{[\mathcal{L}_f,\mathcal{Y}_g]} - \mathrm{i}\frac{\mathcal{M}}{4}\iint g\dddot{f}V_0\,dtdr. \quad (10.67)$$

The additional terms on the right are functionals of the form $\int\int fV_0\,dtdr$ *which vanish on* $\mathcal{N}$ *(and whose Hamiltonian acts of course trivially on* $\mathcal{N}$*).*

Proof. Straightforward computations. □

Chapter 11
Supersymmetric Extensions of the Schrödinger–Virasoro Algebra

Recall the following facts from Chaps. 1 and 2:

- The Schrödinger algebra $\mathfrak{sch}$ is defined as the algebra of projective Lie symmetries of the free Schrödinger equation in $(1+1)$-dimensions, $(-2\mathrm{i}\mathcal{M}\partial_t - \partial_r^2)\psi = 0$.
- Replacing formally $\mathcal{M}$ by ∂_ζ (i.e. taking a Laplace transform with respect to the mass parameter) yields an equation which is formally equivalent to the free Laplace equation $\Delta\psi(t,r,\zeta) = 0$ in *three* dimensions; this way, $\mathfrak{sch}$ is shown to be embedded into the conformal algebra in three dimensions, with its natural action on the space of solutions of $\Delta\psi = 0$.
- Independently from this conformal embedding, $\mathfrak{sch}$ (in the above realization as Lie symmetries of an equation) may be extended into an infinite-dimensional Lie algebra, the Schrödinger–Virasoro algebra, which may in turn also be obtained as a subquotient of an extended Poisson algebra on the torus, namely, $\tilde{\mathcal{A}}(S^1)_{\leq 1}/\tilde{\mathcal{A}}(S^1)_{\leq -\frac{1}{2}}$.

The purpose of this chapter, adapted from [55], is to give generalizations (as systematically as possible) of these constructions in the supersymmetric setting. The analogues of the free Schrödinger/Laplace equations are, in this respect, what we call the *super-Schrödinger model*, see (11.35) and the $(3|2)$-*supersymmetric model*, see (11.32), with algebra of projective Lie symmetries $\tilde{\mathfrak{s}}^{(2)}$, called $(N=2)$-*supersymmetric Schrödinger algebra* (see Proposition 11.5), resp. $\mathfrak{s}^{(2)} \supset \tilde{\mathfrak{s}}^{(2)}$, see Proposition 11.6. As in the non-supersymmetric case, $\tilde{\mathfrak{s}}^{(2)}$ has a semi-direct product structure, $\tilde{\mathfrak{s}}^{(2)} \simeq \mathfrak{osp}(2|2) \ltimes \mathfrak{sh}(2|2)$, where $\mathfrak{osp}(2|2)$ is an orthosymplectic Lie algebra, and $\mathfrak{sh}(2|2)$ a super-Heisenberg algebra, while $\mathfrak{s}^{(2)} \simeq \mathfrak{osp}(2|4)$ is a simple Lie superalgebra.

As in the non-supersymmetric case too, $\tilde{\mathfrak{s}}^{(2)}$ may be seen to have an infinite-dimensional extension, $\mathfrak{sns}^{(2)}$ (called *Schrödinger-Neveu-Schwarz algebra with two supercharges*), which may be realized as a subquotient of an extended Poisson algebra of the super-torus with two odd coordinates, and the action of $\tilde{\mathfrak{s}}^{(2)}$ on

J. Unterberger and C. Roger, *The Schrödinger-Virasoro Algebra*, Theoretical and Mathematical Physics, DOI 10.1007/978-3-642-22717-2_11,

solutions of the super-Schrödinger model may be extended into a representation of $\mathfrak{sns}^{(2)}$, see Proposition 11.24.

Schrödinger-Neveu-Schwarz Lie superalgebras $\mathfrak{sns}^{(N)}$ with N supercharges are introduced in complete generality. They appear as a semi-direct product of Lie algebras of super-contact vector fields with infinite-dimensional nilpotent Lie superalgebras.

Note that supersymmetric extensions of the Schrödinger algebra have been discussed several times in the past [7–9, 26, 33, 35, 36], sometimes in the context of supersymmetric quantum mechanics. Here, we consider the problem from a field-theoretical perspective.

We begin in Sect. 11.1 by recalling some useful facts about the Schrödinger-invariance of the scalar free Schrödinger equation and then give a generalization to its spin-$\frac{1}{2}$ analogue, the Lévy-Leblond equation. By considering the 'mass' as an additional variable, we extend the spinor representation of the Schrödinger algebra $\mathfrak{sch}$ into a representation of $\mathfrak{conf}_3$. As an application, we derive the Schrödinger-covariant two-point spinorial correlation functions. In Sect. 11.2, we combine the free Schrödinger and Lévy-Leblond equations (together with a scalar auxiliary field) into a *super-Schrödinger model*, and show, by using a superfield formalism in 3 + 2 dimensions, that this model has a kinematic supersymmetry algebra with $N = 2$ supercharges. Including then time-inversions, we compute the full dynamical symmetry algebra and prove that it is isomorphic to the Lie algebra of symmetries $\mathfrak{osp}(2|2) \ltimes \mathfrak{sh}(2|2)$ found in several mechanical systems with a finite number of particles. By treating the 'mass' as a coordinate, we obtain a well-known supersymmetric model (see [29]) that we call the (3|2)-*supersymmetric model*. Its dynamical symmetries form the Lie superalgebra $\mathfrak{osp}(2|4)$. The derivation of these results is greatly simplified through the correspondence with Poisson structures and the introduction of several gradings which will be described in detail. In Sect. 11.3, we use a Poisson algebra formalism to construct for every N an infinite-dimensional supersymmetric extension with N supercharges of the Schrödinger algebra that we call *Schrödinger-Neveu-Schwarz algebra* and denote by $\mathfrak{sns}^{(N)}$. At the same time, we give an extension of the differential-operator representation of $\mathfrak{osp}(2|4)$ into a differential-operator representation of $\mathfrak{sns}^{(2)}$. We compute in Sect. 11.4 the two-point correlation functions that are covariant under $\mathfrak{osp}(2|4)$ or under some of its subalgebras. Remarkably, in many instances, the requirement of supersymmetric covariance is enough to allow only a finite number of possible quasiprimary superfields. Our conclusions are given in Sect. 11.5. In Appendix B, we present the details for the calculation of the supersymmetric two-point functions, and collect for easy reference the numerous Lie superalgebras introduced in the paper and their differential-operator realization as Lie symmetries of the (3|2)-supersymmetric model.

The reader may conveniently refer to the anthology on superalgebras [28] for any information on superalgebras.

11.1 On the Dirac-Lévy-Leblond Equation

Throughout this chapter we shall use the following notation: $[A, B]_{\mp} := AB \mp BA$ stand for the commutator and anticommutator, respectively. We shall often simply write $[A, B]$ if it is clear which one should be understood. Furthermore $\{A, B\} := \frac{\partial A}{\partial q}\frac{\partial B}{\partial p} - \frac{\partial A}{\partial p}\frac{\partial B}{\partial q}$ denotes the Poisson bracket or supersymmetric extensions thereof which will be introduced below. We shall use the Einstein summation convention unless explicitly stated otherwise.

In this section we first recall some properties of the one-dimensional free Schrödinger equation before considering a reduction to a system of first-order equations introduced by Lévy-Leblond [76].

Consider the free Schrödinger or diffusion equation

$$\Delta_0 \tilde{\phi} = (2\mathcal{M}\partial_t - \partial_r^2)\tilde{\phi} = 0 \tag{11.1}$$

in one space-dimension, where the Schrödinger operator may be expressed in terms of the generators of $\mathfrak{sch}$ as $\Delta_0 := 2M_0 L_{-1} - Y_{-1/2}^2$. In many situations, it is useful to consider the 'mass' $\mathcal{M}$ as an additional variable such that $\tilde{\phi} = \tilde{\phi}_{\mathcal{M}}(t, r)$. As a general rule (see Chap. 1), we shall denote in this chapter by ζ the variable conjugate to $\mathcal{M}$ via a Fourier-Laplace transformation, and the corresponding wave function by the same letter but without the tilde, here $\phi = \phi(\zeta, t, r)$. In this way, one may show that there is a complex embedding of the Schrödinger algebra into the conformal algebra, viz. $(\mathfrak{sch}_1)_{\mathbb{C}} \subset (\mathfrak{conf}_3)_{\mathbb{C}}$ (see Sect. 2.2) [1]

We illustrate this for the one-dimensional case $d = 1$ in Fig. 11.1, where the root diagram for $(\mathfrak{conf}_3)_{\mathbb{C}} \cong B_2$ is shown. We also give a name to the extra conformal generators. In particular, $\langle N, L_0\rangle$ form a Cartan subalgebra and the eigenvalue of $\mathrm{ad}N$ on any root vector is given by the coordinate along $-e_1$.[2] Furthermore, the conformal invariance of the Schrödinger equation follows from $[\Delta_0, V_-] = 0$ (see [54]). Note that the generator N is related to the generator N_0 introduced in Sect. 6.1 (namely, N is a linear combination of N_0 and L_0).

One of the main applications of the (super-)symmetries studied in this article will be the calculation of *covariant* or *quasi-primary* correlation functions, as defined in Sect. 2.4.

As a specific example, let us consider here n-point functions that are covariant under $(\mathfrak{conf}_3)_{\mathbb{C}}$ or one of its Lie subalgebras, which for our purposes will be either $\mathfrak{sch}_1$ or the maximal *parabolic* subalgebra, introduced in Chap. 2 (and used in Chap. 6)

$$\overline{\mathfrak{sch}} := \mathbb{C}N \ltimes \mathfrak{sch} \tag{11.2}$$

[1] This apparently abstract extension becomes important for the explicit calculation of the two-time correlation function in phase-ordering kinetics [53].

[2] For example, $\mathrm{ad}N(Y_{\frac{1}{2}}) = [N, Y_{\frac{1}{2}}] = -Y_{\frac{1}{2}}$ or $\mathrm{ad}N(Y_{-\frac{1}{2}}) = [N, Y_{-\frac{1}{2}}] = 0$.

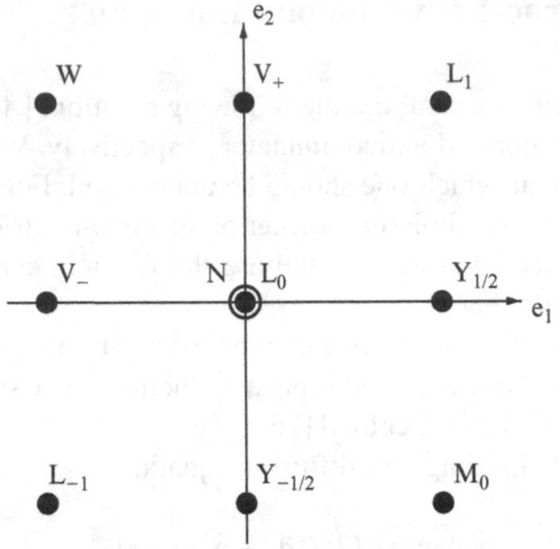

Fig. 11.1 Roots of the three-dimensional complexified conformal Lie algebra $(\mathfrak{conf}_3)_{\mathbb{C}} \cong B_2$ and their relation with the generators of the Schrödinger algebra $\mathfrak{sch}_1$. The double circle in the centre denotes the Cartan subalgebra $\mathfrak{h}$

The generator N is given by

$$N = -t\partial_t + \zeta\partial_\zeta + \nu \tag{11.3}$$

with $\nu = 0$ [54]. But any choice for the value of ν also gives a representation of $\widetilde{\mathfrak{sch}}(1)$, although it does not extend to the whole conformal Lie algebra. So one may consider more generally $\widetilde{\mathfrak{sch}}(1)$-quasiprimary fields characterized by a scaling exponent x and a N*-exponent* ν.

It will turn out later to be useful to work with the variable ζ conjugate to $\mathcal{M}$. If we arrange for $\mathcal{M} = -\frac{1}{2}\partial_\zeta$ through a Laplace transform, see (11.9) below, it is easy to see that the $\mathfrak{sch}$-covariant two-point function under the vector field representation $d\tilde{\pi}_\lambda$ (see Introduction) is given by [54]

$$\begin{aligned}\langle\phi_1\phi_2\rangle &:= \langle\phi_1(\zeta_1, t_1, r_1)\phi_2(\zeta_2, t_2, r_2)\rangle \\ &= \psi_0\delta_{x_1,x_2}(t_1 - t_2)^{-x_1} f\left(\zeta_1 - \zeta_2 + \frac{1}{4}\frac{(r_1 - r_2)^2}{t_1 - t_2}\right)\end{aligned} \tag{11.4}$$

where $x_{1,2}$ are the scaling dimensions of the $\mathfrak{sch}$-quasiprimary fields $\phi_{1,2}$, f is an undetermined scaling function and ψ_0 a normalization constant. If $\phi_{1,2}$ are $\widetilde{\mathfrak{sch}}$-quasiprimary fields with N-exponents $\nu_{1,2}$, then $f(u) = u^{-x_1-\nu_1-\nu_2}$ is fixed[3]. If in addition $\psi_{1,2}$ are $\mathfrak{conf}_3$-quasiprimary, then $f(u) = u^{-x_1}$ and, after an inverse Laplace transform, this two-point function becomes the well-known heat kernel

[3] This fixing of the scaling function through the additional N-covariance prompted us to consider $\widetilde{\mathfrak{sv}}$-primary fields instead of $\mathfrak{sv}$-primary fields in our vertex construction (see Sect. 6.1).

$\langle\tilde{\phi}_1\tilde{\phi}_2\rangle = \phi_0\delta_{x_1,x_2}t^{-x_1}\exp(-r^2/(2\mathscr{M}t))$, together with the causality condition $t > 0$ [54]. The same form for $\langle\tilde{\phi}_1\tilde{\phi}_2\rangle$ also holds true for $\mathfrak{sch}$-quasiprimary fields, since the function f in (11.4) simply gives, after the Laplace transform, a mass-dependent normalization constant ϕ_0.

We now turn to the Dirac-Lévy-Leblond equations (see also Sect. 8.3). They were constructed by Lévy-Leblond [76] by adapting to a non-relativistic setting Dirac's square-root method for finding a relativistically covariant partial differential equation of first order. Consider in d space dimensions a first-order vector wave equation of the form

$$\mathscr{R}\tilde{\Phi} := \left(A\frac{\partial}{\partial t} + \sum_{i=1}^{d} B_i\frac{\partial}{\partial r^i} + \mathscr{M}C\right)\tilde{\Phi} = 0 \tag{11.5}$$

where A, B_i and C are matrices to be determined such that the square of the operator $\mathscr{R}$ is equal to the free Schrödinger operator $\mathscr{R}^2 \stackrel{!}{=} \Delta_0 = 2\mathscr{M}\partial_t - \sum_{i=1}^{d}(\partial_{r^i})^2$. It is easy to see that the matrices A, B_i, C give a representation of a Clifford algebra (with an unconventional metric) in $d+2$ dimensions. Namely, if one sets

$$\begin{aligned} \mathfrak{B}_j &:= \mathrm{i}\sqrt{2}\,B_j;\quad j = 1,\ldots,d \\ \mathfrak{B}_{d+1} &:= A + \frac{1}{2}C,\quad \mathfrak{B}_{d+2} := \mathrm{i}\left(A - \frac{1}{2}C\right) \end{aligned} \tag{11.6}$$

then the condition on $\mathscr{R}$ is equivalent to $[\mathfrak{B}_j, \mathfrak{B}_k]_+ = 2\delta_{j,k}$ for $j,k = 1,\ldots,d+2$.

We are interested in the case $d = 1$. Then the Clifford algebra generated by $\mathfrak{B}_j$, $j = 1,2,3$, has exactly one irreducible representation up to equivalence, which is given for instance by

$$\mathfrak{B}_1 = \sigma^3 = \begin{pmatrix} 1 & 0 \\ 0 & -1 \end{pmatrix},\quad \mathfrak{B}_2 = \sigma^2 = \begin{pmatrix} 0 & \mathrm{i} \\ -\mathrm{i} & 0 \end{pmatrix},\quad \mathfrak{B}_3 = \sigma^1 = \begin{pmatrix} 0 & 1 \\ 1 & 0 \end{pmatrix}. \tag{11.7}$$

Then the wave equation $\mathscr{R}\tilde{\Phi} = \mathscr{R}\begin{pmatrix}\tilde{\psi} \\ \tilde{\phi}\end{pmatrix} = 0$ becomes explicitly, after a rescaling $r \mapsto \sqrt{2}r$

$$\partial_t\tilde{\psi} = \partial_r\tilde{\phi},\quad 2\mathscr{M}\tilde{\phi} = \partial_r\tilde{\psi} \tag{11.8}$$

These are the *Dirac-Lévy-Leblond equations* in one space dimension.

Since the masses $\mathscr{M}$ are by physical convention real and positive, it is convenient to define their conjugate ζ through a Laplace transform

$$\psi(\zeta,t,r) = \int_0^\infty \mathrm{d}\mathscr{M}\, e^{-2\mathscr{M}\zeta}\tilde{\psi}(\mathscr{M};t,r) \tag{11.9}$$

and similarly for ϕ. Then (11.8) become

$$\partial_t \psi = \partial_r \phi, \ \ \partial_\zeta \phi = -\partial_r \psi. \tag{11.10}$$

Actually, it is easy to see that these (11.10) are equivalent to the three-dimensional massless free Dirac equation $\gamma^\mu \partial_\mu \Phi = 0$, where $\partial_\mu = \partial/\partial\xi^\mu$ and ξ^μ with $\mu = 1, 2, 3$ are the coordinates. If we set $t = \frac{1}{2}(\xi^1 + \mathrm{i}\xi^2)$, $\zeta = \frac{1}{2}(\xi^1 - \mathrm{i}\xi^2)$ and finally $r = \xi^3$, and choose the representation $\gamma^\mu = \sigma^\mu$, then we recover indeed (11.10).

The invariance of the free massless Dirac equation under the conformal group is well-known.[4] The generators of $\mathfrak{conf}_3$ act as follows on the spinor field $\Phi = \begin{pmatrix} \psi \\ \phi \end{pmatrix}$, (see again Fig. 11.1)

$$\begin{aligned}
L_{-1} &= -\partial_t, \ Y_{-\frac{1}{2}} = -\partial_r, \ M_0 = \frac{1}{2}\partial_\zeta \\
V_- &= -r\partial_t + 2\zeta\partial_r - \begin{pmatrix} 0 & 1 \\ 0 & 0 \end{pmatrix} \\
N &= -t\partial_t + \zeta\partial_\zeta + \frac{1}{2}\begin{pmatrix} 1 & 0 \\ 0 & -1 \end{pmatrix} \\
Y_{\frac{1}{2}} &= -t\partial_r + \frac{1}{2}r\partial_\zeta - \frac{1}{2}\begin{pmatrix} 0 & 0 \\ 1 & 0 \end{pmatrix} \\
L_0 &= -t\partial_t - \frac{1}{2}r\partial_r - \frac{1}{2}\begin{pmatrix} x & 0 \\ 0 & x+1 \end{pmatrix} \\
W &= -\frac{1}{2}r^2\partial_t + 2\zeta^2\partial_\zeta + 2\zeta r\partial_r + \begin{pmatrix} 2(x+1)\zeta & -r \\ 0 & 2x\zeta \end{pmatrix} \\
V_+ &= -tr\partial_t - \zeta r\partial_\zeta - \frac{1}{2}(r^2 - 4\zeta t)\partial_r - \frac{1}{2}\begin{pmatrix} (2x+1)r & 2t \\ -2\zeta & (2x+1)r \end{pmatrix} \\
L_1 &= -t^2\partial_t - tr\partial_r + \frac{1}{4}r^2\partial_\zeta - \begin{pmatrix} xt & 0 \\ r/2 & (x+1)t \end{pmatrix}
\end{aligned} \tag{11.11}$$

For solutions of (11.10), one has $x = \frac{1}{2}$. As in the case of the vector field representation $d\tilde{\pi}_\lambda$ (see Introduction), arbitrary values of the scaling exponent $x = 2\lambda$ also give a representation of the conformal algebra.

There are three 'translations' $(L_{-1}, Y_{-\frac{1}{2}}, M_0)$, three 'rotations' $(V_-, N, Y_{\frac{1}{2}})$, one 'dilatation' (L_0) and three 'inversions' or special transformations (W, V_+, L_1). It is sometimes useful to work with the generator $D := 2L_0 - N$ (whose differential part is the Euler operator $-t\partial_t - r\partial_r - \zeta\partial_\zeta$) instead of either L_0 or N. We also see that the individual components ψ, ϕ of the spinor Φ have scaling dimensions $x_\psi = x$ and $x_\phi = x + 1$, respectively. If we write the Dirac operator as

---

[4] The Schrödinger-invariance of a free non-relativistic particle of spin S is proven in [45].

$$\mathscr{D} = \frac{1}{\mathrm{i}}\mathscr{R} = \begin{pmatrix} \partial_r & \partial_\zeta \\ \partial_t & -\partial_r \end{pmatrix} \tag{11.12}$$

then the Schrödinger- and also the full conformal invariance of the Dirac-Lévy-Leblond equation $\mathscr{D}\Phi = 0$ follows from the commutators

$$[\mathscr{D}, L_1] = -t\mathscr{D} - \left(x - \frac{1}{2}\right)\begin{pmatrix} 0 & 0 \\ 0 & 1 \end{pmatrix}$$

$$[\mathscr{D}, L_{-1}] = \left[\mathscr{D}, Y_{-\frac{1}{2}}\right] = [\mathscr{D}, V_-] = 0 \tag{11.13}$$

It is clear that dynamical symmetries of the Dirac-Lévy-Leblond equation are obtained only if $x = \frac{1}{2}$. Since $L_1, L_{-1}, Y_{-\frac{1}{2}}, V_-$ generate $(\mathfrak{conf}_3)_{\mathbb{C}}$, as can be seen from the root structure represented in Fig. 11.1, the symmetry under the remaining generators of $(\mathfrak{conf}_3)_{\mathbb{C}}$ follows from the Jacobi identities.

Let $\Phi_i = \begin{pmatrix} \psi_i \\ \phi_i \end{pmatrix}$, $i = 1, 2$ be two quasiprimary spinors under the representation (11.11) of either $\mathfrak{sch}, \widetilde{\mathfrak{sch}}$ or $\mathfrak{conf}_3$, with scaling dimensions $\begin{pmatrix} x_i \\ x_i + 1 \end{pmatrix}$ of the component fields. We now consider the covariant two-point functions; from translation-invariance it is clear that these will only depend on $\zeta = \zeta_1 - \zeta_2, t = t_1 - t_2$ and $r = r_1 - r_2$.

Proposition 11.1. *Suppose Φ_1, Φ_2 are quasiprimary spinors under the representation (11.11) of $\mathfrak{sch}$. Then their two-point functions vanish unless $x_1 = x_2$ or $x_1 = x_2 \pm 1$, in which case they read (ϕ_0, ψ_0 are normalization constants)*

(i) if $x_1 = x_2$, then

$$\begin{aligned} \langle \psi_1 \psi_2 \rangle &= \psi_0 t \left(4\zeta t + r^2\right)^{-x_1 - 1} \\ \langle \psi_1 \phi_2 \rangle = \langle \phi_1 \psi_2 \rangle &= -\frac{1}{2}\psi_0 r \left(4\zeta t + r^2\right)^{-x_1 - 1} \\ \langle \phi_1 \phi_2 \rangle &= \frac{\psi_0}{4}\frac{r^2}{t}\left(4\zeta t + r^2\right)^{-x_1 - 1} + \phi_0 \frac{1}{t}\left(4\zeta t + r^2\right)^{-x_1} \end{aligned} \tag{11.14}$$

(ii) if $x_1 = x_2 + 1$, then

$$\begin{aligned} \langle \psi_1 \psi_2 \rangle = \langle \phi_1 \psi_2 \rangle &= 0 \\ \langle \psi_1 \phi_2 \rangle &= \psi_0 \left(4\zeta t + r^2\right)^{-x_1} \\ \langle \phi_1 \phi_2 \rangle &= -\frac{\psi_0}{2}\frac{r}{t}\left(4\zeta t + r^2\right)^{-x_1} \end{aligned} \tag{11.15}$$

The case $x_1 = x_2 - 1$ is obtained by exchanging Φ_1 with Φ_2.

For brevity, the arguments of the two-point functions in (11.14), (11.15) were suppressed. Let us emphasize that the scaling dimensions of the component fields with a standard Schrödinger form (11.4) ($\langle\psi_1\psi_2\rangle$ and $\langle\phi_1\phi_2\rangle$ in (11.14), and $\langle\psi_1\phi_2\rangle$ in (11.15)) must agree, which is not the case for the other two-point functions which are obtained from them by applying derivative operators.

On the other hand, the covariance under the whole conformal group implies the additional constraint $x_1 = x_2$ (equality of the scaling exponents), and we have

Proposition 11.2. *The non-vanishing two-point functions, $(\mathfrak{conf}_3)_{\mathbb{C}}$-covariant under the representation (11.11), of the fields ψ and ϕ are obtained from (11.14) with $x_1 = x_2$ and the extra condition $\phi_0 = -\psi_0/4$, which gives*

$$\langle\phi_1\phi_2\rangle = -\psi_0\zeta\left(4\zeta t + r^2\right)^{-x_1-1} \tag{11.16}$$

Proof. In proving these two propositions, we merely outline the main ideas since the calculations are straightforward. We begin with Proposition 11.1. Given the obvious invariance under the translations, we first consider the invariance under the special transformation L_1 and use the form (11.11). With the help of dilatation invariance (L_0) and Galilei-invariance ($Y_{\frac{1}{2}}$) this simplifies to

$$(x_1 - x_2)t\langle\psi_1\psi_2\rangle = 0, \quad (x_1 - x_2 - 1)t\langle\psi_1\phi_2\rangle - \frac{1}{2}r\langle\psi_1\psi_2\rangle = 0,$$
$$(x_1 - x_2 + 1)t\langle\phi_1\psi_2\rangle + \frac{1}{2}r\langle\psi_1\psi_2\rangle = 0,$$
$$(x_1 - x_2)t\langle\phi_1\phi_2\rangle + \frac{1}{2}r\langle\psi_1\phi_2\rangle - \frac{1}{2}r\langle\phi_1\psi_2\rangle = 0$$

Considering the first of these equations leads us to distinguish two cases: either (i) $x_1 = x_2$ or (ii) $\langle\psi_1\psi_2\rangle = 0$.

In the first case, we get from the remaining three equations

$$\langle\psi_1\phi_2\rangle = \langle\phi_1\psi_2\rangle = -\frac{r}{2t}\langle\psi_1\psi_2\rangle \tag{11.17}$$

and the covariance under $Y_{\frac{1}{2}}$, N and L_0, respectively, leads to the following system of equations

$$\left(-t\frac{\partial}{\partial r} + \frac{r}{2}\frac{\partial}{\partial\zeta}\right)\langle\psi_1\psi_2\rangle = 0, \quad \left(-t\frac{\partial}{\partial t} + \zeta\frac{\partial}{\partial\zeta} + 1\right)\langle\psi_1\psi_2\rangle = 0,$$
$$\left(-t\frac{\partial}{\partial t} - \frac{r}{2}\frac{\partial}{\partial r} - x_1\right)\langle\psi_1\psi_2\rangle = 0$$

with a unique solution (up to a multiplicative factor) given by the first line of (11.14). Similarly, covariance under the same three generators leads to a system of three linear inhomogeneous equations for $\langle\phi_1\phi_2\rangle$ whose general solution is also given in (11.14).

In the second case, the remaining conditions coming from L_1 are

$$(x_1 - x_2 - 1)t\langle\psi_1\phi_2\rangle = 0, (x_1 - x_2 + 1)t\langle\phi_1\psi_2\rangle = 0,$$

$$(x_1 - x_2)t\langle\phi_1\phi_2\rangle + \frac{r}{2}\left(\langle\psi_1\phi_2\rangle - \langle\phi_1\psi_2\rangle\right) = 0$$

and one of the conditions $x_1 = x_2 \pm 1$ must hold true. Supposing that $x_1 = x_2 + 1$, we get $\langle\phi_1\phi_2\rangle = -\frac{1}{2}(r/t)\langle\psi_1\phi_2\rangle$ and an analogous relation holds (with the first and second field exchanged) in the other case. Again, covariance under $Y_{\frac{1}{2}}, N, L_0$ leads to a system of three linear equations for $\langle\psi_1\phi_2\rangle$ whose general solution in given in (11.15).

To prove Proposition 11.2, it is now sufficient to verify covariance under the generator V_-. Direct calculation shows that (11.14) is compatible with this condition only if $\phi_0 = -\psi_0/4$. On the other hand, compatibility with the second case (11.15) requires that $\psi_0 = 0$. □

Remark. If we come back to the original fields $\tilde{\psi}, \tilde{\phi}$ by inverting the Laplace transform (11.9), the $\overline{\mathfrak{sch}}$-covariant two-point functions of (11.14) take the form

$$\begin{aligned}
\langle\tilde{\psi}_1\tilde{\psi}_2\rangle &= \psi_0'\left(\frac{\mathcal{M}}{t}\right)^{x_1}\exp\left(-\frac{\mathcal{M}}{2}\frac{r^2}{t}\right)\\
\langle\tilde{\psi}_1\tilde{\phi}_2\rangle &= \langle\tilde{\phi}_1\tilde{\psi}_2\rangle = -\psi_0'\frac{r}{2t}\left(\frac{\mathcal{M}}{t}\right)^{x_1}\exp\left(-\frac{\mathcal{M}}{2}\frac{r^2}{t}\right) \qquad (11.18)\\
\langle\tilde{\phi}_1\tilde{\phi}_2\rangle &= \frac{\psi_0'}{4}\frac{r^2}{t}\left(\frac{\mathcal{M}}{t}\right)^{x_1}\exp\left(-\frac{\mathcal{M}}{2}\frac{r^2}{t}\right) + \phi_0'\left(\frac{\mathcal{M}}{t}\right)^{x_1-1}\exp\left(-\frac{\mathcal{M}}{2}\frac{r^2}{t}\right)
\end{aligned}$$

where $\psi_0' = \psi_0/(\Gamma(x_1+1)2^{x_1+1})$, $\phi_0' = \phi_0/(\Gamma(x)2^{x_1})$, and $\Gamma(x)$ is the Gamma function.

Proposition 11.3. *(i) Let f be a solution of the Laplace-transformed Schrödinger equation $(\partial_\zeta\partial_t + \partial_r^2)f = 0$. Then $\Phi = \begin{pmatrix}\psi\\ \phi\end{pmatrix} := \begin{pmatrix}-\partial_\zeta f\\ \partial_r f\end{pmatrix}$ satisfies the Dirac-Lévy-Leblond equations (11.10).*

(ii) Suppose that f_1, f_2 are $\overline{\mathfrak{sch}}(1)$-quasiprimary fields with scaling exponents $x = x_1 = x_2$ and N-exponents $\nu_1 = \nu_2 = -\frac{1}{2}$, and let $\Phi_i := \begin{pmatrix}-\partial_\zeta f_i\\ \partial_r f_i\end{pmatrix}$. Then the covariant two-point function

$$\langle f_1 f_2\rangle = t^{-x}(\zeta + r^2/4t)^{1-x}$$

implies a particular case of (11.14), given by $\psi_0 = -x(x-1)2^{2x+2}$ *and* $\phi_0 = (x-1)2^{2x-1}$.

Both assertions are easily checked by straightforward calculations. Note that (ii) is a priori surprising since the vector field representation $d\tilde{\pi}_{1/4}$ acting on solutions of the free Schrödinger equation is carried over by the transformation $f \to \Phi$ not to the representation (11.11), but to a different representation made up of integro-differential operators.

11.2 Supersymmetry in Three Dimensions and Supersymmetric Schrödinger-Invariance

11.2.1 From $N = 2$ Supersymmetry to the Super-Schrödinger Equation

We begin by recalling the construction of super space-time [29, Lectures 3 & 4] cf. also [18]. Take as n-dimensional space-time $\mathbb{R}^n$ (or, more generally, any n-dimensional Lorentzian manifold). One has a quite general construction of (non-supercommutative) superspace-time $\mathbb{R}^{n|s}$, with s odd coordinates, as the exponential of the Lie superalgebra $V \oplus S$, where the even part V is an n-dimensional vector space, and the s-dimensional odd part S is a spin representation of dimension s of $\mathrm{Spin}(n-1,1)$, provided with non-trivial Lie super-brackets $(f_1, f_2) \in S \times S \mapsto [f_1, f_2]_+ \in V$ which define a $\mathrm{Spin}(n-1,1)$-equivariant pairing $\Gamma : \mathrm{Sym}^2(S) \to V$ from symmetric two-tensors on S into V (see [29], Lecture 3). Super-spacetime $\mathbb{R}^{n|s}$ can then be extended in a natural way into the exponential of the super-Poincaré algebra $(\mathfrak{spin}(n-1,1) \ltimes V) \oplus S$, with the canonical action of $\mathfrak{spin}(n-1,1)$ on V and on S.

Let us make this construction explicit in space-time dimension $n = 3$, which is the only case that we shall study in this paper. Then the minimal spin representation is two-dimensional, so we consider super-spacetime $\mathbb{R}^{3|2}$ with two odd coordinates $\theta = (\theta^1, \theta^2)$. We shall denote by D_{θ^a}, $a = 1, 2$, the associated *left-invariant derivatives*, namely, the left-invariant super-vector fields that coincide with $\partial_{\theta^1}, \partial_{\theta^2}$ when $\theta^1, \theta^2 \equiv 0$. Consider $\mathbb{R}^2$ with the coordinate vector fields $\partial_{y^1}, \partial_{y^2}$ and the associated symmetric two-tensors with components $\partial_{y^{ij}}$, $i, j = 1, 2$. These form a three-dimensional vector space with natural coordinates $y = (y^{11}, y^{12}, y^{22})$ defined by

$$\left[\partial_{y^{cd}}, y^{ab}\right]_- := \delta_{ca}\delta_{db} + \delta_{cb}\delta_{da} \tag{11.19}$$

Then define the map Γ introduced above to be

$$\Gamma(\partial_{\theta^a}, \partial_{\theta^b}) := \partial_{y^{ab}} \tag{11.20}$$

Hence, one has the simple relation $[D_{\theta^a}, D_{\theta^b}]_+ = \partial/\partial y^{ab}$ for the odd generators of $\mathbb{R}^{3|2}$. So, by the Campbell-Hausdorff formula,

$$D_{\theta^a} = \partial_{\theta^a} + \theta^b \partial_{y^{ab}}. \tag{11.21}$$

In this particular case, $\mathfrak{spin}(2,1) \cong \mathfrak{sl}(2,\mathbb{R})$. The usual action of $\mathfrak{gl}(2,\mathbb{R}) \supset \mathfrak{spin}(2,1)$ on $\mathbb{R}^2$ is given by the two-by-two matrices E_{ab} such that $E_{ab}\partial_{y^c} = -\delta_{ac}\partial_{y^b}$ and extends naturally to the following action on symmetric 2-tensors

$$E_{ab}\partial_{y^{cd}} = -\delta_{ac}\partial_{y^{bd}} - \delta_{ad}\partial_{y^{cb}} = \left[y^{a\bar{a}}\partial_{y^{\bar{a}b}}, \partial_{y^{cd}}\right]_-, \tag{11.22}$$

so E_{ab} is represented by the vector field on $V \oplus S$

$$E_{ab} = y^{a\bar{a}}\partial_{y^{\bar{a}b}} + \theta^a \partial_{\theta^b} \tag{11.23}$$

One may verify that the adjoint action of E_{ab} on the left-covariant derivatives is given by the usual matrix action, namely, $[E_{ab}, D_{\theta^c}] = -\delta_{ac} D_{\theta^b}$.

Consider now a superfield $\Phi(y^{11}, y^{12}, y^{22}; \theta^1, \theta^2)$: we introduce the Lagrangian density

$$\mathcal{L}(\Phi) = \frac{1}{2}\varepsilon^{ab}(D_{\theta^a}\Phi)^*(D_{\theta^b}\Phi) \tag{11.24}$$

where ε^{ab} is the totally antisymmetric two-tensor defined by $\varepsilon^{12} = -\varepsilon^{21} = 1$, $\varepsilon^{11} = \varepsilon^{22} = 0$. It yields the equations of motion

$$\varepsilon^{ab} D_{\theta^a} D_{\theta^b}\Phi = (D_{\theta^1} D_{\theta^2} - D_{\theta^2} D_{\theta^1})\Phi = 0. \tag{11.25}$$

This equation is invariant under even translations $\partial_{y^{ab}}$, and under right-invariant super-derivatives

$$\bar{D}_{\theta^a} = \partial_{\theta^a} - \theta^b \partial_{y^{ab}} \tag{11.26}$$

since these anticommute with the D_{θ^a}. Furthermore, the Lagrangian density is multiplied by $\det(g)$ under the action of $g \in \mathrm{GL}(2,\mathbb{R})$, hence all elements in $\mathfrak{gl}(2,\mathbb{R})$ leave equation (11.25) invariant.

Note that the equations of motion are also invariant under the left-invariant super-derivatives D_{θ^a} since these commute with the coordinate vector fields $\partial_{y^{bc}}$ (this is true for flat space-time manifolds only).

All these translational and rotational symmetries form by linear combinations a Lie superalgebra that we shall call (in the absence of any better name) the *super-Euclidean Lie algebra of* $\mathbb{R}^{3|2}$, and denote by $\mathfrak{se}(3|2)$, viz.

$$\mathfrak{se}(3|2) = \left\langle \partial_{y^{ab}}, D_{\theta^a}, \bar{D}_{\theta^a}, E_{ab}; a, b \in \{1,2\}\right\rangle \tag{11.27}$$

We shall show later that it can be included in a larger Lie super-algebra which is more interesting for our purposes.

Let us look at this more closely by using proper coordinates. The vector fields $\partial/\partial y^{ij}$ are related to the physical-coordinate vector fields by

$$\frac{\partial}{\partial t} = \frac{\partial}{\partial y^{11}}, \frac{\partial}{\partial r} = \frac{\partial}{\partial y^{12}}, \frac{\partial}{\partial \zeta} = \frac{\partial}{\partial y^{22}} \tag{11.28}$$

hence by (11.19) we have $t = 2y^{11}, r = y^{12}, \zeta = 2y^{22}$. We set

$$\Phi(\zeta,t,r;\theta^1,\theta^2) = f(\zeta,t,r) + \theta^1\phi(\zeta,t,r) + \theta^2\psi(\zeta,t,r) + \theta^1\theta^2 g(\zeta,t,r). \tag{11.29}$$

Then the left-invariant superderivatives read

$$D_{\theta^1} = \partial_{\theta^1} + \theta^1\partial_t + \theta^2\partial_r, \; D_{\theta^2} = \partial_{\theta^2} + \theta^1\partial_r + \theta^2\partial_\zeta. \tag{11.30}$$

The equations of motion (11.25) become

$$\left(\partial_{\theta^1}\partial_{\theta^2} + \theta^1\theta^2(\partial_\zeta\partial_t - \partial_r^2) + \theta^1(\partial_{\theta^2}\partial_t - \partial_{\theta^1}\partial_r) + \theta^2(\partial_{\theta^2}\partial_r - \partial_{\theta^1}\partial_\zeta)\right)\Phi = 0 \tag{11.31}$$

which yields the following system of equations in the coordinate fields:

$$\begin{aligned} g &= 0 \\ \partial_r\phi &= \partial_t\psi, \;\; \partial_r\psi = \partial_\zeta\phi \\ (\partial_r^2 - \partial_\zeta\partial_t)f &= 0. \end{aligned} \tag{11.32}$$

We shall call this system the (3|2)-*supersymmetric model*. From the two equations in the second line of (11.32) we recover the Dirac-Lévy-Leblond equations (11.10) after the change of variables $\zeta \mapsto -\zeta$.

Equation (11.32) may be obtained in turn from the action

$$S = \int \mathrm{d}\zeta \; \mathrm{d}t \; \mathrm{d}r \; \mathrm{d}\theta^2 \; \mathrm{d}\theta^1 \; \mathscr{L}(\Phi) = \int \mathrm{d}\zeta \; \mathrm{d}t \; \mathrm{d}r \; L(f,\phi,\psi,g) \tag{11.33}$$

where

$$L(f,\phi,\psi,g) = f^*(\partial_\zeta\partial_t - \partial_r^2)f + \phi^*(\partial_t\psi - \partial_r\phi) + \psi^*(\partial_\zeta\phi - \partial_r\psi) + g^*g. \tag{11.34}$$

Now consider the field $\Phi = (f,\psi,\phi,g)$ as the Laplace transform $\Phi = \int \mathrm{d}\mathscr{M}\, e^{2\mathscr{M}\zeta}\, \tilde{\Phi}_{\mathscr{M}}$ of the field $\tilde{\Phi}_{\mathscr{M}}$ with respect to ζ, so that the derivative operator ∂_ζ corresponds to the multiplication by twice the mass coordinate $2\mathscr{M}$. The equations of motion (11.32) then read as follows:

Table 11.1 Defining equations of motion of the supersymmetric models. The kinematic and dynamic symmetry algebras (see the text for the definitions) are also listed

Model	(3|2)-supersymmetric	Super-Schrödinger
	$g = 0$	$\tilde{g} = 0$
	$\partial_r \phi = \partial_t \psi$	$\partial_r \tilde{\phi} = \partial_t \tilde{\psi}$
	$\partial_r \psi = \partial_\zeta \phi$	$\partial_r \tilde{\psi} = 2\mathscr{M}\tilde{\phi}$
	$(\partial_r^2 - \partial_\zeta \partial_t) f = 0$	$(\partial_r^2 - 2\mathscr{M}\partial_t)\tilde{f} = 0$
Kinematic algebra	$\mathfrak{se}(3\|2)$	$\mathfrak{sgal}$
Dynamic algebra	$\mathfrak{s}^{(2)} \cong \mathfrak{osp}(2\|4)$	$\tilde{\mathfrak{s}}^{(2)} \cong \mathfrak{osp}(2\|2) \ltimes \mathfrak{sh}(2\|2)$

$$
\begin{aligned}
\tilde{g} &= 0 \\
\partial_r \tilde{\phi} &= \partial_t \tilde{\psi}, \partial_r \tilde{\psi} = 2\mathscr{M}\tilde{\phi} \\
(\partial_r^2 - 2\mathscr{M}\partial_t)\tilde{f} &= 0
\end{aligned}
\tag{11.35}
$$

We shall refer to (11.35) as the *super-Schrödinger model.*

In this context, g or $\tilde{g}$ can be interpreted as an auxiliary field, while (ψ, ϕ) is a spinor field satisfying the Dirac equation in (2+1) dimensions (11.10) and its inverse Laplace transform $(\tilde{\psi}, \tilde{\phi})$ satisfies the Dirac-Lévy-Leblond equation in one space dimension, see (11.8), and $\tilde{f}$ is a solution of the free Schrödinger equation in one space dimension.

Let us now study the kinematic Lie symmetries of the (3|2)-supersymmetric model (11.32) and of the super-Schrödinger model (11.35). For convenience, we collect their definitions in Table 11.1. By definition, *kinematic* symmetries are (super)-translations and (super-)rotations, and also scale transformations, that leave invariant the equations of motion. Generally speaking, the kinematic Lie symmetries of the super-Schrödinger model contained in $\mathfrak{sgal}$ correspond to those symmetries of the (3|2) supersymmetric model such that the associated vector fields do not depend on the coordinate ζ, in other words which leave the 'mass' invariant. Below, we shall also consider the so-called *dynamic* symmetries of the two free-field models which arise when also inversions $t \mapsto -1/t$ are included, and form a strictly larger Lie algebra. We anticipate on later results and already include the dynamic algebras in Table 11.1.

Let us summarize the results obtained so far on the kinematic symmetries of the two supersymmetric models.

Proposition 11.4. *1. The Lie algebra of kinematic Lie symmetries of the* (3|2)*-supersymmetric model (11.32) contains a subalgebra which is isomorphic to* $\mathfrak{se}(3|2)$*. The Lie algebra* $\mathfrak{se}(3|2)$ *has dimension* 11*, and a basis of* $\mathfrak{se}(3|2)$ *in its realization as Lie symmetries is given by the following generators. There are the three even translations*

$$
L_{-1}, Y_{-\frac{1}{2}}, M_0 \tag{11.36}
$$

the four odd translations

$$G^1_{-\frac{1}{2}} = -\frac{1}{2}\left(D_{\theta^1} + \bar{D}_{\theta^1}\right),\ G^2_{-\frac{1}{2}} = -\frac{1}{2}\left(D_{\theta^1} - \bar{D}_{\theta^1}\right),$$
$$\bar{Y}^1_0 = -\frac{1}{2}\left(D_{\theta^2} + \bar{D}_{\theta^2}\right),\ \bar{Y}^2_0 = -\frac{1}{2}\left(D_{\theta^2} - \bar{D}_{\theta^2}\right) \tag{11.37}$$

and the four generators in $\mathfrak{gl}(2,\mathbb{R})$

$$Y_{\frac{1}{2}} = -\frac{1}{2}E_{12}, L_0 = -\frac{1}{2}E_{11} - \frac{x}{2},\ D = -\frac{1}{2}(E_{11}+E_{22}) - x, V_- = -\frac{1}{2}E_{21} \tag{11.38}$$

An explicit realization in terms of differential operators is

$$\begin{aligned}
L_{-1} &= -\partial_t,\ Y_{-\frac{1}{2}} = -\partial_r,\ M_0 = -\frac{1}{2}\partial_\zeta \\
G^1_{-\frac{1}{2}} &= -\partial_{\theta^1},\ G^2_{-\frac{1}{2}} = -\theta^1\partial_t - \theta^2\partial_r \\
\bar{Y}^1_0 &= -\partial_{\theta^2},\ \bar{Y}^2_0 = -\theta^1\partial_r - \theta^2\partial_\zeta \\
Y_{\frac{1}{2}} &= -t\partial_r - \frac{1}{2}r\partial_\zeta - \frac{1}{2}\theta^1\partial_{\theta^2} \\
L_0 &= -t\partial_t - \frac{1}{2}\left(r\partial_r + \theta^1\partial_{\theta^1}\right) - \frac{x}{2} \\
D &= -t\partial_t - r\partial_r - \zeta\partial_\zeta - \frac{1}{2}\left(\theta^1\partial_{\theta^1} + \theta^2\partial_{\theta^2}\right) - x \\
V_- &= -\zeta\partial_r - \frac{1}{2}r\partial_t - \frac{1}{2}\theta^2\partial_{\theta^1}.
\end{aligned} \tag{11.39}$$

Here a scaling dimension x of the superfield Φ has been added such that for $x = 1/2$ the generators L_0 and D (which correspond to the action of non trace-free elements of $\mathfrak{gl}(2,\mathbb{R})$) leave invariant the Lagrangian density. By changing the value of x one finds another realization of $\mathfrak{se}(3|2)$.

2. *The super-galilean Lie subalgebra $\mathfrak{sgal} \subset \mathfrak{se}(3|2)$ of symmetries of the super-Schrödinger model (11.35) is 9-dimensional. Explicitly*

$$\mathfrak{sgal} = \left\langle L_{-1,0}, Y_{\pm\frac{1}{2}}, M_0, G^{1,2}_{-\frac{1}{2}}, \bar{Y}^{1,2}_0 \right\rangle \tag{11.40}$$

We stress the strong asymmetry between the two odd coordinates $\theta^{1,2}$ as they appear in the dilatation generator L_0. This is a consequence of our identification $L_0 = -\frac{1}{2}E_{11} - \frac{1}{2}x$, which is dictated by the requirement that the system exhibit a non-relativistic behaviour with a dynamic exponent $z = 2$. As we shall show in Sect. 11.4, this choice will have important consequences for the calculation of covariant two-point functions. In comparison, in relativistic systems with an extended ($N = 2$) supersymmetry (see e.g. [20, 88, 97]), one needs a dynamic exponent $z = 1$. In our notation, the generator D would then be identified as the generator of dilatations, leading to a complete symmetry between θ^1 and θ^2.

The supersymmetries of the free non-relativistic particle with a fixed mass have been discussed by Beckers *et al.* long ago [7,9] and, as we shall recall in Sect. 11.2.3, $\mathfrak{sgal}$ is a subalgebra of their dynamical algebra $\mathfrak{osp}(2|2) \ltimes \mathfrak{sh}(2|2)$.

Let us give the Lie brackets of these generators for convenience, and also for later use. The three generators $(L_{-1}, Y_{-\frac{1}{2}}, M_0)$ commute with all translations, even or odd. The commutators of the odd translations yield four non-trivial relations:

$$\left[G^1_{-\frac{1}{2}}, G^2_{-\frac{1}{2}}\right]_+ = -L_{-1}, \ \left[\bar{Y}^1_0, \bar{Y}^2_0\right]_+ = -2M_0$$
$$\left[G^1_{-\frac{1}{2}}, \bar{Y}^2_0\right]_+ = \left[G^2_{-\frac{1}{2}}, \bar{Y}^1_0\right]_+ = -Y_{-\frac{1}{2}}. \tag{11.41}$$

The rotations act on left- or right-covariant odd derivatives by the same formula

$$[E_{ab}, D_{\theta^c}] = -\delta_{ac} D_{\theta^b}, \ \left[E_{ab}, \bar{D}_{\theta^c}\right] = -\delta_{ac} \bar{D}_{\theta^b}, \tag{11.42}$$

which gives in our basis

$$\left[L_0, G^{1,2}_{-\frac{1}{2}}\right] = \frac{1}{2} G^{1,2}_{-\frac{1}{2}}, \left[L_0, \bar{Y}^{1,2}_0\right] = 0$$
$$\left[Y_{\frac{1}{2}}, G^{1,2}_{-\frac{1}{2}}\right] = \frac{1}{2} \bar{Y}^{1,2}_0, \ \left[Y_{\frac{1}{2}}, \bar{Y}^{1,2}_0\right] = 0$$
$$\left[V_-, G^{1,2}_{-\frac{1}{2}}\right] = 0, \qquad \left[V_-, \bar{Y}^{1,2}_0\right] = \frac{1}{2} G^{1,2}_{-\frac{1}{2}}. \tag{11.43}$$

Finally, the commutators of elements in $\mathfrak{gl}(2, \mathbb{R})$ may be computed by using the usual bracket of matrices, and brackets between elements in $\mathfrak{gl}(2, \mathbb{R})$ and even translations are obvious.

11.2.2 Dynamic Symmetries of the Super-Schrödinger Model

Let us consider the symmetries of the super-Schrödinger model, starting from the 9-dimensional Lie algebra of symmetries $\mathfrak{sgal}$ that was introduced in Proposition 11.4. This Lie algebra may be enlarged by adding the generator

$$N_0 = -\theta^1 \partial_{\theta^1} - \theta^2 \partial_{\theta^2} + x \tag{11.44}$$

(Euler operator on odd coordinates), together with three special transformations $L_1, G^{1,2}_{\frac{1}{2}}$ that will be defined shortly. First notice that the operators

$$\Delta_0 := 2\mathcal{M}\partial_t - \partial_r^2, \quad \Delta_0'' := \partial_{\theta^1}\partial_{\theta^2}$$
$$\Delta_0' := 2\mathcal{M}\partial_{\theta^1} - \partial_{\theta^2}\partial_r, \quad \bar{\Delta}_0' := \partial_{\theta^1}\partial_r - \partial_{\theta^2}\partial_t \tag{11.45}$$

cancel on solutions of the equations of motion. So

$$L_1 := -\frac{1}{2\mathcal{M}}\left(Y_{\frac{1}{2}}^2 + t^2\Delta_0 + t\theta^1\Delta_0'\right) = -t^2\partial_t - t\left(r\partial_r + \theta^1\partial_{\theta^1}\right)$$
$$-xt - \frac{\mathcal{M}}{2}r^2 - \frac{1}{2}r\theta^1\partial_{\theta^2} \tag{11.46}$$

is also a symmetry of (11.35), extending the special Schrödinger transformation $d\pi_{x/2}(L_1)$ (see Introduction). One obtains two more generators by straightforward computations, namely

$$G_{\frac{1}{2}}^1 := \left[L_1, G_{-\frac{1}{2}}^1\right] = -t\partial_{\theta^1} - \frac{1}{2}r\partial_{\theta^2}$$
$$G_{\frac{1}{2}}^2 := \left[L_1, G_{-\frac{1}{2}}^2\right] = -t\left(\theta^1\partial_t + \theta^2\partial_r\right) - \frac{1}{2}\theta^1 r\partial_r - x\theta^1 - \mathcal{M}r\theta_2 + \frac{1}{2}\theta^1\theta^2\partial_{\theta^2}. \tag{11.47}$$

Proposition 11.5. *The vector space generated by* $\mathfrak{sgal}$ *introduced in Proposition 11.4, together with* N_0 *and the three special transformations* $L_1, G_{\frac{1}{2}}^{1,2}$*, closes into a 13-dimensional Lie superalgebra. We shall call this Lie algebra the* $(N = 2)$*-super-Schrödinger algebra and denote it by* $\tilde{\mathfrak{s}}^{(2)}$*. Explicitly,*

$$\tilde{\mathfrak{s}}^{(2)} = \left\langle L_{\pm1,0}, G_{\pm1/2}^{1,2}, Y_{\pm1/2}, \bar{Y}_0^{1,2}, M_0, N_0\right\rangle \tag{11.48}$$

and the generators are listed in (11.39), (11.44), (11.46), (11.47). See also Appendix B, Sect. 2.

Proof. One may check very easily the following formulas (note that the correcting terms of the type function times $\mathcal{D}$, where $\mathcal{D} = \Delta_0, \Delta_0', \bar{\Delta}_0'$ or Δ_0'', are here for definiteness but yield 0 modulo the equations of motion when commuted against elements of $\mathfrak{se}(3|2)$, so they can be dismissed altogether when computing brackets)

$$M_0 = \left[Y_{\frac{1}{2}}, Y_{-\frac{1}{2}}\right]$$
$$L_{-1} = -\frac{1}{2\mathcal{M}}Y_{-\frac{1}{2}}^2 - \frac{1}{2\mathcal{M}}\Delta_0$$
$$L_0 = -\frac{1}{4\mathcal{M}}\left(Y_{-\frac{1}{2}}Y_{\frac{1}{2}} + Y_{\frac{1}{2}}Y_{-\frac{1}{2}}\right) - \frac{t}{2\mathcal{M}}\Delta_0 - \frac{\theta^1}{4\mathcal{M}}\Delta_0'$$

$$G^1_{-\frac{1}{2}} = \frac{1}{2\mathcal{M}}\bar{Y}_0^1 Y_{-\frac{1}{2}} - \frac{1}{2\mathcal{M}}\Delta'_0$$

$$G^2_{-\frac{1}{2}} = \frac{1}{2\mathcal{M}}\bar{Y}_0^2 Y_{-\frac{1}{2}} - \frac{\theta^1}{2\mathcal{M}}\Delta_0$$

$$G^1_{\frac{1}{2}} = \frac{1}{2\mathcal{M}}\bar{Y}_0^1 Y_{\frac{1}{2}} - \frac{t}{2\mathcal{M}}\Delta'_0$$

$$G^2_{\frac{1}{2}} = \frac{1}{2\mathcal{M}}\bar{Y}_0^2 Y_{\frac{1}{2}} - \frac{t\theta^1}{2\mathcal{M}}\Delta_0$$

$$N_0 = -\frac{1}{4\mathcal{M}}\left(\bar{Y}_0^2\bar{Y}_0^1 - \bar{Y}_0^1\bar{Y}_0^2\right) - \frac{\theta^1}{2\mathcal{M}}\Delta'_0.$$

So it takes only a short time to compute the adjoint action of $G^{1,2}_{\frac{1}{2}}$ on $\mathfrak{se}(3|2)$. On the even translations we have

$$\left[G^{1,2}_{\frac{1}{2}}, L_{-1}\right] = G^{1,2}_{-\frac{1}{2}},\ \left[G^{1,2}_{\frac{1}{2}}, Y_{-\frac{1}{2}}\right] = \frac{1}{2}\bar{Y}_0^{1,2},\ \left[G^{1,2}_{\frac{1}{2}}, M_0\right] = 0. \tag{11.49}$$

By commuting the G-generators we find

$$\left[G^{1,2}_{\frac{1}{2}}, G^{1,2}_{-\frac{1}{2}}\right]_+ = 0,\ \left[G^1_{\frac{1}{2}}, G^2_{-\frac{1}{2}}\right]_+ = -\frac{1}{2}N_0 - L_0,\ \left[G^2_{\frac{1}{2}}, G^1_{-\frac{1}{2}}\right]_+ = \frac{1}{2}N_0 - L_0. \tag{11.50}$$

The action on the odd translations is given by

$$\left[G^{1,2}_{\frac{1}{2}}, \bar{Y}_0^{1,2}\right]_+ = 0,\ \left[G^{1,2}_{\frac{1}{2}}, \bar{Y}_0^{2,1}\right]_+ = -Y_{\frac{1}{2}}. \tag{11.51}$$

Finally,

$$\left[G^{1,2}_{\frac{1}{2}}, Y_{\frac{1}{2}}\right] = 0,\ \left[G^{1,2}_{\frac{1}{2}}, L_0\right] = \frac{1}{2}G^{1,2}_{\frac{1}{2}}. \tag{11.52}$$

The generator N_0 acts diagonally on the generators of $\mathfrak{se}(3|2)$: the eigenvalue of ad N_0 on a generator without upper index is 0, while it is $+1$ (resp. -1) on generators with upper index 1 (resp. 2). Note that this is also true for the action of N_0 on $G^{1,2}_{\frac{1}{2}}$.

The proof may now be finished by verifying that $[G^i_{\frac{1}{2}}, G^i_{\frac{1}{2}}]_+ = 0$ (for both $i = 1,2$), $L_1 = -[G^1_{\frac{1}{2}}, G^2_{\frac{1}{2}}]_+$ and $[L_1, G^{1,2}_{\frac{1}{2}}] = 0$. □

Remark. In order to prove the invariance of the equations of motion under $\tilde{\mathfrak{s}}^{(2)}$ it is actually enough to prove the invariance under $Y_{\pm\frac{1}{2}}$ and $\bar{Y}_0^{1,2}$ since all other generators are given (*modulo* the equations of motion) as quadratic expressions in these four generators.

11.2.3 Some Physical Applications

We now briefly recall some earlier results on supersymmetric non-relativistic systems with a dynamic supersymmetry algebra which contains $\mathfrak{osp}(2|2)$.

Beckers *et al.* [7–9] studied the supersymmetric non-relativistic quantum mechanics in one spatial dimension and derived the dynamical Lie superalgebras for any given superpotential W. The largest superalgebras are found for the free particle, the free fall or the harmonic oscillator, where the dynamic algebra is [9]

$$\tilde{\mathfrak{s}}^{(2)} \cong \mathfrak{osp}(2|2) \ltimes \mathfrak{sh}(2|2) \tag{11.53}$$

where $\mathfrak{sh}(2|2)$ is the Heisenberg super-algebra. We explicitly list the correspondence for the harmonic oscillator with total Hamiltonian, see [7]

$$H = H_B + H_F = \frac{1}{2}\left(p^2 + \frac{1}{4}x^2 + \frac{1}{2}\sigma_3\right) \tag{11.54}$$

The $\mathfrak{osp}(2|2)$-subalgebras of symmetries of our (3|2)-supersymmetric model and of the harmonic oscillator in the notation of [7] may be identified by setting

$$\begin{gathered} H_B = L_0, H_F = \frac{1}{2}N_0, \quad C_\pm = \pm \mathrm{i} X_{\mp 1}, \\ Q_+ = G^1_{\frac{1}{2}}, \quad Q_- = -G^2_{-\frac{1}{2}}, \quad S_+ = G^1_{-\frac{1}{2}}, \quad S_- = -G^2_{\frac{1}{2}} \end{gathered} \tag{11.55}$$

while the identification of the symmetries in $\mathfrak{sh}(2|2)$ of both models is given by

$$P_\pm = Y_{\mp\frac{1}{2}}, \quad T_\pm = \frac{\mathrm{i}}{\sqrt{2}}\bar{Y}_0^{1,2}, \quad I = -M_0 \tag{11.56}$$

We remark that the total Hamiltonian corresponds to $H = L_0 + \frac{1}{2}N_0$ in our notation.

Duval and Horvathy [26] systematically constructed supersymmetric extensions with N supercharges of the Schrödinger algebra $\mathfrak{sch}(d)$ as subalgebras of the extended affine orthosymplectic superalgebras. In general, there is only one 'standard' possible type of such extensions, but in two space-dimensions, there is a further 'exotic' superalgebra with a different structure. Relationships with Poisson algebras (see below) are also discussed. While the kind of supersymmetries discussed above [7, 9] belong to the first type, the 'exotic' type arises for example in Chern-Simons matter systems, whose $N = 2$ supersymmetry was first described by Leblanc *et al.* [78].[5] In [26], the supersymmetries of a scalar particle in a Dirac monopole and of a magnetic vortex are also discussed.

[5] In *non*-commutative space-time, extended supersymmetries still persist, but scale- and Galilei-invariance are broken [83].

The uniqueness of $\mathfrak{osp}(2|2)$-supersymmetry constructions has been addressed by Ghosh [36]. Indeed, the generators of the $\mathfrak{osp}(2|2)$ algebra can be represented in two distinct ways in terms of the coordinates of the super-Calogero model. This leads to two distinct types of superhamiltonians, which in the simplest case of N free superoscillators read [36]

$$H_{\pm} = \frac{1}{4}\sum_{i=1}^{N}\left[(p_i^2 + x_i^2) \pm \left(\psi_i^\dagger\psi_i - \psi_i\psi_i^\dagger\right)\right] \tag{11.57}$$

$$\hat{H}_{\pm} = \frac{1}{4}\sum_{i=1}^{N}(p_i^2 + x_i^2) \pm \frac{\gamma_5}{4}\left[N - \mathrm{i}\sum_{i,j=1}^{N}\left(\psi_i^\dagger\psi_j^\dagger + \psi_i\psi_j + \psi_i^\dagger\psi_j - \psi_j^\dagger\psi_i\right)L_{ij}\right] \tag{11.58}$$

where x_i and p_i are bosonic coordinates and momenta, $L_{ij} = x_i p_j - x_j p_i$ are angular momenta, the ψ_i are fermionic variables satisfying $[\psi_i, \psi_j^\dagger]_+ = \delta_{ij}$ and the operator γ_5 anticommutes with the ψ_i. The Hamiltonian $H_\pm$ in (11.57) is identical to the one discussed in [7, 9, 26]. Further examples discussed in [36] include superconformal quantum mechanics and Calogero models but will not be detailed here. Dynamical $\mathfrak{osp}(2|2)$-supersymmetries also occur in the d-dimensional Calogero-Marchioro model [35].

Finally, we mention that the $SU(2)_0$ Wess-Zumino-Witten model has a hidden $\mathfrak{osp}(2|2)_{-2}$ symmetry, with a relationship to logarithmic conformal field-theories [74].

11.2.4 Dynamic Symmetries of the (3|2)*-Supersymmetric Model*

So far, we have considered the mass $\mathscr{M}$ as fixed. Following what has been done for the simple Schrödinger equation, we now relax this condition and ask what happens if $\mathscr{M}$ is treated as a variable [52]. We then add the generators D and V_- to $\tilde{\mathfrak{s}}^{(2)}$ which generates, through commutation with L_1 and $G_{\frac{1}{2}}^{1,2}$, the following new generators

$$\begin{aligned}
V_+ &= 4[L_1, V_-] = -2tr\partial_t - 2\zeta r\partial_\zeta - \left(r^2 + 4\zeta t\right)\partial_r - r\left(\theta^1\partial_{\theta^1} + \theta^2\partial_{\theta^2}\right) \\
&\quad -2t\theta^2\partial_{\theta^1} - 2\zeta\theta^1\partial_{\theta^2} - 2xr \\
W &= [V_+, V_-] = -2\zeta^2\partial_\zeta - 2\zeta\left(r\partial_r + \theta^2\partial_{\theta^2}\right) - \frac{r^2}{2}\partial_t - r\theta^2\partial_{\theta^1} - 2x\zeta \\
\bar{Z}_0^1 &= \left[G_{\frac{1}{2}}^1, V_-\right] = -\frac{1}{2}\left(\zeta\partial_{\theta^2} + \frac{1}{2}r\partial_{\theta^1}\right) \\
\bar{Z}_0^2 &= \left[G_{\frac{1}{2}}^2, V_-\right] = -\frac{1}{2}\left(\zeta(\theta^2\partial_\zeta + \theta^1\partial_r) + \frac{1}{2}\theta^2 r\partial_r + \frac{1}{2}r\theta^1\partial_t + \frac{1}{2}\theta^1\theta^2\partial_{\theta^1} + x\theta^2\right).
\end{aligned} \tag{11.59}$$

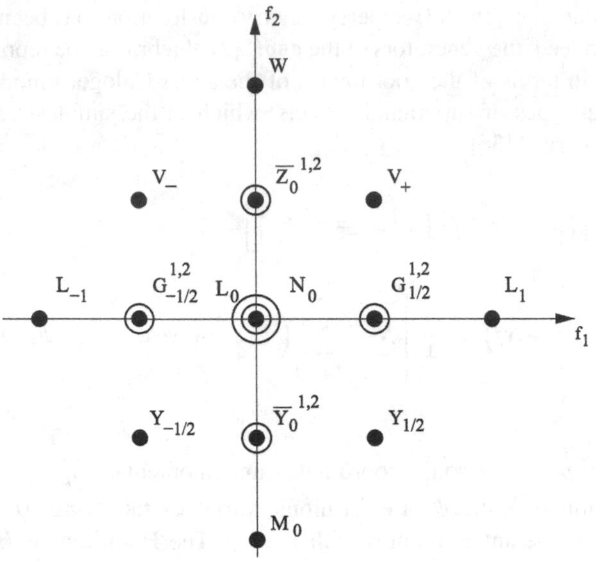

Fig. 11.2 Root vectors of the complexified Lie superalgebra $\mathfrak{s}^{(2)} \cong \mathfrak{osp}(2|4)$. The double circles indicate the presence of two generators corresponding to opposite values of the root projection along α, while the triple circle in the centre corresponds to the Cartan subalgebra $\mathfrak{h}$ (see Proposition 11.14)

Proposition 11.6. *The 19-dimensional vector space*

$$\mathfrak{s}^{(2)} = \left\langle L_{\pm 1,0}, Y_{\pm\frac{1}{2}}, M_0, D, N_0, G^{1,2}_{\pm\frac{1}{2}}, \bar{Y}^{1,2}_0, V_\pm, W, \bar{Z}^{1,2}_0 \right\rangle \tag{11.60}$$

closes as a Lie superalgebra and leaves invariant the equations of motion (11.32) of the $(3|2)$*-supersymmetric model.*

We shall prove this in a simple way in Sect. 11.2.6, by establishing a correspondence between $\mathfrak{s}^{(2)}$ and a Lie subalgebra of a Poisson algebra. This will also show that $\mathfrak{s}^{(2)}$ is isomorphic to the Lie superalgebra $\mathfrak{osp}(2|4)$ - hence one may in the end abandon the notation $\mathfrak{s}^{(2)}$ altogether. The root diagramme of $\mathfrak{osp}(2|4)$ is shown in Fig. 11.2 and the correspondence of the roots with the generators of $\mathfrak{s}^{(2)}$ is made explicit.

11.2.5 First Correspondence with Poisson Structures: The Super-Schrödinger Model

We shall give in this subsection a much simpler-looking presentation of $\tilde{\mathfrak{s}}^{(2)}$ by embedding it into the Poisson algebra of superfunctions on a supermanifold, the Lie bracket of $\tilde{\mathfrak{s}}^{(2)}$ corresponding to the Poisson bracket of the superfunctions. In

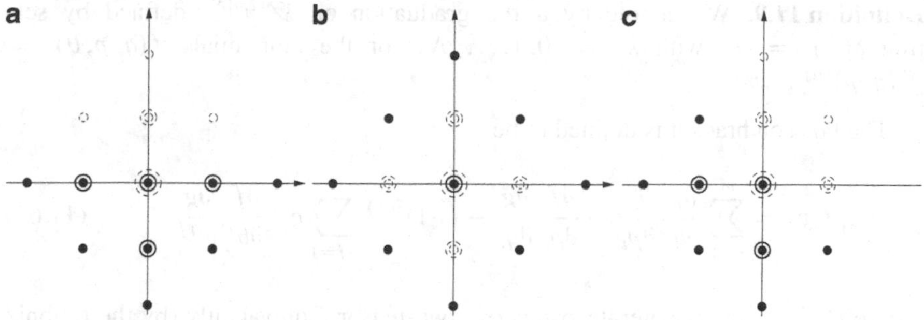

Fig. 11.3 Root vectors of several Lie subalgebras of $\mathfrak{osp}(2|4)$, arranged in the same way as in Fig. 11.2. The full dots and circles give the generators in the respective subalgebra whereas the open dots and broken circles merely stand for the remaining generators of $\mathfrak{osp}(2|4)$. Note that in each case only two generators of the Cartan subalgebra $\mathfrak{h}$ are retained. The subalgebras are **(a)** $\tilde{\mathfrak{s}}^{(2)}$, **(b)** $(\mathfrak{conf}_3)_{\mathbb{C}}$ and **(c)** $\mathfrak{se}(3|2)$

Fig. 11.3a, we show how $\tilde{\mathfrak{s}}^{(2)}$ sits inside $\mathfrak{s}^{(2)} \cong \mathfrak{osp}(2|4)$. For comparison, we display in Fig. 11.3b the even subalgebra $(\mathfrak{conf}_3)_{\mathbb{C}}$ and in Fig. 11.3c the superalgebra $\mathfrak{se}(3|2)$. We see that both $\mathbb{C}D \oplus \tilde{\mathfrak{s}}^{(2)}$ and $\mathbb{C}N_0 \oplus \mathfrak{se}(3|2)$ are maximal Lie subalgebras of $\mathfrak{osp}(2|4)$.

We first need some definitions (see [21], [43], [79] for details).

Definition 11.7. A commutative associative algebra $\mathcal{A}$ is a *Poisson algebra* if there exists a Lie bracket $\{\ ,\ \} : \mathcal{A} \times \mathcal{A} \to \mathcal{A}$ (called Poisson bracket) which is *compatible* with the associative product $f, g \mapsto fg$, that is to say, such that the so-called *Leibniz identity* holds true

$$\{fg, h\} = f\{g, h\} + g\{f, h\}, \quad \forall f, g, h \in \mathcal{A}. \tag{11.61}$$

This definition is naturally superizable and leads to the notion of a *super-Poisson algebra*. Standard examples are Poisson or super-Poisson algebras of smooth functions on supermanifolds:q

Definition 11.8. Let $\mathcal{P}^{(2m|N)}$ be the super-Poisson algebra of functions on the $(2m|N)$-supertorus.

We shall mainly be interested in the cases $m = 1, 2$ and $N = 0, 1, 2$. As an associative algebra, $\mathcal{P}^{(2m|N)}$ may be written as the tensor product

$$\mathcal{P}^{(2m|N)} = \mathcal{P}^{(2m|0)} \otimes \Lambda(\mathbb{R}^N) \tag{11.62}$$

where $\Lambda(\mathbb{R}^N)$ is the Grassmann algebra in the anticommuting variables $\theta^1, \ldots, \theta^N$, and $\mathcal{P}^{(2m|0)}$ is the associative algebra generated by the functions $(q_i, q_i^{-1}, p_i, p_i^{-1})$, $i = 1, \ldots, m$, corresponding to finite Fourier series. (Note that the Poisson algebra of *smooth* functions on the $(2|N)$-supertorus is a kind of completion of $\mathcal{P}^{(2|N)}$.)

Definition 11.9. We denote by δ the graduation on $\mathcal{P}^{(2m|N)}$ defined by setting $\delta(f) = k$, with $k = 0, 1, \ldots, N$, for the monomials $f(q, p, \theta) = f_0(q, p)\theta^{i_1} \ldots \theta^{i_k}$.

The Poisson bracket is defined to be

$$\{f, g\} = \sum_{i=1}^{m} \frac{\partial f}{\partial q_i}\frac{\partial g}{\partial p_i} - \frac{\partial f}{\partial p_i}\frac{\partial g}{\partial q_i} - (-1)^{\delta(f)} \sum_{i,j=1}^{N} \eta^{i\,j} \frac{\partial f}{\partial \theta^i}\frac{\partial g}{\partial \theta^j} \tag{11.63}$$

where η^{ij} is a non-degenerate *symmetric* two-tensor. Equivalently (by the Leibniz identity), it may be defined through the relations

$$\{q_i, p_j\} = \delta_{i,j}, \ \{q_i, \theta^j\} = 0, \ \{p_i, \theta^j\} = 0, \ \{\theta^i, \theta^j\} = \eta^{ij}. \tag{11.64}$$

We warn the reader who is not familiar with Poisson structures in a supersymmetric setting against the familiar idea that the Poisson bracket of two functions should be obtained in a more or less straightforward way from their products. One has for instance in the case $N = 1$

$$(\theta^1)^2 = 0, \quad \{\theta^1, \theta^1\} = \eta^{11} \neq 0 \tag{11.65}$$

which might look a little confusing at first.

It is a well-known fact (see [44] for instance) that the Schrödinger Lie algebra $\mathfrak{sch}$ generated by $X_{\pm 1,0}, Y_{\pm\frac{1}{2}}, M_0$ is isomorphic to the Lie algebra of polynomials in $(q, p) = (q_1, p_1)$ with degree ≤ 2: an explicit isomorphism is given by

$$L_{-1} \to \frac{1}{2}q^2, \ L_0 \to -\frac{1}{2}qp, \ L_1 \to \frac{1}{2}p^2,$$
$$Y_{-\frac{1}{2}} \to q, \ Y_{\frac{1}{2}} \to -p, \ M_0 \to 1. \tag{11.66}$$

In particular, the Lie subalgebra $\langle L_{-1}, L_0, L_1 \rangle$ of quadratic polynomials in $\mathcal{P}^{(2|0)}$ is isomorphic to the Lie algebra $\mathfrak{sp}(2, \mathbb{R})$ of linear infinitesimal canonical transformations of $\mathbb{R}^2$, which is a mere reformulation of the canonical isomorphism $\mathfrak{sl}(2, \mathbb{R}) \cong \mathfrak{sp}(2, \mathbb{R})$ (see Sect. 11.2.6 for an extension of this result).

We now give a natural extension of this isomorphism to a supersymmetric setting. In what follows we take $m = 1, N = 2$ and $\{\theta^1, \theta^1\} = \{\theta^2, \theta^2\} = 0, \{\theta^1, \theta^2\} = 2$.

Definition 11.10. We denote by $\mathcal{P}^{(2|2)}_{\leq 2} \subset \mathcal{P}^{(2|2)}$ the Lie algebra of superfunctions that are polynomials in $(p, q, \theta^1, \theta^2)$ of degree ≤ 2.

Proposition 11.11. *One has an isomorphism from* $\tilde{\mathfrak{s}}^{(2)}$ *to the Lie algebra* $\mathcal{P}^{(2|2)}_{\leq 2}$ *given explicitly as*

$$
\begin{aligned}
&L_{-1} \to \frac{1}{2}q^2,\ L_0 \to -\frac{1}{2}qp,\ L_1 \to \frac{1}{2}p^2,\ Y_{-\frac{1}{2}} \to q,\ Y_{\frac{1}{2}} \to -p,\ M_0 \to 1\\
&N_0 \to -\frac{1}{2}\theta^1\theta^2\\
&\bar{Y}_0^1 \to -\theta^1,\ \bar{Y}_0^2 \to \theta^2\\
&G^1_{-\frac{1}{2}} \to -\frac{1}{2}q\theta^1,\ G^2_{-\frac{1}{2}} \to \frac{1}{2}q\theta^2\\
&G^1_{\frac{1}{2}} \to \frac{1}{2}p\theta^1,\ G^2_{\frac{1}{2}} \to -\frac{1}{2}p\theta^2.
\end{aligned} \tag{11.67}
$$

Remark. An equivalent statement of this result and its extension to higher spatial dimensions was given in [26], (4.10). This Lie isomorphism allows a rapid computation of Lie brackets in $\tilde{\mathfrak{s}}^{(2)}$.

Proof. The subalgebra of $\mathscr{P}^{(2|2)}_{\leq 2}$ made up of the monomials of degree 0 in p decomposes as a four-dimensional commutative algebra $\langle q^2, q, 1, \theta^1\theta^2\rangle$ (for the even part), plus four odd generators $\theta^i, q\theta^i$, $i = 1, 2$. One may easily check the identification with the 3 even translations and the 'super-Euler operator' N_0, plus the 4 odd translations of $\mathfrak{se}(3|2)$, see Fig. 11.2.

Then the two allowed rotations, L_0 and $Y_{\frac{1}{2}}$, form together with the translations a nine-dimensional algebra that is also easily checked to be isomorphic to its image in $\mathscr{P}^{(2|2)}_{\leq 2}$.

Finally, one sees immediately that the quadratic expressions (appearing just before Proposition 3.2 and inside its proof) that give $L_1, G^{1,2}_{\frac{1}{2}}$ in terms of $Y, \bar{Y}$ also hold in the associative algebra $\mathscr{P}^{(2|2)}_{\leq 2}$ with the suggested identification (actually, this is also true for the generators $N_0, L_{-1}, L_0, G^{1,2}_{-\frac{1}{2}}$, so one may still reduce the number of verifications.) □

Let us finish this paragraph by coming back to the original $N = 2$ supersymmetry algebra (see Sect. 11.2.1). Suppose we want to consider only left-invariant odd translations $D_{\theta^1} = -G^1_{-\frac{1}{2}} - G^2_{-\frac{1}{2}}$ and $D_{\theta^2} = -\bar{Y}_0^1 - \bar{Y}_0^2$. It is then natural to consider the vector space

$$
\tilde{\mathfrak{s}}^{(1)} := \left\langle L_{-1}, L_0, L_1, Y_{-\frac{1}{2}}, Y_{\frac{1}{2}}, M_0, G^1_{-\frac{1}{2}} + G^2_{-\frac{1}{2}}, G^1_{\frac{1}{2}} + G^2_{\frac{1}{2}}, \bar{Y}_0^1 + \bar{Y}_0^2 \right\rangle \tag{11.68}
$$

and to ask whether this is a Lie subalgebra of $\tilde{\mathfrak{s}}^{(2)}$. The answer is yes[6] and this is best proved by using the Poisson algebra formulation. Since restricting to this subalgebra amounts to considering functions that depend only on p, q and $\tilde{\theta} := (\theta^1 - \theta^2)/(2\mathrm{i})$, with $(\tilde{\theta})^2 = 0$ and $\{\tilde{\theta}, \tilde{\theta}\} = 1$, $\tilde{\mathfrak{s}}^{(1)}$ can be seen as the Lie

[6] An isomorphic Lie superalgebra was first constructed by Gauntlett *et al.* [33].

algebra of polynomials of degree ≤ 2 in $\mathcal{P}^{(2|1)}$: it sits inside $\mathcal{P}^{(2|1)}$ just in the same way as $\tilde{\mathfrak{s}}^{(2)}$ sits inside $\mathcal{P}^{(2|2)}$. Of course, the conjugate algebra obtained by taking the same linear combinations, but with a minus sign instead (that is, generated by $L_{\pm 1,0}, Y_{\pm\frac{1}{2}}, G^1_{\frac{1}{2}} - G^2_{\frac{1}{2}}$ and the right-invariant odd translations $\bar{D}_{\theta^1} = G^2_{-\frac{1}{2}} - G^1_{-\frac{1}{2}}$ and $\bar{D}_{\theta^2} = \bar{Y}^2_0 - \bar{Y}^1_0$) is isomorphic to $\tilde{\mathfrak{s}}^{(1)}$. The commutation relations of $\tilde{\mathfrak{s}}^{(1)}$ are again illustrated in Fig. 11.3a, where the four double circles of a pair of generators should be replaced by a single generator. We shall consider this algebra once again in Sect. 11.4.

11.2.6 Second Correspondence with Poisson Structures: The Case of $\mathfrak{osp}(2|4)$, or the (3|2)-Supersymmetric Model

We shall prove in this subsection Proposition 11.6 by giving an embedding of the vector space $\mathfrak{s}^{(2)}$ into a Poisson algebra, from which the fact that $\mathfrak{s}^{(2)}$ closes as a Lie algebra becomes self-evident.

Let us first recall the definition of the orthosymplectic superalgebras.

Definition 11.12 (orthosymplectic algebra). Let $n, m = 1, 2, \ldots$ The *Lie superalgebra* $\mathfrak{osp}(n|2m)$ is the set of linear vector fields in the coordinates $x^1, \ldots, x^{2m}, \theta^1, \ldots, \theta^n$ preserving the 2-form $\sum_{i=1}^m \mathrm{d}x^i \wedge \mathrm{d}x^{m+i} + \sum_{j=1}^n (\mathrm{d}\theta^i)^2$.

In the following proposition, we recall the folklore result which states that the Lie superalgebra $\mathfrak{osp}(2|2m)$ may be embedded into a super-Poisson algebra of functions on the $(2m|2)$-supertorus, and detail the root structure in this very convenient embedding.

Definition 11.13. We denote by $\mathcal{P}^{(2m|2)}_{(2)}$ the Lie subalgebra of quadratic polynomials in the super-Poisson algebra $\mathcal{P}^{(2m|2)}$ on the $(2m|2)$-supertorus.

Proposition 11.14. *Equip the super-Poisson algebra* $\mathcal{P}^{(2m|2)}$ *with the super-Poisson bracket* $\{q_i, p_j\} = \delta_{i,j}$, $\{\theta^1, \theta^2\} = 2$, *and consider its Lie subalgebra* $\mathcal{P}^{(2m|2)}_{(2)} \subset \mathcal{P}^{(2m|2)}$. *Then*

1. *The Lie algebra* $\mathcal{P}^{(2m|2)}_{(2)}$ *is isomorphic to* $\mathfrak{osp}(2|2m)$.
2. *Using this isomorphism, a Cartan subalgebra of* $\mathfrak{osp}(2|2m)$ *is given by* $p_i q_i$ $(i = 1, \ldots, 2m)$ *and* $\theta^1\theta^2$. *Let* $(f_1, \ldots, f_{2m}, \alpha)$ *be the dual basis. Then the root-space decomposition is given by*

$$\begin{aligned}\mathfrak{osp}(2|2m) = \mathfrak{osp}(2|2m)_0 &\oplus \mathfrak{osp}(2|2m)_{f_i - f_j} \oplus \mathfrak{osp}(2|2m)_{\pm(f_i+f_j)} \\ &\oplus \mathfrak{osp}(2|2m)_{\pm 2f_i} \\ &\oplus \mathfrak{osp}(2|2m)_{\pm f_i + 2\alpha} \oplus \mathfrak{osp}(2|2m)_{\pm f_i - 2\alpha}\ (i \neq j)\end{aligned} \tag{11.69}$$

Except $\mathfrak{osp}(2|2m)_0$ which is equal to the Cartan subalgebra, all other root-spaces are one-dimensional, and

$$\begin{aligned}
\mathfrak{osp}(2|2m)_{f_i-f_j} &= \langle p_i q_j \rangle, \\
\mathfrak{osp}(2|2m)_{f_i+f_j} &= \langle p_i p_j \rangle, \ \ \mathfrak{osp}(2|2m)_{-(f_i+f_j)} = \langle q_i q_j \rangle \\
\mathfrak{osp}(2|2m)_{2f_i} &= \langle p_i^2 \rangle, \ \ \mathfrak{osp}(2|2m)_{-2f_i} = \langle q_i^2 \rangle \\
\mathfrak{osp}(2|2m)_{f_i+2\alpha} &= \langle p_i \theta^1 \rangle, \ \ \mathfrak{osp}(2|2m)_{-f_i+2\alpha} = \langle q_i \theta^1 \rangle \\
\mathfrak{osp}(2|2m)_{f_i-2\alpha} &= \langle p_i \theta^2 \rangle, \ \ \mathfrak{osp}(2|2m)_{-f_i-2\alpha} = \langle q_i \theta^2 \rangle. \qquad (11.70)
\end{aligned}$$

Proof. Straightforward. □

The root structure is illustrated in Fig. 11.2 in the case $m = 2$. We may now finally state the last ingredient for proving Proposition 11.6.

Proposition 11.15. *1. The linear application $\tilde{\mathfrak{s}}^{(2)} \to \mathfrak{osp}(2|4)$ defined on generators by*

$$\begin{aligned}
& L_{-1} \to \frac{1}{2} q_1^2, \ L_0 \to -\frac{1}{2} q_1 p_1, \ L_1 \to \frac{1}{2} p_1^2, \\
& Y_{-\frac{1}{2}} \to q_1 q_2, \ Y_{\frac{1}{2}} \to -p_1 q_2, \ M_0 \to q_2^2 \\
& N_0 \to \frac{1}{2} \theta^1 \theta^2 \\
& \bar{Y}_0^1 \to -q_2 \theta^1, \ \bar{Y}_0^2 \to q_2 \theta^2 \qquad (11.71) \\
& G^1_{-\frac{1}{2}} \to -\frac{1}{2} q_1 \theta^1, \ G^1_{\frac{1}{2}} \to \frac{1}{2} p_1 \theta^1, \ G^2_{-\frac{1}{2}} \to \frac{1}{2} q_1 \theta^2, \ G^2_{\frac{1}{2}} \to -\frac{1}{2} p_1 \theta^2,
\end{aligned}$$

where $i = 1, 2$, is a Lie algebra morphism and gives an embedding of $\tilde{\mathfrak{s}}^{(2)}$ into $\mathfrak{osp}(2|4)$.

2. This application can be extended into a Lie algebra isomorphism from $\mathfrak{s}^{(2)}$ onto $\mathfrak{osp}(2|4)$ by putting

$$\begin{aligned}
& D \to -\frac{1}{2}(q_1 p_1 + q_2 p_2), \quad V_+ \to p_1 p_2, \quad W \to \frac{1}{4} p_2^2, \\
& V_- \to -\frac{1}{4} q_1 p_2, \quad \bar{Z}_0^1 \to \frac{1}{8} p_2 \theta^1, \quad \bar{Z}_0^2 \to -\frac{1}{8} p_2 \theta^2. \qquad (11.72)
\end{aligned}$$

Proof. The first part is an immediate consequence of Proposition 11.11. One merely needs to replace q, p by q_1, p_1 and then make all generators quadratic in the variables $p_1, p_2, q_1, q_2, \theta^1, \theta^2$ by multiplying with the appropriate power of q_2.

We now turn to the second part. The root diagram of $\mathfrak{osp}(2|4)$ in Fig. 11.2 helps to understand. First $\langle \bar{Z}_0^{1,2}, G_{\frac{1}{2}}^{1,2}; W, V_+, L_1\rangle$ form a Lie algebra of dimension 7 that is isomorphic to $\langle p_2\theta^{1,2}, p_1\theta^{1,2}, p_2^2, p_1p_2, p_1^2\rangle$: in particular, the even part $\langle W, V_+, L_1\rangle$ is commutative and commutes with the 4 other generators; brackets of the odd generators $\bar{Z}_0^{1,2}, G_{\frac{1}{2}}^{1,2}$ yield the whole vector space $\langle W, V_+, L_1\rangle$. Note that part of these computations (commutators of $L_1, G_{\frac{1}{2}}^{1,2}$) come from the preceding subsection, the rest must be checked explicitly. So all there remains to be done is to check for the adjoint action of $\bar{Z}_{1,2}^0, G_{\frac{1}{2}}^{1,2}$ on $\mathfrak{se}(3|2)$. We already computed the action of $G_{\frac{1}{2}}^{1,2}$ on (even or odd) translations; in particular, $G_{\frac{1}{2}}^{1,2}$ preserves this subspace. On the other hand, commutators of $G_{\frac{1}{2}}^{1,2}$ with rotations $V_-, L_0, N_0, Y_{\frac{1}{2}}$ yield linear combinations of $\bar{Z}_0^{1,2}$ and $G_{\frac{1}{2}}^{1,2}$: by definition,

$$\left[G_{\frac{1}{2}}^{1,2}, V_-\right] = \bar{Z}_0^{1,2} \tag{11.73}$$

while other commutators $[G_{\frac{1}{2}}^{1,2}, L_0] = \frac{1}{2}G_{\frac{1}{2}}^{1,2}, [G_{\frac{1}{2}}^1, N_0] = -G_{\frac{1}{2}}^1, [G_{\frac{1}{2}}^2, N_0] = G_{\frac{1}{2}}^2, [G_{\frac{1}{2}}^{1,2}, Y_{\frac{1}{2}}] = 0$ are already known. Now the symmetry $t \leftrightarrow \zeta, \theta^1 \leftrightarrow \theta^2$ preserves $\mathfrak{se}(3|2)$ and sends $G_{\frac{1}{2}}^{1,2}$ into $2\bar{Z}_0^{1,2}$, and corresponds to the symmetry $p \leftrightarrow q$ on $\mathfrak{osp}(2|4) \cong \mathscr{P}_{(2)}^{(4|2)}$, so the action of $\bar{Z}_0^{1,2}$ on the rotation-translation symmetry algebra is the right one. Finally, since W, V_+ and L_1 are given by commutators of $G_{\frac{1}{2}}^{1,2}$ and $\bar{Z}_0^2$, and the commutators of D with the other generators are easily checked to be correct, we are done. □

In Sect. 11.4 we shall consider two-point functions that are covariant under the vector space $\tilde{\mathfrak{s}}_1^{(2)} = \langle L_{-1}, G_{-\frac{1}{2}}^{1,2}, L_0, N_0, G_{\frac{1}{2}}^{1,2}, L_1\rangle \subset \tilde{\mathfrak{s}}^{(2)}$ (actually $\tilde{\mathfrak{s}}_1^{(2)} \subset \tilde{\mathfrak{s}}^{(2)}$ is made of symmetries of the super-Schrödinger model). On the root diagram Fig. 11.2, the generators of $\tilde{\mathfrak{s}}_1^{(2)}$ are all on the f_1-axis, hence (as one sees easily) $\tilde{\mathfrak{s}}_1^{(2)}$ is a Lie algebra. The following proposition gives several equivalent characterizations of $\tilde{\mathfrak{s}}_1^{(2)}$. We omit the easy proof.

Proposition 11.16. *1. The embedding $\tilde{\mathfrak{s}}^{(2)} \subset \mathscr{P}_{\leq 2}^{(2|2)}$ of (11.67) in Proposition 11.11 maps $\tilde{\mathfrak{s}}_1^{(2)}$ onto $\mathscr{P}_{(2)}^{(2|2)}$. Hence, by Proposition 11.14, $\tilde{\mathfrak{s}}_1^{(2)} \cong \mathfrak{osp}(2|2)$.*

2. *The Poisson bracket on $\mathscr{P}_{\leq 2}^{(2|2)}$ (see Proposition 11.11) is of degree -1 with respect to the graduation $\widetilde{\deg}$ of $\mathscr{P}^{(2|2)}$ defined by $\widetilde{\deg}(q) = \widetilde{\deg}(p) = \widetilde{\deg}(\theta^i) = \frac{1}{2}$: in other words, $\widetilde{\deg}\{f, g\} = \widetilde{\deg}(f) + \widetilde{\deg}(g) - 1$ for $f, g \in \mathscr{P}^{(2|2)}$. Hence the set $\{X \in \mathscr{P}^{(2|2)} \mid \widetilde{\deg}(X) = 1\} \cong \tilde{\mathfrak{s}}_1^{(2)}$ is a Lie subalgebra of $\mathscr{P}^{(2|2)}$.*
3. *The Poisson bracket on $\mathscr{P}_{(2)}^{(4|2)}$ (see Proposition 11.14) is of degree -1 with respect to the graduation* deg *of $\mathscr{P}^{(4|2)}$ defined by* $\deg(q_1) = \deg(p_1) = \deg(\theta^1) =$

$\deg(\theta^2) = \frac{1}{2}, \deg(p_2) = 1, \deg(q_2) = 0$. *Hence the set* $\{X \in \mathcal{P}^{(4|2)}_{(2)} \mid \deg(X) = 1\} \cong \tilde{\mathfrak{s}}_1^{(2)} \oplus \mathbb{R}D$ *is a Lie subalgebra of* $\mathcal{P}^{(4|2)}$.

Note that points 1 and 2 use the first correspondence (see (11.67) in Proposition 11.11) while point 3 uses the second correspondence (see (11.71), (11.72) in Proposition 11.15).

11.3 Extended Super-Schrödinger Transformations

We shall be looking in this section for infinite-dimensional extensions of various Lie algebras of Schrödinger type ($\mathfrak{sch}, \tilde{\mathfrak{s}}^{(1)}, \tilde{\mathfrak{s}}^{(2)}, \tilde{\mathfrak{s}}_1^{(2)} \cong \mathfrak{osp}(2|2)$) that we introduced until now. Note that the Lie superalgebra $\mathfrak{s}^{(2)} \cong \mathfrak{osp}(2|4)$ was purposely not included in this list, nor could $\mathfrak{conf}_3$ be included: it seems that there is a 'no-go theorem' preventing this kind of embedding of Schrödinger-type algebras into infinite-dimensional Virasoro-like algebras to extend to an embedding of the whole conformal-type Lie algebra (see Sect. 2.3).

In the preceding section, we saw that all Schrödinger or super-Schrödinger or conformal or 'super-conformal' Lie symmetry algebras could be embedded in different ways into some Poisson algebra or super-algebra $\mathcal{P}^{(n|N)}$.

We shall extend the Schrödinger-type Lie algebras by embedding them in a totally different way into some of the following 'extended' Poisson algebras, where, roughly speaking, one is allowed to consider the square-root of the coordinate p (see Sect. 2.2).

This whole section is really a generalization of the results of Chap. 2 to a supersymmetric setting, in exactly the same spirit.

Definition 11.17. The *extended Poisson algebra* $\widetilde{\mathcal{P}}^{(2|N)}$ is the associative algebra of super-functions

$$f(p,q;\theta) := f(p,q;\theta^1,\ldots,\theta^N) = \sum_{i\in\frac{1}{2}\mathbb{Z}} \sum_{j\in\mathbb{Z}} \sum_{k=1}^{N} \sum_{i_1<\ldots<i_k} c_{i,j,i_1,\ldots,i_k} p^i q^j \theta^{i_1} \ldots \theta^{i_k} \tag{11.74}$$

with usual multiplication and Poisson bracket defined by

$$\{f,g\} := \frac{\partial f}{\partial q}\frac{\partial g}{\partial p} - \frac{\partial f}{\partial p}\frac{\partial g}{\partial q} - (-1)^{\delta(f)} \sum_{i=1}^{N} \partial_{\theta^i} f \, \partial_{\theta^i} g \tag{11.75}$$

with the graduation $\delta : \widetilde{\mathcal{P}}^{(2|N)} \to \mathbb{N}$ defined as a natural extension of Definition 11.9 (see Sect. 11.2.5) on the monomials by

$$\delta(f(p,q)\theta^{i_1} \ldots \theta^{i_k}) := k. \tag{11.76}$$

The Poisson bracket may be defined more loosely by setting $\{q, p\} = 1$, $\{\theta^i, \theta^j\} = \delta^{i,j}$ and applying the Leibniz identity.

Definition 11.18. We denote by $\mathfrak{gra} : \widetilde{\mathscr{P}}^{(2|N)} \to \{0, \frac{1}{2}, 1, \ldots\}$ the graduation (called *grade*) on the associative algebra $\widetilde{\mathscr{P}}^{(2|N)}$ defined by

$$\mathfrak{gra}\left(q^n p^m \theta^{i_1} \ldots \theta^{i_k}\right) := m + k/2 \tag{11.77}$$

on monomials.

This graduation may be defined more simply by setting $\mathfrak{gra}(q) = 0$, $\mathfrak{gra}(p) = 1$, $\mathfrak{gra}(\theta^i) = \frac{1}{2}$. Note that it is closely related but clearly different from the graduations deg, $\widetilde{\deg}$ defined on untwisted Poisson algebras in Proposition 11.16.

Definition 11.19. We denote by $\widetilde{\mathscr{P}}^{(2|N)}_{\leq\kappa}$, $\kappa \in \frac{1}{2}\mathbb{Z}$ (resp. $\widetilde{\mathscr{P}}^{(2|N)}_{(\kappa)}$) the vector subspace of $\widetilde{\mathscr{P}}^{(2|N)}$ consisting of all elements of grade $\leq \kappa$ (resp. of grade equal to κ).

Since the Poisson bracket is of grade -1 (as was the case for deg and $\widetilde{\deg}$) it is clear that $\widetilde{\mathscr{P}}^{(2|N)}_{\leq\kappa}$ (resp. $\widetilde{\mathscr{P}}^{(2|N)}_{(\kappa)}$) is a Lie algebra if and only if $\kappa \leq 1$ (resp. $\kappa = 1$).

It is also easy to check, by the same considerations, that $\widetilde{\mathscr{P}}^{(2|N)}_{\leq\kappa}$ $(\kappa \leq \frac{1}{2})$ is a (proper) Lie ideal of $\widetilde{\mathscr{P}}^{(2|N)}_{\leq 1}$, so one may consider the resulting quotient algebra. In the following, we shall restrict to the case $\kappa = -\frac{1}{2}$:

Definition 11.20 (*Schrödinger-Neveu-Schwarz algebra* $\mathfrak{sns}^{(N)}$)**.** Let $N \geq 0$. Then the N-Schrödinger-Neveu-Schwarz Lie algebra is

$$\mathfrak{sns}^{(N)} := \widetilde{\mathscr{P}}^{(2|N)}_{\leq 1} / \widetilde{\mathscr{P}}^{(2|N)}_{\leq -1/2} \tag{11.78}$$

The choice for the name is by reference to the case $N = 1$ (see below).

11.3.1 Elementary Examples

Let us study in this subsection the simplest examples $N = 0, 1$.

- $\underline{N = 0}$.
  The Lie algebra $\mathfrak{sns}^{(0)}$ is generated by (images in the quotient $\widetilde{\mathscr{P}}^{(2|0)}_{\leq 1} / \widetilde{\mathscr{P}}^{(2|0)}_{\leq -1/2}$) of the fields L, Y, M defined by

$$\mathscr{L}_\phi = \phi(q)p, \mathscr{Y}_\phi = \phi(q)p^{\frac{1}{2}}, \mathscr{M}_\phi = \phi(q). \tag{11.79}$$

By computing the commutators in the quotient, we see that $\mathfrak{sns}^{(0)} = \mathfrak{sv}$ is the Schrödinger–Virasoro algebra $\mathfrak{sv}$, with mode expansion $L_n = q^{n+1}p$, $Y_m = q^{m+\frac{1}{2}}p^{\frac{1}{2}}$, $M_n = q^n$ (where $n \in \mathbb{Z}$, $m \in \mathbb{Z} + \frac{1}{2}$). One retrieves, in slightly different notations, the isomorphism

$$\mathfrak{sv} \simeq \tilde{\mathscr{A}}\left(S^1\right)_{\leq 1} / \tilde{\mathscr{A}}\left(S^1\right)_{\leq -\frac{1}{2}}, \tag{11.80}$$

see Proposition 2.8, which allowed us to see $\mathfrak{sv}$ as a subquotient of the extended Poisson algebra on the torus. Each of these three fields $A = L, Y$ or M has a mode expansion of the form $A_n = q^{n+\varepsilon}p^\varepsilon$. We may rewrite this as $A_{\lambda-\varepsilon} = q^\lambda p^\varepsilon$ with $\lambda \in \mathbb{Z} + \varepsilon$ and see that the shift ε in the indices of the generators (with respect to the power of q) is equal to the opposite of the power of p. This will also hold true for any value of N.

It is important to understand that successive 'commutators' $\{\mathscr{Y}_\phi, \mathscr{M}_\psi\}, \{\mathscr{Y}_{\phi_1}, \{\mathscr{Y}_{\phi_2}, \mathscr{M}_\psi\}\}, \ldots$ in $\widetilde{\mathscr{P}}^{(2|0)}_{\leq 1}$ are generally non-zero and yield ultimately the whole algebra $\widetilde{\mathscr{P}}^{(2|0)}_{\leq -\frac{1}{2}}$. This is due to the fact that derivatives of $p^{\frac{1}{2}}$ give p to power $-\frac{1}{2}, -\frac{3}{2}, \ldots$, unlike derivatives of *integer* positive powers of p, which cancel after a finite time and give only polynomials in p.

The algebraic structure of $\mathfrak{sv}$ is as follows, see Chap. 1 for a more mathematically-minded terminology in terms of tensor-density modules for $\mathrm{Vect}(S^1)$. If one considers the generators L_n, Y_m and M_n as the components of associated conserved currents L, Y and M, then L is a Virasoro field, while Y, M are primary with respect to L, with conformal dimensions $\frac{3}{2}$, respectively 1.

Note also that the conformal dimension of the ε-shifted field A^ε ($A^\varepsilon = L, Y, M$) with mode expansion $A^\varepsilon_n = q^{n+\varepsilon}p^\varepsilon$ ($\varepsilon = 0, \frac{1}{2}, 1$) is equal to $1 + \varepsilon$. This fact is also a general one (see Sect. 11.3.2 below).

For later use, we collect in Tables 11.2 and 11.3 the conformal dimensions and grades of the generators of $\mathfrak{sns}^{(N)}$, with $N = 0, 1, 2$.

- <u>$N = 1$.</u>
The Lie algebra $\mathfrak{sns}^{(1)}$ is generated by (images in the quotient) of the *even* functions $\mathscr{L}_\phi = \phi(q)p, \mathscr{Y}_\phi = \phi(q)p^{\frac{1}{2}}, \mathscr{M}_\phi = \phi(q)$, and of the *odd* functions $\mathfrak{g}_\phi = \phi(q)\theta^1 p^{\frac{1}{2}}, \bar{\mathscr{Y}}_\phi = \phi(q)\theta^1, \bar{\mathscr{M}}_\phi = \phi(q)\theta^1 p^{-\frac{1}{2}}$. We use the same notation as in the case $N = 0$ for the mode expansions $L_n = q^{n+1}p, M_n = q^n, \bar{Y}_n = q^n\theta^1$ ($n \in \mathbb{Z}$), $Y_m = q^{m+\frac{1}{2}}p^{\frac{1}{2}}, G_m = q^{m+\frac{1}{2}}p^{\frac{1}{2}}\theta^1, \bar{M}_m = q^{m-\frac{1}{2}}p^{-\frac{1}{2}}\theta^1$ ($m \in \frac{1}{2} + \mathbb{Z}$), with the same shift in the indices, equal to the opposite of the power in p.

We have a semi-direct product structure $\mathfrak{sns}^{(1)} = \mathfrak{ns} \ltimes \mathfrak{h}^{(1)}$, where

$$\mathfrak{ns} := \langle L, G \rangle \tag{11.81}$$

is isomorphic to the Neveu-Schwarz algebra [89] with a vanishing central charge, and

$$\mathfrak{h}^{(1)} = \big\langle (Y, \bar{Y}), (M, \bar{M}) \big\rangle. \tag{11.82}$$

The commutators of G with these fields read in mode expansion (where we identify the Poisson bracket with an (anti)commutator)

$$[G_n, Y_m] = \frac{1}{2}(n-m)\bar{Y}_{n+m}, \quad [G_n, \bar{Y}_m] = Y_{n+m}$$
$$[G_n, M_m] = -\frac{1}{2}m\bar{M}_{n+m}, \quad [G_n, \bar{M}_m] = M_{n+m}. \tag{11.83}$$

The Lie algebra $\mathfrak{h}^{(1)}$ is two-step nilpotent, which is obvious from the definition of the quotient: the only non-trivial brackets are between elements Y_ϕ and $\bar{\mathscr{Y}}_\phi$ of grade $\frac{1}{2}$ and give elements $\mathscr{M}_\phi$ or $\bar{\mathscr{M}}_\phi$ of grade 0. Explicitly, we have:

$$[Y_n, Y_m] = \frac{1}{2}(n-m)M_{n+m}, \quad [\bar{Y}_n, \bar{Y}_m] = M_{n+m}, \quad [Y_n, \bar{Y}_m] = -\frac{1}{2}m\bar{M}_{n+m}. \tag{11.84}$$

The fields (L, G), $(Y, \bar{Y})$ and $(M, \bar{M})$ can be seen as supersymmetric doublets of conformal fields with conformal dimensions $(2, \frac{3}{2})$, $(\frac{3}{2}, 1)$, $(1, \frac{1}{2})$, see also Table 11.2. Once again, the conformal dimension of any of those fields is equal to the power of p plus one. The grades of the fields are given by, see Table 11.3

$$\mathfrak{gra}(\mathscr{L}_\phi) = \mathfrak{gra}(\mathscr{G}_\phi) = 1, \ \mathfrak{gra}(\mathscr{Y}_\phi) = \mathfrak{gra}(\bar{\mathscr{Y}}_\phi) = \frac{1}{2},$$
$$\mathfrak{gra}(\mathscr{M}_\phi) = \mathfrak{gra}(\bar{\mathscr{M}}_\phi) = 0. \tag{11.85}$$

11.3.2 General Case

We shall actually mainly be interested in the case $N = 2$, but the algebra $\mathfrak{sns}^{(2)}$ is quite large and one needs new insight to study it properly. So let us consider first the main features of the general case.

Table 11.2 Conformal dimensions cdim of the generators of the three supersymmetric extensions $\mathfrak{sv}$, $\mathfrak{sns}^{(1)}$ and $\mathfrak{sns}^{(2)}$ of the one-dimensional Schrödinger algebra $\mathfrak{sch}$

	cdim	Pair	Impair
$\mathfrak{sv}$	2	L	
	3/2	Y	
	1	M	
$\mathfrak{sns}^{(1)}$	2	L	
	3/2	Y	G
	1	M	$\bar{Y}$
	1/2		$\bar{M}$
$\mathfrak{sns}^{(2)}$	2	L	
	3/2	Y	G^1, G^2
	1	M, N	$\bar{Y}^1, \bar{Y}^2$
	1/2	P	$\bar{M}^1, \bar{M}^2$
	0	Q	

Table 11.3 Grades $\mathfrak{gra}$ of the generators of the three supersymmetric extensions $\mathfrak{sv}$, $\mathfrak{sns}^{(1)}$ and $\mathfrak{sns}^{(2)}$ of the one-dimensional Schrödinger algebra $\mathfrak{sch}$

	$\mathfrak{gra}$	Pair	Impair
$\mathfrak{sv}$	1	L	
	1/2	Y	
	0	M	
$\mathfrak{sns}^{(1)}$	1	L	G
	1/2	Y	$\bar{Y}$
	0	M	$\bar{M}$
$\mathfrak{sns}^{(2)}$	1	L, N	G^1, G^2
	1/2	Y, P	$\bar{Y}^1, \bar{Y}^2$
	0	M, Q	$\bar{M}^1, \bar{M}^2$

By considering the grading $\mathfrak{gra}$, one sees immediately that $\mathfrak{sns}^{(N)}$ has a semi-direct product structure

$$\mathfrak{sns}^{(N)} = \mathfrak{g}^{(N)} \ltimes \mathfrak{h}^{(N)} \tag{11.86}$$

where the Lie algebra $\mathfrak{g}^{(N)}$ contains the elements of grade one and the nilpotent algebra $\mathfrak{h}^{(N)}$ contains the elements of grade $\frac{1}{2}$ or 0. The algebra $\mathfrak{g}^{(N)}$ has been studied by Leites and Shchepochkina [79] as one of the 'stringy' superalgebras, namely, the superalgebra $\mathfrak{k}(1|N)$ of supercontact vector fields on the supercircle $S^{(1|N)}$. Let us just mention that $\mathfrak{g}^{(N)}$ shows up as a geometric object, namely, as the superalgebra of vector fields preserving the (kernel of the) 1-form $\mathrm{d}q + \sum_{i=1}^{N} \theta^i \mathrm{d}\theta^i$. Recall also that a supercontact vector field X can be obtained from its *generating function* $f = f(q, \theta^1, \ldots, \theta^N)$ by putting

$$X_f = -\left(1 - \frac{1}{2}E\right)(f)\partial_q - \frac{1}{2}\partial_q f E - (-1)^{\delta(f)} \sum_{i=1}^{N} \partial_{\theta^i} f \partial_{\theta^i}, \tag{11.87}$$

where $E := \sum_{i=1}^{N} \theta^i \partial_{\theta^i}$ is the Euler operator for odd coordinates, and δ is the eigenvalue of E for homogeneous superfunctions as defined in (11.76). Then one has

$$\left[X_f, X_g\right] = X_{\{f,g\}_{\mathfrak{k}(1|N)}} \tag{11.88}$$

where $[\, , \,]$ is the usual Lie bracket of vector fields, and the contact bracket $\{\, , \,\}_{\mathfrak{k}(1|N)}$ is given by

$$\{f, g\}_{\mathfrak{k}(1|N)} := -\left(1 - \frac{1}{2}E\right)(f)\partial_q g + \partial_q f \left(1 - \frac{1}{2}E\right)(g) - (-1)^{\delta(f)} \sum_{i=1}^{N} \partial_{\theta^i} f \partial_{\theta^i} g. \tag{11.89}$$

Proposition 11.21. *The Lie algebras $\mathfrak{g}^{(N)}$ and $\mathfrak{k}(1|N)$ are isomorphic.*

Proof. Let $f = f(q,\theta)$ and $g = g(q,\theta)$ be two E-homogeneous superfunctions. Then

$$\tilde{f}(q,p,\theta) = f(q,\theta)\cdot p^{1-\delta(f)/2}, \quad \tilde{g}(q,p,\theta) = g(q,\theta)\cdot p^{1-\delta(g)/2} \tag{11.90}$$

belong to the subalgebra of elements of grade one in $\widetilde{\mathcal{P}}^{(2|N)}$. Formula (11.75) for the Lie bracket of $\widetilde{\mathcal{P}}^{(2|N)}$ entails

$$\left\{\tilde{f},\tilde{g}\right\}(q,p,\theta) = \left[\left(1-\frac{\delta(g)}{2}\right)(\partial_q f)g - \left(1-\frac{\delta(f)}{2}\right)f(\partial_q g)\right]p^{1-\frac{\delta(f)+\delta(g)}{2}}$$
$$-(-1)^{\delta(f)}\left[\sum_{i=1}^{N}\partial_{\theta^i} f\,\partial_{\theta^i} g\right]p^{2-\frac{\delta(f)+\delta(g)}{2}} \tag{11.91}$$

while formula (11.89) for the contact bracket yields

$$\widetilde{\{f,g\}_{\mathfrak{k}(1|N)}}(q,p,\theta) = -\left[\left(1-\frac{1}{2}\delta(f)\right)f(\partial_q g) - \left(1-\frac{1}{2}\delta(g)\right)(\partial_q f)g\right]$$
$$\times p^{1-\frac{1}{2}(\delta(f)+\delta(g))} - (-1)^{\delta(f)}\left[\sum_{i=1}^{N}\partial_{\theta^i} f\,\partial_{\theta^i} g\right]$$
$$\times p^{1-\frac{1}{2}(\delta(f)+\delta(g)-2)}.$$

Hence

$$\{\tilde{f},\tilde{g}\}_{\widetilde{\mathcal{P}}^{(2|N)}} = \widetilde{\{f,g\}_{\mathfrak{k}(1|N)}}. \tag{11.92}$$

So the assignment $f \to \tilde{f}$ according to (11.90) defines indeed a Lie algebra isomorphism from $\mathfrak{k}(1|N)$ onto $\mathfrak{g}^{(N)}$. □

The application $f \to \tilde{f}$ just constructed may be extended in the following natural way.

Proposition 11.22. *Assign to any superfunction $f(q,\theta)$ on $S^{(1|N)}$ the following superfunctions in the Poisson superalgebra $\widetilde{\mathcal{P}}^{(2|N)}$:*

$$f^{(\alpha)}(q,p,\theta) := f(q,\theta)\cdot p^{\alpha-\delta(f)/2}, \quad \alpha\in\frac{1}{2}\mathbb{Z} \tag{11.93}$$

so that, in particular, $f^{(1)} = \tilde{f}$ as defined in (11.90). Then $f \to f^{(\alpha)}$ defines a linear *isomorphism from the algebra of superfunctions on $S^{(1|N)}$ into the vector space of superfunctions in $\widetilde{\mathcal{P}}^{(2|N)}$ with grade α, and the Lie bracket (11.75) on the Poisson algebra may be written in terms of the superfunctions on $S^{(1|N)}$ in the following way: let f,g be two E-homogeneous functions on $S^{(1|N)}$,*

$$\{f^{(\alpha)}, g^{(\beta)}\}_{\widetilde{\mathcal{P}}^{(2|N)}}$$

$$= \left(-\left(\alpha - \frac{1}{2}E\right)(f)\partial_q g - \partial_q f\left(\beta - \frac{1}{2}E\right)(g) - (-1)^{\delta(f)}\sum_{i=1}^{N}\partial_{\theta^i} f\,\partial_{\theta^i} g\right)^{(\alpha+\beta-1)}. \tag{11.94}$$

Proof. Similar to the proof of proposition 11.21. □

Coming back to $\mathfrak{sns}^{(N)}$, we restrict to the values $\alpha = 1, \frac{1}{2}, 0$. Put $f_n(q) = q^{n+1}$ ($n \in \mathbb{Z}$) and $g_m(q) = q^{m+\alpha-|I|/2}\theta^I$, where $I = \{i_1 < \cdots < i_k\} \subset \{1, \ldots, N\}$ and $\theta^I := \theta^{i_1} \wedge \cdots \wedge \theta^{i_k}$, and $m \in \mathbb{Z} - \alpha + |I|/2$. Then

$$\begin{aligned}
\{f_n^{(1)}, g_m^{(\alpha)}\}_{\widetilde{\mathcal{P}}^{(2|N)}} &= \left\{q^{n+1}p, q^{m+\alpha-|I|/2}\theta^I p^{\alpha-|I|/2}\right\}_{\mathcal{P}^{(2|N)}} \\
&= \left[-(m+\alpha-|I|/2) + \left(\alpha - \frac{1}{2}|I|\right)(n+1)\right] q^{n+m+\alpha-|I|/2}\theta^I \\
&\quad \times p^{\alpha-|I|/2} \\
&= ((\alpha - |I|/2)n - m) g_{n+m}^{(\alpha)},
\end{aligned} \tag{11.95}$$

so the $\tilde{f}_n = f_n^{(1)}$ may be considered as the components of a centerless Virasoro field L, while the $g_m^{(\alpha)}$ are the components of a *primary* field Z_α^I, with conformal dimension $1 + \alpha - |I|/2$, in the sense of [10].

Note also that, as in the cases $N = 0, 1$ studied in Sect. 11.3.1, the conformal dimension cdim of each field is equal to the power of p plus one, and the shift in the indices (with respect to the power of q) is equal to the opposite of the power of p.

This formula (11.94) yields an operator-product formula which is an extension of a formula due to Kac and Cheng [65]. Consider the formal field in $\mathcal{P}^{(N)}$

$$\Theta_I^{(\alpha)}(z) = \sum_{n\in\mathbb{Z}} (t^n \theta^I p^{\alpha-|I|/2}) z^{-n-1}, \tag{11.96}$$

with $I = \{i_1, \cdots, i_k\} \subset \{1, \ldots, N\}$ and $\theta^I = \theta^{i_1} \wedge \cdots \wedge \theta^{i_k}$. If I, J are two families of indices, we denote by $I \bullet J$ the concatenation of I and J. Note that the fields $\Theta_I^{(1)}$ can be seen as a set of formal generators of $\mathfrak{k}(1|N)$ by the isomorphism $\mathfrak{g}^{(N)} \cong \mathfrak{k}(1|N)$, with (almost) the same notations as in [43], Chap. IX. Then one has the following OPE's in $\widetilde{\mathcal{P}}^{(2|N)}$:

Proposition 11.23. *The following OPE hold true in* $\widetilde{\mathcal{P}}^{(2|N)}$:

$$\Theta_I^{(\alpha)}(z) \cdot \Theta_J^{(\beta)}(w) \sim (|I| + |J| - 2(\alpha+\beta)) \frac{\Theta_{I\bullet J}^{(\alpha+\beta-1)}(w)}{(z-w)^2}$$

$$+(|I|-2\alpha)\frac{\partial_w \Theta^{(\alpha+\beta-1)}_{I\bullet J}(w)}{z-w}$$
$$+(-1)^{|I|}\sum_{i=1}^{N}\frac{(\partial\Theta_I/\partial\theta_i\ \wedge\ \partial\Theta_J/\partial\theta_i)^{(\alpha+\beta-1)}(w)}{z-w}. \qquad (11.97)$$

Restricting to $\alpha = 0, \frac{1}{2}, 1$ and setting to zero in the above expansion all fields $\Theta^{(\alpha)}_I$ with $\alpha < 0$, along with their derivatives, one obtains a set of OPE for the fields in the Schrödinger-Neveu-Schwarz algebra $\mathfrak{sns}^{(N)}$.

Note that restricting simply to $\alpha = 1$ yields the OPE of the fields in $\mathfrak{k}(1|N)$ as in the above-cited article by Kac and Cheng [65].

11.3.3 Study of the Case $N = 2$

As follows from the preceding subsection, the superalgebra $\mathfrak{sns}^{(2)}$ is generated by the fields Z^I_α where $I = \emptyset$, $\{1\}$, $\{2\}$ or $\{1,2\}$ and $\alpha = 0, \frac{1}{2}$ or 1. Set

$$L = Z^{\emptyset}_1,\ \ G^{1,2} = \frac{1}{\sqrt{2}}\left(Z^{\{1\}}_1 \pm \mathrm{i}Z^{\{2\}}_1\right),\ \ N = \mathrm{i}\sqrt{2}Z^{\{1,2\}}_1 \qquad (11.98)$$

for generators of grade one,

$$Y = \sqrt{2}Z^{\emptyset}_{\frac{1}{2}},\ \ \bar{Y}^{1,2} = \pm Z^{\{1\}}_{\frac{1}{2}} + \mathrm{i}Z^{\{2\}}_{\frac{1}{2}},\ \ P = 2\mathrm{i}Z^{\{1,2\}}_{\frac{1}{2}} \qquad (11.99)$$

for generators of grade $\frac{1}{2}$, and

$$M = Z^{\emptyset}_0,\ \ \bar{M}^{1,2} = \frac{1}{2\sqrt{2}}\left(\mp Z^{\{1\}}_0 + \mathrm{i}Z^{\{2\}}_0\right),\ \ Q = \frac{\mathrm{i}}{\sqrt{2}}Z^{\{1,2\}}_0 \qquad (11.100)$$

for generators of grade 0. Their conformal dimensions are listed in Table 11.2.

Then the superalgebra $\mathfrak{sns}^{(2)}$ is isomorphic to $\mathfrak{k}(1|2) \ltimes \mathfrak{h}^{(2)}$, with

$$\mathfrak{k}(1|2) \cong \langle L, G^{1,2}, N\rangle$$
$$\mathfrak{h}^{(2)} \cong \langle Y, \bar{Y}^{1,2}, P\rangle \oplus \langle M, \bar{M}^{1,2}, Q\rangle \qquad (11.101)$$

The fields in the first line of (11.101) are of grade 1, while the three first fields in the second line have grade $\frac{1}{2}$ and the three other grade 0.

Put $\theta = (\theta^1 + \mathrm{i}\theta^2)/\sqrt{2}, \bar{\theta} = (\theta^1 - \mathrm{i}\theta^2)/\sqrt{2}$, so that $\{\theta,\theta\} = \{\bar{\theta},\bar{\theta}\} = 0, \{\theta,\bar{\theta}\} = 1$ and $\theta\bar{\theta} = -\mathrm{i}\sqrt{2}\theta^1\theta^2$ (this change of basis is motivated by a need of coherence with Sect. 11.2, see Proposition 11.24 below): then these generators are given by the images in $\mathfrak{sns}^{(2)} = \widetilde{\mathscr{P}}^{(2|2)}_{\le 1} / \widetilde{\mathscr{P}}^{(2|2)}_{\le -\frac{1}{2}}$ of

$$\begin{aligned}
&\mathscr{L}_\phi = \phi(q)p,\ \mathscr{G}^1_\phi = \phi(q)\theta p^{\frac{1}{2}},\ \mathscr{G}^2_\phi = -\phi(q)\bar\theta p^{\frac{1}{2}},\ \mathscr{N}_\phi = -\theta\bar\theta\phi(q)\\
&\bar{\mathscr{Y}}^1_\phi = \sqrt{2}\phi(q)\theta,\ \bar{\mathscr{Y}}^2_\phi = -\sqrt{2}\phi(q)\bar\theta,\ \mathscr{Y}_\phi = \sqrt{2}\phi(q)p^{\frac{1}{2}},\ \mathscr{P}_\phi = -\sqrt{2}\theta\bar\theta\phi(q)p^{-\frac{1}{2}}\\
&\mathscr{M}_\phi = \phi(q),\ \bar{\mathscr{M}}^1_\phi = \frac{1}{2}\phi(q)\theta p^{-\frac{1}{2}},\ \bar{\mathscr{M}}^2_\phi = -\frac{1}{2}\phi(q)\bar\theta p^{-\frac{1}{2}},\ \mathscr{Q}_\phi = -\frac{1}{2}\theta\bar\theta\phi(q)p^{-1}.
\end{aligned} \tag{11.102}$$

Commutators in the Lie superalgebra $\mathfrak{k}(1|2) \cong \langle L, G^{1,2}, N\rangle$ are given as follows:

$$\{\mathscr{L}_\phi, \mathscr{L}_\psi\} = \mathscr{L}_{\phi'\psi-\phi\psi'},\ \left\{\mathscr{L}_\phi, \mathscr{G}^{1,2}_\psi\right\} = \mathscr{G}^{1,2}_{\frac{1}{2}\phi'\psi-\phi\psi'},\ \{\mathscr{L}_\phi, \mathscr{N}_\psi\} = \mathscr{N}_{-\phi\psi'} \tag{11.103}$$

(in other words, $G^{1,2}$ have conformal dimension $\frac{3}{2}$, and N conformal dimension 1);

$$\begin{aligned}
\left\{\mathscr{G}^i_\phi, \mathscr{G}^i_\psi\right\} &= 0,\ i = 1,2;\ \{\mathscr{N}_\phi, \mathscr{N}_\psi\} = 0\\
\left\{\mathscr{G}^1_\phi, \mathscr{G}^2_\psi\right\} &= -\mathscr{L}_{\phi\psi} - \mathscr{N}_{\frac{1}{2}\phi'\psi-\phi\psi'}\\
\left\{\mathscr{G}^{1,2}_\phi, \mathscr{N}_\psi\right\} &= \mp\mathscr{G}^{1,2}_{\phi\psi}.
\end{aligned} \tag{11.104}$$

Note that $\mathfrak{sns}^{(2)}$ is generated (as a Lie algebra) by the fields $L, G^{1,2}$ and Y since one has the formula

$$\mathscr{N}_{\frac{1}{2}\phi'\psi-\phi\psi'} = \left\{\mathscr{G}^1_\phi, \mathscr{G}^2_\psi\right\} - \mathscr{L}_{\phi\psi} \tag{11.105}$$

for the missing generators of grade 1;

$$\bar{\mathscr{Y}}^{1,2}_{\frac{1}{2}(\phi'\psi-\phi\psi')} = \left\{\mathscr{G}^{1,2}_\phi, \mathscr{Y}_\psi\right\},\ -\frac{1}{2}\mathscr{P}_{\phi\psi'} = \left\{\mathscr{G}^1_\phi, \bar{\mathscr{Y}}^2_\psi\right\} + \mathscr{Y}_{\phi\psi} \tag{11.106}$$

for the missing generators of grade $\frac{1}{2}$; and

$$\begin{aligned}
&\mathscr{M}_{\frac{1}{2}(\phi'\psi-\phi\psi')} = \{\mathscr{Y}_\phi, \mathscr{Y}_\psi\},\ \bar{\mathscr{M}}^{1,2}_{\phi\psi'} = \left\{\mathscr{G}^{1,2}_\phi, \mathscr{M}_\psi\right\} = \frac{1}{2}\left\{\mathscr{Y}_\phi, \bar{\mathscr{Y}}^{1,2}_\psi\right\},\\
&2\mathscr{Q}_{(\phi'\psi+\phi\psi')} = \{\mathscr{Y}_\phi, \mathscr{P}_\psi\}
\end{aligned} \tag{11.107}$$

for the generators of grade 0.

Proposition 11.24. *1. The subspace* $\mathscr{R} := \langle\mathscr{M}_\phi - \mathscr{Q}_{\phi'}, \bar{\mathscr{M}}^1_\phi\rangle$, *(with* $\phi'(t) = \mathrm{d}\phi(t)/\mathrm{d}t$*) is an ideal of* $\mathfrak{sns}^{(2)}$ *strictly included in the ideal of elements of grade zero.*

2. *The quotient Lie algebra* $\mathfrak{sns}^{(2)}/\mathscr{R}$ *has a realization in terms of differential operators of first order that extends the representation of* $\tilde{\mathfrak{s}}^{(2)}$ *given e.g. in appendix, Sect. 13.2 : the formulas read (in decreasing order of conformal dimensions)*

$$
\begin{aligned}
& -\mathscr{L}_\phi \to \phi(t)\partial_t + \frac{1}{2}\phi'(t)(r\partial_r + \theta^1\partial_{\theta^1}) + \frac{x}{2}\phi'(t) + \frac{1}{4}\mathscr{M}\phi''(t)r^2 \\
& \qquad + \frac{1}{4}\phi''(t)r\theta^1\partial_{\theta^2} - \mathscr{Y}_\phi \to \phi(t)\partial_r + \mathscr{M}\phi'(t)r + \frac{1}{2}\phi'(t)\theta^1\partial_{\theta^2} \\
& -\mathscr{M}_\phi \to \mathscr{M}\phi(t) \\
& -\mathscr{N}_\phi \to \phi(t)\left(\theta^1\partial_{\theta^1} + \theta^2\partial_{\theta^2} - x\right) - \frac{\mathscr{M}}{4}\phi'(t)r^2 + \frac{1}{4}\phi'(t)r\theta^1\partial_{\theta^2} \\
& -\mathscr{P}_\phi \to \phi(t)\left(\theta^1\partial_{\theta^2} - 2\mathscr{M}r\right) \\
& -\mathscr{Q}_\phi \to \mathscr{M}\int\phi
\end{aligned}
\tag{11.108}
$$

for the even generators, and

$$
\begin{aligned}
& -\mathscr{G}^2_\phi \to \phi(t)\left(\theta^1\partial_t + \theta^2\partial_r\right) + \phi'(t)\left(\frac{1}{2}\theta^1 r\partial_r + x\theta^1 + \mathscr{M}r\theta^2 - \frac{1}{2}\theta^1\theta^2\partial_{\theta^2}\right) \\
& \qquad + \frac{\mathscr{M}}{2}\phi''(t)r^2\theta^1 \\
& -\mathscr{G}^1_\phi \to \phi(t)\partial_{\theta^1} + \frac{1}{2}\phi'(t)r\partial_{\theta^2} \\
& -\bar{\mathscr{Y}}^1_\phi \to \phi(t)\partial_{\theta^2} \\
& -\bar{\mathscr{Y}}^2_\phi \to \phi(t)\left(\theta^1\partial_r + 2\mathscr{M}\theta^2\right) + 2\mathscr{M}\phi'(t)r\theta^1 \\
& -\bar{\mathscr{M}}^1_\phi \to 0 \\
& -\bar{\mathscr{M}}^2_\phi \to \mathscr{M}\phi(t)\theta^1
\end{aligned}
\tag{11.109}
$$

for the odd generators. Their conformal dimensions are listed in Table 11.2 and their grades in Table 11.3.

Proof. 1. Since $\mathfrak{sns}^{(2)}$ is generated as a Lie algebra by the fields $L, G^{1,2}$ and Y, one only needs to check that $[\mathscr{L}_\phi, \mathscr{R}] \subset \mathscr{R}$, $[\mathscr{G}^{1,2}_\phi, \mathscr{R}] \subset \mathscr{R}$ and $[\mathscr{Y}_\phi, \mathscr{R}] \subset \mathscr{R}$ for any ϕ. Then straightforward computations show that

$$[\mathscr{L}_\phi, \mathscr{M}_\psi - \mathscr{Q}_{\psi'}] = -\mathscr{M}_{\phi\psi'} + \mathscr{Q}_{(\phi\psi')'} \tag{11.110}$$

$$[\mathscr{G}^i_\phi, \mathscr{M}_\psi - \mathscr{Q}_{\psi'}] = -2\delta_{i,1}\bar{\mathscr{M}}^1_{\phi\psi'} \tag{11.111}$$

$$[\mathscr{G}^i_\phi, \bar{\mathscr{M}}^1_\psi] = -\frac{1}{2}\delta_{i,2}(\mathscr{M}_{\phi\psi} - \mathscr{Q}_{(\phi\psi)'}) \tag{11.112}$$

while $[\mathscr{Y}_\phi, \mathscr{M}_\psi - \mathscr{Q}_{\psi'}] = [\mathscr{Y}_\phi, \bar{\mathscr{M}}^1_\psi] = 0$ by the definition of $\mathfrak{sns}^{(2)}$ as a quotient.

2. This is a matter of straightforward but tedious calculations. □

Remarks. 1. Each of the above generators is homogeneous with respect to an $\mathbb{R}^3$-valued graduation for which t, r, θ^1 are independent measure units and $[\mathscr{M}] \equiv [t/r^2], [\theta^2] \equiv [\theta^1 r/t]$.

2. One may read (up to an overall translation) the conformal dimensions of the fields by putting $[t] = -1$, $[r] = [\theta^1] = -\frac{1}{2}$, $[\theta^2] = [\mathscr{M}] = 0$.
3. Consider the two distinct embeddings $\tilde{\mathfrak{s}}^{(2)} \subset \mathfrak{sns}^{(2)}$ and $\tilde{\mathfrak{s}}^{(2)} \subset \mathscr{P}^{(2|2)}$ with respective graduations $\mathfrak{gra}$ (see definition 11.18) and deg (defined in Proposition 11.16). Then both graduations coincide on $\tilde{\mathfrak{s}}^{(2)}$. In particular, the Lie subalgebra $\tilde{\mathfrak{s}}^{(2)}_1 \cong \mathfrak{osp}(2|2) \subset \tilde{\mathfrak{s}}^{(2)}$ may be defined either as the set of elements X of $\tilde{\mathfrak{s}}^{(2)}$ with $\deg X = 1$ or else as the set of elements with $\mathfrak{gra} X = 1$, depending on whether one looks at $\tilde{\mathfrak{s}}^{(2)}$ as sitting inside $\mathscr{P}^{(2|2)}$ or inside $\tilde{\mathscr{P}}^{(2|2)}$.
4. When we reconsider the four operators (11.45) for the supersymmetric equations of motion

$$\Delta_0 = 2M_0 L_{-1} - Y^2_{-\frac{1}{2}}, \quad \Delta'_0 = 2M_0 G^1_{-\frac{1}{2}} - Y_{-\frac{1}{2}} \bar{Y}^1_0,$$
$$\bar{\Delta}'_0 = Y_{-\frac{1}{2}} G^1_{-\frac{1}{2}} - L_{-1} \bar{Y}^1_0, \quad \Delta''_0 = G^1_{-1/2} \bar{Y}^1_0 \tag{11.113}$$

and use the Poisson algebra representation (11.102) of $\mathfrak{sns}^{(2)}$, then $\Delta_0 = \Delta'_0 = \bar{\Delta}'_0 = \Delta''_0 = 0$. The consequences of this observation remain to be explored.

11.4 Two-Point Functions

We shall compute in this section the two-point functions $\langle \Phi_1 \Phi_2 \rangle$ that are covariant under some of the Lie subalgebras of $\mathfrak{osp}(2|4)$ introduced previously. Consider the two superfields

$$\Phi_1 = \Phi_1(t_1, r_1, \theta_1, \bar{\theta}_1) = f_1(t_1, r_1) + \phi_1(t_1, r_1)\theta_1 + \bar{\phi}_1(t_1, r_1)\bar{\theta}_1 + g_1(t_1, r_1)\theta_1\bar{\theta}_1$$
$$\Phi_2 = \Phi_2(t_2, r_2, \theta_2, \bar{\theta}_2) = f_2(t_2, r_2) + \phi_2(t_2, r_2)\theta_2 + \bar{\phi}_2(t_2, r_2)\bar{\theta}_2 + g_2(t_2, r_2)\theta_2\bar{\theta}_2 \tag{11.114}$$

with respective masses and scaling dimensions $(\mathscr{M}_1, x_1)$ and $(\mathscr{M}_2, x_2)$. With respect to (11.29), we made a simple change of notations. The Grassmann variable previously denoted by θ^1 is now called θ and the Grassmann variable θ^2 is now called $\bar{\theta}$. The *lower* indices of the Grassmann variables now refer to the first and second superfield, respectively. The two-point function is

$$\mathscr{C}(t_1, r_1, \theta_1, \bar{\theta}_1; t_2, r_2, \theta_2, \bar{\theta}_2) := \langle \Phi_1(t_1, r_1, \theta_1, \bar{\theta}_1) \Phi_2(t_2, r_2, \theta_2, \bar{\theta}_2) \rangle. \qquad (11.115)$$

Since we shall often have invariance under translations in either space-time or in superspace, we shall use the following abbreviations

$$t := t_1 - t_2, \quad r := r_1 - r_2, \quad \theta := \theta_1 - \theta_2, \quad \bar{\theta} := \bar{\theta}_1 - \bar{\theta}_2 \qquad (11.116)$$

The generators needed for the following calculations are collected in appendix, Sect. 13.2.

Proposition 11.25. *(see Appendix B, Sect. 1.5) The* $\mathfrak{osp}(2|4)$*-covariant two-point function is, where the constraints* $x := x_1 = x_2$ *and* $\mathscr{M} := \mathscr{M}_1 = -\mathscr{M}_2$ *hold true and* c_2 *is a normalization constant*

$$\mathscr{C} = c_2 \delta_{x,\frac{1}{2}} t^{-\frac{1}{2}} \exp\left(-\frac{\mathscr{M}}{2}\frac{r^2}{t}\right)\left(\bar{\theta} - \frac{r}{2t}\theta\right) \qquad (11.117)$$

In striking contrast with the usual 'relativistic' $N = 2$ superconformal theory, see e.g. [20, 88, 97], we find that covariance under a finite-dimensional Lie algebra is enough to fix the scaling dimension of the quasi-primary fields. We have already pointed out that this surprising result can be traced back to our non-relativistic identification of the dilatation generator L_0 as $-\frac{1}{2}E_{11} - \frac{1}{2}x$ in Proposition 11.4.

It is quite illuminating to see how the result (11.117) is modified when one considers two-point functions that are only covariant under a subalgebra $\mathfrak{g}$ of $\mathfrak{osp}(2|4)$. We shall consider the following four cases and refer to Fig. 11.3 for an illustration how these algebras are embedded into $\mathfrak{osp}(2|4)$.

1. $\mathfrak{g} = \tilde{\mathfrak{s}}^{(1)}$, which describes invariance under an $N = 1$ superextension of the Schrödinger algebra.
2. $\mathfrak{g} = \tilde{\mathfrak{s}}^{(2)}$, which describes invariance under an $N = 2$ superextension of the Schrödinger algebra.
3. $\mathfrak{g} = \mathfrak{osp}(2|2)$, where, as compared to the previous case $\mathfrak{g} = \tilde{\mathfrak{s}}^{(2)}$, invariance under spatial translations is left out, which opens prospects for a future application to non-relativistic supersymmetric systems with a boundary.
4. $\mathfrak{g} = \mathfrak{se}(3|2)$, for which time-inversions are left out.

From the cases 2 and 4 together, the proof of the proposition 11.25 will be obvious.

Proposition 11.26. *The non-vanishing two-point function,* $\tilde{\mathfrak{s}}^{(1)}$*-covariant under the representation (11.39), (11.44), (11.46), (11.47), of the superfields* $\Phi_{1,2}$ *of the form (11.114) is given by*

$$\mathscr{C} = \delta_{x_1,x_2} \delta_{\mathscr{M}_1+\mathscr{M}_2,0} \left[c_1 \mathscr{C}_1 + c_2 \mathscr{C}_2\right] \qquad (11.118)$$

where $c_{1,2}$ *are constants and*

$$\mathscr{C}_1(t,r;\theta_1,\bar{\theta}_1,\theta_2,\bar{\theta}_2) = t^{-x_1}\exp\left(-\frac{\mathscr{M}_1}{2}\frac{r^2}{t}\right)\left\{1+\frac{1}{t}\left(-x_1+\frac{\mathscr{M}_1}{2}\frac{r^2}{t}\right)\theta_1\theta_2 + 2\mathscr{M}_1\bar{\theta}_1\bar{\theta}_2 - \mathscr{M}_1\frac{r}{t}\left(\theta_1\bar{\theta}_2-\theta_2\bar{\theta}_1\right) - \frac{1}{t}\mathscr{M}_1(2x_1-1)\theta_1\theta_2\bar{\theta}_1\bar{\theta}_2\right\} \tag{11.119}$$

$$\mathscr{C}_2(t,r;\theta_1,\bar{\theta}_1,\theta_2,\bar{\theta}_2) = t^{-x_1}e^{-\mathscr{M}_1r^2/2t}\left\{-\frac{r}{2t}(\theta_1-\theta_2)+(\bar{\theta}_1-\bar{\theta}_2)+\frac{1}{t}\left(\frac{1}{2}-x_1\right)\left(\bar{\theta}_1\theta_1\theta_2-\bar{\theta}_2\theta_1\theta_2\right)\right\}. \tag{11.120}$$

The proof is given in Appendix B, Sect. 1.1.

For ordinary quasiprimary superfields with fixed masses, the two-point function $\mathscr{C} = c_1\mathscr{C}_1 + c_2\mathscr{C}_2$ reads as follows, where we suppress the obvious arguments (t,r) and also the constraints $x := x_1 = x_2$ and $\mathscr{M} := \mathscr{M}_1 = -\mathscr{M}_2$:

$$\begin{aligned}
\langle f_1 f_2\rangle &= c_1\, t^{-x} e^{-\frac{\mathscr{M}r^2}{2t}} \\
\langle \phi_1\phi_2\rangle &= c_1\left(-x+\frac{\mathscr{M}}{2}\frac{r^2}{t}\right) t^{-x-1}e^{-\frac{\mathscr{M}r^2}{2t}} \\
\langle \bar{\phi}_1\bar{\phi}_2\rangle &= 2\mathscr{M}c_1\, t^{-x}e^{-\frac{\mathscr{M}r^2}{2t}} \qquad (11.121)\\
\langle \phi_1\bar{\phi}_2\rangle &= \langle \bar{\phi}_1\phi_2\rangle = -c_1\,\mathscr{M}r\, t^{-x-1}e^{-\frac{\mathscr{M}r^2}{2t}} \\
\langle g_1 g_2\rangle &= -c_1\,\mathscr{M}(2x-1)t^{-x-1}e^{-\frac{\mathscr{M}r^2}{2t}} \\
\langle f_1\phi_2\rangle &= -\langle \phi_1 f_2\rangle = c_2\,\frac{r}{2}\, t^{-x-1}e^{-\frac{\mathscr{M}r^2}{2t}} \\
\langle f_1\bar{\phi}_2\rangle &= -\langle \bar{\phi}_1 f_2\rangle = -c_2\, t^{-x}e^{-\frac{\mathscr{M}r^2}{2t}} \\
\langle \bar{\phi}_1 g_2\rangle &= \langle g_1\phi_2\rangle = -c_2\left(\frac{1}{2}-x\right)t^{-x-1}e^{-\frac{\mathscr{M}r^2}{2t}} \\
\langle \phi_1\bar{\phi}_1\rangle &= \langle \phi_2\bar{\phi}_2\rangle = \langle \phi_1 g_2\rangle = \langle g_1\bar{\phi}_2\rangle = 0.
\end{aligned}$$

Corollary. *Any $\tilde{\mathfrak{s}}^{(2)}$-covariant two-point function has the following form, where $x = x_1 = x_2$ and $\mathscr{M} = \mathscr{M}_1 = -\mathscr{M}_2$ and $c_{1,2}$ are normalization constants*

$$\mathscr{C} = c_1\delta_{x,0}\,\delta_{\mathscr{M},0} + c_2\delta_{x,\frac{1}{2}}\,t^{-\frac{1}{2}}\exp\left(-\frac{\mathscr{M}}{2}\frac{r^2}{t}\right)\left(\bar{\theta}-\frac{r}{2t}\theta\right) \tag{11.122}$$

For the proof see Appendix B, Sect. 1.2. We emphasize that covariance under $\tilde{\mathfrak{s}}^{(2)}$ is already enough to fix x to be either 0 or $\frac{1}{2}$. The contrast with $\tilde{\mathfrak{s}}^{(1)}$ comes from the fact that $\tilde{\mathfrak{s}}^{(1)}$ does not contain the generator N_0, while $\tilde{\mathfrak{s}}^{(2)}$ does. The only non-vanishing two-point functions of the superfield components are

$$\begin{aligned}
\langle f_1 f_2 \rangle &= \delta_{x,0}\delta_{\mathcal{M},0}\, c_1 \\
\langle f_1 \phi_2 \rangle &= -\langle \phi_1 f_2 \rangle = \delta_{x,\frac{1}{2}}\, c_2 \frac{r}{2}\, t^{-3/2} \exp\left(-\frac{\mathcal{M}}{2}\frac{r^2}{t}\right) \\
\langle f_1 \bar{\phi}_2 \rangle &= -\langle \bar{\phi}_1 f_2 \rangle = -\delta_{x,\frac{1}{2}}\, c_2\, t^{-1/2} \exp\left(-\frac{\mathcal{M}}{2}\frac{r^2}{t}\right).
\end{aligned} \tag{11.123}$$

We now study the case of covariance under the Lie algebra

$$\mathfrak{g} = \left\langle L_{-1,0,1}, G^1_{\pm\frac{1}{2}}, G^2_{\pm\frac{1}{2}}, N_0 \right\rangle \cong \mathfrak{osp}(2|2), \tag{11.124}$$

see Sect. 2 in Appendix B for the explicit formulas. We point out that neither space-translations nor phase-shifts are included in $\mathfrak{osp}(2|2)$, so that the two-point functions will in general depend on both space coordinates $r_{1,2}$, and there will in general be no constraint on the masses $\mathcal{M}_{1,2}$. On the other hand, time-translations and odd translations are included, so that $\mathscr{C}$ will only depend on $t = t_1 - t_2$ and $\theta = \theta_1 - \theta_2$. From a physical point of view, the absence of the requirement of spatial translation-invariance means that the results might be used to describe the kinetics of a supersymmetric model close to a boundary surface, especially for semi-infinite systems [6, 50, 100], cf. also [17, 63].

Proposition 11.27. *There exist non-vanishing, $\mathfrak{osp}(2|2)$-covariant two-point functions, of quasiprimary superfields Φ_i of the form (11.114) with scaling dimensions x_i, $i = 1, 2$, only in the three cases $x_1 + x_2 = 0, 1$ or 2. Then the two-point functions $\mathscr{C}$ are given as follows.*

(i) if $x_1 + x_2 = 0$, then necessarily $x_1 = x_2 = 0$, $\mathcal{M}_1 = \mathcal{M}_2 = 0$ and

$$\mathscr{C} = a_0 \tag{11.125}$$

where a_0 is a constant.

(ii) if $x_1 + x_2 = 1$, then

$$\mathscr{C} = t^{-1/2}\left\{\left(\bar{\theta}_1 - \frac{r_1}{2t}(\theta_1 - \theta_2)\right) h_1 + \sqrt{\frac{\mathcal{M}_1}{\mathcal{M}_2}}\left(\bar{\theta}_2 - \frac{r_2}{2t}(\theta_1 - \theta_2)\right) h_2\right\} \tag{11.126}$$

where

$$h_1 = \left(\frac{r_1 r_2}{t}\right)^{1/2} \left(\frac{r_1}{r_2}\right)^{-(x_1-x_2)/2} \left(\alpha J_\mu\left(\frac{\mathscr{M} r_1 r_2}{t}\right) + \beta J_{-\mu}\left(\frac{\mathscr{M} r_1 r_2}{t}\right)\right)$$
$$\exp\left[-\frac{\mathscr{M}_1}{2}\frac{r_1^2}{t} + \frac{\mathscr{M}_2}{2}\frac{r_2^2}{t}\right]$$
$$h_2 = \left(\frac{r_1 r_2}{t}\right)^{1/2} \left(\frac{r_1}{r_2}\right)^{-(x_1-x_2)/2} \left(\alpha J_{\mu+1}\left(\frac{\mathscr{M} r_1 r_2}{t}\right) - \beta J_{-\mu-1}\left(\frac{\mathscr{M} r_1 r_2}{t}\right)\right)$$
$$\exp\left[-\frac{\mathscr{M}_1}{2}\frac{r_1^2}{t} + \frac{\mathscr{M}_2}{2}\frac{r_2^2}{t}\right] \tag{11.127}$$

and

$$\mathscr{M}^2 = \mathscr{M}_1 \mathscr{M}_2, \quad 1 + 2\mu = x_2 - x_1 \tag{11.128}$$

while $J_\mu(x)$ are Bessel functions and α, β are arbitrary constants.

(iii) if $x_1 + x_2 = 2$, then

$$\mathscr{C} = (\theta_1 - \theta_2)\left(\bar\theta_1 - \frac{r_1}{r_2}\bar\theta_2\right) B + \bar\theta_1\bar\theta_2 D \tag{11.129}$$

where

$$B = t^{-3/2}\left(\frac{r_1^2}{t}\right)^{-(x_1-1/2)} h\left(\frac{r_1 r_2}{t}\right) \exp\left[-\frac{\mathscr{M}_1}{2}\frac{r_1^2}{t} + \frac{\mathscr{M}_2}{2}\frac{r_2^2}{t}\right]$$
$$D = \frac{2t}{r_2} B \tag{11.130}$$

and where h is an arbitrary function.

Again, the proof in given in Appendix B (see Sect. 1.3). A few comments are in order.

1. In applications, one usually considers either (a) response functions which in a standard field-theoretical setting may be written as a correlator $\langle\phi\widetilde{\phi}\rangle$ of an order-parameter field ϕ with a 'mass' $\mathscr{M}_\phi \geq 0$ and a conjugate response field $\widetilde{\phi}$ whose 'mass' is non-positive $\mathscr{M}_{\widetilde{\phi}} \leq 0$ [51, 99] or else (b) for purely imaginary masses $\mathscr{M} = im$, correlators $\langle\phi\phi^*\rangle$ of a field and its complex conjugate [50].
2. The two supercharges essentially fix the admissible values of the sum of the scaling dimensions $x_1 + x_2$, which is a consequence of covariance under the supersymmetry generator N_0.
3. For systems which are covariant under the scalar Schrödinger generators $X_{\pm1,0}$ only, it is known that for scalar quasiprimary fields $\phi_{1,2}$ [50]

$$\langle\phi_1(t_1, r_1)\phi_2(t_2, r_2)\rangle = \delta_{x_1,x_2} t^{-x_1} h\left(\frac{r_1 r_2}{t_1 - t_2}\right) \exp\left(-\frac{\mathscr{M}_1}{2}\frac{r_1^2}{t_1 - t_2} + \frac{\mathscr{M}_2}{2}\frac{r_2^2}{t_1 - t_2}\right) \tag{11.131}$$

where h is an arbitrary function. At first sight, there appears some similarity of (11.131) with the result (11.130) obtained for $x_1 + x_2 = 2$ in that a scaling function of a single variable remains arbitrary, but already for $x_1 + x_2 = 1$ the very form of the scaling function (11.127) is completely distinct. The main difference between (11.131) and Proposition 11.27 is that here we have a condition on the *sum* of the scaling dimensions, whereas (11.131) rather fixes the *difference* $x_1 - x_2 = 0$.

Proposition 11.28. *(see Appendix B, Sect. 1.4) The non-vanishing two-point function which is covariant under* $\mathfrak{se}(3|2)$ *is, where* $x = x_1 + x_2$ *and* c_2 *and* d_0 *are normalization constants*

$$\begin{aligned}\mathscr{C} &= c_2\, \delta_{x,1} \delta_{\mathscr{M}_1+\mathscr{M}_2,0}\, t^{-1/2} \exp\left(-\frac{\mathscr{M}_1}{2}\frac{r^2}{t}\right)\left(\bar{\theta} - \frac{r}{2t}\theta\right)\\ &\quad + d_0\, \delta_{\mathscr{M}_1+\mathscr{M}_2,0}\, \mathscr{M}_1^{(x-1)/2} t^{-(x+1)/2} \exp\left(-\frac{\mathscr{M}_1}{2}\frac{r^2}{t}\right)\theta\bar{\theta}\end{aligned} \tag{11.132}$$

This makes it clear that one needs both $N = 2$ supercharges and the time-inversions $t \mapsto -1/t$ in order to obtain a finite list of possibilities for the scaling dimension $x = x_1 + x_2$.

Appendix A
Appendix to Chapter 6

Let $\Phi_i, i = 1, 2, \ldots$ be $\overline{\mathfrak{sch}}$-quasi-primary fields as in Sect. 6.1. The general problem we address in this Appendix (and solve in some generality for $n = 2, 3$) is: what is the most general n-point function $\langle 0 \mid \Phi_1(t_1, r_1, \zeta_1) \ldots \Phi_n(t_n, r_n, \zeta_n) \mid 0 \rangle$ compatible with the constraints coming from symmetries?

It has been solved in general (see [50, 54, 55] and [5], Appendix B) for scalar massive $\mathfrak{sch}$-quasi-primary fields, i.e. for fields such that the representation ρ of $\mathfrak{sv}_0 = \langle L_0 \rangle \ltimes \langle Y_{\frac{1}{2}}, M_1 \rangle$ is one-dimensional, namely $\rho(L_0) = -\lambda$ (where $x := 2\lambda$ is the scaling exponent of the field, see Preface) and $\rho(Y_{\frac{1}{2}}) = 0$. Note that in the whole discussion, the value of $\rho(M_1)$ is irrelevant since M_1 does not belong to $\overline{\mathfrak{sch}}$. Let us recall the results for two- and three-point functions. In the following proposition, we also consider the natural extension to scalar $\overline{\mathfrak{sch}}$-quasi-primary fields:

Definition A.1 (*quasi-primary fields*).
A *scalar* (λ, λ')*-quasi-primary field* is a ρ-$\overline{\mathfrak{sch}}$-quasi-primary field for which ρ is scalar, with $\rho(L_0) = -\lambda$, $\rho(N_0) = -\lambda'$ and $\rho(Y_{\frac{1}{2}}) = 0$.

When speaking of two-point functions, we shall generally use the notation $t = t_1 - t_2$, $r = r_1 - r_2$, $\zeta = \zeta_1 - \zeta_2$. The notations $u = r^2/2t$, $\xi = \zeta - u = \zeta - r^2/2t$ will also show up frequently.

Note quite generally that the Bargmann superselection rule (due to the covariance under the phase shift M_0), as mentioned in the Preface, forbids scalar massive fields $\Phi_1, \ldots, \Phi_n$ with total mass $\mathcal{M} = \mathcal{M}_1 + \ldots + \mathcal{M}_n$ different from 0 to have a non-zero n-point function.

Proposition A.2. *(i) Let $\Phi_{1,2}$ be two scalar $\mathfrak{sch}$-quasi-primary fields with scaling exponents $\lambda_{1,2}$. Then their two-point function $\mathscr{C} = \langle \Phi_1(t_1, r_1, \zeta_1) \Phi_2(t_2, r_2, \zeta_2) \rangle$ vanishes except if $\lambda_1 = \lambda_2 =: \lambda$, in which case it is equal to*

$$\mathscr{C} = t^{-2\lambda} f\left(\zeta - \frac{r^2}{2t}\right) \tag{A.1}$$

J. Unterberger and C. Roger, *The Schrödinger-Virasoro Algebra*, Theoretical and Mathematical Physics, DOI 10.1007/978-3-642-22717-2,

where f is an arbitrary scaling function. The inverse Laplace transform with respect to ζ gives (up to the multiplication by an arbitrary function of the mass) for fields with the same mass $\mathcal{M}$ a generalized heat kernel,

$$\mathcal{C} = g(\mathcal{M}) t^{-2\lambda} e^{-\mathcal{M}\frac{r^2}{2t}}. \tag{A.2}$$

(ii) Suppose furthermore that $\Phi_{1,2}$ are $(\lambda_i, \lambda_i') - \overline{\mathfrak{sch}}$-quasi-primary, $i = 1, 2$, with $\lambda_1 = \lambda_2 =: \lambda$ (otherwise $\mathcal{C}$ vanishes). Then the two-point function is fixed (up to a constant),

$$\mathcal{C} = t^{-2\lambda}(\zeta - \frac{r^2}{2t})^{-\frac{\lambda_1' + \lambda_2'}{2}}. \tag{A.3}$$

The inverse Laplace transform with respect to ζ of this function yields (up to a constant)

$$\mathcal{C} = \mathcal{M}^{\frac{\lambda_1' + \lambda_2'}{2} - 1} t^{-2\lambda} e^{-\mathcal{M}\frac{r^2}{2t}}. \tag{A.4}$$

(iii) Let $\Phi_{1,2,3}$ be three scalar $\mathfrak{sch}$-quasi-primary fields with scaling exponents $\lambda_{1,2,3}$. Then (as already recalled in Sect. 2.4)

$$\begin{aligned}\mathcal{C} &= \langle \Phi_1(t_1, r_1; \mathcal{M}_1)\Phi_2(t_2, r_2; \mathcal{M}_2)\Phi_3(t_3, r_3; \mathcal{M}_3)\rangle \\ &= \delta(\mathcal{M}_1 + \mathcal{M}_2 + \mathcal{M}_3) t_{12}^{\lambda_3 - \lambda_1 - \lambda_2} t_{23}^{\lambda_1 - \lambda_2 - \lambda_3} t_{31}^{\lambda_2 - \lambda_3 - \lambda_1} \\ &\exp\left[-\frac{\mathcal{M}_1}{2}\frac{r_{13}^2}{t_{13}} - \frac{\mathcal{M}_2}{2}\frac{r_{23}^2}{t_{23}}\right] F\left(\frac{(r_{13}t_{23} - r_{23}t_{13})^2}{t_{12}t_{23}t_{13}}\right)\end{aligned} \tag{A.5}$$

where F is an arbitrary scaling function.

Note that the N_0-symmetry constraint is necessary to fix (up to a constant) even the two-point function in the variables (t, r, ζ), contrary to the more rigid case of conformal invariance which fixes two- and three-point functions. That is the reason why we consider fields that are covariant under the extended Schrödinger or Schrödinger–Virasoro algebra.

The non-scalar fields considered below are actually the most general possible for finite-dimensional representations ρ (see discussion before Theorem 6.1), since one does not consider $\rho(M_1)$.

Theorem A.1 (two-point functions for non-scalar fields).
Consider two d-dimensional representations ρ_i, $i = 1, 2$ of $\overline{\mathfrak{sv}}_0 = (\langle L_0\rangle \oplus \langle N_0\rangle) \ltimes \langle Y_{\frac{1}{2}}, M_1\rangle$ indexed by the parameters $\lambda_{1,2}, \lambda_{1,2}', \alpha_{1,2}$ such that

$$\rho_i(L_0) = -\lambda_i \mathrm{Id} + \frac{1}{2}\sum_{\mu=0}^{d-1} \mu E_{\mu,\mu},\ \rho_i(N_0) = -\lambda_i' \mathrm{Id} - \sum_{\mu=0}^{d-1} \mu E_{\mu,\mu} \tag{A.6}$$

$$\rho_i\left(Y_{\frac{1}{2}}\right) = \alpha_i \sum_{\mu=0}^{d-2} E_{\mu,\mu+1} \tag{A.7}$$

Let $\Phi_i = (\Phi_i^\mu(t,r,\zeta))_{\mu=0,\dots,d-1}$, be ρ_i-quasiprimary fields, $i = 1,2$. Then their two-point functions $\mathscr{C}^{\mu,\nu} = \langle \Phi_1^{(\mu)}(t_1,r_1,\zeta_1)\Phi_2^\nu(t_2,r_2,\zeta_2)\rangle$ vanish unless $2(\lambda_1-\lambda_2)$ is an integer. Supposing that $\lambda_1 = \lambda_2$, they may be expressed in terms of d arbitrary parameters $c_0,\dots,c_{d-1}$ as follows:

$$\mathscr{C}^{\mu,\nu} = t^{-\lambda+\frac{\mu+\nu}{2}} \sum_{\delta=\max(\mu,\nu)}^{d-1} c_\delta \frac{\alpha_1^{\delta-\mu}\alpha_2^{\delta-\nu}}{(\delta-\mu)!(\delta-\nu)!}\left(\frac{r^2}{t}\right)^{\delta-\frac{\mu+\nu}{2}}\left(\zeta-\frac{r^2}{2t}\right)^{-(\frac{\lambda'}{2}+\delta)} \tag{A.8}$$

where $\lambda = \lambda_1+\lambda_2$ (= $2\lambda_1$ here) and $\lambda' = \lambda_1'+\lambda_2'$.

Remark. The assumption $\lambda_1 = \lambda_2$ is no restriction of generality: assuming that $\Delta := 2(\lambda_1-\lambda_2)$ is (say) a positive integer implies a shift in the index μ with respect to ν in formula (A.8) and restricts the number of unknown constants. By working through the proof of this Theorem, it is possible to see that the $\mathscr{C}^{\mu,\nu}$ vanish for $\max(\mu,\nu) > d-1-\Delta$ (hence all of them vanish if $\Delta \geq d$) and that the other components depend on $d-\Delta$ coefficients.

Proof. First of all, invariance under translations $\rho(L_{-1}) = -\partial_t, \rho(Y_{-\frac{1}{2}}) = -\partial_r$, $\rho(M_0) = -\partial_\zeta$ implies that $\mathscr{C}^{\mu,\nu}$ is a function of the differences of coordinates t,r,ζ only. Set $\lambda = \lambda_1+\lambda_2, \lambda' = \lambda_1'+\lambda_2'$; we do not assume anything on $\lambda_1-\lambda_2$ for the moment. Let us write the action of $\rho(L_0) = -t\partial_t - \frac{1}{2}r\partial_r + \rho_1(L_0)\otimes \mathrm{Id} + \mathrm{Id}\otimes\rho_2(L_0)$ on $\mathscr{C}^{\mu,\nu}$. By definition, one has

$$\left(t\partial_t + \frac{1}{2}r\partial_r\right)\mathscr{C}^{\mu,\nu} = \left(-\lambda+\frac{\mu+\nu}{2}\right)\mathscr{C}^{\mu,\nu} \tag{A.9}$$

hence

$$\mathscr{C}^{\mu,\nu} = f^{\mu,\nu}(\zeta,u)t^{-\lambda+\frac{\mu+\nu}{2}} \tag{A.10}$$

where $u := \frac{r^2}{2t}$. Then invariance under $\rho(Y_{\frac{1}{2}}) = -t\partial_r - r\partial_\zeta + \rho_1(Y_{\frac{1}{2}})\otimes\mathrm{Id} + \mathrm{Id}\otimes\rho_2(Y_{\frac{1}{2}})$ implies

$$(t\partial_r + r\partial_\zeta)\mathscr{C}^{\mu,\nu} = \rho_1\left(Y_{\frac{1}{2}}\right)_l^\mu \mathscr{C}^{l,\nu} + \rho_2\left(Y_{\frac{1}{2}}\right)_l^\nu \mathscr{C}^{\mu,l} \tag{A.11}$$

$$= \alpha_1\mathscr{C}^{\mu+1,\nu} + \alpha_2\mathscr{C}^{\mu,\nu+1} \tag{A.12}$$

hence

$$\sqrt{2u}(\partial_u+\partial_\zeta)f^{\mu,\nu}(\zeta,u) = \alpha_1 f^{\mu+1,\nu}(\zeta,u) + \alpha_2 f^{\mu,\nu+1}(\zeta,u). \tag{A.13}$$

The solutions of the homogeneous equation associated with (A.13) are the functions of $\xi := \zeta - u$. In the new set of coordinates (ξ, u), (A.13) reads as

$$\sqrt{2u}\partial_u f^{\mu,\nu}(\xi,u) = \alpha_1 f^{\mu+1,\nu}(\xi,u) + \alpha_2 f^{\mu,\nu+1}(\xi,u). \qquad \text{(A.14)}$$

These coupled equations are easily solved. First, $\partial_u f^{d-1,d-1}(\xi,u) = 0$, hence $g^{d-1,d-1} := f^{d-1,d-1}$ is a function of ξ only. It is clear by decreasing induction on μ and ν that the general solution may be expressed in terms of d^2 undetermined functions $g^{\mu,\nu}(\xi)$, $0 \le \mu, \nu \le d-1$, through the relations

$$f^{\mu,\nu}(\xi,u) = \alpha_1 \int \frac{f^{\mu+1,\nu}(\xi,u)}{\sqrt{2u}}\, du \,+\, \alpha_2 \int \frac{f^{\mu,\nu+1}(\xi,u)}{\sqrt{2u}}\, du \,+ g^{\mu,\nu}(\xi) \qquad \text{(A.15)}$$

Let us now use covariance under $\rho(N_0) \equiv -r\partial_r - 2\zeta\partial_\zeta + \rho_1(N_0) \otimes \mathrm{Id} + \mathrm{Id} \otimes \rho_2(N_0)$: one gets

$$2(u\partial_u + \xi\partial_\xi) f^{\mu,\nu}(\xi,u) = -(\lambda' + \mu + \nu) f^{\mu,\nu}(\xi,u)$$

hence

$$f^{\mu,\nu}(\xi,u) := \xi^{-\frac{\lambda'+\mu+\nu}{2}} f_0^{\mu,\nu}(\frac{\xi}{u});$$

this implies immediately $g^{d-1,d-1}(\xi) = \xi^{-(\frac{\lambda'}{2}+d-1)}$ up to a multiplicative constant. Then $\int \frac{\xi^{-\frac{\lambda'+\mu+\nu+1}{2}} f_0^{\mu+1,\nu}(\frac{\xi}{u})}{\sqrt{2u}}\, du$ is homogeneous of degree $-(\lambda' + \mu + \nu)$ with respect to $2(u\partial_u + \xi\partial_\xi)$, hence the defining relations (A.15) are compatible with covariance under $\rho(N_0)$, provided that $g^{\mu,\nu}(\xi) = \xi^{-\frac{\lambda'+\mu+\nu}{2}}$ up to a constant.

Covariance under $\rho(L_1) = -\sum_{i=1}^{2}(t_i^2\partial_{t_i} + t_i r_i \partial_{r_i} + \frac{1}{2}r_i^2\partial_{\zeta_i}) + (2t_1\rho_1(L_0) + r_1\rho_1(Y_{\frac{1}{2}})) \otimes \mathrm{Id} + \mathrm{Id} \otimes (2t_2\rho_2(L_0) + r_2\rho_2(Y_{\frac{1}{2}}))$ is seen to be equivalent (after some easy computations) to the coupled equations

$$(t^2\partial_t + tr\partial_r + \frac{1}{2}r^2\partial_\zeta)\mathscr{C}^{\mu,\nu}(t,r,\zeta) = 2t\rho_1(L_0)^\mu_l \mathscr{C}^{l,\nu} + r\rho_1(Y_{\frac{1}{2}})^\mu_l \mathscr{C}^{l,\nu} \qquad \text{(A.16)}$$

Using the above Ansatz (A.10) yields

$$\left[\left(\frac{\nu-\mu}{2} + \lambda_1 - \lambda_2\right) + u\partial_u\right] f^{\mu,\nu}(\xi,u) = \alpha_1\sqrt{2u} f^{\mu+1,\nu}(\xi,u). \qquad \text{(A.17)}$$

Applying this relation to $f^{d-1,d-1} = c_{d-1}.\xi^{-(\frac{\lambda'}{2}+d-1)}$ gives $c_{d-1} = 0$ unless $\lambda_1 = \lambda_2$, which we assume from now on. Let us compute $f^{d-2,d-1}$ and $f^{d-1,d-2}$ before we attempt the general case; one may set $c_{d-1} = 1$ for the moment. Then $f^{d-2,d-1}(\xi,u) = (\alpha_1\sqrt{2u} + c\sqrt{\xi})\xi^{-(\frac{\lambda'}{2}+d-1)}$ must satisfy $(\frac{1}{2} + u\partial_u) f^{d-2,d-1}(\xi,u) = \alpha_1\sqrt{2u}\xi^{-(\frac{\lambda'}{2}+d-1)}$. The function $\mu_1\sqrt{2u}\xi^{-(\frac{\lambda'}{2}+d-1)}$

is indeed a solution of this equation, and any other solution will be a linear combination of this with some function $u^{-\frac{1}{2}}h(\xi)$, hence $c = 0$ and $f^{d-2,d-1}$ is totally determined by $f^{d-1,d-1}$. On the other hand, $f^{d-1,d-2}(\xi,u) = (\alpha_2\sqrt{2u} + c\sqrt{\xi})\xi^{-(\frac{\lambda'}{2}+d-1)}$ must satisfy $(-\frac{1}{2}+u\partial_u)f^{d-1,d-2}(\xi,u) = \alpha_1\sqrt{2u}\xi^{-(\frac{\lambda'}{2}+d-1)}$. The general solution of this equation is

$$f^{d-1,d-2}(\xi,u) = \sqrt{2u}h(\xi) + \alpha_1\sqrt{2u}(\ln u)\xi^{-(\frac{\lambda'}{2}+d-1)}. \tag{A.18}$$

Both Ansätze are clearly compatible if and only if $c = 0$.

Let us now prove the general case by decreasing induction on $\max(\mu,\nu)$. Assume formula (A.8) of the Theorem has been proved for $\max(\mu,\nu) > M$. Then formula (A.14) gives $f^{M,M}$ up to an undetermined function $g^{M,M}(\xi)$ which is proportional to $\xi^{-\lambda'-M}$ due to covariance with respect to $\rho(N_0)$; it is compatible with formula (A.8) and formula (A.17). One may now go down or left along a line or a row: if for instance all $f^{M-i,M}$, $i < I$ have been found to agree with (A.8), then formula (A.14) again gives $f^{M+I,M}$, in accordance with (A.8), up to an undetermined function $g^{M+I,M}(\xi)$. Compatibility with covariance under $\rho(L_1)$ (formula (A.17)) gives $(\frac{I}{2} + u\partial_u)g^{M+I,M}(\xi) = 0$, hence $g^{M+I,M} = 0$ as soon as $I > 0$. □

Let us now turn to the computation of the general three-point function for *scalar* quasi-primary fields.

Theorem A.2. *Let Φ_i, $i = 1,2,3$ be (λ_i,λ_i')-quasi-primary fields. Then their general three-point function $\mathscr{C}(t_i,r_i,\zeta_i) = \langle\Phi_1(t_1,r_1,\zeta_1)\Phi_2(t_2,r_2,\zeta_2)\Phi_3(t_3,r_3,\zeta_3)\rangle$ may be written as*

$$\mathscr{C} = t_{12}^{-\lambda_1-\lambda_2+\lambda_3}t_{23}^{-\lambda_2-\lambda_3+\lambda_1}t_{13}^{-\lambda_1-\lambda_3+\lambda_2}\xi_{12}^{\frac{1}{2}(-\lambda_1'-\lambda_2'+\lambda_3')}\xi_{23}^{\frac{1}{2}(-\lambda_2'-\lambda_3'+\lambda_1')}\xi_{13}^{\frac{1}{2}(-\lambda_1'-\lambda_3'+\lambda_2')}$$
$$\cdot F(\xi_{12},\xi_{13},\xi_{23}) \tag{A.19}$$

where F is any function of $\xi_{12} := \zeta_{12} - r_{12}^2/2t_{12}, \xi_{13} := \zeta_{13} - r_{13}^2/2t_{13}, \xi_{23} := \zeta_{23} - r_{23}^2/2t_{23}$ which is homogeneous of degree zero, i.e.

$$(\xi_{12}\partial_{\xi_{12}} + \xi_{13}\partial_{\xi_{13}} + \xi_{23}\partial_{\xi_{23}})F = 0. \tag{A.20}$$

Remark. In the case $\lambda_i' = 2\lambda_i$, F constant, one retrieves the standard result for the three-point function in 3d conformal field theory, with a Lorentzian pseudo-distance given (in light-cone coordinates) by $d^2((t_i,r_i,\zeta_i),(t_j,r_j,\zeta_j)) = t_{ij}\zeta_{ij} - r_{ij}^2/2$. The explicit connection between the n-point functions in the Schrödinger/conformal cases has been made in [54] and in [56]. In the last reference, an explicit computation of the three-point function in the dual *mass* coordinates $\mathscr{M}_i$, $i = 1,2,3$ is given – assuming covariance under the whole conformal group – in the case when $\lambda_1 = \lambda_2$, $\mathscr{M}_1 = \mathscr{M}_2$, $r_1 = r_2$. The general result is a combination of two confluent hypergeometric functions. Note that in the present case, $\lambda_i' \neq 2\lambda_i$ in general, but this leads simply to a different time-dependent pre-factor.

Proof. Set $r = r_1 - r_3$, $r' = r_2 - r_3$ and similarly for t, t' and ζ, ζ'. The covariance under the action of L_0, $Y_{\frac{1}{2}}$, N_0 and L_1 yields respectively

$$\left(\sum_i t_i \partial_{t_i} + \frac{1}{2} \sum_i r_i \partial_{r_i} + \lambda \right) \mathscr{C} = 0, \tag{A.21}$$

$$\left(\sum_i t_i \partial_{r_i} + r_i \partial_{\zeta_i} \right) \mathscr{C} = 0, \tag{A.22}$$

$$\left(\sum_i r_i \partial_{r_i} + 2\zeta_i \partial_{\zeta_i} + \lambda' \right) \mathscr{C} = 0, \tag{A.23}$$

$$\left(\sum_i t_i^2 \partial_{t_i} + t_i r_i \partial_{r_i} + \frac{1}{2} r_i^2 \partial_{\zeta_i} + 2\lambda_i t_i \right) \mathscr{C} = 0 \tag{A.24}$$

with $\rho_i(L_0) = -\lambda_i$, $\rho_i(N_0) = -\lambda_i'$, and $\lambda := \sum_i \lambda_i$, $\lambda' = \sum_i \lambda_i'$. The function

$$\mathscr{C} = t_{12}^{-\lambda_1-\lambda_2+\lambda_3} t_{23}^{-\lambda_2-\lambda_3+\lambda_1} t_{13}^{-\lambda_1-\lambda_3+\lambda_2} \xi_{12}^{\frac{1}{2}(-\lambda_1'-\lambda_2'+\lambda_3')} \xi_{23}^{\frac{1}{2}(-\lambda_2'-\lambda_3'+\lambda_1')} \xi_{13}^{\frac{1}{2}(-\lambda_1'-\lambda_3'+\lambda_2')} \tag{A.25}$$

is a particular solution of this system of equations. Hence the general solution is given by $\mathscr{C}_0(t_i, r_i, \zeta_i)\mathscr{C}(t_i, r_i, \zeta_i)$, where $\mathscr{C}_0$ is any solution of the homogeneous system obtained by setting $\lambda_i, \lambda_i' = 0$. By taking an appropriate linear combination of (A.21) and (A.23), one gets

$$(E + E')\mathscr{C}_0 = 0, \tag{A.26}$$

where $E = \sum_i t_i \partial_{t_i} + r_i \partial_{r_i} + \zeta_i \partial_{\zeta_i}$ is the Euler operator in the variables t_i, r_i, ζ_i and similarly for E'. An appropriate linear combination of (A.21), (A.22) and (A.24) gives

$$\left(t^2 \partial_t + t'^2 \partial_{t'} + t r \partial_r + t' r' \partial_{r'} + \frac{1}{2} r^2 \partial_\zeta + \frac{1}{2} r'^2 \partial_{\zeta'} \right) \mathscr{C}_0 = 0 \tag{A.27}$$

Equation (A.22) is equivalent to saying that

$$\mathscr{C}_0 := \mathscr{C}_1(t, t', \rho, \xi, \xi') \tag{A.28}$$

where

$$\rho = \frac{r}{t} - \frac{r'}{t'}, \xi = \zeta - r^2/2t, \xi' = \zeta' - r'^2/2t'. \tag{A.29}$$

Equation (A.21) may then be rewritten

$$\left(t\partial_t + t'\partial_{t'} - \frac{1}{2}\rho\partial_\rho\right)\mathscr{C}_1 = 0 \tag{A.30}$$

hence

$$\mathscr{C}_1 := \mathscr{C}_2(\tau, \tau', \xi, \xi') \tag{A.31}$$

where

$$\tau = \rho^2 t, \tau' = \rho^2 t'. \tag{A.32}$$

Equation (A.27) reads now simply

$$(\tau^2\partial_\tau + \tau'^2\partial_{\tau'})\mathscr{C}_2 = 0, \tag{A.33}$$

with general solution

$$\mathscr{C}_2 := \mathscr{C}_3\left(\frac{1}{\tau} - \frac{1}{\tau'}, \xi, \xi'\right). \tag{A.34}$$

Set $\xi_{ij} := \zeta_{ij} - r_{ij}^2/2t_{ij}$. Then

$$\frac{1}{2}\left(\frac{1}{\tau} - \frac{1}{\tau'}\right) = \xi_{12} + \xi_{23} + \xi_{31} \tag{A.35}$$

Hence the final result by taking into account the equation $(E + E')\mathscr{C}_0 = 0$. □

Appendix B
Appendix to Chapter 11

Section B.1 supplies proofs for the results on covariant supersymmetric two-point functions listed in Sect. 11.4.
Section B.2 is a table of the Lie superalgebras introduced in Chap. 11.

B.1 Supersymmetric Two-Point Functions

B.1.1 $\tilde{\mathfrak{s}}^{(1)}$-Covariant Two-Point Functions

We prove the formulas (11.118) and (11.119) of Proposition 11.26 for the two independent $\tilde{\mathfrak{s}}^{(1)}$-covariant two-point functions.

Let Φ_1 and Φ_2 two superfields with respective masses and dimensions $(\mathcal{M}_1, x_1)$, $(\mathcal{M}_2, x_2)$ as in Chapter 11, Sect. 4, and let $\mathscr{C}(t_1, r_1, \theta_1, \bar{\theta}_1; t_2, r_2, \theta_2, \bar{\theta}_2) := \langle \Phi_1(t_1, r_1, \theta_1)\Phi_2(t_2, r_2, \theta_2)\rangle$ be the associated two-point function. One assumes that $\mathscr{C}$ is covariant under the Lie symmetry representation (see e.g. Sect. 13.2 for a list of the generators) of the 'chiral' superalgebra $\tilde{\mathfrak{s}}^{(1)}$ generated by $L_{-1}, L_0, L_1, Y_{\pm\frac{1}{2}}, G_{-\frac{1}{2}} := G^1_{-\frac{1}{2}} + G^2_{-\frac{1}{2}}, \bar{Y}_0 := \bar{Y}_0^1 + \bar{Y}_0^2$ and M_0.

Because of the invariance under time- and space-translations $L_{-1}, Y_{-\frac{1}{2}}$ and under the mass generator M_0, the two-point function $\mathscr{C}$ depends on time and space only through the coordinates $t := t_1 - t_2$ and $r := r_1 - r_2$, and one can assume that Φ_1 and Φ_2 have opposite masses. We set $\mathcal{M} = \mathcal{M}_1 = -\mathcal{M}_2$.

Covariance under L_0 of the two-point function C gives

$$\left(t\partial_t + \frac{1}{2}r\partial_r + \frac{1}{2}(\theta_1\partial_1 + \theta_2\partial_2) + \frac{1}{2}(x_1 + x_2)\right)\mathscr{C}(t, r, \theta_1, \bar{\theta}_1, \theta_2, \bar{\theta}_2) = 0. \quad (B.1)$$

Covariance under $Y_{\frac{1}{2}}$ gives

J. Unterberger and C. Roger, *The Schrödinger-Virasoro Algebra*, Theoretical and Mathematical Physics, DOI 10.1007/978-3-642-22717-2,

$$\left(t\partial_r + \mathscr{M}r + \frac{1}{2}(\theta_1\bar{\partial}_1 + \theta_2\bar{\partial}_2)\right)\mathscr{C}(t, r, \theta_1, \bar{\theta}_1, \theta_2, \bar{\theta}_2) = 0. \tag{B.2}$$

Covariance under L_1 entails

$$\left(t^2\partial_t + tr\partial_r + t\theta_1\partial_1 + tx_1 + \frac{1}{2}\mathscr{M}r^2 + \frac{1}{2}r\theta_1\bar{\partial}_1\right)\mathscr{C}(t, r, \theta_1, \bar{\theta}_1, \theta_2, \bar{\theta}_2) = 0. \tag{B.3}$$

Covariance under $G_{-\frac{1}{2}}$ yields

$$\left(\partial_1 + \partial_2 + \theta\partial_t + \bar{\theta}\partial_r\right)\mathscr{C}(t, r, \theta_1, \bar{\theta}_1, \theta_2, \bar{\theta}_2) = 0. \tag{B.4}$$

Finally, covariance under $\bar{Y}_0$ yields

$$\left(\bar{\partial}_1 + \bar{\partial}_2 + (\theta_1 - \theta_2)\partial_r + 2\mathscr{M}(\bar{\theta}_1 - \bar{\theta}_2)\right)\mathscr{C}(t, r, \theta_1, \bar{\theta}_1, \theta_2, \bar{\theta}_2) = 0. \tag{B.5}$$

In general, the two-point function may be written as

$$\begin{aligned}\mathscr{C}(t, r, \theta_1, \bar{\theta}_1, \theta_2, \bar{\theta}_2) &= A(t,r) + B_i(t,r)\theta_i + \bar{B}_i(t,r)\bar{\theta}_i \\ &\quad + C_{12}\theta_1\theta_2 + C_{\bar{1}\bar{2}}\bar{\theta}_1\bar{\theta}_2 + C_{1\bar{1}}\theta_1\bar{\theta}_1 + C_{2\bar{2}}\theta_2\bar{\theta}_2 + C_{1\bar{2}}\theta_1\bar{\theta}_2 \\ &\quad + C_{2\bar{1}}\theta_2\bar{\theta}_1 + D_1(t,r)\bar{\theta}_1\theta_1\theta_2 \\ &\quad + D_2(t,r)\bar{\theta}_2\theta_1\theta_2 + \bar{D}_1(t,r)\theta_1\bar{\theta}_1\bar{\theta}_2 \\ &\quad + \bar{D}_2(t,r)\theta_2\bar{\theta}_1\bar{\theta}_2 + E(t,r)\theta_1\theta_2\bar{\theta}_1\bar{\theta}_2.\end{aligned} \tag{B.6}$$

where i is summed over $i = 1, 2$.

Covariance under $G_{-\frac{1}{2}}$ (B.4) gives the following system of linearly independent equations:

$$B_1 = -B_2 \tag{B.7}$$

$$C_{12} = \partial_t A \tag{B.8}$$

$$C_{1\bar{1}} + C_{2\bar{1}} =: -\partial_r A,\ C_{1\bar{2}} + C_{2\bar{2}} = \partial_r A \tag{B.9}$$

$$\partial_r(\bar{B}_1 + \bar{B}_2) + (\bar{D}_1 + \bar{D}_2) = 0 \tag{B.10}$$

$$\partial_t\bar{B}_1 - \partial_r B_1 - D_1 = 0 \tag{B.11}$$

$$\partial_t\bar{B}_2 + \partial_r B_1 - D_2 = 0 \tag{B.12}$$

$$\partial_t C_{\bar{1}\bar{2}} - \partial_r C_{1\bar{1}} - \partial_r C_{1\bar{2}} - E = 0 \tag{B.13}$$

$$\partial_t(\bar{D}_1 + \bar{D}_2) + \partial_r(D_1 + D_2) = 0 \tag{B.14}$$

Covariance under $\bar{Y}_0$ gives the following system of linearly independent equations:

$$\bar{B}_1 = -\bar{B}_2 \tag{B.15}$$

$$\partial_r A = C_{1\bar{1}} + C_{1\bar{2}}, -\partial_r A = C_{2\bar{1}} + C_{2\bar{2}} \tag{B.16}$$

$$2\mathscr{M} A = C_{\bar{1}\bar{2}} \tag{B.17}$$

$$\partial_r(B_1 + B_2) + (D_1 + D_2) = 0 \tag{B.18}$$

$$\partial_r \bar{B}_1 - 2\mathscr{M} B_1 + \bar{D}_1 = 0 \tag{B.19}$$

$$-\partial_r \bar{B}_2 + 2\mathscr{M} B_2 - \bar{D}_2 = 0 \tag{B.20}$$

$$\partial_r C_{1\bar{1}} + \partial_r C_{2\bar{1}} + 2\mathscr{M} C_{12} - E = 0 \tag{B.21}$$

$$\partial_r(\bar{D}_1 + \bar{D}_2) + 2\mathscr{M}(D_1 + D_2) = 0 \tag{B.22}$$

Combining these relations, we can express all the coefficients of $\mathscr{C}$ in terms of $A, B_1, \bar{B}_1, D_1, \bar{D}_1$ and $\Gamma := C_{1\bar{1}} = C_{2\bar{2}}$ through the obvious relations $B_1 + B_2 = \bar{B}_1 + \bar{B}_2 = D_1 + D_2 = \bar{D}_1 + \bar{D}_2 = 0$ and the (less obvious) relations

$$C_{2\bar{1}} = -\partial_r A - \Gamma,\ C_{1\bar{2}} = \partial_r A - \Gamma,\ C_{12} = \partial_t A,\ C_{\bar{1}\bar{2}} = 2\mathscr{M} A. \tag{B.23}$$

There remain only three supplementary equations : (B.11), (B.19) and

$$(2\mathscr{M}\partial_t - \partial_r^2)A - E = 0. \tag{B.24}$$

Recall that the only solution (up to scalar multiplication) of the equations

$$(t\partial_r + \mathscr{M} r)F(t,r) = (t\partial_t + \frac{1}{2} r\partial_r + \lambda)F(t,r) = 0 \tag{B.25}$$

is $F(t,r) = \mathscr{F}_\lambda(t,r) := t^{-\lambda} e^{-\mathscr{M} r^2/2t}$, which one might call a 'Schrödinger quasiprimary function'. Looking now at the consequences of L_0- and $Y_{-\frac{1}{2}}$-covariance, one understands easily that the coefficients in $\mathscr{C}$ of the polynomials in $\theta_i, \bar{\theta}_i$ that depend on θ_1, θ_2 only through $\theta_1\bar{\theta}_1$ and $\theta_2\bar{\theta}_2$ are Schrödinger quasiprimary functions, namely:

$$\begin{aligned} A &= a\mathscr{F}_{\frac{1}{2}(x_1+x_2)}\ , \quad \bar{B}_1 = \bar{b}\mathscr{F}_{\frac{1}{2}(x_1+x_2)} \\ \bar{D}_1 &= \bar{d}\mathscr{F}_{\frac{1}{2}(x_1+x_2)+\frac{1}{2}}\ , \quad \Gamma = \gamma\mathscr{F}_{\frac{1}{2}(x_1+x_2)+\frac{1}{2}} \\ & \qquad\qquad E = e\mathscr{F}_{\frac{1}{2}(x_1+x_2)+1} \end{aligned} \tag{B.26}$$

with yet undetermined constants $a, \bar{b}, \bar{d}, \gamma, e$. This, together with the previous relations, allows one to express all the coefficients of $\mathscr{C}$ in terms of these constants, since all other coefficients are derived directly from $A, \bar{B}_1, \bar{D}_1, \Gamma$ and E. Equation (B.24) gives

$$e = -a\mathscr{M}(x_1 + x_2 - 1). \tag{B.27}$$

Finally, it remains to check covariance under L_1, which gives constraints on the scaling dimensions of the Schrödinger quasiprimary coefficients, namely: $x_1 = x_2$ unless $a = \bar{b} = 0$; $\gamma = \bar{d} = 0$ (otherwise we would have simultaneously $x_1 - x_2 = 1$ and $x_1 - x_2 = -1$). In order to get a non-zero solution, we have to put in the constraint $x_1 = x_2 =: x$, and find $\mathscr{C} = a\mathscr{C}_1 + \bar{b}\mathscr{C}_2$.

One then checks that all supplementary relations coming from $Y_{-\frac{1}{2}}$, L_0- and L_1-covariance are already satisfied. □

B.1.2 $\tilde{\mathfrak{s}}^{(2)}$-Covariant Two-Point Functions

Starting from an $\tilde{\mathfrak{s}}^{(1)}$-covariant two-point function $\mathscr{C} = a\mathscr{C}_1 + \bar{b}\mathscr{C}_2$, all there is to do is to postulate invariance of $\mathscr{C}$ under the vector field $G^1_{-\frac{1}{2}} = -\partial_1 - \partial_2$. We find that either $\bar{b} = 0$ and then also $x = 0$ and $\mathscr{M} = 0$, or else $a = 0$ and furthermore $x = \frac{1}{2}$ which establishes (11.122). □

B.1.3 $\mathfrak{osp}(2|2)$-Covariant Two-Point Functions

Here, we prove Proposition 11.27. From the definition of $\mathfrak{osp}(2|2) = \langle L_{\pm 1,0}, G^{1,2}_{\pm\frac{1}{2}}, N_0\rangle$ we see that time-translations $L_{-1} = -\partial_{t_1} - \partial_{t_2}$ and odd translations $G^1_{-\frac{1}{2}} = -\partial_{\theta_1} - \partial_{\theta_2}$ are included, hence $\mathscr{C}$ will only depend on $t = t_1 - t_2$ and $\theta = \theta_1 - \theta_2$. From the explicit differential-operator representation (11.39), (11.44), (11.46), (11.47) we obtain the following covariance conditions for $\mathscr{C} = \mathscr{C}(t, r_1, r_2; \theta, \bar{\theta}_1, \bar{\theta}_2)$

$$
\begin{aligned}
-L_0\mathscr{C} &= \left[t\partial_t + \frac{1}{2}(r_1\partial_{r_1} + r_2\partial_{r_2} + \theta\partial_\theta) + \frac{x_1 + x_2}{2}\right]\mathscr{C} = 0 \\
-L_1\mathscr{C} &= \left[t^2\partial_t + t\theta\partial_\theta + tr_1\partial_{r_1} + x_1 t + \frac{1}{2}\left(\mathscr{M}_1 r_1^2 + \mathscr{M}_2 r_2^2\right) + \frac{r_1}{2}\theta\partial_{\bar{\theta}_1}\right]\mathscr{C} = 0 \\
-G^1_{\frac{1}{2}}\mathscr{C} &= \left[t\partial_\theta + \frac{1}{2}\left(r_1\partial_{\bar{\theta}_1} + r_2\partial_{\bar{\theta}_2}\right)\right]\mathscr{C} = 0 \\
-G^2_{-\frac{1}{2}}\mathscr{C} &= \left[\theta\partial_t + \partial_{r_1}\bar{\theta}_1 + \partial_{r_2}\bar{\theta}_2\right]\mathscr{C} = 0 \\
-G^2_{\frac{1}{2}}\mathscr{C} &= \left[t\theta\partial_t + t\partial_{r_1}\bar{\theta}_1 + \frac{1}{2}\left((r_1\partial_{r_1} + 2x_1)\theta\right.\right. \\
&\quad \left.\left. + 2\left(\mathscr{M}_1 r_1\bar{\theta}_1 + \mathscr{M}_2 r_2\bar{\theta}_2\right) - \theta\bar{\theta}_1\partial_{\bar{\theta}_1}\right)\right]\mathscr{C} = 0 \\
-N_0\mathscr{C} &= \left[\theta\partial_\theta + \bar{\theta}_1\partial_{\bar{\theta}_1} + \bar{\theta}_2\partial_{\bar{\theta}_2} - x_1 - x_2\right]\mathscr{C} = 0
\end{aligned} \tag{B.28}
$$

The solutions of this system of equations can be written in the form

$$\mathscr{C} = A + \theta A_0 + \bar{\theta}_1 A_1 + \bar{\theta}_2 A_2 + \theta\bar{\theta}_1 B_1 + \theta\bar{\theta}_2 B_2 + \theta\bar{\theta}_1\bar{\theta}_2 C + \bar{\theta}_1\bar{\theta}_2 D \quad \text{(B.29)}$$

where the functions $A = A(t, r_1, r_2), \ldots$ depend on the variables t, r_1, r_2 and are to be determined. In what follows, the arguments of these functions will usually be suppressed.

First, we consider the condition $G^1_{\frac{1}{2}}\mathscr{C} = 0$ which together with (B.29) leads to the following equations

$$\begin{aligned} tA_0 + \frac{r_1}{2}A_1 + \frac{r_2}{2}A_2 &= 0 \\ tB_1 - \frac{1}{2}r_2 D &= 0 \\ tB_2 + \frac{1}{2}r_1 D &= 0 \\ C &= 0 \end{aligned} \quad \text{(B.30)}$$

Next, we use the condition $N_0\mathscr{C} = 0$, which together with (B.29) leads to

$$\begin{aligned} (x_1 + x_2)A &= 0 \\ (1 - x_1 - x_2)A_i &= 0 \\ (2 - x_1 - x_2)B_j &= 0 \\ (2 - x_1 - x_2)D &= 0 \end{aligned} \quad \text{(B.31)}$$

for $i = 0, 1, 2$ and $j = 1, 2$. Therefore, we have to distinguish the three cases $x_1 + x_2 = 0, 1, 2$, respectively.

We begin with the case (i) $x_1 + x_2 = 0$. Then $A_i = B_j = C = D = 0$. From $G^2_{-\frac{1}{2}}\mathscr{C} = \left[\theta\partial_t + \bar{\theta}_1\partial_{r_1} + \bar{\theta}_2\partial_{r_2}\right]A = 0$ it follows that $A = a_0$ is a constant. Furthermore, the covariance $L_1\mathscr{C} = 0$ implies $x_1 = x_2 = 0$ and $\mathscr{M}_1 = \mathscr{M}_2 = 0$.

Next, we consider the case (ii) $x_1 + x_2 = 1$. Then $A = B_i = C = D = 0$ and it remains to find $A_{1,2}$, whereas A_0 is given by the first of (B.30). From the condition $G^2_{-\frac{1}{2}}\mathscr{C} = 0$, we have

$$\partial_t A_1 = \partial_{r_1} A_0, \quad \partial_t A_2 = \partial_{r_2} A_0, \quad \partial_{r_1} A_2 = \partial_{r_2} A_1 \quad \text{(B.32)}$$

From the condition $G^2_{\frac{1}{2}}\mathscr{C} = 0$, we find

$$\left(t\partial_t + \frac{1}{2}r_1\partial_{r_1} + \left(x_1 - \frac{1}{2}\right)\right) A_1 = (t\partial_{r_1} + \mathscr{M}_1 r_1)\, A_0$$
$$\left(t\partial_t + \frac{1}{2}r_1\partial_{r_1} + x_1\right) A_2 = \mathscr{M}_2 r_2 A_0$$
$$(t\partial_{r_1} + \mathscr{M}_1 r_1)\, A_2 = \mathscr{M}_2 r_2 A_1 \tag{B.33}$$

Dilatation-covariance $L_0\mathscr{C} = 0$ gives

$$\left(t\partial_t + \frac{1}{2}\left(r_1\partial_{r_1} + r_2\partial_{r_2}\right) + \frac{1}{2}(1 + x_1 + x_2)\right) A_0 = 0$$
$$\left(t\partial_t + \frac{1}{2}\left(r_1\partial_{r_1} + r_2\partial_{r_2}\right) + \frac{1}{2}(x_1 + x_2)\right) A_{1,2} = 0 \tag{B.34}$$

and finally, covariance under the special transformations $L_1\mathscr{C} = 0$ leads to

$$\left(t\left(r_1\partial_{r_1} - r_2\partial_{r_2}\right) + t(1 + x_1 - x_2) + \left(\mathscr{M}_1 r_1^2 + \mathscr{M}_2 r_2^2\right)\right) A_0 + r_1 A_1 = 0$$
$$\left(t\left(r_1\partial_{r_1} - r_2\partial_{r_2}\right) + t(x_1 - x_2) + \left(\mathscr{M}_1 r_1^2 + \mathscr{M}_2 r_2^2\right)\right) A_{1,2} = 0 \tag{B.35}$$

To solve (B.30), (B.32), (B.33), (B.34), (B.35), we use that $x_1 + x_2 = 1$ and have the scaling ansatz

$$A_{1,2} = t^{-1/2}\mathscr{A}_{1,2}(u_1, u_2), \quad A_0 = t^{-1}\mathscr{A}_0(u_1, u_2) \tag{B.36}$$

where $u_i = r_i/\sqrt{t}$, $i = 1, 2$. Then (B.30) becomes $\mathscr{A}_0 + \frac{1}{2}u_1\mathscr{A}_1 + \frac{1}{2}u_2\mathscr{A}_2 = 0$. On the other hand, (B.35) gives

$$\mathscr{A}_{1,2}(u_1, u_2) = h_{1,2}(u_1 u_2)\, u_1^{x_2 - x_1} \exp\left[-\frac{\mathscr{M}_1}{2}u_1^2 + \frac{\mathscr{M}_2}{2}u_2^2\right] \tag{B.37}$$

It is now easily seen that the remaining equations all reduce to the following system of equations for the two functions $h_{1,2}(v)$

$$\frac{\mathrm{d}h_1(v)}{\mathrm{d}v} = -\mathscr{M}_1 h_2(v)$$
$$\frac{\mathrm{d}h_2(v)}{\mathrm{d}v} = \mathscr{M}_2 h_1(v) + (x_1 - x_2)\frac{1}{v}h_2(v) \tag{B.38}$$

The general solution of these equations is found with standard techniques

$$h_1(v) = \alpha'\left(\frac{\mathscr{M}v}{2}\right)^{-\mu} J_\mu(\mathscr{M}v) + \beta'\left(\frac{\mathscr{M}}{2v}\right)^{\mu} J_{-\mu}(\mathscr{M}v)$$
$$h_2(v) = \sqrt{\frac{\mathscr{M}_2}{\mathscr{M}_1}}\left[\alpha'\left(\frac{\mathscr{M}v}{2}\right)^{-\mu} J_{\mu+1}(\mathscr{M}v) - \beta'\left(\frac{\mathscr{M}}{2v}\right)^{\mu} J_{-\mu-1}(\mathscr{M}v)\right] \tag{B.39}$$

where we used (11.128), J_μ is a Bessel function and α', β' are arbitrary constants. Combination with (B.29) establishes the second part of the assertion.

Finally, we consider the third case (iii) $x_1 + x_2 = 2$. Then $A = A_j = C = 0$ and we still have to find $B_{1,2}$ and D. Going through the covariance conditions, we obtain the following system of equations

$$D = \frac{2t}{r_2} B_1, \quad D = -\frac{2t}{r_1} B_2 \tag{B.40}$$

and

$$\begin{aligned}
\partial_{r_1} B_2 - \partial_{r_2} B_1 &= \partial_t D \\
\left(t\partial_t + \frac{1}{2}\left(r_1\partial_{r_1} + r_2\partial_{r_2}\right) + 1 \right) D &= 0 \\
\left(t\partial_t + \frac{1}{2}\left(r_1\partial_{r_1} + r_2\partial_{r_2}\right) + \frac{3}{2} \right) B_{1,2} &= 0 \\
\left(t^2\partial_t + tr_1\partial_{r_1} + t(x_1+1) + \frac{1}{2}\mathcal{M}_1 r_1^2 + \frac{1}{2}\mathcal{M}_2 r_2^2 \right) B_1 &= 0 \\
\left(t^2\partial_t + tr_1\partial_{r_1} + tx_1 + \frac{1}{2}\mathcal{M}_1 r_1^2 + \frac{1}{2}\mathcal{M}_2 r_2^2 \right) B_2 &= 0 \\
\left(t^2\partial_t + tr_1\partial_{r_1} + tx_1 + \frac{1}{2}\mathcal{M}_1 r_1^2 + \frac{1}{2}\mathcal{M}_2 r_2^2 \right) D &= 0 \\
\left(t\partial_t + \frac{1}{2} r_1\partial_{r_1} + x_1 - \frac{1}{2} \right) D - \left(t\partial_{r_1} + \mathcal{M}_1 r_1\right) B_2 + \mathcal{M}_2 r_2 B_1 &= 0
\end{aligned} \tag{B.41}$$

We see that $B_2 = -(r_1/r_2)B_1$ and it further follows that (B.41) can be reduced to the system

$$\begin{aligned}
\left(t\partial_t + \frac{1}{2}\left(r_1\partial_{r_1} + r_2\partial_{r_2}\right) + \frac{3}{2} \right) B_1 &= 0 \\
\left(t^2\partial_t + tr_1\partial_{r_1} + t(x_1+1) + \frac{1}{2}\mathcal{M}_1 r_1^2 + \frac{1}{2}\mathcal{M}_2 r_2^2 \right) B_1 &= 0
\end{aligned} \tag{B.42}$$

with the general solution

$$B_1 = t^{-3/2} f\left(\frac{r_1 r_2}{t}\right) \left(\frac{r_1^2}{t}\right)^{\frac{1}{2}-x_1} \exp\left[-\frac{\mathcal{M}_1}{2}\frac{r_1^2}{t} + \frac{\mathcal{M}_2}{2}\frac{r_2^2}{t} \right] \tag{B.43}$$

where $f = f(v)$ is an arbitrary function. We have hence found the function $B = B_1$. Combining this with (B.29) then yields the last part of the assertion. □

B.1.4 $\mathfrak{se}(3|2)$-Covariant Two-Point Functions

In order to prove Proposition 11.28, we first observe that because of the covariance under the generators $L_{-1}, Y_{-\frac{1}{2}}, M_0, G^1_{-\frac{1}{2}}$ and $\bar{Y}^1_0$, we have

$$\mathscr{C} = \delta(\mathscr{M}_1 + \mathscr{M}_2) G(t, r, \mathscr{M}_1, \theta, \bar{\theta}) \tag{B.44}$$

where the notation of (11.116) was used. The remaining six conditions become

$$\left[t\partial_t + \frac{1}{2} r\partial_r + \frac{1}{2}\theta\partial_\theta + \frac{x_1 + x_2}{2} \right] G = 0 \tag{B.45}$$

$$\left[t\partial_r + \mathscr{M}_1 r + \frac{1}{2}\theta\partial_{\bar{\theta}} \right] G = 0 \tag{B.46}$$

$$\left[-\partial_{\mathscr{M}_1}\partial_r + r\partial_t + \bar{\theta}\partial_\theta \right] G = 0 \tag{B.47}$$

$$\left[t\partial_t + r\partial_r + \frac{1}{2}\theta\partial_\theta + \frac{1}{2}\bar{\theta}\partial_{\bar{\theta}} - \mathscr{M}_1\partial_{\mathscr{M}_1} + (x_1 + x_2 - 1) \right] G = 0 \tag{B.48}$$

$$\left[\theta\partial_t + \bar{\theta}\partial_r \right] G = 0 \tag{B.49}$$

$$\left[\theta\partial_r + 2\mathscr{M}_1\bar{\theta} \right] G = 0 \tag{B.50}$$

These are readily solved through the expansion

$$G = A + \theta B + \bar{\theta} C + \theta\bar{\theta} D \tag{B.51}$$

where $A = A(t, r, \mathscr{M}_1)$ and so on. Now, from (B.49) we have $\partial_t A = \partial_r A = \partial_t C - \partial_r B = 0$. Similarly, from (B.50) we find $2\mathscr{M}_1 A = 0$ and $\partial_r C = 2\mathscr{M}_1 B$.

First, we consider the coefficient A. From (B.45) it follows that $x_1 + x_2 = 0$ and from (B.48) it can be seen that $(\mathscr{M}_1\partial_{\mathscr{M}_1} + 1)A = 0$, hence $A = a_0/\mathscr{M}_1$. Because of $2\mathscr{M}_1 A = 0$ as derived above it follows $a_0 = 0$.

Next, we find C from (B.45) and (B.46) which give $(t\partial_t + \frac{1}{2}r\partial_t + \frac{x}{2})C = 0$ and $(t\partial_r + \mathscr{M}_1 r)C = 0$ with the result $C = c(\mathscr{M}_1)t^{-x/2}\exp\left(-\mathscr{M}_1 r^2/(2t)\right)$. From the above relation $\partial_r C = 2\mathscr{M}_1 B$ it follows that $B = -r/(2t)C$ and the relation $\partial_t C - \partial_r B = 0$ derived before then implies $x = x_1 + x_2 = 1$. Hence the terms parametrized jointly by B and C reads $(\bar{\theta} - \theta r/t)c(\mathscr{M}_1)t^{-1/2}e^{-\mathscr{M}_1 r^2/(2t)}$. Its covariance under V_- and D (B.47), (B.48) leads to $c(\mathscr{M}_1) = c_2 = \text{cste.}$.

Finally, it remains to find D, which completely decouples from the other coefficients. From (B.45), (B.46), (B.47), (B.48) we have, with $x = x_1 + x_2$

$$\left[t\partial_t + \frac{1}{2} r\partial_r + \frac{1}{2}(x + 1) \right] D = 0$$

$$\left[t\partial_r + \mathscr{M}_1 r \right] D = 0$$

$$[-\partial_{\mathscr{M}_1}\partial_r + r\partial_t]\,D = 0 \tag{B.52}$$
$$[t\partial_t + r\partial_r - \mathscr{M}_1\partial_{\mathscr{M}_1} + x]\,D = 0$$

whose general solution is

$$D = d_0 \mathscr{M}_1^{(x-1)/2} t^{-(x+1)/2} \exp\left(-\frac{\mathscr{M}_1}{2}\frac{r^2}{t}\right) \tag{B.53}$$

which proves the assertion. □

B.1.5 $\mathfrak{osp}(2|4)$-Covariant Two-Point Functions

In order to prove Proposition 11.25 it is enough to observe that $\mathfrak{osp}(2|4)$ includes both $\tilde{\mathfrak{s}}^{(2)}$ and $\mathfrak{se}(3|2)$, hence $\mathfrak{osp}(2|4)$-covariant two-point functions must be also covariant under these subalgebras. The assertion follows immediately by comparing (11.122) and (11.132). □

B.2 Table of Lie Superalgebras

In order to help the reader find his way through the numerous Lie superalgebras defined all along Chap. 11, we recall here briefly their definitions and collect the formulas for the realization of $\mathfrak{osp}(2|4)$ as Lie symmetries of the $(3|2)$-supersymmetric model.

The super-Euclidean Lie algebra of $\mathbb{R}^{3|2}$ is

$$\mathfrak{se}(3|2) = \left\langle L_{-1,0}, Y_{\pm\frac{1}{2}}, M_0, D, V_-, G^{1,2}_{-\frac{1}{2}}, \bar{Y}_0^{1,2}\right\rangle \tag{B.54}$$

whose commutator relations are given at the end of Sect. 3.1 (see the root diagram on Fig. 11.3c). From this, the super-Galilean Lie algebra $\mathfrak{sgal} \subset \mathfrak{se}(3|2)$ is obtained by fixing the mass

$$\mathfrak{sgal} = \left\langle L_{-1,0}, Y_{\pm\frac{1}{2}}, M_0, G^{1,2}_{-\frac{1}{2}}, \bar{Y}_0^{1,2}\right\rangle \tag{B.55}$$

The super-Schrödinger algebras with $N = 1$ or $N = 2$ supercharges are called $\tilde{\mathfrak{s}}^{(1)}$ and $\tilde{\mathfrak{s}}^{(2)}$ and read

$$\tilde{\mathfrak{s}}^{(1)} = \left\langle L_{\pm1,0}, Y_{\pm\frac{1}{2}}, M_0, G^1_{-\frac{1}{2}} + G^2_{-\frac{1}{2}}, G^1_{\frac{1}{2}} + G^2_{\frac{1}{2}}, \bar{Y}^1_0 + \bar{Y}^2_0\right\rangle \tag{B.56}$$

and

$$\tilde{\mathfrak{s}}^{(2)} = \left\langle L_{\pm 1,0}, Y_{\pm\frac{1}{2}}, M_0, G^{1,2}_{\pm\frac{1}{2}}, \bar{Y}^{1,2}_0, N_0 \right\rangle \cong \mathfrak{osp}(2|2) \ltimes \mathfrak{sh}(2|2) \tag{B.57}$$

The commutators of $\tilde{\mathfrak{s}}^{(2)}$ are coherent with the root diagram of Fig. 11.3a and those of $\tilde{\mathfrak{s}}^{(1)} \subset \tilde{\mathfrak{s}}^{(2)}$ follow immediately. Finally, all these Lie superalgebras can be embedded into the Lie superalgebra $\mathfrak{s}^{(2)}$

$$\mathfrak{s}^{(2)} = \left\langle L_{\pm 1,0}, Y_{\pm\frac{1}{2}}, M_0, D, N_0, G^{1,2}_{\pm\frac{1}{2}}, \bar{Y}^{1,2}_0, V_\pm, W, \bar{Z}^{1,2}_0 \right\rangle \cong \mathfrak{osp}(2|4), \tag{B.58}$$

see Fig. 11.2 for the root diagram. This is the largest dynamical symmetry algebra of the (3|2)-supersymmetric model with equations of motion (11.32). To make the connection with the infinite-dimensional Lie superalgebras introduced in Chapter 11, Sect. 4, let us mention that the Lie algebra

$$\tilde{\mathfrak{s}}^{(2)}_1 = \left\langle L_{\pm 1,0}, N_0, G^{1,2}_{\pm\frac{1}{2}} \right\rangle \cong \mathfrak{osp}(2|2) \tag{B.59}$$

is the subalgebra of $\tilde{\mathfrak{s}}^{(2)}$ made up of all grade-one elements, with the identification of $\tilde{\mathfrak{s}}^{(2)}$ as a subalgebra of $\mathfrak{sns}^{(2)}/\mathscr{R}$ given in Proposition 11.24.

Let us finally give explicit formulas for the realization of $\mathfrak{osp}(2|4)$ as Lie symmetries of the (3|2)-supersymmetric model, using the notation of Sects. 3 and 4. In formulas (B7) through (B21), the indices n range through $-1, 0, 1$ while $m = \pm\frac{1}{2}$. Note that these formulas are compatible with those of Proposition 11.24 if one substitutes $2\mathscr{M}$ for ∂_ζ.

$$\begin{aligned} L_n &= -t^{n+1}\partial_t - \frac{n+1}{2}t^n\left(r\partial_r + \theta^1\partial_{\theta^1}\right) - \frac{(n+1)x}{2}t^n - \frac{n(n+1)}{8}t^{n-1}r^2\partial_\zeta \\ &\quad - \frac{n(n+1)}{4}t^{n-1}r\theta^1\partial_{\theta^2} \end{aligned} \tag{B.60}$$

$$Y_m = -t^{m+1/2}\partial_r - \frac{1}{2}\left(m+\frac{1}{2}\right)t^{m-1/2}r\partial_\zeta - \frac{1}{2}\left(m+\frac{1}{2}\right)t^{m-1/2}\theta^1\partial_{\theta^2} \tag{B.61}$$

$$M_0 = -\frac{1}{2}\partial_\zeta \tag{B.62}$$

$$D = -t\partial_t - \zeta\partial_\zeta - r\partial_r - \frac{1}{2}\left(\theta^1\partial_{\theta^1} + \theta^2\partial_{\theta^2}\right) - x \tag{B.63}$$

$$N_0 = -\theta^1\partial_{\theta^1} - \theta^2\partial_{\theta^2} + x \tag{B.64}$$

$$G^1_m = -t^{m+1/2}\partial_{\theta^1} - \frac{1}{2}\left(m+\frac{1}{2}\right)t^{m-1/2}r\partial_{\theta^2} \tag{B.65}$$

$$G^2_m = -t^{m+1/2}\left(\theta^1\partial_t + \theta^2\partial_r\right) - \left(m+\frac{1}{2}\right)t^{m-1/2}$$

$$\left(\frac{1}{2}\theta^1 r\partial_r+\frac{1}{2}r\theta^2\partial_\zeta-\frac{1}{2}\theta^1\theta^2\partial_{\theta^2}+x\theta^1\right)-\frac{1}{4}\left(m^2-\frac{1}{4}\right)t^{m-3/2}r^2\theta^1\partial_\zeta \quad \text{(B.66)}$$

$$\bar{Y}_0^1 = -\partial_{\theta^2} \quad \text{(B.67)}$$

$$\bar{Y}_0^2 = -\theta^1\partial_r - \theta^2\partial_\zeta \quad \text{(B.68)}$$

$$V_- = -\frac{1}{2}r\partial_t - \zeta\partial_r - \frac{1}{2}\theta^2\partial_{\theta^1} \quad \text{(B.69)}$$

$$V_+ = -2tr\partial_t - 2\zeta r\partial_\zeta - (r^2+4\zeta t)\partial_r - r(\theta^1\partial_{\theta^1}+\theta^2\partial_{\theta^2}) - 2t\theta^2\partial_{\theta^1} - 2\zeta\theta^1\partial_{\theta^2} - 2xr \quad \text{(B.70)}$$

$$W = -2\zeta^2\partial_\zeta - 2\zeta(r\partial_r+\theta^2\partial_{\theta^2}) - \frac{r^2}{2}\partial_t - r\theta^2\partial_{\theta^1} - 2x\zeta \quad \text{(B.71)}$$

$$\bar{Z}_0^1 = -\frac{1}{2}\left(\zeta\partial_{\theta^2}+\frac{1}{2}r\partial_{\theta^1}\right) \quad \text{(B.72)}$$

$$\bar{Z}_0^2 = -\frac{1}{2}\left(\zeta(\theta^2\partial_\zeta+\theta^1\partial_r)+\frac{1}{2}\theta^2 r\partial_r+\frac{1}{2}r\theta^1\partial_t+\frac{1}{2}\theta^1\theta^2\partial_{\theta^1}+x\theta^2\right). \quad \text{(B.73)}$$

References

1. M. Abramowitz, A. Stegun, M. Danos, J. Rafelski, *Handbook of Mathematical Functions* (Harri Deutsch, Germany, 1984).
2. C. Albert, P. Molino, *Pseudogroupes de Lie Transitifs. I* (Hermann, Paris, 1984).
3. J. Balog, L. Fehér, L. Palla, Coadjoint orbits of the Virasoro algebra and the global Liouville equation, Int. J. Mod. Phys. **A13**, 315–362 (1998).
4. V. Bargmann, On unitary ray representations of continuous groups, Ann. Math. **59**(2), 1–46 (1954).
5. F. Baumann, S. Stoimenov, M. Henkel, Local scale invariances in the bosonic contact and pair-contact processes, J. Phys. **A39**, 4095–4118 (2006).
6. F. Baumann, M. Pleimling, Out-of-equilibrium proerties of the semi-infinite kinetic spherical model, J. Phys. **A39**, 1981–1999 (2006).
7. J. Beckers, V. Hussin, Dynamical supersymmetries of the harmonic oscillator, Phys. Lett. **118A**, 319–321 (1986).
8. J. Beckers, N. Debergh, ($N = 2$)-extended supersymmetries and Clifford algebras, Helv. Phys. Acta **64**, 24–47 (1991).
9. J. Beckers, N. Debergh, A.G. Nikitin, On supersymmetries in nonrelativistic quantum mechanics, J. Math. Phys. **33**, 152–160 (1992).
10. A.A. Belavin, A.M. Polyakov, A.B. Zamolodchikov, Infinite conformal symmetry in two-dimensional quantum field theory, Nucl. Phys. **B241**, 333–380 (1984).
11. F.A. Berezin, M.A. Shubin, *The Schrödinger Equation* (Kluwer, Dordrecht, 1991).
12. J. de Boer, L. Feher, Wakimoto realizations of current algebras: an explicit construction, Comm. Math. Phys. **180**, 759–793 (1997).
13. A. Bohm, A. Mostafazadeh, H. Koizumi, Q. Niu, J. Zwanziger, *The Geometric Phase in Quantum Systems*, Texts and Monographs in Physics (Springer, New York, 2003).
14. G. Burdet, M. Perrin, P. Sorba, About the non-relativistic structure of the conformalalgebra, Comm. Math. Phys. **34**, 85–90 (1973).
15. E. Cartan, Sur les variétés à connexion affine, et la théorie de la relativité généralisée (deuxième partie), Ann. Sci. Ecole Norm. Sup. **42**(3), 17–88 (1925).
16. F. Constantinescu, H.F. De Groote, *Geometrische und Algebraische Methoden der Physik: Supermannigfaltigkeiten und Virasoro-Algebren* (Teubner, Germany, 1994).
17. M. de Crombrugghe, V. Rittenberg, Supersymmetric quantum mechanics, Ann. Phys. **151**, 99–126 (1983).
18. P. Deligne, P. Etingof, D. Freed, A. Jeffrey, *Quantum Fields and Strings: A Course for Mathematicians* (AMS, Providence, Rhode Island, 1999).
19. P. Di Francesco, C. Itzykson, J.-B. Zuber, Classical W-algebras, Comm. Math. Phys. **140**(3), 543–567 (1991).

J. Unterberger and C. Roger, *The Schrödinger-Virasoro Algebra*, Theoretical and Mathematical Physics, DOI 10.1007/978-3-642-22717-2,

20. F.A. Dolan, H. Osborn, Superconformal symmetry, correlation functions and the operator product expansion, Nucl. Phys. **B629**, 3–73 (2002).
21. J.-P. Dufour, N.T. Zung, *Poisson Structures and Their Normal Forms* (Birkhäuser, Switzerland, 2005).
22. C. Duval, H.P. Künzle, Minimal gravitational coupling in the Newtonian theory and the covariant Schrödinger equation, Gen. Relat. Gravit. **16**(4) (1984).
23. C. Duval. Nonrelativistic conformal symmetries and Bargmann structures, in: Conformal groups and related symmetries: physical results and mathematical background, Springer Lecture Notes in Physics **261** (1986).
24. C. Duval, G. Gibbons, P. Horváthy, Celestial mechanics, conformal structures, and gravitational waves, Phys. Rev. **D43**(12) (1991).
25. C. Duval, On Galilean isometries, Class. Quantum Grav. **10**, 2217–2221 (1993).
26. C. Duval, P. Horvathy, On Schrödinger superalgebras, J. Math. Phys. **35**, 2516–2538 (1994).
27. A. Erdelyi, W. Magnus, F. Oberhettinger, F. Tricomi, H. Bateman, *Higher Transcendental Functions* (McGraw-Hill, New York, 1953).
28. Frappat L. Sciarrino A., Sorba P, *Dictionary on Lie Algebras and Superalgebras* (Academic Press, New York, 2000).
29. D. Freed, *Five Lectures on Supersymmetry* (AMS, Providence, 1999).
30. D. Friedan, E. Martinec, S. Shenker, Conformal invariance, supersymmetry and string theory, Nucl. Phys. **B271**, 93–165 (1986).
31. D.B. Fuks, *Cohomology of Infinite-Dimensional Lie Algebras*, Contemporary Soviet Mathematics (Consultants Bureau, New York, 1986).
32. W.I. Fushchich, W.M. Shtelen, N.I. Serov, *Symmetry Analysis and Exact Solutions of Equations of Nonlinear Mathematical Physics* (Kluwer, Dordrecht, 1993).
33. J.P. Gauntlett, J. Gomis, P.K. Townsend, Supersymmetry and the physical phase-space formulation of spinning particles, Phys. Lett. **248B**, 288–294 (1990).
34. I.M. Gelfand, G.E. Shilov, *Generalized Functions*, Vol. 1 (Academic Press, New York, 1964).
35. P.K. Ghosh, Extended superconformal symmetry and Calogero-Marchioro model, J. Phys. **A34**, 5583–5592 (2001).
36. P.K. Ghosh. Super-Calogero model with OSp(2|2) supersymmetry: is the construction unique? Nucl. Phys. **B681**, 359–373 (2004).
37. H. Goldstein, *Classical Mechanics, Series in Advances Physics* (Addison-Wesley, Reading, Massachusetts, 1959).
38. I.S. Gradshteyn, I.M. Ryzhik, *Table of Integrals, Series, and Products* (Academic Press, New York, 1980).
39. P.Ja. Grozman, Classification of bilinear invariant operators on tensor fields, Funktsional. Anal. i Prilozhen. **14**(2), 58–59 (1980).
40. M.F. Guasti, H. Moya-Cessa, Solution of the Schrödinger equation for time-dependent 1D harmonic oscillators using the orthogonal functions invariant, J. Phys. **A36**, 2069–2076 (2003).
41. M.F. Guasti, H. Moya-Cessa, Coherent states for the time-dependent harmonic oscillator: the step function, Phys. Lett. A **311**, 1–5 (2003).
42. L. Guieu, Sur la géométrie des orbites de la représentation coadjointe du groupe de Bott-Virasoro, Ph-D thesis, Université Aix-Marseille I, 1994.
43. L. Guieu, C. Roger, *L'Algèbre et le Groupe de Virasoro: Aspects Géométriques et Algébriques, Généralisations* (Publications CRM, Montreal, 2007).
44. V. Guillemin, S. Sternberg, *Symplectic Techniques in Physics* (Cambridge University Press, Cambridge, 1984).
45. C.R. Hagen, Scale and conformal transformations in Galilean-covariant field-theory, Phys. Rev. **D5**, 377 (1972).
46. G. Hagedorn, Raising and lowering operators for semiclassical wave packets, Ann. Phys. **269**(1), 77–104 (1998).
47. R.S. Hamilton, The inverse function theorem of Nash and Moser, Bull. Am. Math. Soc. (N.S.) **7**(1), 65–222 (1982).

48. E. Hansen, *A Table of Series and Products* (Prentice-Hall, New Jersey, 1975).
49. G. Hector, U. Hirsch, Introduction to the geometry of foliations, Part B. Foliations of codimension one, Aspects of Mathematics, E3 (1987).
50. M. Henkel, Schrödinger invariance and strongly anisotropic critical systems, J. Stat. Phys. **75**, 1023 (1994).
51. M. Henkel, Phenomenology of local scale invariance: from conformal invariance to dynamical scaling, Nucl. Phys. **B641**, 405 (2002).
52. M. Henkel, M. Pleimling, Local scale invariance as dynamical space-time symmetry in phase-ordering kinetics, Phys. Rev. **E68**, 065101(R) (2003).
53. M. Henkel, A. Picone, M. Pleimling, Two-time autocorrelation function in phase-ordering kinetics from local scale-invariance, Europhys. Lett. **68**, 191–197 (2004).
54. M. Henkel, J. Unterberger, Schrödinger invariance and space-time symmetries, Nucl. Phys. **B660**, 407 (2003).
55. M. Henkel, J. Unterberger, Supersymmetric extensions of Schrödinger invariance, Nucl. Phys. **B746**, 155–201 (2006).
56. M. Henkel, A. Picone, M. Pleimling, Europhys. Lett. **68**, 191 (2004).
57. M. Henkel, H. Hinrichsen, S. Lübeck, *Nonequilibrium Phase Transitions, vol. 1: Absorbing phase transitions* (Springer, New York, 2010).
58. M. Henkel, M. Pleimling, *Nonequilibrium Phase Transitions, vol. 2: Ageing and Dynamical Scaling Far from Equilibrum* (Springer, New York, 2010).
59. J. Hoppe, P. Schaller, Infinitely many version of $SU(\infty)$, Phys. Lett. **B237**(3-4), 407–410 (1990).
60. Humphreys, J.E, *Introduction to Lie Algebras and Representation Theory* (Springer, New York, 1972).
61. E. Inönü, E.P. Wigner, On the contraction of groups and their representations, Proc. Nat. Acad. Sci. U.S.A. **39**, 510–524 (1953).
62. A. Joye, Geometric and mathematical aspects of the adiabatic theorem of quantum mechanics, Ph.D. thesis, Ecole Polytechnique Fédérale de Lausanne, 1992.
63. G. Junker, *Supersymmetric Methods in Quantum and Statistical Physics* (Springer, Heidelberg, 1996).
64. V.G. Kac, *Vertex Algebras for Beginners* (AMS, Providence, 1998).
65. V.G. Kac, S.-J. Cheng, A new $N = 6$ superconformal algebra, Comm. Math. Phys. **186**(1), 219–231 (1997).
66. V.G. Kac, A.K. Raina, *Bombay Lectures on Highest Weight Representations of Infinite Dimensional Lie Algebras* (World Scientific, Singapore, 1987).
67. M. Kaku, *Quantum Field Theory. A Modern Introduction* (The Clarendon Press, Oxford University Press, New York, 1993).
68. B. Khesin, R. Wendt, The geometry of infinite-dimensional Lie groups, Series of Modern Surveys in Mathematics, Vol. 51 (Springer, New York, 2008).
69. B.A. Khesin, O.S. Kravchenko, A central extension of the algebra of pseudodifferential symbols (translated from Russian), Funct. Anal Appl. **25**(2), 152–154 (1991).
70. A.A. Kirillov, Infinite-dimensional Lie groups: their orbits, invariants and representations. The geometry of moments, Lect. Note Math. **970**, 101–123 (1982).
71. A.A. Kirillov, Merits and demerits of the orbit method, Bull. Am. Math. Soc. (N.S.) **36**(4), 433–488 (1999).
72. A.W. Knapp, *Representation Theory of Semisimple Groups: An Overview Based on Examples* (Princeton University Press, Princeton, 1986).
73. S. Kobayashi, K. Nomizu, *Foundations of differential geometry, Vol. 1* (Wiley, New York, 1963).
74. I.I. Kogan, A. Nichols, $SU(2)_0$ and $OSp(2|2)_{-2}$ WZNW models: two current algebras, one logarithmic CFT, Int. J. Mod. Phys. **A17**, 2615–2643 (2002).
75. V.F. Lazutkin, T.F. Pankratova, Normal forms and versal deformations for Hill's equation. Funct. Anal. Appl. **9**(4), 306–311 (1975).

76. J.-M. Lévy-Leblond. Nonrelativistic particles and wave equations, Comm. Math. Phys. **6**, 286 (1967).
77. P.G.L. Leach, H.R. Lewis, A direct approach to finding exact invariants for one-dimensional time-dependent classical Hamiltonians, J. Math. Phys **23**(12), 2371–2374 (1982).
78. M. Leblanc, G. Lozano, H. Min, Extended superconformal Galilean symmetry in Chern-Simons matter systems, Ann. Phys. **219**, 328–348 (1992).
79. D. Leites, I. Shchepochkina, The classification of the simple Lie superalgebras of vector fields, preprint Max-Planck Institute Bonn MPIM2003-28.
80. H.R. Lewis, W.B. Riesenfeld, An exact quantum theory of the time-dependent harmonic oscillator and of a charged particle in a time-dependent electromagnetic field, J. Math. Phys. **10**(8), 1458–1473 (1969).
81. J. Li, Y. Su, Representations of the Schrödinger–Virasoro algebras, J. Math. Phys. **49**(5), (2008).
82. J. Li, Y. Su, Irreducible weight modules over the twisted Schrödinger–Virasoro algebra, Acta Math. Sin. (Engl. Ser.) **25**(4), 531–536 (2009).
83. G. Lozano, O. Piguet, F.A. Schaposnik, L. Sourrouille, Nonrelativistic supersymmetry in noncommutative space, Phys. Lett. **B630**, 108–114 (2005).
84. W. Magnus, S. Winkler, *Hill's Equation* (Wiley, New York, 1966).
85. C. Martin, A. Piard, Classification of the indecomposable bounded admissible modules over the Virasoro Lie algebra with weight spaces of dimension not exceeding two, Comm. Math. Phys. **150**, 465–493 (1992).
86. O. Mathieu, Classification of Harish-Chandra modules over the Virasoro Lie algebra, Invent. Math. **107**, 225–234 (1992).
87. R.V. Moody, A. Pianzola, *Lie Algebras with Triangular Decompositions* (Wiley, New York, 1995).
88. J. Nagi, Logarithmic primary fields in conformal and superconformal field theory, Nucl. Phys. **B722**, 249–265 (2005).
89. A. Neveu, J.H. Schwarz, Factorizable dual model of pions, Nucl. Phys. **31**, 86 (1971).
90. U. Niederer, Maximal kinematical invariance group of free Schrödinger equation, Helv. Phys. Acta **45**(5), 802–810 (1972).
91. U. Niederer, The maximal kinematical invariance groups of Schrödinger equations with arbitrary potentials, Helv. Phys. Act. **47**, 167–172 (1974).
92. V. Ovsienko, C. Roger, Generalizations of Virasoro group and Virasoro algebra through extensions by modules of tensor-densities on S^1, Indag. Math. (N.S.) **9**(2), 277–288 (1998).
93. V. Ovsienko, C. Roger, Deforming the Lie algebra of vector fields on S^1 inside the Poisson algebra on $\dot{T}^* S^1$, Comm. Math. Phys. **198**, 97–110 (1998).
94. V. Ovsienko, C. Roger, Looped cotangent Virasoro algebra and non-linear integrable systems in dimension $2+1$, Comm. Math. Phys. **273**(2), 357–378 (2007).
95. V. Ovsienko, S. Tabachnikov, What is ... the Schwarzian derivative? Notices Am. Math. Soc. **56**(1), 34–36 (2009).
96. P.B.E. Padilla, Ermarkov-Lewis dynamic invariants with some applications, Master Thesis, Institutot de Fisica (Guanajuato, Mexico), available on arXiv:math-ph/0002005 (2000).
97. J.-H. Park, Superconformal symmetry in three dimensions, J. Math. Phys. **41**, 7129–7161 (2000).
98. M. Perroud, Projective representations of the Schrödinger group, Helv. Phys. Acta **50**, 233 (1977).
99. A. Picone, M. Henkel, Local scale invariance and ageing in noisy systems, Nucl. Phys. **B688**, 217 (2004).
100. M. Pleimling, F. Iglói, Nonequilibrium critical dynamics in inhomogeneous systems, Phys. Rev. **B71**, 094424 (2005).
101. E. Ramos, C.-H. Sah, R.E. Shrock, Algebras of diffeomorphisms of the N-torus, J. Math. Phys. **31**(8), 1805–1816 (1990).
102. J.R. Ray, J.L. Reid, Invariants for forced time-dependent oscillators and generalizations, Phys. Rev. A **26**(2), 1042–1047 (1982).

103. A.G. Reiman, M.A. Semenov-Tyan-Shanskii, Hamiltonian structure of Kadomtsev-Petivashvili type equations, Zapiski Nauchnykh Seminarov Leningradskogo Otdeleniya Matematicheskogo Instituta im. V.A. Steklova AN SSSR **133**, 212–227 (1984).
104. C. Roger, Déformations algébriques et applications à la physique, Gaz. Math. **49**, 75–94 (1991).
105. C. Roger, Sur les origines du cocycle de Virasoro. (French) [On the origins of the Virasoro cocycle] Confluentes Math. 2 (2010), no. 3, 313–332.
106. C. Roger, J. Unterberger, The Schrödinger–Virasoro Lie group and algebra: representation theory and cohomological study, Ann. Henri Poincaré **7**, 1477–1529 (2006).
107. C. Roger, J. Unterberger, A Hamiltonian action of the Schrödinger-Virasoro algebra on a space of periodic time-dependent Schrödinger operators in $(1+1)$-dimensions. J. Nonlinear Math. Phys. 17 (2010), no. 3, 257–279.
108. I.M. Singer, S. Sternberg, The infinite groups of Lie and Cartan. I. The transitive groups, J. Analyse Math. **15**, 1–114 (1965).
109. J.-M. Souriau, *Structure des Systèmes Dynamiques*, Maîtrises de mathématiques (Dunod, Paris, 1970).
110. R.F. Streater, A.S. Wightman, *PCT, spin and statistics and all that*, Advanced Book Classics (Addison-Wesley, California, 1989).
111. H. Tamanoi, Elliptic genera and vertex operator super-algebras, Lect. Notes Math. 1704 (1999).
112. S. Tan, X. Zhang, Automorphisms and Verma modules for generalized Schrödinger–Virasoro algebras, J. Algebra **322**(4), 1379–1394 (2009).
113. J. Unterberger, The Schrödinger–Virasoro Lie algebra: a mathematical structure between conformal field theory and non-equilibrium dynamics, J. Phys. Conference Series **40**, 156 (2006).
114. J. Unterberger, On vertex algebra representations of the Schrödinger–Virasoro algebra, Nucl. Phys **B823**(3), 320–371 (2009).
115. S. Weinberg, *The Quantum Theory of Fields, vol. 1* (Cambridge University Press, Cambridge, 1996).

Index

J. Unterberger and C. Roger, *The Schrödinger-Virasoro Algebra*, Theoretical and Mathematical Physics, DOI 10.1007/978-3-642-22717-2,